TRANSFORMATIVE SUSTAINABILITY EDUCATION

This book lays out the principles and practices of transformative sustainability education using a relational way of thinking and being.

Elizabeth A. Lange advocates for a new approach to environmental and sustainability education, that of rethinking the Western way of knowing and being and engendering a frank discussion about the societal elements that are generating climate, environmental, economic, and social issues. Highlighting the importance of Indigenous and life-giving cultures, the book covers educational theory, transformation stories of adult learners, social and economic critique, and visions of changemakers. Each chapter also has a strong pedagogical element, with entry points for learners and embodied practices and examples of taking action at micro/meso/macro levels woven throughout. Overall, this book enacts a relational approach to transformative sustainability education that draws from post humanist theory, process thought, relational ontology, decolonization theory, Indigenous philosophy, and a spirituality that builds a sense of sacred towards the living world.

Written in an imaginative, storytelling manner, this book will be a great resource for formal and nonformal environmental and sustainability educators.

Elizabeth A. Lange is Honorary and Adjunct Fellow at the Institute for Sustainable Futures of the University of Technology Sydney, Australia. At three Canadian universities, she has served as a professor in adult and lifelong education with over 40 years' experience as a formal and nonformal transformative educator. She is a scholar of transformative learning, sustainability education, and transcultural learning, winning awards for her research.

Research and Teaching in Environmental Studies

This series brings together international educators and researchers working from a variety of perspectives to explore and present best practice for research and teaching in environmental studies.

Given the urgency of environmental problems, our approach to the research and teaching of environmental studies is crucial. Reflecting on examples of success and failure within the field, this collection showcases authors from a diverse range of environmental disciplines including climate change, environmental communication and sustainable development. Lessons learned from interdisciplinary and transdisciplinary research are presented, as well as teaching and classroom methodology for specific countries and disciplines.

Theatre Pedagogy in the Era of Climate Crisis
Edited by Conrad Alexandrowicz and David Fancy

Institutionalizing Interdisciplinarity and Transdisciplinarity
Collaboration across Cultures and Communities
Edited by Bianca Vienni Baptista and Julie Thompson Klein

Interdisciplinary Research on Climate and Energy Decision Making
30 Years of Research on Global Change
Edited by M. Granger Morgan

Transformative Sustainability Education
Reimagining Our Future
Elizabeth A. Lange

For more information about this series, please visit: www.routledge.com/Research-and-Teaching-in-Environmental-Studies/book-series/RTES

TRANSFORMATIVE SUSTAINABILITY EDUCATION

Reimagining Our Future

Elizabeth A. Lange

Routledge
Taylor & Francis Group
LONDON AND NEW YORK

earthscan
from Routledge

Designed cover image: "Renewal" by Canadian artist Winona Doram Sunshyneart. Reproduced with permission.

First published 2023
by Routledge
4 Park Square, Milton Park, Abingdon, Oxon OX14 4RN

and by Routledge
605 Third Avenue, New York, NY 10158

Routledge is an imprint of the Taylor & Francis Group, an informa business

British Library Cataloguing-in-Publication Data
A catalogue record for this book is available from the British Library

ISBN: 978-0-367-74704-6 (hbk)
ISBN: 978-0-367-74706-0 (pbk)
ISBN: 978-1-003-15964-3 (ebk)

DOI: 10.4324/9781003159643

Typeset in Bembo
by Apex CoVantage, LLC

Honouring My Ancestors

I offer my gratitude to my four grandparents and my loving parents Eric and Sylvia Lange. Together, all their lived examples of sustainability, love of the land, hopes and dreams for a better life, stories and convictions, set me on this path.

Honouring My Present

I offer my deep gratitude to my husband, Ian Waugh, for his longstanding patience and support, through a dissertation and now this book. He is a partner of shared values as a biologist, environmental educator, and manager of protected areas. We have shared a life journey together advocating for such things as intact forests, salmon conservation as well as for education programmes and social policies that create a society of justice and ecological integrity. We have shared in the care for/of four children, and now grandchildren. The intentions of our shared community involvements have been for creating a safe and regenerative future for all living beings on this tender, exquisite planet.

Honouring My Descendants—the Future

I offer my gratitude to Blaire and Erin, Jenny and
Carly, their partners,
and our little granddaughter, Zaylee, as well as those
not yet named but hoped for.
May the possibilities herein inspire you and your generations
towards life-giving ways that are a worthy future for you
and all your/our descendants.

CONTENTS

FIGURES

BOXES

PREFACE

We can no longer write or speak from nowhere to abstract audiences.[1]

Welcome to this book, a dialogue hopefully engaging you in (re)considering the history of, and a vision for, sustainability education and sustainable societies. This book takes a transformative approach to environmental and sustainability education (ESE).

I write from within my relations, my Life matrix. I am Elizabeth A. Lange, the daughter of Eric Lange and Sylvia Newman Lange, the granddaughter of Emil Lange and Christina Pohl Lange and of Alexander Newman and Susanna Ertman

Newman, all from Alberta, Canada. I am the great granddaughter of Wilhelm Neumann and Rosalie Gerst Neumann, Matthias Erdmann and Carolina Kaddatz Erdmann, Friedrich Lange and Julianna Kottke Lange, and Gottlieb Pohl and Emilie Eckert Pohl, all from Volhynia. My people are called the "Wandering Volhynians."

My Germanic ancestors originated in the Eastern German Plains: Silesia for my grandmothers and Brandenburg and Saxony for my grandfathers. Constant war including mass death, poverty, and famine led them to look elsewhere in the late 1700s and early 1800s. They wanted to live lives of sustainable self-reliance with economic freedom to own farmlands rather than live under a feudal master. They wanted freedom of religion (Protestantism) and freedom from war.

Before Germany was united into a nation-state and European borders solidified, all my great grandparents moved to Volhynia, west of Kiev in Ukraine, although it was under Russian rule at the time. Germanic peasants had been invited by Russia's Catherine the Great, of German origins, to help put the Russian steppe under cultivation, despite local resistance. Several generations later, the political winds turned against them. The promises of freedom for their religion, language, and education, as well as freedom from war, ended. Repression began. This coincided with invitations from Canadian immigration agents promising "free land" to help put the Canadian plains under cultivation, despite local resistance.

My maternal great grandparents as well as paternal grandparents left between the turn of the century and just prior to World War I. Over the two world wars, family members who remained in Europe and Ukraine lost their land, were shot on sight, pressed into war service, or loaded onto railcars and taken deep into Siberia, where they died, or we lost touch with them. Only one great aunt escaped Siberia as a stowaway under a railcar for thousands of miles, eventually finding her way to Canada, with the help of my grandfather. My grandparents worked hard to bring their remaining siblings to Canada, until they could find no others.

Thus, we are plains people, living on three plains: the German Plain, Russian Plain, then the Canadian Plains. My people have been "settlers" for several generations. They believed the imperial/colonial mythologies that the lands to which they were invited were *terra nullius* or "empty." This land in Canada became transformative for our family, fostering deep love.

I grew up on the beautiful land of the Plains Cree, where the grassland plains meet the boreal forest, fed by rushing rivers from the Rocky Mountain glaciers, notably the Columbia Icefield. These mountains were known to local Indigenous peoples as the "Shining Mountains," given their glistening, snowy peaks.[2] They were sacred lands, as they became to me, where I could hear their large voices on the wind. These mountains form the spine of North America, the continental divide of what is called "Turtle Island" by Indigenous[3] peoples.

Our ecoregion is a transition zone called the Aspen Parkland,[4] an area of very fertile black loam soil that holds moisture well, some of the finest in the world. It features softly undulating hills and hummocks, dotted with small lakes and wetlands.

FIGURE 0.1 The "Shining Ones"—Rocky Mountains

Credit: Ian Waugh photo. Reprinted with permission.

The more northern areas are populated with boreal forest of black and Engelmann spruce, lodgepole and white pine, fir, larch/tamarack, juniper, buffalo berry, and kinnikinnick with ravens, gray jays, loons, wood frogs, rainbow trout, and whitefish, extending into the subalpine areas of the Rockies. The Aspen Parkland area is primarily populated with aspen, oval leaves trembling and rattling in the soft winds accompanied with the balsamy scents of poplar trees. Together with white birch and Manitoba maple, they merge into long and short grass prairie. This is all interspersed with willows and the many berries we pick such as strawberries, blueberry, cranberry, Saskatoons, gooseberry, currants, and chokecherry. The muted fragrances of wild rose, red-osier dogwood, wild mint, and honeysuckle grace us in this understated landscape. The sky holds crows, magpies, blue jays, chickadees, many kinds of sparrows, vireos, woodpeckers, grosbeaks, sandhill cranes, many kinds of ducks, and of course Canada geese, the symphonic sounds of "home." These are my relations, the human and nonhuman species that have been part of my matrix of Life.

Alberta, my Canadian province of birth, is big sky country. Given the diminutive size of trees and general flatness, the horizon is visible in all directions. We all watch the clouds, especially on hot summer days, as they billow in volume all day towards dramatic thunder and lightening storms each evening, refreshing the air and nourishing the earth with rain. In the deep cold of winters, down to minus 40°

C, the dry snow is driven horizontally across the prairies, whistling, penetrating our many layers of clothing, swirling into nooks and crannies. As a child, I loved to watch the lazy flakes of early season snow in the evening darkness as it got caught in the streetlight beam in front of our home, slowly building into layers measured by the foot—snow we could build and shape into caves and figures. This snow quiets the world, muffling any sound, creating a sense of contemplation and sacredness, the spirit of winter.

The Plains Cree people call themselves *Nêhiyawak* (neh-HEE-oh-wuk), or the "four direction people," says Elder Jimmy O'Chiese.[5] Their original territory was immense, from the subarctic south into what is now the Northern United States, encompassing the Canadian provinces of British Columbia, Alberta, and Saskatchewan. The Cree have been here over 12,000 years, since the last Ice Age glaciers retreated, supported by archeological evidence.[6] They know their land intimately. It is a storied land where each feature has a legend connected to it that has been carefully passed down, orally, from generation to generation. While the Sarcee (*Tsuu T'ina*), Beaver (Woodland or *Dane-zaa*) Cree, Blackfoot (*Siksika*), and Stoney-Nakoda (Assiniboine or *Sihábi*) moved continuously through the area where I lived, it was predominately the land of the Plains Cree. Overall, the Cree are the largest and most widely spread Indigenous group in what is now Canada.[7]

The Plains Cree were nomadic in that they followed regular seasonal migration routes for food, primarily following the immense herds of buffalo that fed upon the grasslands and Aspen Parkland forests and foothills until the 1880s. They shared the land with robust numbers of coyotes, fox, lynx, bobcats, cougars, wolves, beaver, muskrats, moose, grizzly and black bear, mink, elk, mule and white-tailed deer, and even caribou in high country. While they lived sustainably, considered Old

FIGURE 0.2 Cree Camp on the Prairie, South of Vermilion, September 1871

Credit: Charles Horetzky/Library and Archives Canada/Sandford Fleming fonds/c-005181. Reprinted with permission.

Sustainability,[8] there were lean times as with any people. Then, the numbers of large mammals and birds began to plummet with sustained contact.

The disruption that occurred through European contact shifted Cree traditional movements, technology, and eventually their economy over the space of several hundred years. Initially participating in the fur trade as trappers and negotiators, their way of life altered, becoming more focussed on trading forts, where they received supplies in exchange for pelts. I lived for most of my life in what had been Fort Edmonton. The subgroup of Plains Cree in this area originally moved within the watershed of the North Saskatchewan River, a bioregional form of habitation.

The virtual elimination of the buffalo by settlers to "clear the plains"[9] as well as deliberate attacks on Indigenous well-being and economy, including introduced disease and denigration of cultural and spiritual history, left the Plains Cree population decimated, sick, and starving. They signed Treaty 6 with "the (British) Crown" for the welfare of their people. They believed in the process of mutuality and the many promises given in exchange for most of their land. They understood they would receive benefits, as is expressed in the Cree language, for "as long as the sun shines, the rivers flow, and the grass grows."[10] The intrusions upon their reserved land continues to this day, from mining and forestry to pipelines. Beyond this brief introduction, their story, a story of survival and now resurgence, is not mine to tell. So, I point you towards the many Indigenous fiction and nonfiction authors and scholars.[11]

Upon research, I realized my maternal great/grandparents obtained land that was promised to the Beaver Cree as part of Treaty Six. They settled on *Papaschase* Cree territory, a group who had moved to Fort Edmonton during the fur trade. The reserve never materialized, given opposition, and was sold to incoming settlers.

My paternal grandparents obtained Plains Cree land that the federal government gave to the railroad to sell as they wish. It was *Maskwacîs* Cree territory, not far from a significant spiritual area that was an Indigenous gathering place of midsummer festivities, before they would disperse into smaller groups for winter survival. It is also not far from Wetaskiwin,[12] which derives from the Cree words *wîtaskiwinihk spatinow*, meaning "the hills were peace was made" between the Cree and Blackfoot in 1871. These hills are one place I go to make some peace with my settler family history.

The underbelly of my family story is this displacement of various nomadic peoples. My people came to Canada with great hope and dreams, leaving behind the troubles and tragic memories of the "Old World." As homesteaders who cleared the trees and brush, literally "breaking" the land, they lived close to the land with only the help of horses. My dad was the horse whisperer in the family, keeping the unruly ones pulling the plows, harrows, and wagons. Their survival was based on hours of backbreaking labour, carrying water, building fires, stooking grain by hand, hauling grain for milling, smoking meat, feeding chickens and geese, weeding the vast garden, hanging meat and dairy down the well for refrigeration, and making all things from scratch—soap, medicines, and whittled butter paddles.

Rows of canned preserves lined basement shelves from berries, fish, and vegetables to jams, jellies, and pickles, beside mounds of potatoes, carrots, cabbages, and hanging dried herbs. Staples like sugar or apples came from only 8 hours away, although a local beekeeper provided large tubs of honey.

In winter, howling snow filtered in through the roof to settle on their thick goose down quilts under which multiple children huddled. In summer, they ran barefoot. It was so quiet, my Dad said, that they could hear the first mechanical washing machine 2 miles away. Each of the many children felt important by contributing to the tasks of sustenance. My paternal grandparents felt deeply privileged to own land with the freedom to live their convictions. They felt "masters of their own destiny."[13] My grandfather's handshake was his bond, and both him and my Dad were known for their solid values, including honesty, fairness, and generosity. They lived sustainably within the mutuality of community. There were few uses for money.[14]

Both sides of my family settled close to each other, comforted by shared language and religion with extended families all supporting one another in harsh circumstances. For instance, my Great Grandmother was accompanied by my Grandmother, who together acted as midwives. My Grandmother's sisters helped each other during the time of new babies.

My Dad and his siblings remember these days with great happiness. I and my cousins remember playing in the haystacks and forests, feeding the pigs from the slop pail or throwing hay to the cattle, traipsing out to the outhouse, riding horses, and pumping water from the well until modern utilities, such as running water, electric lights, flush toilets, and phones, appeared. In the outskirts of the city, my

FIGURE 0.3 Lange Homestead

Credit: Eric Lange photo.

mother continued the big garden and rows of preserves that she learned from her mother, aided by new modern conveniences. This was the sustainability I learned as a child, seeing both its virtues and its limitations.

Under this (heroic) pioneer story and the hope that sailed with my people across the sea was their participation in land theft, cultural genocide, and systemic impoverishment, still ongoing. Yet, we are all "treaty people."[15] All those in Canada are bound together by the promises of fair and peaceful coexistence, established by the treaties. The mutuality of coexistence and the original promises offered did not materialize anywhere on Turtle Island.[16]

This remains unfinished business. It is time to honour the terms of co-existence. Globally, those from colonial countries and those from settler countries have an opportunity to forge relationships of trust and dialogue around right relationship with Indigenous peoples, wherever we are on this beautiful planet. This is, in part, one motivation for this book—to honour our human, nonhuman, and more-than-human relations and our histories in rightful and just ways. Who are you? Who do you come from? What is/has been your matrix of Life? What does right relationship mean where you are?

In my work at the University of Alberta, I was privileged to meet Dr Claudine Louis. She is *Maskwacîs* Cree. Through many conversations, we came to realize that my people settled only a few miles from where her people now lived on reserve. She shared with me the living memory and experiences of her people. Soon, I heard stories from within my own family, of exchanges between the peoples, moccasins for bread or fish for money. Our family stories have become interwoven through the land over the last 100 years. Both hunted, fished, and picked berries and searched for medicines on land that was considered "common" at the time. My dad mourned when the land that was providing sustenance was "enclosed" by being purchased, even though they themselves were participating in this private property relation. Slowly, the routes that Claudine's people used, which passed by my family and their farms, were purchased and put under cultivation. As a child, I remember being chased off newly enclosed land with a gun when we were berry picking with parents and grandparents, forced to dump our berries.

I grew up hearing horrendous racist stereotypes about Indigenous people. Simultaneously, based on the "fatness of the land," I saw my family transform from peasant farmers to modernized professionals with all the accoutrements of being modern—modern educations, modern homes, lives of comfort, and, for those who urbanized, increasing disconnection from the land. As a petrostate since the 1940s, Big Oil touched everything in our province, including most jobs and the funding of social programmes and education, including universities. While my grandparents wanted their descendants to have an easier life, some of this is surely not what they had in mind. This is, in part, another motivation for this book—to return to a more fulsome understanding of sustainability and sustainable living, blending what I learned from my ancestors together with newly emerging technology, ideas, and relational ways of thinking. It is a reclaiming of kinship with the Land that we, as newcomers, share with the Peoples original to this Land.

Claudine shared with me that she and her family respect the courage and faith with which my ancestors came to this land. Alongside my grief at their many losses and tragedies, I continue to be deeply humbled by her and her family's generosity of spirit. Similarly, Elder Joseph, who would grant me my spiritual name conveying my "true face," suggested "that was then, this is now . . . let's find a way to move forward with mutual respect."

Thus, one aspect of this book is honouring Indigenous people's generosity, their teachings on relationality, and their impact globally as well as on Canadian societal development.[17] Throughout this book, I draw on many Indigenous teachers and scholars with whom I have had the privilege of working, listening to, or reading their powerful words. Claudine has called me a "translator," which I hope I am living into. Yet, I must be clear that I am only sharing oral knowledge with you on the basis of what I have been learning and have been given permission to share. I am also only sharing knowledge written for the public, through the best extent of my understanding, which is certainly modest and not from an Indigenous place of knowing. Rather, it is from a place of struggling to understand, particularly how to be in good relations. Throughout the book, I cite and point you to global Indigenous leaders, authors, and scholars who speak for themselves. I also hope to convey my experience walking a decolonizing pathway of unlearning and relearning, needed for both settler and colonizer descendants.

Like her educational work with Indigenous women, Claudine challenged me to uncover my own ancestry, older ways of knowing/being, relational ways that were lost, leading me for decades on a long and winding journey. I am still excavating the meaning of the Land that my people originated from, their stories, and their ways of being in relation. I now know that the gifts of my people are as farmers, philosophers, spiritual leaders, educators, blacksmiths and metalsmiths, as well as artists, which I see throughout my extended family as well as in my own children. My father, grandfather, and other family members revere(d) trees, just as our ancient ancestors did. It was the Roman Empire that colonized my people and destroyed their old history, including their spirituality and ways of life.[18] When Charlemagne was crowned the first Emperor (*Kaiser*) of the realm (*Reich*) in 800 CE by the Roman Catholic Pope, becoming the first Holy Roman Emperor, he deliberately dismantled all the ancient tribal ruling systems and destroyed the "pagan" worship sites. He acquired a vast collection of Germanic songs, stories, runes, and practices, which he ordered to be destroyed at his death.[19]

Exploring Indigenous worldviews has helped me in the process of remembering my own People. Three other Teachers, particularly my dear friend Nova Poirier but also Diane *Kahontakwas* Longboat, Daniel Foor, and Grandmother Flordemayo, have taught me different ways of knowing and ways to search for seeds of possibilities in deep time, past the living memory of ancestors. I am from the Orphaned Ones, a cultural orphan, one who has been called into the role of

Rememberer. I am using all my senses and intelligence to understand what my own People lost, who left the land in which they were indigenous. It was another of my Teachers, Mayan shaman Martín Prechtel, who illustrated the importance of the creative convergence of a land, people, and spirit, from which unique cosmologies, languages, and knowledges flow. In his workshops, he clearly illustrated the vast destructiveness of "Empire" over human history, while helping us relearn Old Ways.

Over the 40 years I have been a transformative educator, initially as a high school teacher, community educator, and then professor, I experienced several deep transformations, understanding the pain and liminality of the transformative learning process firsthand. My first transformative experience was as a young adult when I travelled to Peru on a study tour and came face-to-face with the reality of my middle class, white cocoon juxtaposed to rich, ancient cultures like the Incas. I witnessed Peruvian popular leaders carry out powerful social analyses of their society. When I asked the typical question, "what can I do?," I was told, "don't come here to help us. Go back to the belly of the beast. Challenge and change it. That is what will help us."

My educational work in Canada expanded to "development education" then "global education" in multiple educational contexts, from church basements to government boardrooms. I co-led a travel seminar to El Salvador, overcoming the rest of my romanticism about development. We witnessed the vitality of life among Salvadoreans, routinely putting their lives on the line for justice. I came to appreciate social experiments of commoning,[20] in small, humble communities, showing a new way.[21] This led to my scholarly study of international development, the Nicaraguan Revolution, and the role of popular education in imagining and transforming society, based on the work of Paulo Freire. Travelling to many "development projects" around the world, including in India, Nepal, and Thailand, I saw the damage and dependencies created alongside good intentions. Later trips to Costa Rica and Cuba illustrated more clearly cultures of *buen vivir* or living well, leading me to question many more assumptions and possible shapes for low-carbon, sustainable societies. Working with immigrant and refugee groups, I witnessed the process of how Westernism both colonizes and grants freedoms. I helped facilitate and then witnessed the thinking about what to preserve from cultures of origin and what to accept from Canadian culture, whether in parenting, new relations with local ecosystems, or transcultural leadership.

During this time, neoliberalism came to Alberta, the first place these ideas took root in Canada. Already part of the social justice movement, we took to the streets protesting cuts to public education, public health, the social welfare system, public services like utilities, higher education, environmental legislation and watchdogs, and nonprofit and volunteer organizations. The world I grew up in and the values of a fair and just society, including the protection of people from the vagaries of capitalism through a social safety net established after the Great Depression of the 1930s, were torn apart by successive governments. It was a traumatizing time. Some of my daughters' earliest memories are holding placards on the legislature steps. They became familiar with debate, discussion of ideas, and social analysis. As community educators, we took on demystifying the neoliberal rhetoric in many professional and organizational circles.

Given the magnitude of the assault on our society, we began to realize that the social, the economic, and the environmental were profoundly interconnected. Development theory morphed into postdevelopment and sustainability theorizing. As so many lost their jobs and professions during structural adjustment policies, I began offering courses and workshops on sustainability, entitled *Transforming Your Working and Living*, finding a way forward in the complex and highly contentious context that is Alberta.[22] We started the Fireweed Institute for Sustainable Living so that a range of potential educators could offer learning opportunities on various facets of sustainability.

Then, as a professor of transformative learning in the adult education and lifelong learning field, I engaged undergraduate and graduate students on globalization, neoliberalism, and sustainability topics, always starting from their own experiences. To deepen understanding further, together with one of my daughters on an international development placement in Ecuador, we travelled and witnessed the pristine Amazon rainforest, traditional Indigenous lifeways, the threats from Big Oil, mining, and many other corporate ventures, as well as deliberate Indigenous alternatives to oil. We saw the extractivism first hand, similar in many ways to the extractivism in Alberta, only bolder and more violent. My other daughter has now taken up work in ocean conservation and coral reef restoration from our new habitation on Vancouver Island, the lands of the Coast Salish, a place of Pacific salmon and Big Trees—cedar, Douglas fir, and Sitka spruce. My partner works in salmon conservation, a species challenged by overfishing but now by climate change, with implications for local whale species. Currently, I am working with our municipality, designing a Climate Action Plan with an equity lens in collaboration with the local Coast Salish and educating around it for community engagement. I am working to help save remarkable old-growth and second-growth forest stands close to us from industrial extractivism and residential development. These are elements of our Great Work. What is your Great Work?

As you will see, much of what you read in successive chapters reflects facets of my lived experience and my learning journey. Overall, this book aims to trace the big arc of history that has led to this moment of epochal shift. It then offers a more focussed history of environmentalism and environmental education, concentrating on founders, key voices, and strands of practice, within political, economic, and cultural realities. It traces the many antecedents to sustainability as well as the dialogues about and approaches to sustainability education. Profiling the ideological struggles in the education field illustrates the crises *of* and *in* education, as well as the backlash and barriers to environmental and sustainability education. It presents the theory and practice of Transformative Sustainability Education, as one pathway to reclaiming education and reimagining our future.

Reclaiming education is about bringing humans back into harmony with the Earth, for our collective survival. Transformative Sustainability Education offers a *way of thinking, a way of knowing*, and a *way of being* that stretches beyond the Western model of education. It is based on the emerging science of relationality, that the universe is connected as an unbroken wholeness of flowing energy. General principles and pedagogical practices offered here are, in part, drawn from nonwestern cultural systems and ancient knowledges as well as the New Science. Together, these new stories are like seeds that hold potential of life-giving possibility. For educators, engaging these life-giving possibilities is an embrace of our Great Work in this important historical moment.

Now, after nearly 40 years of personal and scholarly exploration, I have arrived back to where I began. My thoughts and beingness have been fostered by "all my relations"—the soil that built my body, the fresh Rocky Mountain water that has coursed through my veins, the trees, birds, and animals that were my daily companions, ancestral stories that kept me wondering especially about wisdom, the Stories embedded in the Land that entered my dreams, intuitive memories carried in my genes and blood—this is the real/reality behind these chapters.

As philosopher Bayo Akomolafe suggests, many of us around the world have been colonized, but we must not presume that any one "other" way of being and knowing has "the" answers, including for relational living and sustainable societies. He says,

> Colonization isn't so much the extraction of wealth from a subdued land to the conqueror's, as it is the shrinking of experience to fit the perspectival frames of the conqueror; it is the imposition of a single way of knowing, a single regime of noticing, a single apparatus for meaning-making—so that what is truly lost isn't gold and corn and [people] as such, but ways of making sense of the vibrancy around us. What is truly lost . . . is the freedom to think otherwise. . . . Decolonizing ourselves is more about . . . reclaiming our imaginaries, our hopes.[23]

We are all struggling to ask the best questions and to look in fruitful places for new frames of meaning making and principles for sustainable living, uniquely manifested in each place. May what I share from my meaning making over decades and the hope I still hold help you a little way on your journey. As I am pondering and writing, I am walking step by step into a new way of being, planting seeds for a new social imaginary. I hope you will join me.

Notes and References

Chapter line drawing by Canadian artist Erin Marie Lange Christensen. www.erinchristensen.com Reproduced with permission.

1 Gustavo Esteva and Madhu Suri Prakash, *Grassroots Postmodernism: Remaking the Soil of Cultures* (London: Zed Books, 1998), p. 7.
2 Chief John Snow, *These Mountains Are Our Sacred Places* (Toronto: Samuel Stevens, 1977).
3 The term Indigenous refers to the original inhabitants of a particular place. George Sefa Dei calls "Indigenousness" as the knowledge consciousness arising locally and in

association with the long-term occupancy of a place. See George Sefa Dei, Budd Hall, and Dorothy Rosenberg (Eds.), *Indigenous Knowledges in Global Context: Multiple Readings of our World* (Toronto: University of Toronto Press), p. 72. Marie Battiste reminds us that these are not the terms that the People call themselves.
Terms like Indigenous, Original, or Aboriginal are capitalized as these refer to a formal identity. I use these terms in referring to Indigenous people globally. When quoting Indigenous scholars, I use their specific tribal or other affiliation. See Marie Battiste, *Decolonizing Education: Nourishing the Learning Spirit* (Saskatoon: Purich Publishing, 2013), p. 13.

4 W. L. Strong and K. R. Leggat, *Ecoregions of Alberta* (Edmonton: Alberta Lands, Forestry and Wildlife, 1992).

5 As told to Claudine Louis, April 11, 2008, personal communication. Claudine Louis, *KIWEYTOTAHK ISKWEW ISKOTAYOW: Returning to the Women Fire* (Edmonton: University of Education, 2008), p. 3.

6 Robert McGhee, "Prehistory," in *The Canadian Encyclopedia* (Historica Canada, July 2016). Article published February 24, 2010; Last Edited July 29, 2016. www.thecanadianencyclopedia.ca/en/article/prehistory.

7 Richard Preston, "Nêhiyawak (Cree)," in *The Canadian Encyclopedia* (Historica Canada, May 2021). www.thecanadianencyclopedia.ca/en/article/cree.

8 The concept of Old Sustainability comes from Hayden Washington, *Demystifying Sustainability: Towards Real Solutions* (London: Earthscan from Routledge, 2015).

9 James Daschuk, *Clearing the Plains: Disease, Politics of Starvation, and the Loss of Aboriginal Life* (Regina: University of Regina Press, 2013).

10 Claudine Louis, personal communication, September 28, 2022.

11 Indigenous authors from what is now Canada or who are/have been working in the Canadian context include: Thomas King, Richard Wagamese, Lee Maracle, Tanya Tagaq, Eden Robinson, Joseph Boyden, Marie Battiste, Jeannette Armstrong, Leanne Simpson, Drew Hayden Taylor, Tomson Highway, Basil Johnson, Tanya Talaga, Daniel Paul, Bev Sellars, Eve Tuck, Gregory Lowan-Trudeau, and Waubgeshig Rice, just to name a few. Many others are cited throughout this book.

12 See https://www.wetaskiwin.ca/699/History-of-Wetaskiwin for a recounting of this history via Indigenous sources.

13 This phrase, of the same time period, comes from Moses Coady, *Masters of Their Own Destiney: The Story of the Antigonish Movement* (New York: Harpers & Brothers, 1939).

14 More of this story is in Elizabeth A. Lange, "Riverspeaking: The Spiralling of Transformative and Restorative Learning toward Kinship Ethics," in Peter Blaze Corcoran, Joseph Weakland and Arjen Wals (Eds.), *Envisioning Futures for Environmental and Sustainability Education* (Wageningen, Netherlands: Wageningen Academic Publishers, 2017), pp. 33–44. https://doi.org/10.39201978-90-8686-846-9_1

15 Roger Epp, *We Are All Treaty People: Prairie Essays* (Edmonton: The University of Alberta Press, 2008), p. xi. Epp says that the words "on these prairies, we are all treaty people" were first used by Muriel Lee, an Ojibwa-Cree.

16 Turtle Island is the traditional name for North America.

17 One general source for this thought is Jack Weatherford, *Indian Givers: How Native Americans Transformed the World* (New York: Three Rivers Press, 1988/2010). See also John Ralston Saul, *A Fair Country: Telling Truths about Canada* (Toronto: Viking Canada, 2008) who asserts Canadians are more Aboriginal than European.

18 Tacitus, *Agricola and the Germania* (New York: Penguin Classics, 2010).

19 Richard Tarnas, *The Passion of the Western Mind* (New York: Ballantine Books, 1991), p. 108.

20 Commoning is a concept explained in much more detail in the last chapter. Briefly, it means the collaborating and sharing to meet the daily needs of individuals and communities within their ecosystems. See David Bollier and Silke Helfrich, *Free, Fair, and Alive: The Insurgent Power of the Commons* (Gabriola Island, BC: New Society Press, 2019).

21 See my articles: Elizabeth A. Lange Christensen, Learning to Work Collectively: Women Artisans in El Salvador. *Connections - Newsletter of the Centre of International Alternatives*, Edmonton, September, 1994.
Elizabeth A. Lange Christensen, Learning to Work Collectively. *Update - National Newsletter of Ten Days for World Development: Ecumenical Coalition for Development Education*, Toronto, Volume 2, Number 4, Summer 1994.
Elizabeth A. Lange Christensen, Individualism and Community in El Salvador and Canada. *Building Bridges: A Change for Children Publication*, Edmonton, May, 1994.

22 For a description, see Elizabeth A. Lange, "Transformative and Restorative Learning: A Vital Dialectic for Sustainable Societies," *Adult Education Quarterly*, Vol. 54, No. 2 (2004), pp. 121–139. https://doi.org/10.1177/071713603260276.

23 Bayo Akomolafe, *Decolonizing Ourselves*, Blog, August 3, 2021. www.bayoakomolafe.net/tag/posts. Inclusive language added.

ACKNOWLEDGEMENTS

It takes a community, not only to raise a child, but also to write a book. I wish to acknowledge numerous academic colleagues for their assistance along the way.

I wish to honour the unique thinking of and my past collaborations with Joy Kcenia Polanco O'Neil and Katie E Ross. They have challenged and inspired me, leading to fruitful work. A special thanks is offered to Cynthia Mitchell for the honorary fellowship at UTS and her shared passion for sustainable futures.

My good friend, Donna Chovanec, who loved the collaborative process of writing and sharing ideas, is deeply missed. Edward Taylor, Tara Fenwick, Peter Mayo, Susie Brigham, Jerry Kachur, Bruce Spencer, Chad Hoggan, Fergal Finnegan, Toh Swee-Hin, Sue Scott, Terry Carson, Kazim Bachus, and Carlos Torres have always supported my work and offered me opportunities to share it. I honour you.

I owe gratitude to many additional writing collaborators over the years—Tiffany Prete, Claudine Louis, Catherine Baillie Abidi, Basia Solarz, Susan Young, Betsy Jardine, Evelyn Tastsoglou, Kathleen Kevany, Aaron Chubb, Shibao Guo, Jenny Kelly, Cora Weber-Pillwax, and Lynette Schultz. I would also like to thank many others, such as Shirley Walters, Dip Kapoor, Walter Leal Filho, Aliki Nicolaides, Marcella Milana, Anne Zimmer, Tadej Kosmerl, Peter Blaze Corcoran, Joseph Weakland, and Arjen Wals, Duane Massing, Alison Mathie, Carol Kasworm, Alexis Kokkos, Mary Drinkwater, and many others for invitations to share in research projects, publications, keynotes, or journal support.

To all those thinkers and writers whose work has inspired me over so many years—many in the following pages—I say thank you. The global sustainability field, including sustainability education, is a vast field with pockets diffused in many other fields in every region of the world. Thus, I am aware and express regret that there will be many who are not referenced, but not for lack of importance. I take responsibility for these absences and any errors in this text. I wish to thank all those friends and colleagues who reviewed earlier drafts. Much appreciation is sent your way.

I give thanks to all my many students and workshop participants over the years. You have all helped keep the process of educating and learning to act in the world very real and very inspirational.

I also wish to thank my spiritual teachers whose training and collaborations have been helping me decolonize my mind and spirit. You have shown me new pathways forward. I thank Nova Poirier for opening up a whole new world of shamanic practices for me. Her friendship has kept my footing on the path of my life work. I give thanks to Claudine Louis for offering to be my Teacher and guiding my ever-deepening understandings related to Indigenous knowledge, ritual and ceremony, for sponsoring me in the naming ceremony, and for encouraging my journey to know my ancestry and ancestors. I thank Yvonne Chiu, Monique Nutter, and Funke Olokode and all the Multicultural Health Brokers for their trust in my facilitation helping navigate the complex challenges of their transcultural community. As ever, they are an inspiring example of creating a justice-seeking, beauty-making, compassion-engendering community, working on their trauma transformatively. It has been a deep privilege working with you all! Thanks to Martín Pretchel, Diane *Kahontakwas* Longboat, Daniel Foor, Grandmother Flordemayo, and many others for your amazing workshops, helping me locate missing pieces—knowing much more is still to be discovered.

I wish to give thanks as well to the Environment and Sustainability publishing team at Routledge, particularly Annabelle Harris for accepting my book and all those who shepherded my book through the publishing stages. I have appreciated your guidance and patience throughout. Thanks to all the reviewers for useful suggestions and guidance.

I give thanks to my partner Ian and our family who witnessed the daily struggle and rhythm of writing, for their patience, encouragement, and humour throughout this venture. I also give thanks to my daughter Erin Lange Christensen for her lovely line drawings that express the theme of seeds throughout the book. Thanks as well to Winona Doram for designing a cover image that flows like the vitality surging through the living world.

I give thanks to all my teachers in the natural world and the Lands I have sunk my roots into. I honour you daily. It is a privilege to awake each day and witness the sheer exuberance, abundance, delicacy, and vitality that envelope us in a loving cosmos. I give thanks to all those who not only watch over me, but more importantly, our world.

1

SEEDING LIFE-GIVING CULTURES

Unearthing Seeds of Life

> The history of words, of language, of perceptions, is an urgent venture for all of us who are committed to life.[1]

Seeds of Hope for the Planet

Earth is *alive*—a mysterious living organism. Earth is sustainable by nature, with an inherent ability to sustain Life indefinitely. This is what is meant by *sustainability*—sustaining Life indefinitely.[2] All we need to do as human communities is respect, honour, and cooperate with these processes. Such human communities are

DOI: 10.4324/9781003159643-1

life-giving. Transformative sustainability education is learning to live, work, and be in a life-giving way.[3]

Doing so, however, is no easy task as it goes against most of the operating assumptions that drive modern Western societies. Moreover, Western societies have been exporting their thought and practices around the globe for 500 years, crushing other ways of thinking and being among cultures who have been sustainable for millennia. The challenge is transforming Western society—the Modern way of thinking and its destructive socio-economic practices—towards a way of being that respects Life, meaning all life forms and processes. The challenge is unearthing seeds of life-giving ways of thinking and being that can be both planted now and carried into the future. No small task—but we must take the long view. This is vital for long-term human habitation on Earth.

Undertaking this collective journey to lifegivingness requires *hope*. Hope is not optimism, and it does not mean we should avoid grief, anxiety, and rage. Hope does not mean executing the usual planning strategies for changing the world. As Vaclav Havel, poet and past president of Czechoslovakia, has said, "Hope is a dimension of the spirit . . . [Hope] is not the conviction that something will turn out well, but the certainty that something makes sense, regardless of how it turns out."[4] We are being called to be humans of depth, humans with great spirit—remade in the fires of transformation. As educators, we are being called to reclaim education and reimagine our future. Further, we are called to active hope—no matter what is to come. Holding hope, even in the small window of time left to mitigate the most catastrophic scenarios, is vital because it is the right thing to do. Whatever happens, we know the planet will survive in some form—that *does* offer hope. Life is persistent.

Epochal Shift: Which Pathway?

Naming our historical moment is important—knowing from where we have come and the big forces shaping our everyday life. Many agree that we are in an epochal shift, from the Modern era that developed over the last 800 years[5] to another, currently unnamed, era. Largely, our era has been ahistorical, where history has been much less relevant to the determinations and decisions of daily life. We generally live for ourselves, not for posterity, our place in history remembered by generations of descendants.[6] We often do not know our family histories or ancestors beyond living memory. Why?

The Modern has been all about "the now"![7] Looking backward has been antipathy for cultures where anything over 5 years is passé. There has been disregard for anything "old" or "dated." In the race towards "progress," we are inundated with trends, what is "cutting edge" in ideas, "new and improved" and "fashion forward" in material consumption, "youthful" and "cool" in appearance, and "the latest" in technology. In feeding this addiction for the new, and the machine that drives it, we have lost our way.

In fact, we are in danger of losing what it means to be human. The logical trajectory of Modernism has led to genetic and geoengineering, artificial intelligence, and robotics, with an increasingly blurred line between machine and natural world, including humans. This "confluence of infotech and biotech" can easily replace not only human labour but also humans, with Machines offering more "advantages" of digital "connectivity and updateability."[8] Through pattern recognition and biochemical algorithms, robots are now able to outperform humans in assessing "the emotions and desires of other people."[9] In these ways and more, artificial intelligence has the potential to surpass human intelligence. Some say it has the possibility of a new hybrid form never seen before, as promoted by the 1% elites.[10] Others pose the ethical question about a tech-driven future: is *this* what we want?

In all of this, the so-called objectivity of the "hard" sciences and technology has not been troubled by morals and ethics. Moral valuations are based on so-called "nebulous" criteria, the job of the "softer" disciplines. This avoidance is further supported by the notion of progress, assuming change is towards increasingly optimal states. As sociologist Max Weber[11] suggested, we are now locked into an iron cage of technological imperative, creating more and more technology, simply because we can. At the moment, the consequences are someone else's problem, namely our descendants.

We are in danger of losing the purpose of the human species. Purpose has historically been the concern of educators, among others. Nonwestern cultures and pre-modern cultures have often understood the learning goals of humanity to be developing "wisdom, understanding of the natural order, and living in harmony with it."[12] Yet, the 18th-century Enlightenment *philosophes* considered themselves making a dramatic leap in human knowledge, reaching towards new heights of possibilities for freedom. They believed in the superiority of Modern science which could grant control over, and thus freedom from, the vagaries of nature as well as the unsavoury aspects of human nature. Enlightenment philosophy was about remaking humans and societies. It was about conquering religious dogma, superstitious myths, and other human illusions, using rational, analytic reason. The world was no longer understood as one of soul and matter, but mind over matter.[13] We lost our humility in the order of Life. Wisdom and the pathways to wisdom became obsolete.

Humanity is always in danger of losing the path towards higher qualities and what it means to live a good life. Pointing to this pathway has been the *raison d'etre* of most spiritual traditions. In the Modern world, humans do not necessarily need to be dependent on gods and sacred values. Their search has been to find morality within, as part of the dignity of humanism. Humans became the measure of all things and the source of all value,[14] profound anthropocentrism. Even so, humans were considered to have no ultimate purpose or destiny within this toughminded Modern scepticism.[15] Rather, humans were expected to be self-determining and self-governing in pursuit of individual autonomy, with their individual consciousness as the gauge for a moral life. With the very term *enlightenment*, philosopher Immanuel Kant expected that individuals would use their courage to move beyond immaturity, which he defined as relying on guardians and guides to do their thinking.[16] It eventually became the Modern task of individuals to determine how to stand on their own, be true to themselves, and pursue individual self-fulfilment.[17] The question of "who am I?" has plagued Modern generations, having lost the markers of real identity as well as the depth of being human, now only skimming the surface.

In the Western system, humanity no longer opts for moderation or communal duty. While the Protestant work ethic initially supported disciplined work at a sacred vocation while promoting economic thrift, which indirectly generated "capital," this morality was lost.[18] In a profound ethical reversal, individual pursuit of monetary gain became "ethical" by encouraging humans to compete for their share of the market—as labourers, professionals, owners, and entrepreneurs—thereby contributing to a nation's wealth production. This morphed into greed, consumption, and self-aggrandizement. Veneration of the ultrawealthy includes a deliberate blindness to their corrupt and nefarious means of enrichment, as well as the obscene concentration of wealth on the backs of working, and now disposable, people. With unparalleled power to manipulate global systems and future trajectories, the 1% undertake a space race away from Earth and the issues in which they are implicated. Meanwhile, without a pervasive sense of the communal, we grasp onto polarizing ideologies; vacant dreams; and political, economic, and religious fundamentalisms, as a steadying force.

We have lost more expansive understandings of what it means to be "a people." Now we live in a "society"[19] which in the West is primarily an economic entity, supported by national boundaries and representative democracy. The basic premise of Western societies is unlimited, undeterred economic growth. Since the formal dissolution of colonies, economic growth and industrialization have been touted as the route to a mature Modern society, ostensibly "freeing" people from their ancestral locales and extended family to become wage workers in the capitalist machine. At the personal level, we hope to be able to report back to families that we are "doing well" in this machine. While some have gained materially in the process, we tend to have lost our deepest human roots and most meaningful identities.

The hungry search in the West for more ways to make money has led to neoliberal policies which have progressively removed public goods—such as health, education, pensions, utilities including heat and water, and other basic elements—out of the social welfare net into profit-making. Several generations of neoliberal schooling have taught younger generations the "virtues" of entrepreneurship and self-marketing, rendering the vulnerability of the gig economy acceptable. The social contract between workers and company owners, in terms of reciprocal benefits and responsibilities, has been broken. Trumpeting worker "flexibility" and "independence," business has asserted the constant need to grow profit margins. Further, the voracious appetite for unlimited profits has led companies into more and more remote areas of land and sea for mining, forestry, and fossil fuels, despoiling pristine habitat and disrupting the planet's living processes. We have lost a moral economy and the ancient knowledge of "dwelling in place."[20]

We stand at a precipice. The stakes are high. Just as other empires have fallen across human history, many predict a decline and disassembly of Modern civilization, rather than a technological utopia.[21] As Jeremy Lent recounts, many studies have shown a link between periods of climate change and civilization decline, often from resource contamination and depletion.[22] The human pathos is captured by Ronald Wright in describing the demise of the Easter Islanders, "The people who felled the last tree could see it was the last, could know with complete certainty that there would never be another. And they felled it anyway."[23] The renewed rate of global deforestation is one example that indicates we are well on this path.

Significant climate disruption will create a Dark Age,[24] ostensibly the first phase of this epochal shift. Many systems will begin to unravel,[25] although the overall pace and scenarios are under debate. We see early signs already, from the doubling and intensity of natural disasters, rise in global epidemics, precipitous decline in biodiversity, and the scarcity stimulating war and environmental refugees. We know there is a hard climate deadline with little time left for addressing climate change drivers.

Those alive today are *the* transitional generations. We have a historic choice to make, in terms of how we respond to these factors pressing in on us. This choice will determine the significance of our individual lives as well as our collective future as humanity, part of our Great Work.[26] There is much we *can* do, which is what this book is about. While we might think we are only one . . . in reality, we are the many. We, especially educators, can facilitate the emerging, while the old is crumbling.

The Great Transformation

The antidotes to Modernism are seeds that lie in multiple soils. We can find some seeds in the New Science, particularly quantum physics and systems biology, with their new understandings of reality emerging over the last 100 years. We can also dust off or dig into the earth for human history through archaeological fragments that remain of life-honouring societies. We can also explore peoples and cultural knowledge systems that have stubbornly persevered in the face of all odds, including attempted cultural and physical genocide. Through colonialism and now neocolonialism, global humanity has been losing traditional cultural systems, languages, and knowledge at the same rate as nonhuman species are being lost.[27] The losses of the ethnosphere and the ecosphere are profoundly connected, as consequences of the same way of thinking.[28] So while time is of the essence on all these fronts, we must not rush in a reactionary way. We must think deeply, feel deeply, and act deeply. Our long-term survival depends on this.

So, what is entailed in this challenge of transformation? It is not just a change in the contents of our thinking, but in the very frames that boundary our thinking and daily lives, shaping and normalizing what we see. Changing the very neural pathways of how we think, the assumptions that shape our perceptions of reality, the ways we assign meaning to life experience, and the ways of being as well as social structures and processes that all emerge from this, is the very definition of transformative learning. Charlene Spretnak calls this "a transformation [on] a grand scale."[29]

Stories of Modernity

To do this, we need new stories. Stories carry our basic understandings and assumptions about reality. It has been said that "The universe is made of stories, not of atoms"[30], underlining the inherent human need for story as our primary meaning-making process. It has been our ancestral process for intergenerational teaching and learning.

What have been our Modern stories? Our Modern story of the universe, or *cosmology*, was inherited from Copernicus, Galileo, and Newton, a heliocentric universe where the planets in our solar system revolve around the sun, all running mechanically in geometric rotation and clockwork predictability. Over the course of Modernity, matter has become considered dead and lifeless, like clock components.

Our Modern story of reality, or *ontology*, has been about atoms, the smallest indivisible particles of matter. This idea of reality, from the early 20th century, is of small building blocks constituting both the natural and social worlds. We seek to reduce things down to their smallest component to understand them. By tinkering with these small components, we are remaking reality.

Our Modern story of knowledge and knowing, or *epistemology*, has been predicated on the scientific way of knowing. Science uses both logic and empiricism, observing the world and experimenting to deduce universal laws. Science and all "respected" disciplines have used mathematics and statistical patterns as the objective language through which these understandings are best conveyed.

Our Modern story of what it means to be human, or *ontos*,[31] is individualism, where the individual right to choose a pattern of life and a set of convictions is protected, balanced with the collective good.[32] Self-consciousness and the ability to reason have been considered the two elements which set humans apart from all other species. Thus, a meaningful life has been about seeking enlightenment through the use of reason. Reason, it is argued, can liberate us from tumultuous emotions, from weaknesses of the body and its appetites, and from the spirit world and magical or superstitious thinking. Gaining control of the thinking mind can generate truth and certainty. Its application through technology can control the capriciousness of natural forces, enhance the human character, and design a new society. In sum, our Modern way of being has been the story of control and mastery, over self and Other.

Our Modern story regarding the values guiding collective human life, or *axiology*, has been about shunning the "uncivilized," "backward" or "yesterday," and adopting success, progress, efficiency, civilization, development, and constant improvement as the Modern ruling values. Under this, the foundation of Modern ethics is about separation, where we must illustrate our distinction from each other as well as the human distinction from all other life forms. We must illustrate our individuality, our uniqueness. Yet, this occurs only within tight bounds, as we also wish to belong, necessitating conformity more than we acknowledge. We come to conceive of each other through categories which most often imply a hierarchy, of more important to less important. Out of this separation logic has emerged colonialism, slavery, caste, oppression, exploitation, patriarchy, ableism, racism, heterosexism, and all other fears and repulsions of the Other.

In sum, the primary Modern story has been one of separation and machine logic. The weight of this logic and the ponderousness of the institutions and structures that operationalize them make it hard to imagine how and where transformation may occur. Interestingly, while science has been one origin of our current crisis, science now is one origin of a newly emerging story, a hopeful seed.

The Paradox of Science

Through science, our human hubris has known no bounds. Nothing in the physical world has been spared our probing. Yet, the scope of science in terms of knowledge production has been very narrow, only what lends itself to the scientific method. French philosopher and sociologist Edgar Morin calls this "blind or narrow intelligence." He says the result has been too many silos of specialized knowledge, leaving the larger questions facing humanity almost untouched.[33]

Paradoxically, it is science which is undergoing a paradigm shift. It is teaching new perspectives on reality, including the aliveness and sensitivities of the living world and the continual expansiveness of the cosmos. This knowledge is late in the day, as survival conditions for so many species and ecosystems are already endangered by the exuberant technological application of science. A leading educator of environmental studies, David Orr, summarizes, "Our science and technology have changed beyond recognition and the Earth is rapidly shifting from the Holocene [our current geological epoch] to something being called the Anthropocene,"[34] the era in which human actions are the key force impacting planetary well-being.[35] It is now recognizable that the Modern age has led us to the brink of unliveability and perhaps, incivility. Yet, this Reality Revolution[36] offers seeds for a life-giving future. It is known as Relationality.

Relationality

Emerging Story

Relationality is not about the nature of interpersonal relations between humans or even between humans and the natural world. From quantum physics, Relationality is the new understanding of the universe as a deeply interwoven unity. From quantum physicist David Bohm,[37] Relationality expresses the perception that the universe is an unbroken wholeness in flowing movement. All is connected in a multidimensional, flowing fabric of energy. From systems biology,

Relationality is the mutual interrelatedness of all living beings and nonliving elements, impacting each other synergistically, within a dynamic network of nested living systems.

We and all things are made entirely of relationships, the basis of reality. We do not have relationships; we *are* our relationships. From the moment we are born, we are in a web of relations that include body relations, social relations, cultural relations, ancestral relations, historical relations, land relations, and ecological relations. No person is self-contained but is embedded relationally from their very first breath. This is our matrix of life. This matrix shapes who we are while also giving us the ability to develop a separate sense of self from within these relations, a "Nested-I."[38] When these relations change, we change. This insight can change how we live together and how we harmonize with the natural world.

This is a profound transformation of perception, away from a Machine-like universe with isolated bits of matter floating about, with the objective human mind standing above it. Human mind, matter, body, and spirit are all part of the Whole, and one is not more important than the others. We are embodied, embedded, and kin to all other living beings. We matter within our relations. We are home. We do belong.

Through Relationality, we can re-story our understanding of the universe, the nature of reality, our knowledge, and our way of knowing as well as our way of being and lived ethics. Relationality leads to new stories, stories that can free us from the literal dead end of Modernity. Stories are an ancient way of learning that has been relegated to the world of children for too long. Further, we do not recognize our collective stories. Creative stories can teach us about this, about an understanding of Relationality, and nurture an imagination of the impossible made possible.

Inspired by Ancient and Indigenous thought, there are other rich stories which illustrate Relationality and related beliefs of what is real, what is good, what is wise, what is possible, and the importance of the impossible. As Madeleine L'Engle says,

> Most of the time the fact that this fact
> Is impossible doesn't bother me.
> I live by the impossible.
> I find it a good discipline to practice believing
> As many as seven impossible things every morning before breakfast.
> How dull the world would be
> If we limited ourselves to the possible.[39]

Seeds of Hope for Humanity

The work here is to form strong intentions to do the good and important things that will build a hopeful future for humanity over the long term. If we become overwhelmed with the compounding challenges, then we are already defeated. If we feel powerless to make intelligent change, then we are. No matter how well, or poorly, global humanity responds to these great challenges, some of us need to live into relational lifeways with an eye towards surviving and thriving.

We *can* re-weave our societies to be life-giving, staying true to the spirit of each local place. These pages offer a guide for exploring the layers of change that are needed by humanity and for facilitating various transformations, particularly by educators. As philosopher Christian de Quincey says,

> Our only hope is to reinvigorate the language of the body, learn to let our muscles, sinews, blood, and bones sing again in harmony with the wild chorus of the land, sea, and air—to feel again the pulse of natural kinship. We need to open the vital channels between our worlds—even our written words—our bodies, and the articulate flesh of the world.[40]

We can become Re-memberers, as we remember how it was to feed the processes of Earth and the living beings around us, sustaining life, ours included.[41] We can reclaim those practices which have been viewed as impediments to Modernity. There is no pre-established roadmap, only signposts that emerge out of the fog as we travel near. We will learn through the process of living into the future, testing pathways, experimenting, and finding travelling companions, much as Life teaches us. Each group of educators and learners, committed to the common good, dedicated to continual learning, taking steps in a good direction, are living courageously and creatively.[42] They are holding gently the seeds of hope, looking for welcoming places to receive and nurture them, an important step into a life-giving future.

We now stand in a liminal zone—between the Modern era and glimmers of a new era, sometimes called the postmodern or the post-post.[43] With a transformed knowing, being, and doing, lies our hope for a future. The new era is calling us—ideas for responding lay within. It is truly a privilege to be an educator in this historical moment.

Notes and References

1 Gustavo Esteva, "Regenerating People's Space," *Alternatives*, Vol. 12, No. 198 (1988), p. 140.
2 I am capitalizing the word Life, where I am referring to the principle of Life.
3 This thought is inspired by Fritjof Capra in his Capra course and in his book, Fritjof Capra and Pier Luigi Luisi, *The Systems View of Life* (Cambridge: Cambridge University Press, 2015).
4 Vaclav Havel cited in Frances Westley, Brenda Zimmerman, and Michael Quinn Patton, *Getting to Maybe* (Toronto, ON: Vintage Canada, 2007).
5 Richard Tarnas suggests that Western cultural history can be separated into the classical, medieval, and modern worldviews, with Ancient Greece considered the "dawn of Western civilization" and the Western intellectual tradition. I am suggesting the Modern era as we know it began to take shape in the crisis of the late Middle Ages, including the little Ice Age, famine, plague, war as well as political, economic, and religious changes, around 1250 CE. See Richard Tarnas, *The Passion of the Western Mind: Understanding the Ideas That Have Shaped Our World View* (New York: Ballantine Books, 1991).

6 Charles Taylor discusses this in *The Malaise of Modernity* (Concord, ON: House of Anansi Press, 1991).
7 The term *Modern* comes from the Latin *modo*, or that which is "just now." Albert Borgmann, *Crossing the Postmodern Divide* (Chicago: University of Chicago Press, 1992).
8 Yuval Noah Harari, *21 Lessons for the 21st Century* (London: Jonathan Cape, 2018), p. 22.
9 Harari, *Lessons for the 21st Century*, p. 21.
10 Jeremy Lent, *The Patterning Instinct* (New York: Prometheus Books, 2017), p. 422
11 Max Weber, *The Protestant Ethic and the Spirit of Capitalism* (London: Routledge, 1930).
12 Fritjof Capra, *The Tao of Physics* (Boulder, CA: Shambala, 2010), p. 335.
13 Rosemary Ruether, *Gaia & God* (New York: HarperSanFrancisco, 1992).
14 Elizabeth A. Lange, "Transforming Transformative Learning through Sustainability and the New Science," in Taylor Edward, Cranton Patricia, and Associates (Eds.), *The Handbook of Transformative Learning* (San Francisco: Jossey-Bass Publishers, 2012), pp. 195–211.
15 Richard Tarnas, *The Passion of the Western Mind* (New York: Ballantine Books, 1991).
16 Immanuel Kant, "An Answer to the Question: What is Enlightenment?" in Immanuel Kant (Ed.), *Perpetual Peace and Other Essays on Politics, History, and Moral Practice* (Ted Humphrey, Trans.) (Indianapolis: Hackett, 1983; original work published in 1784).
17 For more on this, see philosopher Charles Taylor, *Sources of the Self* (Cambridge, MA: Harvard University Press, 1989).
18 For more on this, see sociologist Max Weber, *The Protestant Ethic and the Spirit of Capitalism* (London: Routledge, 1930).
19 Highly debated by theoretical schools, society is most commonly defined as "a group of people who interact in a defined territory and share a culture." In John Macionis and Linda Gerber, *Sociology*, Eighth Edition (Toronto: Pearson, 2014), p. 639.
20 The concept of the moral economy is from historian E. P. Thompson. Gustavo Esteva and Madhu Suri Prakash discuss dwelling in *Grassroots Post-Modernism* (London: Zed Books, 1998).
21 See James Hansen, *Storms of My Grandchildren* (New York: Bloomsbury, 2009); Naomi Oreskes and Erik Conway, *The Collapse of Western Civilization* (New York: Columbia University Press, 2014); and James Howard Kunstler, *The Long Emergency* (New York: Grove Press, 2005).
22 Lent, *The Patterning Instinct*, p. 411.
23 Ronald Wright, *A Short History of Progress* (Toronto: Anansi Press, 2004), p. 60.
24 Jane Jacobs, *Dark Age Ahead* (Toronto: Random House Canada, 2004).
25 Jacobs, *Dark Age Ahead*.
26 The notion of Great Work is from Thomas Berry, *The Great Work* (New York: Bell Tower, 1999).
27 This phenomenon has been well tracked by Wade Davis in his books *The Wayfinders: Why Ancient Wisdom Matters in the Modern World* (Toronto: House of Anansi Press, 2009) and *Light at the End of the World* (Madeira Park, BC: Douglas and McIntyre, 2001).
28 Davis, *Light at the Edge of the World*, p. 5.
29 Charlene Spretnak, *Relational Reality: New Discoveries of Interrelatedness That Are Transforming the World* (Topsham, ME: Green Horizon Books, 2011), p. 6.
30 Muriel Rukeyser cited in Tom Hampson and Loretta Whalen, *Tales of the Heart: Affective Approaches to Global Education* (New York: Friendship Press, 1991), p. viii.
31 Chris Beeman and Sean Blenkinsop, "Environmental End Game: *Ontos*," *Environmental Education Research* (2019), pp, 1479–1490. https://doi.org/10.1080/13504622.2019.1663792.
32 Charles Taylor discusses this further in *The Malaise of Modernity*.
33 Edgar Morin, *On Complexity* (Cresskill, NJ: Hampton Press Inc, 2008).
34 David Orr, "Foreword," in Bob Jickling and Stephen Sterling (Eds.), *Post-Sustainability and Environmental Education* (Cham, Switzerland: Palgrave Macmillan, 2017), p. ix.
35 Carl Folke, "Respecting Planetary Boundaries and Reconnecting to the Biosphere," in World Watch Institute (Ed.), *State of the World 2013: Is Sustainability Possible?* (Washington, DC: Island Press, 2013).

36 See Charlene Spretnak, *Relational Reality* (Topsham, ME: Green Horizon Books, 2011).
37 David Bohm, *Wholeness and the Implicate Order* (London: Routledge, 1980).
38 David Bollier and Silke Helfrich, *Free, Fair and Alive: The Insurgent Power of the Commons* (Gabriola Island, BC: New Society Press, 2019).
39 Madeleine L'Engle, *The Irrational Season: The Crosswick Journals* (New York: Open Road Media, 2016), p. 18.
40 Christian de Quincey in *Radical Nature: The Soul of Matter* (Rochester: Park Street Press, 2010), p. 104 and *Radical Knowing: Understanding Consciousness through Relationship* (Rochester: Park Street Press, 2005), p. 33.
41 See Esteva and Prakash, *Grassroots Post-Modernism.*
42 This statement is inspired by Margaret Mead's quote: "Never doubt that a small group of thoughtful, committed, citizens can change the world. Indeed, it is the only thing that ever has."
43 Since the early 20th century with atomic physics, there has been a significant rupture in Modern philosophical thinking, deconstructing the grand narratives and technologies of the Modern era, including common educational beliefs. Since postmodernism and the rise of new technologies such as bio- and genetic engineering, blurring the lines of the human and Machine, post-post thinking is moving into posthumanism and an ontological turn, concerned primarily with "being" than "knowing." See Patti Lather, "Top Ten+ List: (Re)thinking Ontology in (Post)Qualitative Research," *Cultural Studies↔Critical Methodologies*, Vol. 16, No. 2 (2016), pp. 125–131; Elizabeth St. Pierre, "The Posts Continue: Becoming," *International Journal of Qualitative Studies in Education*, Vol. 26, No. 6 (2013), pp. 646–657; and Karen Barad, *Meeting the Universe Halfway Quantum Physics and the Entanglement of Matter and Meaning* (Durham, NC: Duke University Press, 2007).

2

HOW DID WE GET HERE?

Historical Context—Part One

> History is a reservoir of human creativeness.
> Without the perpetual rediscovery and
> reinterpretation of history,
> without free access to that reservoir,
> the life of any single generation would be
> but a trickle of water in a desert.[1]

Western civilization and the Modern age have been under scrutiny not only for their failings, but also for their redemption. This chapter aims to hopscotch briefly through some key events in the history of the West and some elements that have

DOI: 10.4324/9781003159643-2

created the Modern age. As Mark Twain has said, "Every civilization carries the seeds of its own destruction, and the same cycle shows in them all."[2] By sketching a very broad historical arc, the origins of the crisis we now face are more visible. Similarly, as Lewis Mumford explains at the outset, revisiting interpretations of history can also reveal a reservoir of potential. That which lays outside EuroAmerican activity and Western thinking is particularly instructive.

As transformative educators in this epochal shift, it is necessary to be students of history and social change as well as of learning processes. This chapter firstly offers an orthodox understanding of the general eras comprising Western civilization and then, more specifically, the Modern age, all currently under examination by transdisciplinary scholars.[3]

While environmentalism is described as starting in the late 19th and early 20th centuries, environmental studies scholar Ramachandra Guha[4] argues that the origins are much further back. Thus, this chapter, secondly, describes the colonial age as one key origin of our current crises. By tracing the spread of death that accompanied colonialism and then the early Industrial age, along with some historical resistances to them, the patterns we witness now are clearly a continuation of this history. These patterns shaped the blind spots of the international development movement as well as influenced the First and Second Waves of environmentalism as it became a self-aware movement.

A historical era can be conceptualized as constellation of individual consciousness (attitudes, beliefs, values), social consciousness (shared symbols, social organization, and collective identities), material interests and practices (including the dominant political economic system, changing technology, and daily material habits) as well as cultural consciousness (prevalent ideas and perceptions of reality). These constantly changing patterns shape and are shaped by each other within specific historical conditions. A Great Transformation necessarily addresses all these facets.

Scope of Western Civilization

> Today the Western mind appears to be undergoing an epochal transformation, of a magnitude perhaps comparable to any in our civilization's history. I believe we can participate intelligently in that transformation only to the extent to which we are historically informed.[5]

Human presence on this Earth is nested within a much bigger cosmological story. Western civilization is just one of many civilizational stories of the human journey on Earth. Western civilization has largely been located on the European continent; however, early influences on what became "the West" came from far beyond, including Persia, India, and China.[6] "The West" now includes European ethnically derived populations, the Western political economic system of capitalism, and representative liberal democracy as they developed and spread their influence on every other continent.

While Western civilization is generally traced from the Classical periods of Greece and Rome, it is useful to briefly discuss the Neolithic era, from approximately 3500 BCE to 1700 BCE,[7] or what is called Old Europe.[8] The Neolithic era preceded, as well as simultaneously existed with, the Bronze Age, later giving way to the Iron Age. The Neolithic era saw the expansion of small-scale agriculture into Europe from the Mediterranean. Archaeologist Marija Gimbutas found little evidence of extensive fortifications or violence over this time, indicating peaceful and egalitarian lifestyles in family-based communities. A Mother Goddess or Female Principle dominated most Neolithic spiritualities, expressing reverence for the mysterious cyclical processes of birth, death, and regeneration. Evidence of this Goddess worship has been found throughout Paleolithic and Neolithic habitations for over 20,000 years.[9] The significance of the Neolithic era for our future is expanded in Chapter 8.

Around 4000 BCE, horse-riding pastoralists and warrior nomads from the Eurasian steppes invaded Europe.[10] These Indo-Europeans became some of the primary ancestors of current-day Europeans. They brought with them systems of patriarchy and patriliny, hierarchy and stratification, weapons and war, heroism and warriorship,[11] establishing the preconditions for contemporary Western culture.

The following thumbnail traces key turning points in Western civilization and more specifically, the Modern age. All dates and categories are approximate, provided just to illustrate a general historical arc of how we got to where we are.

- Early features of the West emerge from the *Classical* era (500 BCE to 500 CE), both Classical Greece, and then Classical and Imperial Rome. The Greek pantheon of gods which had been part of the Ancient Greek mythic vision began to shift towards philosophy and science.[12] The powerful concepts that began to populate the Western world view included: objective rationalism; dualism; democracy; humanism; individualism and heroism; the contest between idealism and empiricism, the universal ideas of Beauty, Truth, Justice, and the Good; and an understanding of the universe as mathematically organized.

  With the rise of the Roman empire, which carried on the Greek intellectual framework and colonial expansion, came the development of "law, political administration, and military strategy."[13] Scholarship during this time was extensive as were academic institutions and libraries. "Considered as a single entity, classical Greco-Roman civilization arose, flourished, and declined in the course of a thousand years."[14]
- During the late Roman Empire and early *Middle Ages*, Christianity arose as a dominant religious framework. With the conversion of Roman emperor Constantine and the adoption of Christianity as the state religion, he was committed to the spread of Christianity. Under Charlemagne, Christianity was enforced in the furthest reaches of the empire. Charlemagne ordered that the indigenous European mythologies, stories, poems, and music he had collected

be destroyed upon his death, as they were an "impediment to the propagation of the Christian faith."[15]

Christianity shaped the Middle Ages, from 500 to 1450 CE, the era between the Classical and Renaissance eras. The Christian world view and cosmology "permeat[ed] the Western cultural psyche," including ideas of truth through religious revelation; a divine salvific plan to be taken "to the ends of the Earth;" a linear and progressive sense of history; a personal relationship to the Divine; dualistic thinking (sinful humanity/omnipotent God, light/darkness, spirit/matter, good/evil); a monotheistic cosmic hierarchy (the Great Chain of Being); the transformation of the Mother Goddess into Virgin Mary; the rise of universities and early disciplines within a rigidly theological framework; and a modified theocracy where European state leaders and thinkers were subordinate to religious institutions and leaders (i.e. Holy Roman Empire).[16] The distinction of "East" and "West" evolved from the Western Latin-speaking Roman Catholic Empire and the Eastern Byzantine Empire of Greek-speaking Orthodox Christians. This later evolved into Western Europe and Eastern Europe. During the late Middle Ages, from 1000 to 1450 CE, foundations were laid for Early Modernism.[17]

- The era of the *Renaissance and Reformation* (1450–1560 CE) is often considered *Early Modernism*, in terms of constituting the modern world view. With the rediscovery of classical humanism from the Greco-Roman eras after the Medieval Age, "Petrarch began the reeducation of Europe."[18] This led to "a rebirth of culture, creativity, and human greatness."[19] The classics were integrated into Christian thinking, and there was a revaluing of human dignity and reason over revelation.

  In part, the Renaissance was a cultural revolution. It was shaped in part by the empirical accuracy of art and sculpture, offered by Michelangelo and Leonardo da Vinci for instance. Shakespeare penetrated human psychology and the human condition through literature. However, it was the personal angst of the Augustinian monk Martin Luther and his zeal for reforming the corrupt practices within Christianity which cracked the unity of Christendom and catalysed the Protestant Reformation. His rebellious, self-governing individualism—particularly the right of individual Scriptural exegesis and refusal to recant his understandings, stubbornly asserting the importance of individual conscience—inadvertently undermined Christian dogma and institutional authority, while according primacy to private judgement and individual autonomy. Charles Taylor identifies this as the one of the origins of individualism, contributing to the loss of a coherent and unified moral and ethical horizon.[20]
- The *Scientific Revolution* (1540–1680 CE) led to the modern cosmology of a dead, clockwork universe, from the heliocentrism of Copernicus and Galileo, then Newton. Bacon and many other scientists promoted control of the "thinking form" to probe and command the natural world, now known as the scientific method.[21] They promoted the superiority of scientific knowledge, objectivity,

empiricism, reductionism, universal laws, and linear causality. This "foundationalism" dominated the scientific endeavour for over 500 years, begetting scientism or the belief that science is the only legitimate mode of truth.

This loss of a sacred framework led to the "death of nature," argues Carolyn Merchant.[22] Anthropocentrism became preeminent, where only humans have moral standing, resulting in chauvinism towards other species and those not considered fully human. It also disenchanted the world through instrumental reason, sweeping away "passion," a "sense of higher purpose," and a "heroic dimension to life."[23] New technology such as the magnetic compass, mechanical clock, gunpowder, and the printing press led to significant advances in technology and a reading public. The early colonial enterprise of exploration and conquest began.

- The *Enlightenment* (1687–1776 CE) emerged as *philosophes* embraced human reason as offering the heights of possibility for freedom. In other words, freedom from nature could be won and liberation from all oppressive structures such as Church dogma could be achieved. All the so-called "primitive myths" were eclipsed through a critical scepticism.[24] Explorers and missionaries began to bring back tales of other kinds of societies. Soon, intellectual debates raged in European salons about the natural state of humanity and concepts of freedom, equality, liberty, private property, morality, and the relation of the individual to the social. It was at this time that the West's intellectual heritage gained primacy over other systems of thought, becoming largely self-referencing.[25]

  Rather than seeing the Enlightenment typically as the height of human intellectual achievement, critical theorists now suggest,

  > The Enlightenment has always aimed at liberating men (sic) from fear and establishing their sovereignty. Yet the fully enlightened Earth radiates disaster triumphant. The program of the Enlightenment was the disenchantment of the world.[26]

  In other words, it was a time for the "dissolution of myths . . . extirpation of animism . . . [turning of] nature into mere objectivity." It began the process of alienation: humans alienated from themselves, each other, their labour, products of their labour, and the natural world.[27]
- The *Colonial Empires* (1500s–1900s) and their encounters led to a linear notion of societal development, of which European nations were considered the pinnacle of achievement. The voracious conquering and dividing of continents and intact peoples, as well as the spread of Western ideas and feudal economic relations, including serfdom and slavery, created the origins of continual war, racism, and other forms of division over centuries. Ancient peoples were reconstituted into colonies. All "things" including people were regarded as unlimited raw resources for European enrichment.
- The *Age of Revolution* (1776–1850 CE), including the American, French, and Industrial Revolutions, largely ended the political and economic power of the monarchy and aristocracy in favour of the owners of capital and the captains

of industry. The Protestant work ethic supported economic acquisition as the logical end to disciplined work at a "sacred vocation." Max Weber argued that the religious approval of an acquisitive spirit, accompanied by asceticism and thrift, created individual accumulation of surplus, available as capital.[28] Adam Smith's invisible hand of competitive greed would guide each individual in the pursuit of monetary self-interest, translating into a utilitarian society producing "the most good for the highest number."

The enclosure movement of common lands into private property permanently disrupted geographically bound communities. Dislocated peoples now became wage labourers for the new factories. This trajectory would fork in two directions, into capitalist liberal democracies and into the levellers and diggers who stressed common ownership and redistribution of wealth. These two competing ideologies would dominate the 20th century. Underlying both forks, however, was a privileging of economic progress and a reverence for technology and "the machine,"[29] as the measure and means of national development. Slavery and serfdom would end formally during the 1800s but new oppressive and exploitative practices developed.

- The *Age of Nationalism and Industrialism* (1850–1914 CE) completed unification into European nation-states, with a partial separation of church and state.[30] In North America, it was the Gilded Era of rapid capitalist expansion and urbanization, including the expansion of railroads, coal mining, factory production, papermaking, and shipbuilding. Agricultural settlement further west on the continent was fuelled by mass migration from Europe with continual war on the Indigenous peoples of the Americas. The resulting wealth disparities created vigorous debates on social reform, labour unions, and labour law, as well as concern about extinction of various species. Notions of human rights, representative democracy, constitutionalism, and utopianism evolved.
- The *Late Modern* era (1914–2000s) begins with two World Wars, the rise of fascism and the rise of communism, the dissolution of colonial empires and rise of neocolonialism, the establishment of the United Nations, and then the Cold War.

  The second Industrial Revolution was based on the discovery and rapid expansion of oil and gas, fuelling the new auto industry, air flight, then space exploration. Many new production technologies emerge, reconstituting urban life and geographic mobility.

  After WWII, the rising prosperity of Western nations led to conspicuous consumption and planned obsolescence, causing the pollution, waste, and contamination issues of today. A corresponding rise in "international development" projects from the 1950s and then neoliberal globalization from the 1980s shaped then reshaped global economic and political relations. Fully industrialized nations recalibrated political and economic relations to maintain their superior economic position.

  As a Third Industrial revolution, the rise of the Information age from the 1960s remade production again, this time with digital technology. By the

> 1980s, looking to go digital, manufacturing industries went global. They were seeking cheap, more unorganized labour, favourable tax regimes, and weaker environmental standards for increased profits. Now transnational corporations often exceed nation-states in wealth, political weight, and ideological control. The electronic flow of culture globally, hijacked for consumption, has a homogenizing impact, threatening cultural diversity.

While periodization is slippery, Western civilization and the development of the modern worldview can be said to span over 2,000–2,500 years. The Modern age, generally considered to have arisen during the Renaissance, can be said to span approximately 600–800 years, and the Industrial age for almost 300 years.

Since the end of the 20th century, historian Richard Tarnas asserts that there is a "powerful crescendo" of ideas now occurring, where "virtually every important element of the Western intellectual past is now present and active in one form or another."[31] By teasing out some of these strands of thought and associated practices over the next few chapters, we can unearth life-giving fragments, while leaving behind those that have proven life-denying. In so doing, perhaps we can avoid total civilization collapse while finding routes forward.

Learning from Epochal Shifts and Collapse

> Everyone sees the need of a new principle of life.
> But as always happens in similar crises—some people
> attempt to save the situation by an artificial intensification
> of the very principle which has led to decay.[32]

Many previous epochal transformations—with their knowledge constructs, cultural ways, and political economic systems—preceded and shaped the Western project. Sociologist Allan Johnson suggests that "to understand how social systems work or hold together, we must have a basic understanding of how they change or fall apart."[33]

All ancient societies and empires have fallen, often after 1,000 years, except Egypt and China of over 3,000 years.[34] Since the Agricultural Revolution and use of metals, humankind has been altering the natural world. Most often, this has been to their own long-term detriment.[35]

By looking across multiple civilizations—including the Sumerians, Mayans, Ancient Romans, Minoans in Ancient Greece, the Byzantine Empire, and Easter Island—anthropologist Joseph Tainter[36] suggests three primary reasons for collapse. The *dinosaur model* is a civilization which is a "lumbering colossus" incapable of quick adaptation and response to challenges. A second and related model is the *runaway train* where a civilization is unable to switch directions despite persistent feedback that it is moving inexorably towards catastrophe. The *house of cards* model is a civilization which is "inherently fragile," containing only small reserves of energy and resources to adequately adapt to challenges.

Tainter then proposes a fourth more encompassing explanation. This model is based on assessing the balance of energy flow in relation to a civilization's sociopolitical organizations. Complex societies require much higher amounts of energy and financial expenditures per capita to maintain their large organizational institutions. Therefore, after an optimal point, the amount of energy required to sustain complex societies starts to yield diminishing returns, leading to eventual collapse.[37] This *diminishing returns* model incorporates most attributes of collapse, including resource depletion, loss of soil fertility, deforestation, natural catastrophe, failure to adapt, competition and war between complex societies, invaders, and/or elite mismanagement resulting in loss of political legitimacy. While this is a brief reprise of just one theory of civilizational collapse, it conveys some vulnerabilities that contribute to decline and collapse.

As the initial quote suggests, civilizations tend to intensify the very principles leading to their undoing. Ronald Wright calls this "progress traps." Civilizations do not anticipate or respond adequately as the long-term consequences of their actions become visible.[38]

While the briefest of introductions, studying causes of collapse can assist in thinking about an uncertain future. Scholars, including Jeremy Lent, assert we are in the process of an epochal shift with potential for decline and perhaps collapse.[39] Yet, there is always potential for a breakthrough to another form of social organization, discussed in Chapter 6. All of us here now are the generations that can midwife a life-giving future, including educating about decline factors and more hopeful alternatives.

Seeds of Environmentalism and Justice-Seeking

If we are to educate towards Life-giving possibilities, we must also connect the ecological, social, cultural, educational, economic, and political dots. The following section probes somewhat deeper into colonialism and industrialism, illustrating the interwovenness between ecological ravages, economic thinking, cultural context, social impact, and sociopolitical responses. As Carolyn Merchant asserts,

> Changes in . . . ecology and society . . . were rapid and revolutionary. Only through a historical approach can the magnitude and implications of such changes . . . colonialist and capitalist . . . for the human future be fully appreciated.[40]

We tend to think no previous generation, and no other place, worked to avert social and environmental crises. Nevertheless, each generation responded to the issues as they were emerging from within the Western project. While many heralded the transition to industrial capitalism, others could see troubling consequences. There have been persistent voices critiquing the consequences of the Scientific, Enlightenment, and Industrial revolutions.

Artists and poets often capture best the *zeitgeist* or spirit of a historical period and the critique. For instance John Donne wrote the poem *An Anatomy of the World* in 1611. He was grieving the end of the world as he knew it: the loss of the medieval geocentric cosmology featuring Earth as the centre of the cosmos. He expressed his reservations about the modern view of heliocentrism or the Sun as the centre of our universe.

> The new Philosophy calls all in doubt,
> The Element of fire is quite put out,
> The Sunne is lost, and th'Earth, and no mans wit
> Can well direct him, where to looke for it . . .
> Tis all in peeces, all cohaerence gone.[41]

He expresses well, even in this short section, the sense of undoing that accompanies an epochal shift, which we would do well to heed.

Old Sustainability

There have been historical times when humans *did* live in harmony with Earth. Australian environmental scientist Haydn Washington argues that sustainability is "built on the foundation of how people relate to Nature."[42] For him, cultures that lived in harmony with the natural world, such as Indigenous cultures, are "old sustainability."[43] Their example of "old sustainability underpins the need to discuss 'worldview' and ethics when one talks about 'sustainability.' How do we see our relationship to Nature, are we part of it or are we its masters?"[44]

Indigenous societies have created ways of being that are traditionally in harmony with the cosmos and Earth. As best we know, Neolithic cultures lived by a Feminine Principle that was Life-revering. In both cases, their spiritual ways taught balance with other species, especially maintaining a "sacred balance" of cosmic and Life forces.[45] We have "forgotten the old sustainability and the teachings and wisdom of millennia."[46] We need to gather these forgotten seeds to be replanted so they may flourish once again.

The New Science also supports this perspective. As Canadian scientist and public educator, David Suzuki, argues,

> The heart [of it] is . . . the scientifically supported fact that each of us is quite literally air, water, soil and sunlight, and what cleanses and renews these fundamental elements of life is the web of living things on the planet. Furthermore, as social and spiritual creatures, we need love and spirit if we are to lead rich, full lives. These are the fundamental building blocks of sustainable lives and societies.[47]

Indigenous education scholars are assisting in this task. For instance Indigenous scholar Lewis Williams describes how the confiscation of a sacred mountain protecting a harbour on Māori land contributed to the loss of this wisdom.

> For Māori, an alienation of this nature includes but also goes well beyond a mere physical separation; rather, it represents a deeper alienation from an enduring worldview, a way of being, and ultimately the ability for a fuller experience of life, including one's wairua (spirit) and mana (standing or presence due to a form of spiritual power of authority that is bestowed by the gods).

Such seeds of sustainability can be found in multiple places. Our task is recovery and replanting. This recovery requires an integrated regard for the living world as well as the human world, as a justice-seeking and balance-making. Probing our history to understand the losses is a step in recovery.

Critique of Colonialism as the Spread of Death

The destruction of life, human and nonhuman, during the colonial era is not only staggering but traumatizing. It is a key origin of the environmental and sustainability issues we face today. There are many persistent legacies that flow from colonialism, most notably structural racism and structural inequality between people and within nations as well as between people and nonhuman species, leading to multiple extinctions and permanent damage to the living world.

For Indigenous peoples globally, their epochal shift was forced through the ravages of colonialism, primarily from the flood of European settlement to most other continents. Globally, millions of Indigenous people lost their lives through disease, slaughter, and starvation, as part of deliberate genocide. In North America, *Mi'kmaq* Daniel Paul indicates that the colonization of the Americas resulted in the "annihilation of 70–100 million indigenous souls" as well as the destruction of their thriving civilizations as illustrated by Aaron Carapella in Figure 2.1. He continues,

> Prior to European settlement the Mi'kmaq lived in countries that had developed a culture founded upon three principles: the supremacy of the Great Spirit, respect for Mother Earth, and people power. This instilled in them the benefits of living in a harmonious, healthy, prosperous, and peaceful social environment. The nature of their society, which included sharing and free expression, was so advanced in the establishment of equitable human rights principles that greed and intolerance were all but unknown.[48]

In recounting the horrific actions of the Europeans on the North American continent, Daniel Paul concludes, "we were not the savages."[49]

In Latin America, it was the moral conscience of the 16th-century Spanish priest Bartolomé de las Casas, the first bishop of Chiapas in Mexico/Guatemala, that guided him to faithfully record the atrocities against Indigenous people, including their enslavement, exploitation by the *encomienda* system, and the ongoing destruction of their civilizations throughout Central and South America.[50] "The purpose of all the facts [de las Casas] sets forth is the exposure of the 'sin' of domination, oppression, and injustice that the European was inflicting upon the newly

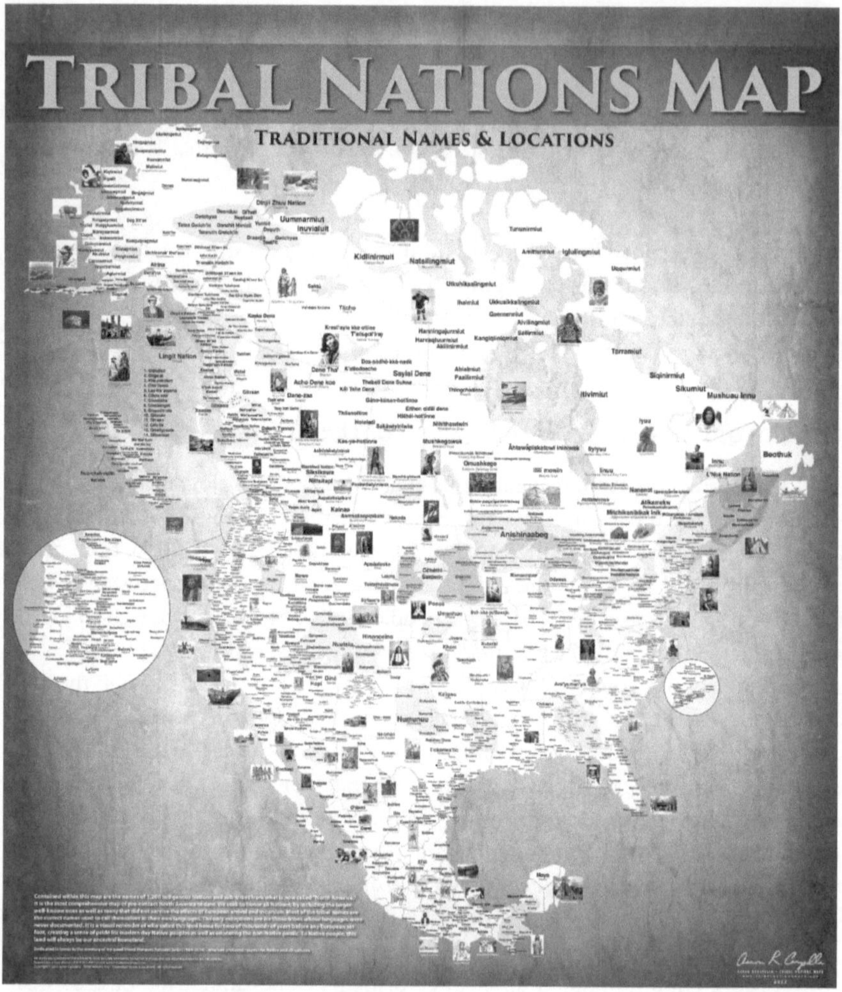

FIGURE 2.1 Tribal Nations Map: Traditional Names and Locations in North America

Credit: Tribal Nations Map 2022: Traditional Names and Locations by Aaron Carapella. Reproduced with permission.

discovered peoples."[51] De la Casas accuses his own nation and church of wholesale slaughter, simply from greed.

> The reason why the Christians have killed and destroyed such an infinite number of souls is that they have been moved by their wish for gold and their desire to enrich themselves in a very short time.[52]

This "get rich quick" scheme has persisted until today. In the 1800s and early 1900s, North American colonists and settlers worked systematically to "clear the plains" of both Indigenous people as well as the buffalo which was their mainstay. Repeatedly, Indigenous populations were forcibly moved entailing much death. They were then confined onto reserves/reservations.[53]

Through the deceit of colonial governments, Indigenous people were not always aware that they were in actuality surrendering the land or what the terms were in colonial minds.[54] They understood they were entering into agreements or "treaties" for sharing the land and its bounty, receiving benefits according to their losses, "for as long as the sun shines and the rivers run."[55] By the late 1800s, *Oglala Lakota* medicine man Black Elk, as well as many other Indigenous leaders, desolately foretold the loss of the Aboriginal way of life. Remembering the slaughter at Wounded Knee, he says,

> When I look back now from this high hill of my old age, I can still see the butchered women and children lying heaped and scattered all along the crooked gulch as plain as when I saw them with eyes still young. And I can see that something else died there in the bloody mud, and was buried in the blizzard. A people's dream died there. It was a beautiful dream. . . . the nation's hoop is broken and scattered. There is no center any longer, and the sacred tree is dead.[56]

Despite multiple debates among Indigenous peoples at the time and their various strategies to address the presence of the "White man,"[57] the growing destruction of the Indigenous way of life became clear. Born in 1899, Chief Dan George of the *Tsleil-Waututh* Coast Salish (Pacific coast) penned important poetry and literature describing the Indigenous way of being in sacred relation with the living world, as well as witnessing the losses.

> Many strange things have happened during my lifetime. Often I could not understand the changes. I have been angered by some, shamed by others, and saddened by many. But nothing can give me a greater feeling of loss than the way nature disappears to make room for people's pleasure. Beyond the reeds of the lake where my cabin stands is a loon pair's nest . . . I no longer go there because the sun shines on an empty nest . . . The evenings are without the laughter of loons . . . Who will bring us the messages from the spirit world when the loons are gone?[58]

Importantly, American anthropologist David Graeber and British comparative archaeologist David Wengrow, in *The Dawn of Everything*, suggest that the Enlightenment debates in European salons were informed by the cogent critique of European society offered by Indigenous leaders, who were "contemptuous of European claims to cultural superiority."[59] They argue instead that it was Indigenous ways of life which provided Europeans with "a sense of social possibility."[60] Conveniently forgotten, Graeber and Wengrow demonstrate that "key Enlightenment thinkers insist[ed] that their ideas of individual liberty and political equality were inspired by

Native American sources and examples."[61] Most irritating to Europeans living in North America, "the Mi'kmaq would constantly assert that they were . . . 'richer' than the French. The French had more material possessions, the Mi'kmaq conceded; but they had other, greater assets: ease, comfort, and time."[62]

In this way, Europeans were confronted with different social ideas, where no one was subjugated to another and where each person was "guaranteed . . . the means to an autonomous life" without "squabbling for advantage."[63] Society "existed not in opposition to but in support of individual freedom"[64] through the sharing of basic resources to ensure food and shelter for all. From this understanding of individual liberty, as Graeber and Wengrow say, flowed "two very different concepts of equality,"[65] either forcing people to behave well towards each other through law or establishing a social framework that evoked this behaviour naturally. They quote Kandiaronk, a brilliant orator and strategist for the *Wendat* Confederacy of Iroquoian-speaking people, in dialogue with Europeans.

> I find it hard to see how you could be much more miserable that you already are. What kind of human, what species of creature, must Europeans be, that they have to be forced to do good, and only refrain from evil because of fear of punishment? . . . we made a decision neither to accept or make use of money. . . . To imagine one can live in the country of money and preserve one's soul is like imagining one could preserve one's life at the bottom of a lake. Money is the father of luxury, lasciviousness, intrigues, trickery, lies, betrayal, insincerity—all of the world's worst behaviour. . . . Over and over I have set forth the qualities that we Wendat believe ought to define humanity—wisdom, reason, equity, etc.—and demonstrated that the existence of separate material interests knocks all these on the head. A man motivated by interest cannot be a man of reason.[66]

Graeber and Wengrow argue that it was the exposure of Europeans to such new ideas regarding social organization, human values, and even processes of rational debate that fed Enlightenment intellectual ideas.[67] While Rousseau disingenuously asserted his ideas were "thought experiments," his ideas clearly derived from Indigenous sources, they say. Eventually, the European notion of a linear progression from hunter-gatherer societies to complex industrial societies would dismiss the importance of the self-aware social organization of Indigenous groups. Rousseau and others infantilized Indigenous people as "innocent child[ren] of nature," thereby avoiding any "meaningful conversation."[68]

Colonialism was the spread of death on other continents as well, particularly through slavery. In the African holocaust, again the land was declared "empty," resources pillaged,[69] and people stolen. Over 400 years, the transatlantic slave trade, the largest slave trade, resulted in the death of between 15 and 28 million Africans, either in the march to the African coast, in the concentration camps awaiting ships, in wretched conditions for one or more months on the ships, or through disease, starvation, dehumanization, overwork, brutality and torture in the "New World."[70] It is now called *the Maafa*, a Kiswahili term meaning "great trauma,"[71] as the "longest and most extensive genocide in human history."[72] The colonial takeover of Africa in a few short years from 1878–1914 is well illustrated in Figure 2.2.

"As the 'dark continent,' Africa has typically been painted with broad-brush strokes, as a place of heathen and uncivilized people, as savage and superstitious, as tribalistic and nepotistic."[73] The legacy from this view and from slavery continues to be "racism, inequality and the vulnerabilities of people of African descent across the globe."[74]

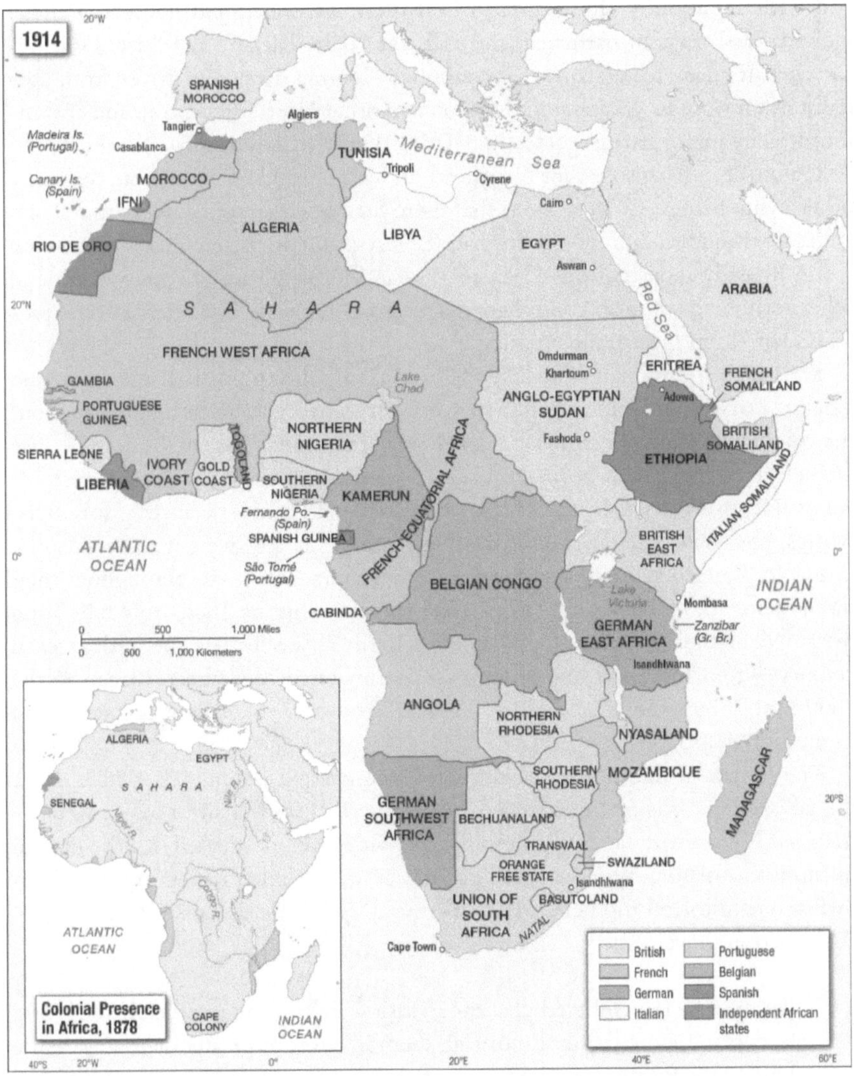

FIGURE 2.2 Colonized Africa (1878–1914)

Credit: Copyright © 2022 Colonized Africa 1878–1914 by *Facing History and Ourselves*. www.facinghistory.org. Reprinted by permission.

Further, slavery is far from over. Modern slavery is responsible for 50 million people who exist in enslaved conditions, according to the International Labour Organization.[75] As Kevin Bales explains in *Disposable People*, the consumer society and need for ever-expanding profits have led to modern slavery that touches most Western lives daily. "This is the new slavery, which focusses on big profits and cheap lives."[76] The central dynamic is little changed.

During the UN International Decade for People of African Descent 2015–2024, there is an urgent call to "dismantle racist structures, reform racist institutions, and confront slavery's legacy of racism together."[77] As "the violence of slavery did not end with abolition,"[78] the goal is to heal the multilayered and transgenerational wounds of slavery. In particular, this calls for "deracializing" and "decolonizing" societies. It means recognizing that continued "looting is a system driven from capitalist institutions in Washington, London and other Northern centres, and accommodated by junior partners across the Third World, including African capitals."[79]

Colonialism has also meant the spread of death, including extinction, for many other living beings. Guha suggests that there have been three recurring moral and cultural critiques and proposed alternatives in response to the ravages of the natural world by early industrialism—back-to-the-land ideas, scientific conservation and other technology/human control approaches, and the ideas of wilderness that preserves land from human intervention.

For instance early naturalists like Carl Linnaeus who originated taxonomy, and Charles Darwin and James Audubon who catalogued natural history throughout the colonies, began noting the extensive losses of flora and fauna in the colonies. By 1650, beavers had largely been eliminated from the East Coast of North America.[80] Across North America, by the 1800s, populations of buffalo, deer, seals, sea otters, whales, pheasants, pigeons, and wolves, to name just a few, were under increasing threat. Explorers like Lewis and Clark recorded abundant species throughout their travels across the United States from 1804 to 1806. But, by 1886, only 540 bison remained of the estimated 40–70 million bison.[81] Species became extinct each century—from the dodo bird around the perimeter of the Indian Ocean in the 1600s, the Tahiti sandpiper in the 1700s, to the North American passenger pigeon in the late 1800s.

The "myth of resource inexhaustibility," human and nonhuman, which drove the colonial era was first questioned by George Perkins Marsh in his 1864 book *Man and Nature*, with far-reaching impact.[82] Marsh even postulated that the human disruption of natural systems could possibly lead to human extinction, foreshadowing our historical moment. His work was a significant precursor to the global conservation movement.[83]

> [I]t must be remembered, that the British were . . . unquestionably the world leaders in deforestation. Emulating them in lesser or greater degree were the Dutch, the Portuguese, the French, the Belgians and the Germans, European powers who were to all become prime agents of ecological destruction in 'their' colonies.[84]

As Guha argues, environmental issues are not new, but once there is a "perception of an environmental *crisis*,"[85] those seeking to address the crisis, though colonists and the source of the crisis themselves, are heralded as the vanguard. Too often, the human and nonhuman lives at most risk are the prime recipients of degradation and then pawns in the politics. Their stories of loss and resistance tend not to be told, given global hierarchies of importance.

For instance out of self-interest, Victorian big game hunters would establish the first wildlife conservation societies as they saw steep declines in animal populations by their own hands. By 1900, a convention for the preservation of African animals, birds, and fish would be signed and games reserves established.[86] In the United States, the public trust doctrine was enshrined in law, 1842, holding that "wildlife and fish belong to *all* the people."[87] Protected species, game reserves, and fishing and hunting licences would become standard programmes. Guha comments, "If there was indeed a 'crisis of African wildlife,' this crisis was created by the white man's gun and rifle, not the native spear and sling shot."[88] Conservation practices, while aimed at conserving these species for colonists, restrict Indigenous people from having traditional access for their livelihoods, perpetuating ongoing economic and cultural injustice.

Again, in the mid-1800s, worries over two centuries of forest destruction finally led to the adoption of forest management practices in India and Burma. This voracious use of wood for shipbuilding, mining infrastructure, and urban expansion led to forestry circles calling for conservation, as a balance between harvest and regeneration.[89] They would use science, or scientific conservation, to find this balance. In establishing the Forest Service, over a fifth of India's forests were protected, to address the loss of teak and other exotic wood forests used for European consumption.[90]

Scientific forestry originated in late medieval Europe, particularly in Germany, where deforestation was well advanced.[91] As ecological writer J.B. MacKinnon explains,

> In the year 1377, a poet in what is now Germany speaks of entering the Great Wilderness, an unbroken forest that took three days to cross and was home to bison, wild boar, wild horses, wolves, bears, lynxes, and wolverines.[92]

MacKinnon continues,

> Five thousand years ago, Britain was covered in forests that are now pleasingly remembered as 'wildwood.' . . . As much as 50 percent of those woodlands were already gone two thousand years ago. By 1900, Britain's original forest cover had dipped below 5 percent and has since recovered to around 10 percent.[93]

MacKinnon suggests that we have a constantly "shifting baseline syndrome" where the creeping disappearance is not noted across generations. We forget or do not know "what the world used to look like," and we deny who is responsible.[94]

Further, the custodial approaches to forestry conservation leave out all those who had been living in harmony with their ancestral lands. In India, scientific forestry meant that subsistence agricultural communities no longer had access to the forests or lands that gave them life.[95] In some cases, the loss of the ancestral relation between human and forest itself degraded the health of the land from loss of this symbiosis. This has been repeated among peasants, pastoralists, and agriculturalists across the globe for centuries. Not only did subsistence peoples not create the situation, they suffered the loss of their livelihoods as well, as another form of enclosure. Centuries of their ongoing protests and militancy have not been well represented in environmental literature.

Another example of the spread of death has been the loss of soil productivity, soil erosion, sometimes even resulting in less rainfall, which occurred with the global expansion of plantation agriculture, from cotton to sugar cane to tea.[96] By 1800, even in the United States, "destructive land practices . . . were already leaving the great tobacco estates of Virginia in ruins."[97] In Australia, *Bunurong/Tasmanian/Yuin* Bruce Pascoe illustrates the "cultural myopia" of colonials that led to a constant misuse of soils and water as well as importation of foreign plants and animals.

> The fertility encouraged by careful husbandry of the soil was destroyed in just a few seasons. The lush yam pastures of Victoria disappeared as soon as sheep grazed upon them. . . . The English pastoralists weren't to know that the fertility they extolled on first entering the country was the result of careful management. Colonial Australia sought to forget the advanced nature of the Aboriginal society and economy, and this amnesia was entrenched when settlers who arrived after the depopulation of whole districts found no structure more substantial than a windbreak, and no populations that were not humiliated, debased, and diseased.[98]

There is much to learn from original inhabitants and much to question regarding assumptions of European superiority, "levels of civilization," and environmental vanguards. The task is deracializing and decolonizing our thinking when retrieving environmental and cultural history. This remembering is vital context to transform the tenor of current global sustainability debates.

Protesting Industrialism

In the late 1700s and early 1800s, the protest of early industrialism was vibrant on many fronts, the seed of today's environmentalism. The Luddites attacked the mechanized production of textile factories which eliminated them as handcrafters. In their new jobs tending machines, they received poor wages, dangerous working conditions, and pitiful urban living conditions.

> The world rolls around for ever like a mill;
> It grinds out death and life and good and ill;
> It has no purpose, heart or mind or will.[99]

Born in 1818, Karl Marx critiqued the degrading, exploitative conditions of the early industrial era, including the textile mills of Hegel's father. He died before he could fully explicate his ideas on alienation, exploitation, processes of social change as well as socialism and communism. Nevertheless, his analyses of oppression, inequality, and concentration of wealth as permanent features of capitalism remains cogent today, echoing the Indigenous critique. As one writer captures,

> The facts of life for most of us are a dark street, crowds, hurry,
> commonplaceness, loneliness, and
> worse than all, a terrible doubt, which can hardly be named, as to the
> meaning and purpose of life.[100]

Wherever nations were industrializing, severe pollution led to illnesses and deaths as well as calls to avoid wasteful consumption. In England, there were calls for conservation of coal in the late 1800s and early 1900s,[101] as the level of coal smoke from factories, home-cooking hearths, and steam engine trains led to what was called the Big Smoke or London Fog. Pittsburg, too, was known as Smoke City.[102] This is repeated today in China and India with poor air quality leading to significant illnesses and premature deaths.

In the 1800s and early 19th century, the Romantic Movement flowered in the literary work of Goethe, Blake, Wordsworth, and Emerson. They all challenged mechanism and constant human expansion into wilderness areas. They promoted a "back-to-nature" movement against the losses endemic to urbanization, from the loss of fresh air to loss of simple contact. In this context, Henry David Thoreau would publish *Walden* in 1854, documenting his simple and self-reliant life in the forests of Massachusetts. This call would be echoed again in the 1960s and 1970s.

Further, the Romantics asserted that the natural world did not operate mechanically as science presumed, but as a sacred whole. They contested instrumental reason by upholding the power of imagination. They disputed scientific abstraction, aggression, and invasiveness by focussing on the living presence of the natural world. They practiced "passive attentiveness" to hone subtle perceptions of its lessons.[103] Goethe revived the perception of nature as a moving order of patterns and relationships within an organized, harmonious whole. This was an intuitive precursor to current systems and Relationality thinking.

> In nature we never see anything isolated, but everything in connection with something else which is before it, beside it, under it and over it.[104]

In the late 1800s and early 1900s, the Arts and Craft Movement resuscitated traditional handcraft through the ideas of John Ruskin and William Morris. They were protesting sterile modern design and, again, the lifelessness of automated production and ugliness of its products.

> Nothing should be made by man's (sic) labour which is not worth making; or which must be made by labour degrading to the makers.[105]

They decried the "desacralization" of nature in seeing the natural world only as raw materials to be exploited.[106] They revered rustic living, away from railways and factories. Their writings presaged the development of many future environmental organizations protesting the ravages of late industrialism. As Guha argues, early environmentalism often began with these heralds first, who document and condemn the degradation while pointing to alternatives.

The Gilded Age of Industrialism, in the late 1800s, was an era of rapid economic growth. In Victorian England and the American Reconstruction (post-Civil War) period, railway building, coal and other mining, agriculture, industrial mechanization, and factory production dramatically expanded. Trade doubled and then tripled between Western nations, from 1830 to 1880, a phenomenon neoliberal leaders would try to replicate in the late 1990s. During this time, English philosopher John Stuart Mill proposed the notion of a "stationary state" society rather than one of "constant progress," a precursor of the steady-state economics of Herman Daly.[107]

In the late 1800s and early 1900s, the influx of so many European migrants to non-European continents created contentious class relations between the ruling classes of colonists, settler migrants, Indigenous populations, and previously enslaved populations. Both progressivists—promoting social reform based on science, technological innovation, health programmes, and public education—and socialists promoting communal ownership of property and the means of production would offer contrasting ideologies for addressing stark class differences.

Social Reformism

Progressive social reform was advocated in industrializing nations to address the precarious poverty, moral degradation, and squalid living conditions that emerged out of the cauldron of rapid industrialization. Reformists advocated for mass public education, or schooling, to take children off the streets as well as out of orphanages and factories. Schooling could offer character training and moral discipline through nonsectarian religious instruction based on common Christian morals. It could cultivate the mind in overcoming superstitious thinking and fostering scientific thinking. Schools could develop literacy "according to one's station" to build a productive work force. Politically, schools could encourage patriotism and civic virtues, promoting freedom and avoiding tyranny. In the United States, school reformer Horace Mann advocated for a universal, tax-supported, centrally planned national education system that was uniform for all classes across the country. He saw this as part of redistributing wealth and contributing to the dignity and improvement of the "common classes."

Places like Toynbee Hall in London and Hull House in Chicago started by Jane Addams sought to address social needs in impoverished communities, through adult education, social welfare programmes, and union organizing. Adult education was also used to address character training and moral discipline, like sobriety, as people were dislocated from their communal bonds, moving into urban jobs.

This was the original purpose of the YMCA. Policy makers also turned to adult education to build scientific thinking and technical skills for industry, such as in the Mechanics Institutes. Public libraries offered reading material for the "common people" in the cities, and university extension departments used circuit-riding professors to take education out to rural areas. Folk schools, such as in Denmark, offered free, nonformal education to adults with a desire to learn. Folk schools gave the "lower" classes an alternative to the university education of the elites. They also reinvigorated a waning cultural identity through music and the arts as well as building the skills for democracy. The overall rationale was to address hardships and promote learning and industriousness as the pathway to individual success and prosperity. Thus, social unrest could be avoided while supporting industrial growth and democracy.

The feminist movement was part of this larger social reform movement, as women advocated for community programmes to address human needs that resulted from industrialization and immigrant settlement. Programmes such as the Farmer's Institutes and Women's Institutes addressed topics in agricultural science, farm safety, rural isolation, food safety, and family health. Feminism, as "the advocacy of social equality for the sexes, in opposition to patriarchy[108] and sexism,"[109] nurtured the strength and persuasiveness of women's voices in their communities.

In the first wave of feminism, from the mid-1800s to early 20th century, women contested their intellectual, social, and political exclusion. They promoted women's access to employment, healthcare, property ownership, and education. They called for the prohibition of liquor to address violence in family life, crime, and other social ills. The key achievement, at this time, was the expansion of suffrage or the right to vote. The universal suffrage movement asserted women were legal persons and that there should be a vote for every citizen regardless of difference or circumstance. This was inspired by the Enlightenment writings of Olympe de Gouges's *Declaration of the Rights of Woman and the Female Citizen*[110] in 1776, Mary Wollstonecraft's *A Vindication of the Rights of Women* in 1792, and John Stuart Mill's *The Subjection of Women* in 1869.

Further, the loss of women's lives from childbearing, disease, hygiene, home-based violence and accidents, their financial dependence on men, and lack of political power continued to shape demands throughout the 20th century. Thus, the feminist movement was "acting to overturn modern constructions of [both] nature and women as culturally passive and subordinate."[111]

In sum, this is just a glimpse of the massive loss of human and nonhuman life as well as the multiple faces of oppression and exploitation. The impacts of colonialism and industrialism on all life forms, human and nonhuman, are profoundly linked. There have also been persistent protest movements and calls for social reform prior to the 20th century. There have been recurring ideas of back-to-land, handcrafting, nonindustrial ways of living, and "good work." Thus, it is important to remember these foregoing and intertwined streams that eventually lead into the environmental, social justice, and sustainability movements of the late 20th and early 21st centuries.

First Wave Environmentalism: Conservation and Preservation

First Wave environmentalism can be traced from the free-for-all in "natural resource" exploitation and slaughter. In the early 20th century, the continued loss of forests for paper, building materials, furniture and ship masts, the loss of mountains for coal and minerals, and the emptying of oceans for fish and whale oil generated even more voices warning that abundance was not endless. First Wave environmentalism formalized two approaches, that of conservation and that of preservation.

At the turn of the 20th century in the United States, John Muir's writing addressed the "ruthless destruction"[112] of expanding pioneer settlement, agriculture, ranching, and urbanization. He celebrated the "glories" of the natural world for the leisure of "tired, nerve-shaken, over-civilized people."[113] His "wilderness thinking" led to the establishment of the first national parks around the world. The first park in the world was the Bogd Khan Uul National Park (1778) in Mongolia, protecting sacred cultural and natural sites. This distinction is often incorrectly assigned to Yellowstone Park (1872) in the United States. Many colonial countries followed suit: Banff National Park in Canada, Tongariro National Park in New Zealand, and Royal National Park in Australia. A specific conception of wilderness emerged—legally protected land free of any human industry or consumptive use. For Muir, preserving wilderness allowed for "geopietistic mystic experience" in natural areas and could foster a moral code or "wilderness ethic."[114] Muir founded the Sierra Club in 1892, long considered the "most influential" preservation society in American environmentalism.[115]

Aldo Leopold, ecologist and author of *The Sand County Almanac* in 1949, founded the Wilderness Society. Leopold promoted preservation of wilderness spaces and species through evocative writing, but unlike Muir, Leopold was also science trained. At first trained in game and forest management, then settling on a farm in Wisconsin, he called for a "harmonious relation to the land."[116] He advocated a shift from the idea of "earth enslavement" towards "interdependent cooperation."[117] Leopold also advocated for responsible living outside of national parks as well as inside.[118] He "recognized that wild areas could hardly be saved without a wider reorganization of the economy on ecological principles, so that the fruits of nature's use could be more equitably distributed among humans."[119] His work anticipated Second Wave environmentalism.

In the 1920s and 1930s, following from George Perkins Marsh, forester Gifford Pinchot elaborated scientific conservation, using reason and rational planning for addressing depletion. He advocated for the optimal or "wise use" of nature, where the rate of extraction does not exceed the rate of growth. Borrowing ideas from German botanist Dietrich Brandis who established the Forest Department in India,[120] he founded the United States Forest Service. His "definition of conservation [was] the greatest good for the greatest number *for the longest time*,"[121] a variation of utilitarian philosophy. He positioned his work within the progressivism taking hold in the United States.

Within First Wave environmentalism, then, these two approaches of conservation or controlled harvesting and preservation as wilderness reserves emerged in an organized way. However, they are at opposite ends of a philosophical continuum. Conservation is anthropocentric in privileging human needs and utilitarian in determining harvest rates, whereas preservation is biocentric, in that the natural world has inherent value beyond utility to humans.[122] Together, these approaches energized the growing environmental movement as well as expanded the professionalization of environmental activity.

End of Colonialism and Old Social Movements

The Imperial era would formally close with the end of World War I (WWI) in 1918. The monarchies of Russia, Germany, Austria-Hungary, and Turkey would fall. The British and Ottoman Empires were fully dissolved by the 1950s, and the Dutch, Americans, French, and Portuguese eventually withdrew from colonized territories as well. This occurred not only as a growing commitment to democracy, but primarily in response to political independence movements throughout Africa, Asia, and Latin America.

In India, Gandhi's nonviolent resistance strategies that overthrew British colonialism inspired many anticolonial movements. His nonviolent tactics of passive resistance, truth telling, and spiritual practices continue to impact social movements today. Gandhi's life of "voluntary simplicity" and reverence for the "simplicity of rural life"[123] illustrated his principles for regenerating and re-empowering India's village life. Gandhi held a vision of "a confederation of self-governing, self-reliant, self-employed people living in village communities, deriving their right livelihood from the products of their homesteads."[124] He maintained that village assemblies needed to retain the power to decide what to export and what to import. This, he said, is how Indian people had lived for millennia in harmony with Earth. He called this *swadeshi*, meaning a "home economy" based on local self-sufficiency. The German Green Party and many others would borrow liberally from Gandhi's ideas[125] as do many environmental protest movements using nonviolent direct-action. The simple living and village life encouraged by Gandhi would inform many back-to-the-land movements of the 1960s as well as the anti-globalization movement in the 1990s.[126]

Early in the 20th century, Western nations shifted increasingly towards bureaucratic organization of work, industrial division of labour, waged labour, and labour mobility. Advanced industrialism emerged, with the use of more complex machinery and innovation of production processes, to achieve economic efficiencies.[127]

As advanced industrialism developed, so did the first wave of 20th-century social movements, defined as organized groups of people joined in collective action towards change-oriented goals using extra- or non-institutional means.[128] Social movements are one legitimate avenue of social protest available in liberal democracies when institutional, legislative, or juridical avenues are not available or not (solely) effective. In other words, there are times when pressure is required from outside the polity, as people "take it to the streets" for their voices to be seen and heard.

While craft and labour union formation and use of strikes began in the 1700s, there was a new surge of unionism in the early 20th century, partly inspired by the ideals of the socialist movement and Marx's *Communist Manifesto*, and partly by pragmatic philosophy which guided social reformism. As mass production industrialism expanded, so did labour organizing among both skilled and unskilled labour to "improve the lot of working people."[129] The dangers of industrialism and the excesses of capitalism and capitalists needed to be tamed, encouraging more vigorous union organizing.

Increasing militancy led to a spectrum of tactics, from sit-down strikes to mass demonstrations. Particularly after the Russian Revolution of 1917 and the Great Crash and Depression from 1929, a new "social contract"[130] was established through unionism. Workers would receive higher wages and reliable benefits including pensions, defended by strong unions. Business would receive reliable, educated, trained, and healthy workers who participated in innovating manufacturing processes to keep prices lower. This social contract was understood as acting in the interests of the whole community where industries were located. Unionism eventually normalized the 8-hour workday, 5-day work week, worker health protection, workplace safety standards, collective bargaining processes, and minority group access to jobs. This social contract held until shareholder capitalism became widespread in the 1980s.[131]

Origin of the United Nations

After two devastating world wars as well as the loss of life during the Holocaust—where 6 million Jews and 5 million non-Jews lost their lives—the United Nations was formed in 1945. Its goal was to maintain global peace and security, protect human rights, offer humanitarian aid, rule on issues of international law, provide peacekeeping in areas of active war, and promote international cooperation on issues of mutual concern. With only 51 member states at the outset, its membership grew rapidly as previous colonies gained independence, with close to 200 member states now.

To further the international rule of law and maintain stable economic systems, the 1947 General Agreement on Tariffs and Trade (GATT) and the 1944 Bretton Woods agreement created a stable international trading system for the period of 1945–1965. It established a fixed currency exchange system using the gold standard. It also founded both the International Monetary Fund and World Bank for short- and long-term international financial assistance.[132]

In the context of the atrocities committed during World War II, the Human Rights Commission of United Nations, formed in 1945, came together to discuss an International Bill of Rights to ensure that the barbarism of the Holocaust would not transpire again. Under the leadership of Eleanor Roosevelt, members of the commission constructed the Universal Declaration of Human Rights (UDHR), defining a set of basic human rights and freedoms as a framework that articulated globally universal ideas and ethical behaviour. Human rights are universal guarantees protecting individuals and groups against actions and omissions that interfere with their fundamental freedoms, entitlements, and human dignity. Such entitlements exist as shared international norms of human morality, justified both as moral norms or *natural rights* supported by reason, and as *legal rights* entrenched in international law. Now codified in international, regional, and national legal systems, they constitute a set of standards against which people and governing bodies are held accountable. The fulfilment of commitments under international human rights treaties is monitored by independent expert committees, who also help clarify the meaning of human rights in specific situations.

Today, the various agencies and councils of the UN address issues of global concern, from climate change, sustainability, education, to all other social, economic, and military issues of international concern. Most importantly, the UN uses its moral force to obtain agreements and monitor adherence, while navigating significant political, economic, and cultural diversity across member nations.

Rise of Social Welfare States

It was the Great Crash of the stock market in 1929 and the subsequent Great Depression of the 1930s that led to the development of social welfarism in Western nations. It was clear that the economic depression and the poverty and hunger it created were not the fault of individuals. Rather, it has been a recurring aspect of capitalism, either as market corrections or corruption, system rigging, and price fixing by monopolies and trusts, a phenomenon foreseen by economist Adam Smith.

In the interwar period, economist John Maynard Keynes advocated for governments to intervene in the market during difficult times, like the Depression, to equitably allocate resources and ensure full employment. He advised spending on national infrastructure which could jumpstart the economy and help nations escape a cycle of economic depression.

To protect citizens from mass poverty and deaths, Western nations began to invest in a social welfare state that provided an economic floor so that basic needs would be met when an individual fell on hard times, especially during such a system correction. The social welfare state was also intended to co-opt the socialism movement of the 1920s and 1930s, which was demanding the means of production be socially owned and under worker control.

President Roosevelt's "New Deal," reproduced in many Western countries, would come to include state-provided compensatory services for those disadvantaged by the economic system, such as social assistance for those unable to work, unemployment insurance for those temporarily unemployed, old age insurance and pensions to avoid elder poverty, and family allowances either universal or for low-income families. It also provided universal basic services outside of the market system, using a not-for-profit system that was managed by the state. This included subsidized mass schooling and, in some countries, free or highly subsidized medical care.

From the 1960s, in an era of optimism and affluence, the social safety net expanded in many nations to cover additional basic needs, such as subsidized utilities, public subsidized housing, student loans, free universities, agricultural subsidies, and resource marketing boards to provide income reliability for producers. This was actively supported by many movements before and after WWII, including labour unions, social gospel movements, farmer's movements, women's movements, student's movements, and the cooperative movements comprising credit cooperatives, producer cooperatives, and consumer cooperatives.

First Global Existential Threat: The Atomic Age

World War II (WWII) marked the beginning of the Atomic Age after the nuclear bombing of Japan, which killed over 200,000 people. Along with their nuclear capacity, the United States rose as a key economic power, becoming the world's banker, initially through its assistance in the reconstruction of Europe. This era saw another massive industrial expansion, redirecting the technological capacity developed during the war into mass production of consumer goods. By 1950, the world had become dependent on oil as its primary energy source, replacing coal. The 1950s and 1960s became a time of prosperity for most Western nations, including more compulsory years of schooling and enhanced accessibility of higher education.

International Development as Industrialism

During decolonization, Western nations felt the "white man's burden" or, rather, the moral duty to spread industrial mechanisms of prosperity to all decolonizing nations, particularly the "Third World."[133] The 1947 American-funded Marshall Plan for the reconstruction of Europe morphed into international development funding by the Truman government. In his inauguration speech of 1949, he asserted, "greater production is the key to prosperity and peace"[134] for newly independent nations. While development work was touted as international goodwill, it was also an expansion of the American and European spheres of political and economic influence, to offset the influence of the growing Union of Soviet Socialist Republics (USSR).

Further, through expanding literacy and public media, now including TV, the West was receiving images of hunger and suffering in other parts of the world. Yet, it was how the West understood "poverty" that would become problematic. Western capitalism, democracy, and industrial development were considered the only acceptable model for "world development." Thus, communities that had been self-sufficient for millennia, now suffering the impacts of colonialism, were considered impoverished and in need of development.

The first UN Development decade, from 1960 to 1970, involved long-term development funds for nations and short-term aid for industrial projects. This was intended to remedy what was considered a temporary state of underdevelopment. Development assistance could move "underdeveloped nations" towards the Rostowian "take-off" stage to join the prosperous world capitalist economy, a fast-tracked mimicry of Europe and the United States.[135] Through the United Nations, every industrialized country agreed to commit 1% of their Gross National Product (GNP) towards international aid. Aid-funded modernization projects assisted not only industrial growth but also social growth in "making humans modern," largely through schooling and higher education.

In postwar industrialized nations, scientific professionalism particularly in engineering and chemistry, grew into a cult of experts and technocrats tasked with solving economic and social "problems." Corporate foundations and industrial research laboratories were "strategic resources to be mobilized in the [American] quest for world dominance,"[136] best seen in the Cold War "race to space." This was also evident in the spread of the Green Revolution, with seeds bred to withstand chemical dependency and mechanized agriculture.[137]

This was the time of the "company man" in new corporate bureaucracies, in the expansion of white-collar salesmen and functionaries whose allegiance to company or organization was primary. Fordism and the principles of mass production and factory line rationalization, followed by Taylorism with the principles of scientific management, would reshape many work lives. This period also saw the expansion of the public sector and the needed civil service in support of the social welfare state.

Shaped by this rapid expansion of income, consumerism and comfort, the middle class in the West grew. Their lives soon included shopping centres, suburban homes, car culture, offices in steel skyscrapers, and radio/TV/movies for popular entertainment. During the Cold War between the United States and USSR in the 1950s, critics of industrial capitalism were silenced. In the United States, the McCarthy era was on the constant lookout for subversive and treasonous activity, particularly socialist sympathies. The FBI rapidly expanded their internal investigation capacities. A reinvigorated competitive individualism, rising entrepreneurial spirit, and growth in the technocratic class helped provide a bulwark against socialism fears. Progressive ideas became suspect, shaping American ideology for decades well beyond the end of the Cold War.

Second Wave Environmentalism

By the late 1950s and early 1960s, the consequences of unmitigated industrial capitalism were becoming apparent to the everyday eye. In addition to lived experience, science itself was revealing profound damage to the natural world.

The second wave of environmentalism is characterized by expanding public awareness and single-issue legislation. The following issues cascaded within a short period of time.

- Persistent smog in all major urban cities, no longer from coal but combined industrial and automotive emissions, were creating significant health issues, deaths, and airport closures. Air pollution studies that began in the 1950s led to a Clean Air Act in England and a 1961 international congress.
- With increased urbanization, the burning of open garbage heaps was no longer acceptable. Research in the early 1950s established requirements for sanitary landfills, leading to the solid waste management policies of the 1960s.
- Rachel Carson in her 1962 book *Silent Spring* addressed the damaging effects of agricultural chemical toxins on species such as birds, leading to the banning of DDT and limits on other agricultural toxins.
- In 1965, Ralph Nader questioned the hand-in-glove relationship between government and industry, particularly in the car industry. He had a profound influence on car safety, from seat belts to air bags, and eventually on pollution from cars.
- Continued atomic testing and dumping led to mass protests from the late 1950s to 1970s, against French atomic dumping in the Mediterranean Sea as well as French and American nuclear testing on Amchitka Island and various Pacific Islands, contaminating people, land, and ocean. Ten Nobel Prize winners including Albert Einstein and Bertrand Russell sign an anti-nuclear manifesto in 1955.
- Partial meltdowns and other accidents in nuclear plants occur in Canada, the United States, and Europe. Helen Caldicott, an Australian physician, called attention to the insanity of nuclear testing and nuclear power, establishing *Physicians for Social Responsibility* among other organizations, winning the 1985 Nobel Peace Prize.
- In 1967, the oil tanker SS Torrey Canyon ran aground off British and French shores, spilling an estimated 100,000 litres of oil. An oil drilling platform off the American Californian coast exploded in 1969, spilling crude which washed up 35 miles of beach, killing untold numbers of ocean life and wildlife.
- Rivers began burning, such as the Cuyahoga River fire in Ohio, 1969, due to industrial dumping of toxins and liquid fossil fuels. Other rivers and the Great Lakes become clogged with raw sewage, textile dyes, detergents, and other waste, forming sludge in many waterways and killing marine, aquatic, and terrestrial life.
- Leaded gasoline in the 1950s was shown to pose grave health risks, including human poisoning, cardiovascular damage, as well as neurological damage in children. Chemist Alice Hamilton begins a campaign against General Motors, but it would take over 50 years to fully ban lead from gasoline.
- In 1966, a coal mine waste pile collapses in Aberfan, Wales, burying a primary school and surrounding homes killing 116 children and 28 adults.

- As early as 1968, meteorologists warn of rising CO_2 levels in the atmosphere.
- The construction of massive dams throughout the world, without concern for flood damage to natural environments or cultural history, led to the creation of UN World Heritage protected sites programme in 1964 and World Heritage Convention in 1972.
- Animal testing in laboratories provokes outrage and changes in practice.
- The use of anti-nausea drugs for pregnant women, particularly thalidomide, causes significant birth defects in 46 countries.

These and other dramatic incidents in the 1950s and 1960s led to important environmental legislation nationally and internationally regarding water and air quality, noise, waste disposal, ocean and waterway dumping, toxic substances, wilderness preservation, animal welfare, endangered species, and more appropriate industrial siting. Many of these environmental campaigns were founded on *reform environmentalism*, which recognized that "humankind is part of nature," and therefore the

FIGURE 2.3 Apollo 8—Earthrise

Credit: NASA. Reprint permitted. www.nasa.gov/image-feature/apollo-8-earthrise

health of ecosystems and human populations are connected.[138] However, there was still only limited critique on the social and economic origins of environmental issues.[139]

In 1965, Adlai Stevenson gave a widely quoted speech to the UN using the metaphor of "spaceship Earth," providing a mental image of a finite planet.[140] Then, in 1968, the Apollo 8 astronauts took the first photos of Earthrise (portrayed in Figure 2.3), and in 1972, the Apollo 17 astronauts took the Blue Marble photo of Earth, an image that would become iconic to the environmental movement and a symbol of common global destiny.

These were just some of the scientific and activist precursors and discrete campaigns that led to a more formal organizing of the environmental movement in the third wave.[141]

The Hippy Generation and New Social Movements of the 1960s

In the midst of growing turmoil, the young generation of Baby Boomers, born in the United States after WWII, began to challenge the ethics of the so-called "American Establishment," who as the dominant elites, controlled much of public life. To them, the dominant culture was a culture of decline, illustrated by a growing malaise and mediocrity. They rejected the repression of the McCarthy anti-communist era and the need for ongoing war by the military industrial complex, from Korea to Vietnam. They profiled the industrial destruction of the environment and resisted the American military draft for the Vietnam War. They spurned the planned obsolescence and throwaway goods of the consumer society as well as the grind of boring, disciplined, factory-style work.

Many young people began to consider themselves a counterculture vanguard, wanting to build a new society out of the destructiveness, alienation, and conformity elicited by the modern industrial era and the Establishment who controlled it. The hippie movement became known simply as the "Great Refusal" to cooperate.[142] The "hip ethics" were to "drop out, do your own thing, be cool, blow the mind of, and freak out" those who *were* cooperating.[143]

As one of the first generations to attend public school for 12 years of their childhood and adolescence, who had not necessarily experienced the world wars and the economic insecurity of the depression, a chasm emerged between them and the older generations called the "generation gap."[144] Yet, they were also the first generation who grew up learning to hide under their school desks in the event of a nuclear war, such as during the Cuban Missile Crisis,[145] the closest the world had come to nuclear war.

Both folk music and the new rock music—culminating in the Woodstock festival attracting up to half a million young people in 1969—expressed their critiques, demands, and yearnings. Their carefree clothing evoked "beat-hip-bohemianism."[146] For the first time, despite clothing bans, women freely wore pants as well as miniskirts. Bras and girdles became symbols of male oppression. Jeans were a symbol of rebellion as well as solidarity with working people. Being "natural" was signified by the long hair and headbands of both male and female, in contrast to the crew cuts and pin curls of the 1950s.

There were three streams within the hippie generation. The first carried a vision of a society founded on love, peace, and equality across race, gender, and class, a new version of the Romantic movement.[147] The hippies created many varieties of communal living, back-to-land, and back-to-nature movements, including commitments to organic food production, farmer's markets, and food cooperatives to bypass supermarkets. The alternative health movement and underground press developed. Fair trade, initially pioneered by Mennonite and Quaker groups, became part of the radical student movements as "trade not aid."[148]

New drugs such as marijuana and LSD as well as sexual freedom aided by "the Pill" augmented this significant cultural revolution. Generally, hippies were white children of relative privilege, rejecting "middle class comforts."[149] Drawing from new psychology, transcendental meditation, and Eastern spirituality, they attempted to deal with ego, arrogance, jealousy, and other human weaknesses, often within communal living experiments. They were anticipating the Age of Aquarius, an astrological age featuring an expansion of human consciousness and enlightenment. This was the precursor of the New Age movement.

The second stream in hippie culture was the New Left, which included radical political activists, most often university students, inspired by Marxism, Critical Theory, Feminist Theory, and other social criticism of industrialism, capitalism, technical rationality, and patriarchy, especially Herbert Marcuse's *One-Dimensional Man*.[150] New Social Movements emerged at this time, including the peace/anti-war movement, nuclear disarmament movement, second wave feminist movement, the anti-poverty movement, and the gay rights movement.

Some of these movements were still concerned with class issues and capitalism akin to the Old Social Movements. They stood in solidarity with "Third World" independence and revolutionary movements. They protested repressive violence and the destabilization of Third World nations by Western democracies.[151] Many young people charged that the stability of capitalist markets was more important to Western nations than the fostering of democracy. They saw the origins of many authoritarian regimes in the Cold War politics of global manipulation.

While the Old Social Movements had been working class, the New Social Movements were generally middle class. The New Social Movements addressed identity politics as well as the meaning systems of modern society, stimulated by a sense of a threatened lifeworld and ecosystem.[152] They were more loosely organized than the formal membership system of the Old Social Movements. They were able to quickly mobilize "conscience constituencies" largely through use of new media like TV, phones, and radio.[153] They staged sits-ins, love-ins, be-ins, die-ins, and teach-ins. They occupied business buildings, banks, government buildings, and university offices, one demonstration ending in the shooting of four students at Kent State University, Ohio, in 1970 portrayed in Figure 2.4.

FIGURE 2.4 Kent State Standoff Between Students and Riot Police

Credit: Bettmann/Getty Images. Reprinted with permission.

The most significant culmination was the 1968 mass protests in Paris, begun by students demanding university democratization. Instead, they experienced a repressive police response. Their protest quickly expanded to general strikes and factory occupations throughout France. This led to global protest and a call for an end to industrial capitalism, consumerism, Western imperialism, traditional politics, patriarchy, and alienating work. In many global cities, urban movements would use guerrilla tactics to target installations, such as nuclear or military,[154] or they would initiate longer term actions, such as the 20-year Greenham Common Women's Peace Camp blockading a UK military base storing cruise missiles.

The third stream within the hippy generation was the militant working class and poor, including the civil rights or Black Power movement and the Native Indian rights or Red Power movement in the United States. Starting in the late 1940s to the 1960s, sit-ins, demonstrations, marches, boycotts, and other forms of nonviolent civil disobedience were used to challenge American racial segregation in housing, bussing, schooling, voting, and laws. After a decade of tumult, President John F. Kennedy, 46 years old, was assassinated in 1963 before he could pass the Civil Rights Act. However, civil rights legislation was passed in 1965. His brother Robert, 42 years old, also a progressive liberal, would be assassinated in 1968. The eloquent leader of the Civil Rights Movement, Martin Luther King Jr, was assassinated in 1968, only 39 years old. As peaceful protests and nonviolent strategies of Black and White allies were met with escalating violence—beating, jackbooting, gassing, and shootings—the Black Panthers organized the militant part of the Black liberation movement.[155] Their 39-year-leader Malcolm X was assassinated by 1965.

The other part of the civil rights movement was the Indigenous Red Power movement in North America, in part inspired by the writing of Vine Deloria Jr in the late 1960s. The movement emerged when the American government withdrew Indigenous recognition, rights, and land guaranteed under the treaty agreements through the Relocation Act of 1956. Further, Native people were still not allowed to vote, defined as "wards of the state." The American Indian Movement (AIM) was formed in 1968 to fight for the civil rights of Indigenous people. They formed a network of support organizations fighting for health, political, land, and educational rights. In 1969, AIM occupied Alcatraz Island, an empty jail, claiming the land as a treaty right and using it as an example of the living conditions on reservations from broken treaties, the first of many such occupations.

The International Indian Treaty Council was formed at Standing Rock in North and South Dakota, 1974, as a voice for Indigenous people globally. This led to a "renaissance" of Native Studies programmes, the taking back of many sacred sites, and the nonviolent protest of projects such as dams and pipelines which desecrated land, water, and cultural sites. As part of reclaiming Indigenous culture after generations of erasure, Indigenous-operated schools and "survival schools" were opened.[156] Indigenous people globally would increasingly become known as the "guardians" of the Earth, both land and water, related to their sacred Original Instructions[157] and Traditional Ecological Knowledge (TEK).

Reflections

We know every civilization carries its own seeds of destruction, working itself into a progress trap, now evident for Western civilization. Cogent critiques of the West have developed from multiple origins, both external and internal.

Within Western civilization, the shift to the modern era was a reaction against the strictures of medieval societies—the burden of tradition, the dogmatism of the Roman Catholic Church, and the obligations of communal life that subordinated individuals, often to oppressive and corrupt authorities.

Through various revolutions, a new cosmology developed of a heliocentric universe that was mathematically and mechanically ordered. From Christianity, history was infused with a sense of progressiveness towards an ideal state. The ideal state of modernity is one that includes liberation, equality, and prosperity. Many of these ideas were nurtured by contact with Indigenous peoples and their various forms of society. A new ontology developed of individual freedom and autonomy, premised on the human dignity explicitly valued in humanism. A new axiology developed of the right of individual conscience and a social contract between nation-states and their citizens. A new epistemology developed on the basis of objective rationalism and universal scientific laws. Over the time of these changes, elite education would eventually shift to mass public schooling, building the capacity for modern individuals to participate in a capitalist economy and democratic polity.

The colonial era spread death to vast numbers of humans as well as nonhuman life. Later, the modern era would become based on industrial capitalism, with additional human and environmental costs. A variety of movements emerged: the Luddites, the Romantic movement, the Arts and Crafts movement, the antislavery and postcolonial movements, the socialist and trade union movements, social reformism, and then the environmental preservationist and conservation wise use movements of the First Wave environmentalism. Under this tumult, however, there has been the persistent belief in the superiority of the West.

After WWII and the intensification of industrialism, we see the rise of the New Left, civil rights movements, and the New Social Movements—including the environmental, feminist, social justice, gay rights, and peace movements—all key players contesting the consequences of modernism and industrialism. Second Wave environmentalism is characterized by this widening of public awareness and single-theme advocacy and legislation.

Yet, the environmental and social justice movements stand on the broad shoulders of dissent since the beginnings of the colonial era and the French, American, and Industrial Revolutions. In the 20th century, the hippies became the next "cultural vanguard," effecting a change in modern consciousness and ethics. They facilitated a rising awareness of technocratic capitalism, patriarchy, environmental desecration, and consumptive materialism, prompting a revaluing of equality, justice, and environmental integrity.[158]

While it was romantic and naïve utopianism in many ways, the hippies questioned the formerly unquestionable, shifting values to the more liberal end of the spectrum. While many baby boomers eventually picked up briefcases and made their living in the mainstream financial and corporate sectors, contradicting earlier commitments, many others took up work in the expanding social professions and social welfare sectors, or they started businesses and cooperatives predicated on a vision of a more peaceful, non-exploitative world. As Timothy Miller claims, "the most enduring legacy of the counterculture is the role that it played in awakening public concern about the environmental crisis of the late twentieth century."[159] They had created political space for the new environmental movement, both reform environmentalism as well as countercultural environmentalism.[160]

The second part of the historical context is recounted in the next chapter, Chapter 3, which traces social and environmental history from the 1970s, as context which impacts the approaches, debates, and challenges in environmental and sustainability education today. Chapter 3 addresses the third and fourth waves of the environmental movement, the emergence of sustainable development and the notion of sustainable societies, as well as the neoliberal backlash and climate crisis.

Notes and References

Chapter line drawing by Canadian artist Erin Marie Lange Christensen. www.erinchristensen.com. Reproduced with permission.

1 Lewis Mumford. Cited in Andrew Jamison and Ron Eyerman, *Seeds of the Sixties* (Berkeley: University of California Press, 1944), p. ii.

2 Mark Twain, *Mark Twain in Eruption* (New York: Harper & Brothers, 1940).
3 Transdisciplinarity here means research that does not originate in any one discipline but crosses many disciplinary boundaries creating holistic or integrated questions and research approach. See Basarab Nicolescu (Ed.), *Transdisciplinarity: Theory and Practice* (Cresskill, NJ: Hampton Press, 2008).
4 Ramachandra Guha, *Environmentalism: A Global History* (London: Longman, 2000).
5 Richard Tarnas, *The Passion of the Western Mind: Understanding the Ideas That Have Shaped Our World View* (New York: Ballantine Books, 1991).
6 David Graeber and David Wengrow, *The Dawn of Everything: A New History of Humanity* (Toronto: Signal McClelland & Stewart, 2021).
7 BCE means Before Common Era previously known as BC. CE means Common Era previously known as AD. This change is inclusive of all faith and cultural traditions.
8 Dates are from Bernard Grun, *The Timetables of History* (New York: A Touchstone Book, 1991).
9 Marija Gimbutas, *The Living Goddesses* (Berkeley: University of California Press, 1999).
10 Riane Eisler, *The Chalice and the Blade: Our History, Our Future* (New York: HarperSanFrancisco, 1987).
11 Eisler, *The Chalice and the Blade*, pp. 42–58.
12 Tarnas, *The Passion of the Western Mind.*
13 Tarnas, *The Passion of the Western Mind*, p. 74.
14 Tarnas, *The Passion of the Western Mind*, p. 89.
15 Tarnas, *The Passion of the Western Mind*, p. 108.
16 Tarnas, *The Passion of the Western Mind*, pp. 89–131. See also Jean Markale, *The Great Goddess* (Rochester, VT: Inner Traditions International, 1999).
17 Tarnas, *The Passion of the Western Mind*, pp. 191–221.
18 Tarnas, *The Passion of the Western Mind,* p. 209.
19 Tarnas, *The Passion of the Western Mind*, p. 209.
20 Charles Taylor, *The Malaise of Modernity* (Toronto: House of Anansi Press, 1991).
21 Charlene Spretnak, *The Resurgence of the Real: Body, Nature, and Place in a Hypermodern World* (New York: Routledge, 1999).
22 Carolyn Merchant, *The Death of Nature: Women, Ecology, and the Scientific Revolution* (New York: Harper & Row, 1980).
23 Taylor, *The Malaise of Modernity*, pp. 2–5.
24 Tarnas, *The Passion of the Western Mind.*
25 Graeber and Wengrow, *The Dawn of Everything.*
26 Theodor Adorno and Max Horkheimer, *Dialectic of Enlightenment* (New York: Herder & Herder, 1947/1979), pp. 3–24. Exclusive language retained.
27 Adorno and Horkheimer, *Dialectic of Enlightenment*, p. 3–24.
28 Max Weber, *The Protestant Ethic and the Spirit of Capitalism* (London: Routledge, 1930).
29 Guha, *Environmentalism*, p. 126.
30 Ronald Johnstone, *Religion in Society* (Upper Saddle River, NJ: Pearson Prentice Hall, 2004), pp. 136–137.
31 Tarnas, *The Passion of the Western Mind*, pp. 411, 402.
32 Ortega as cited in Benjamin Barber, "Jihad vs McWorld," *The Atlantic Monthly*, March 1992, p. 62.
33 Allan Johnson, *The Blackwell Dictionary of Sociology* (Cambridge, MA: Blackwell Publishers, 1995), p. 255.
34 Ronald Wright, *A Short History of Progress* (Toronto, ON: Anansi Press, 2004), p. 107.
35 Malcolm Swan, "Forerunners of Environmental Education," in Noel McInnis and Don Albrecht (Eds.), *What Makes Education Environmental?* (Washington DC and Louisville, KT: Environmental Educators Inc and Data Courier Inc, 1975), pp. 4–20.
36 Joseph Tainter, *The Collapse of Complex Societies* (Cambridge: Cambridge University Press, 2017).
37 Tainter, *The Collapse of Complex Societies*, pp. 91–94.
38 Wright, *A Short History of Progress*, p. 108.

39 Jeremy Lent, *The Patterning Instinct: A Cultural History of Humanity's Search for Meaning* (Amherst, NY: Prometheus Books, 2017). See also Tainter, *The Collapse of Complex Societies*; Jared Diamond, *Collapse: How Societies Choose to Fail or Succeed* (New York: Penguin Books, 2005); Naomi Oreskes and Erik Conway, *The Collapse of Western Civilization: A View from the Future* (New York: Columbia University Press, 2014); John Michael Greer, *Dark Age America: Climate Change, Cultural Collapse and the Hard Future Ahead* (Gabriola Island, BC: New Society Publishers, 2016); among others.
40 Carolyn Merchant, *Ecological Revolutions: Nature, Gender, and Science in New England* (Chapel Hill/London: The University of North Carolina Press, 1980), p. 1.
41 Albert Borgman, *Crossing the Postmodern Divide* (Chicago: University of Chicago Press, 1992), p. 22.
42 Washington, *Demystifying Sustainability*, p. 7.
43 Washington, *Demystifying Sustainability*, pp. 6–16.
44 Washington, *Demystifying Sustainability*, p. 7.
45 David Suzuki with Amanda McConnell, *The Sacred Balance: Rediscovering our Place in Nature* (Vancouver: GreyStone Books, 1997). Cited in Washington, *Demystifying Sustainability*, p. 8.
46 Washington, *Demystifying Sustainability*, p. 8.
47 Suzuki with McConnell, *The Sacred Balance*, p. viii.
48 Daniel Paul, *We Were Not the Savages: Collision between European and Native American Civilizations* (Halifax: Fernwood Publishing, 2006), p. 7.
49 Paul, *We Were Not the Savages*.
50 www.britannica.com/biography/Bartolome-de-Las-Casas.
51 www.britannica.com/biography/Bartolome-de-Las-Casas.
52 www.britannica.com/biography/Bartolome-de-Las-Casas.
53 James Daschuk, *Clearing the Plains: Disease, Politics of Starvation and the Loss of Aboriginal Life* (Regina: University of Regina Press, 2013).
54 Sheldon Krasowski, *No Surrender: The Land Remains Indigenous* (Regina: University of Regina Press, 2019).
55 See Krasowski, *No Surrender*. For a wry account, see Thomas King, *The Inconvenient Indian: A Curious Account of Native People in North America* (Toronto: Anchor Canada, 2012).
56 John Neihardt, *Black Elk Speaks* (Lincoln: University of Nebraska Press, 2014), p. 169.
57 Neihardt, *Black Elk Speaks*.
58 Chief Dan George, *My Spirit Soars* (Surrey: Hancock House Publishers, 1982), p. 17.
59 David Graeber and David Wengrow, *The Dawn of Everything: A New History of Humanity* (New York: Signal of Penguin Random House Canada, 2021), p. 52.
60 Graeber and Wengrow, *The Dawn of Everything*, p. 37.
61 Graeber and Wengrow, *The Dawn of Everything*, p. 37.
62 Graeber and Wengrow, *The Dawn of Everything*, p. 38.
63 Graeber and Wengrow, *The Dawn of Everything*, p. 48.
64 Graeber and Wengrow, *The Dawn of Everything*, p. 48.
65 Graeber and Wengrow, *The Dawn of Everything*, p. 48.
66 Graeber and Wengrow, *The Dawn of Everything*, pp. 54–56.
67 Graeber and Wengrow, *The Dawn of Everything*, p. 61.
68 Graeber and Wengrow, *The Dawn of Everything*, p. 74.
69 See Patrick Bond, *Looting Africa: The Economics of Exploitation* (Pietermaritzburg: University of KwaZulu-Natal Press and London: Zed Press, 2006).
70 www.un.org/en/events/slaveryremembranceday/.
71 UNESCO, *Healing the Wounds of Slave Trade & Slavery* (Brighton: The UNESCO Slave Route Project/GHFP Research Institute, 2021), p. 2.
72 UNESCO, *Healing the Wounds of Slave Trade & Slavery*, p. 2.
73 Bond, *Looting Africa*, p. x.
74 www.un.org/en/events/slaveryremembranceday/confronting-slavery-legacy-of-racism-together-in-covid19.shtml.

75 International Labour Organization, Walk Free, and UN International Organization on Migration. *Global Estimates of Modern Slavery Forced Labour and Forced Marriage*, September 2022. See press release at https://www.ilo.org/global/about-the-ilo/newsroom/news/WCMS_855019/lang--en/index.htm. See report at https://www.ilo.org/wcmsp5/groups/public/---ed_norm/---ipec/documents/publication/wcms_854733.pdf.
76 Kevin Bales, *Disposable People: New Slavery in the Global Economy* (Berkeley: University of California Press, 1999), p. 4.
77 See www.un.org/en/events/slaveryremembranceday/confronting-slavery-legacy-of-racism-together-in-covid19.shtml.
78 file:///C:/Users/EAL/Desktop/UNESCO-GHFP_2020_Healing-the-Wounds-of-Slavey_Desk-Review_Report.pdf.
79 Bond, *Looting Africa*, p. xiii.
80 Robert Brown, "A Conservation Timeline: Milestones of the Model's Evolution," *The Wildlife Professional*, Fall 2010. www.wildlifedepartment.com/aboutodwc/A%20Conservation%20Timeline[1].pdf.
81 Brown, " A Conservation Timeline."
82 Swan, "Forerunners of Environmental Education," p. 5.
83 Discussed in Jacobus Du Pisani, "Sustainable Development – Historical Roots of the Concept," *Environmental Sciences*, Vol. 3, No. 2 (2006), pp. 83–96. https://doi.org/10.1080/15693430600688831, p. 86.
84 Guha, *Environmentalism*, p. 4. Emphasis added.
85 Guha, *Environmentalism*, p. 5.
86 Guha, *Environmentalism*, p. 45.
87 Brown, *A Conservation Timeline.*
88 Guha, *Environmentalism*, p. 47.
89 Discussed in Pisani, "Sustainable Development – Historical Roots of the Concept."
90 Guha, *Environmentalism*, p. 26.
91 See J. B. MacKinnon, *The Once and Future World: Nature As It Was, As It Is, and As It Could Be* (Toronto: Vintage Canada, 2013).
92 MacKinnon, *The Once and Future World*, p. 45.
93 MacKinnon, *The Once and Future World*, p. 91.
94 MacKinnon, *The Once and Future World*, p. 19.
95 Guha, *Environmentalism*, p. 39.
96 Guha, *Environmentalism*, p. 26.
97 Swan, "Forerunners of Environmental Education," p. 5.
98 Bruce Pascoe, *Dark Emu: Aboriginal Australia and the Birth of Agriculture* (Melbourne: Scribe, 2018), p. 11.
99 James Thomson (1874), *The City of the Dreadful Night*, in E. P. Thompson (2011). *William Morris* (Oakland, CA: PM Press and Merlin Press, 2011), p. 128.
100 Mark Rutherford in Thompson, *William Morris*, p. 142.
101 Discussed in Du Pisani, "Sustainable Development – Historical Roots of the Concept."
102 J. R. McNeill, *Something New Under the Sun: An Environmental History of the Twentieth-Century World* (New York: W. W. Norton & Company, 2000), pp. 64–66.
103 In Charlene Spretnak, *The Resurgence of the Real: Body, Nature, and Place in a Hypermodern World* (New York: Routledge, 1999), p. 139.
104 www.brainyquote.com/quotes/johann_wolfgang_von_goeth_393498.
105 William Morris, "Art, Socialism and Environment," in Laurence Coupe (Ed.), *The Green Studies Reader: From Romanticism to Ecocriticism* (London: Routledge, 2000), pp. 32–36.
106 Guha, *Environmentalism*, p. 13.
107 Discussed in Du Pisani, "Sustainable Development – Historical Roots of the Concept," p. 86.
108 Patriarchy means the form of social organization in which wealth, power, privilege, and inheritance accrue to males over females. John Macionis and Linda Gerber, *Sociology*, Eighth Canadian Edition (Toronto: Pearson Prentice Hall, 2014), p. 312.

109 John Macionis and Linda Gerber, *Sociology*, Fifth Canadian Edition (Toronto: Pearson Prentice Hall, 2005), p. 328.
110 Nikki Van der Gaag, *The No-Nonsense Guide to Women's Rights* (Toronto: New Internationalist Press and Between the Lines, 2004), p. 22.
111 Merchant, *The Death of Nature*, p. xvi.
112 John Muir in Guha, *Environmentalism*, p. 50.
113 John Muir in Guha, *Environmentalism*, p. 53.
114 Linda Graber, *Wilderness as Sacred Space* (Washington, DC: The Association of American Geographers, 1976), p. 50, in Guha, *Environmentalism*, p. 86.
115 Guha, *Environmentalism*, p. 50.
116 Aldo Leopold in Guha, *Environmentalism*, p. 56.
117 Aldo Leopold in Guha, *Environmentalism*, p. 56.
118 Guha, *Environmentalism*, pp. 57–58.
119 Guha, *Environmentalism*, p. 58.
120 Guha, *Environmentalism*, p. 8.
121 Guha, *Environmentalism*, p. 31. Emphasis in original.
122 From Heather Farley and Zachary Smith, *Sustainability: If It's Everything, Is It Nothing?* (London: Routledge, 2020), pp. 16–17.
123 Guha, *Environmentalism*, p. 20.
124 Satish Kumar, "Gandhi's Swadeshi: The Economics of Permanence," in Jerry Mander and Edward Goldsmith (Eds.), *The Case against the Global Economy: And For a Turn Toward the Local*, pp. 418–424 (San Francisco: Sierra Club Books, 1996), p. 419.
125 Guha, *Environmentalism*, p. 8.
126 See Jerry Mander and Edward Goldsmith (Eds.), *The Case Against the Global Economy: And For a Turn Toward the Local* (San Francisco: Sierra Club Books, 1996).
127 Tony Watson, *Sociology, Work and Industry* (London: Routledge, 1995), p. 97.
128 Paraphrased from David Snow, Sarah Soule, and Hanspeter Kriesi, *The Blackwell Companion to Social Movements* (Malden, MA: Blackwell Publishing, 2007), p. 6.
129 Snow et al., *The Blackwell Companion to Social Movements*, p. 563.
130 A social contract is an agreement between citizens and their state, where citizens surrender some freedoms to the state for their protection in other ways, so that all parties receive social benefits. Allan Johnson, *The Blackwell Dictionary of Sociology* (Oxford: Blackwell Publishing, 1995), p. 258.
131 Josh Freedman and Michael Lind, "The Past and Future of America's Social Contract," *The Atlantic*, December 19, 2013. www.theatlantic.com/business/archive/2013/12/the-past-and-future-of-americas-social-contract/282511/.
132 Patricia Marchak, *The Integrated Circus: The New Right and The Restructuring of Global Markets* (Montreal: McGill-Queen's University Press, 1991).
133 The Western industrialized nations were considered to comprise the First World, and the Second World was considered the nations within the Soviet communist sphere of influence. The Third World was originally named for the group of nations who remained nonaligned during the Cold War.
134 Guha, *Environmentalism*, p. 64. See also Gustavo Sachs, "Development," in Wolfgang Sachs (Ed.), *The Development Dictionary: A Guide to Knowledge as Power* (London: Zed Books and Johannesburg: Witwatersrand University Press, 1992), pp. 6–25.
135 In 1960, Walt Rostow contributed a stage theory that guided the policies of the American administration in the early 1960s. All nations would need to pass through the five stages of traditional society, preconditions for take-off, take-off, road to maturity, and age of high mass consumption. He assumed that international aid and modern technology could accelerate the process that had taken centuries in Europe and the United States. In George Rist, *The History of Development: From Western Origins to Global Faith* (London: Zed Books, 1997), pp. 93–99.
136 Andrew Jamison and Ron Eyerman, *Seeds of the Sixties* (Berkeley, CA: University of California Press, 1994), p. 5.

137 The Green Revolution was the application of science to breed high-yield crops and grow them with the application of machinery and industrial chemicals, such as pesticides, insecticides, fungicides, and herbicides.
138 Christopher Rootes, "Environmental Movements," in David Snow, Sarah Soule, and Hanspeter Kriesi (Eds.), *The Blackwell Companion to Social Movements* (Malden, MA: Blackwell Publishing, 2007), p. 612. See also Colin Soskolne (Ed.), *Sustaining Life on Earth: Environmental and Human Health through Global Governance* (Lanham, MD: Lexington Books, 2008).
139 Rootes, "Environmental Movement," p. 613.
140 Swan, "Forerunners of Environmental Education," p. 14.
141 Rootes, *Environmental Movements*, p. 612.
142 Timothy Miller, *The Hippies and American Values* (Knoxville, TN: The University of Tennessee Press, 1991), p. 8.
143 Miller, *The Hippies and American Values*, p. 5.
144 Miller, *The Hippies and American Values*, p. 9.
145 The Cuban Missile Crisis was a standoff between the United States and the Soviet Union for 13 nail-biting days in 1962, when the Soviets installed nuclear missiles on Cuban territory, just 100 miles from the United States. John F. Kennedy ordered a naval blockade of Cuba. The issue was solved when Soviet leader Nikita Khrushchev offered to remove the missiles if the United States agreed not to invade Cuba. However, it was the closest the world had come to nuclear war. From this point, while nuclear proliferation continued for some years, the Nuclear Proliferation Treaty was signed in 1968 among 18 nations to restrict development, spread of nuclear weapons, and size of arsenals, as part of disarmament. The treaty is reviewed and renegotiated regularly. See www.un.org>wmd>npt.
146 Miller, *The Hippies and American Values*, p. 9.
147 Miller, *The Hippies and American Values*, p. 11.
148 In addition to Timothy Miller see R. Bahro, *Avoiding Social and Ecological Disaster: The Politics of World Transformation* (Bath, UK: Gateway Books) and Mathias Finger, "New Social Movements and Their Implications for Adult Education," *Adult Education Quarterly*, Vol. 40, No. 1 (1989).
149 Miller, *The Hippies and American Values*, p. 15.
150 Herbert Marcuse, *One-Dimensional Man: Studies in the Ideology of Advanced Industrial Society* (Boston: Beacon Press, 1964).
151 Rod Bantjes, *Social Movements in a Global Context* (Toronto: Canadian Scholars' Press Inc., 2007), pp. 68–69.
152 Michael Welton, "Social Revolutionary Learning: The New Social Movements as Learning Sites," *Adult Education Quarterly*, Vol. 43, No. 3 (1993), p. 153.
153 Bantjes, *Social Movements in a Global Context*, p. 71.
154 Bantjes, *Social Movements in a Global Context*, p. 85.
155 Bantjes, *Social Movements in a Global Context*, p. 82.
156 Zinn Education Project, *Native American Activism: 1960s to Present*. www.zinnedproject.org.
157 "Original instructions refer to the many diverse teachings, lessons, and ethics expressed in the origin stories and oral traditions of Indigenous Peoples." Melissa Nelson (Ed.), *Original Instructions: Indigenous Teachings for a Sustainable Future* (Rochester, VM: Bear & Company, 2008), p. 2.
158 Miller, *The Hippies and American Values*, pp. 125–126.
159 Miller, *The Hippies and American Values*, p. 137.
160 Countercultural environmentalism is "part cultural expression, part social dissatisfaction, part search for new environmental values." Christopher Rootes, "Environmental Movements," in David Snow, Sarah Soule and Hanspeter Kriesi (Eds.), *The Blackwell Companion to Social Movements* (Malden, MA: Blackwell Publishing, 2007), pp. 613, 614.

3

WAVES OF ENVIRONMENTALISM, DEVELOPMENT, AND BACKLASH

Historical Context—Part Two

> But for creatures who live a mere 70 years or so, the study of the past, distant, and recent, is required to know what the range of possibilities includes, and to know what might endure.[1]

The history of the modern environmental movement is typically recounted from the 1970s. However, as Chapter 2 explored, a much broader context is needed for identifying the possibilities that can take us beyond the current environmental and social justice crises. The Modern era has been an ahistorical era, shunning the

DOI: 10.4324/9781003159643-3

past while privileging the "now." Yet, without historical context, as historian J.R. MacNeill asserts earlier, we cannot fully ascertain the significance of current events or what holds promise for truly sustainable societies.

Chapter 2 discussed the scope and historical arc of Western civilization from Classical Greece and Rome, particularly the origins of the Western worldview, vital for recognizing the progress trap in which we are embroiled. This history is also vital for recognizing the epochal shift currently underway, with potential for either civilization collapse or breakthrough.

Chapter 2 also very briefly traced the Modern era, from early to late stages, illustrating early seeds of environmentalism and justice-seeking. It focussed on the colonial era, as the spread of death, as well as previously overlooked critiques from those outside the Western project that perhaps we now have ears to hear.

Radical[2] and reformist protests of industrial capitalism within the West have been numerous and linked, especially as part of both the Old and New social movements. The shifting political economic context after WWII—from the rise of social welfare states, internationalism, development polices, Fordist industrial policies and unionism, a shifting social contract within Western nation-states, to a profound cultural revolution via the counterculture (hippy) generation—shaped state-level and popular responses to emerging environmental and social justice issues.

This chapter, Chapter 3, continues this story from 1970, tracing a larger context. Importantly, this chapter examines the intertwining of emerging environmental concerns and Third and Fourth Wave environmentalism, alongside the 5- and 10-year themes in international development efforts. Even though environmentalism and international development were addressed separately in distinct social movements throughout the 1970s and 1980s, eventually they would merge under the conceptual frameworks of sustainability and sustainable development. The tracing of this larger historical context and interwovenness of movements reveal the need for wholistic and comprehensive approaches within environmental and sustainability education, as well as points to seeds of possibilities beyond the current moment.

Rise of Environmental Movements

> Earth Day is a commitment to make life better,
> not just bigger and faster;
> to provide real rather than rhetorical solutions.
> It is a day to re-examine the ethic of individual progress
> at mankind's (sic) expense.
> It is a day to challenge the corporate and governmental leaders who promise change,
> but who shortchange the necessary programs.
> It is a day for looking beyond tomorrow.
> April 22 seeks a future worth living.
> April 22 seeks a future.[3]

This description of Earth Day is taken from the 1970 advertisement (as seen in Figure 3.1). Earth Day has long been considered the beginning of the environmental movement, as an *organized movement*, which bound single-issue campaigns and the growing number of groups together into a concerted effort. In response to the intensification of environmental and human catastrophes, there was a surge of symbolic and material resources which galvanized a much larger number of participants into social movements. Environmental damage was witnessed not only personally but also nightly on the TV news, with the advent and diffusion of TV. Images of famine and war also entered homes and minds, growing a sense of common humanity. Scientific knowledge was becoming more available to the public, explaining the causes and seriousness of issues.

These new realities coalesced into the first international Earth Day on April 22 (the northern spring equinox) held in New York, 1970. Spearheaded by a Wisconsin senator, Gaylord Nelson, and a small group of activist organizers,[4] the purpose was to hold a National Environmental Teach-In in New York, as the lead event of many decentralized events across the country (see Figure 3.2). Throughout

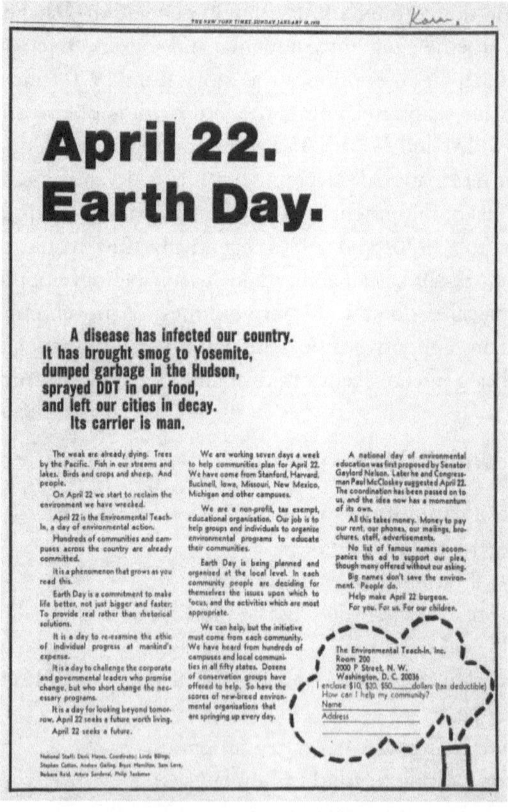

FIGURE 3.1 Earth Day Advertisement 1970

Credit: Wisconsin Historical Society WHI-122404. Reprinted with permission.

FIGURE 3.2 Denis Hayes and Staff Members of the Earth Day National Environmental Teach-In, Washington DC, April 24, 1970

Credit: Photo by John Olson/The Life Picture Collection/Shutterstock. Reprinted by permission.

the United States, an estimated 20 million people from colleges, universities, high schools, and the general public arrived at locations throughout the country to march, speak, and attend teach-ins.

These Earth Day events were covered extensively by the media, in a positive and supportive way (see Figure 3.3). A public consciousness was emerging that environmental protection would be necessary for human survival. Earth Day captured this growing consciousness and provided a space for voice.

The first Earth Day generated momentum for a flurry of activity in the United States. The establishment of the Environmental Protection Agency as well as many new pieces of environmental legislation occurred within just a few months. Yale law professor, Charles Reich, in his groundbreaking book, *The Greening of America*, 1970, asserted that many Americans were realizing their way of life was "dealing death."[5] They grasped that there was an American Crisis developing instead of the American Dream, a crisis sponsored by corporations, the state, and elites . . . "The

FIGURE 3.3 New York Times Headline on Earth Day

Establishment." Reich predicted a revolution was in the making, a consciousness revolution that would be facilitated by the 1960s countercultural movement. He suggested it would foster individual change, ultimately leading towards a revolution of political structures.

UN Secretary General U Thant offered international support for annual Earth Days when, in 1971, he developed an ongoing ceremony to ring the peace bell at the United Nations Headquarters in New York. Ten years later, Earth Day spread to Canada, and 20 years later, in 1990, Earth Day went global, mobilizing 200 million people in 141 countries. These global gatherings would pave the way for the pivotal 1992 Rio Earth Summit. Fifty years later in the new millennium, Earth Day now engages over 1 billion people in over 193 countries. It is still working towards protecting the planet from new ravages, while building a global "cohesive, coordinated, and diverse movement."[6]

Environmental Movement Organizations

The environmental movement is now considered one of "the most comprehensive and influential movements of our time."[7] Political and environmental sociologist Christopher Rootes suggests that the environmental movement became the "embryo of the transformative social movement that would be to the 'postindustrial' society, what the working-class movement promised to be for industrial society."[8] Across the Western world, Earth Day organizers now are just one of the many Environmental Movement Organizations (EMOs).

Together, EMOs mobilize masses of people onto the street, create media spectacle, generate new political agendas, and register responses to international discussions. EMOs work outside the system but in public spaces, where moral and political boldness is possible. Social movements such as EMOs choose forms of protest based on attaining simultaneous goals: maximizing their public exposure, building internal solidarity and a sense of collective identity, achieving legitimacy for their demands in the public mind, and persuading political or corporate authorities to meet their demands.[9] They must constantly innovate to remain newsworthy.

Generally, however, the more radical their choices, the more counterproductive to gaining public sympathy. Eventually, the role of EMOs would largely become direct action.[10] They situate themselves at key points of resource extraction, whether confronting whaling boats, pipeline construction sites, forestry sites, and nuclear installations, or calling attention to practices through images, public information, and obstruction (see Figure 3.4). Their role has been to build public support and political pressure through their visibility. At times, this pressure has assisted their leaders or environmental non-government organizations (ENGOs) who might be negotiating behind the scenes.

In the 1970s and 1980s, EMOs were dominated by charismatic, white, middle-class, straight males who used oligarchic organizational structures. Eventually, after a few decades, this would give way to more diverse, adaptable, and fluid organizational systems. During the 1970s, not only did the membership of larger EMOs surge,[11] but thousands of local grassroots groups would form around anti-pollution, anti-nuclear, anti-war, and anti-toxin goals. These grassroots groups were often led by women concerned for their children and community health, such as Lois Gibb in Love Canal near Niagara Falls, 1979, protesting a municipal waste site that later became a toxic waste site upon which housing, a school, and public park was built.[12] Supporting community outrage at such negligence, the rates of disease such as leukaemia and birth defects were revealed by several journalists.

FIGURE 3.4 Greenpeace Takes on Whaling

Credit: Greenpeace International. Reprinted with permission.

Local groups like this learned how to do citizen research and take public action to change local policy, as self-guided citizen education. Their information was based on scientific research such as Rachel Carson's *Silent Spring*; investigative journalism like at Love Canal; or they learned to follow the money, influence, policy, and practices of locally situated corporations and government. These grassroots groups received generous financial and moral support at the time. The large EMOs received substantial donations, sometimes from wealthy individuals and corporate donors, often to hold their own industries accountable.[13]

Environmental Nongovernmental Organizations

National organizations also emerged, now known as environmental nongovernmental organizations (ENGOs). Their role was public education, raising monies, advocating for and reviewing new legislation and programmes, as well as ensuring proper environmental enforcement. During this first decade, nation-wide ENGOs began to meet to divide the work among themselves, rather than compete for funds, adherents, and goal achievement. They were called the Group of Ten who carried significant influence (Sierra Club, Wilderness Society, Environmental Defense Fund, Environmental Policy Institute, Friends of the Earth, Izaak Walton League of America, National Audubon Society, National Parks and Conservation Association, Natural Resources Defense Council, and Defenders of Wildlife).[14] Together, they would strategize, act as lobbyists, hire lawyers, and consult scientists. They "pledged support for each other's actions."[15]

The Group of Ten became organizations with fulltime professional staff, large budgets, and professionalized campaigning. They represented the institutionalization of environmentalism, eventually as part of Third Wave environmentalism. The goal was to forge a long-term environmental agenda and to frame public environmental discourse, with the goal of generating public consensus.[16] Many parallel organizations emerged across Western nations, in North America, Western Europe, Australia, and New Zealand. As reform environmentalism, they work for pragmatic change through negotiated compromise, rather than fundamental systemic change.

Also emerging from the first Earth Day was the first UN Conference on the Human Environment, held in Stockholm in 1972. As the first international discussion on the environment, it would lead to the establishment of the UN Environment Programme (UNEP). This was the start of nations across the world collectively addressing environmental issues that were global in scope—from air quality to ocean health to illegal wildlife trafficking—now more than 50 years old.

UNEP became the central coordinating body for United Nation environmental activities.[17] It assists in identifying issues and coordinating scientific analysis for the purposes of developing global conventions, cooperating in educational programming, and promoting relevant international policies and practices. Over time, the UN Environment Assembly became the world's highest-level decision-making body on the environment.[18] Located in Nairobi, Kenya, with many regional offices, UNEP carries out its current work in seven theme areas: climate change, disasters and conflicts, ecosystem management, environmental governance, chemicals and waste, resource efficiency, and environment under review.[19]

Green Politics

This upwelling in environmental animation from the 1960's student protests, anti-nuclear movement, and environmental organizations led to the growth of Green political parties in the 1980s. It was clear that most established political parties could not stretch enough to embrace an environmental platform. While EMOs were creating pressure for better environmental laws and policies, the goal was to "transform politics itself in new, more truly democratic, environmentally just, and green directions."[20] Forming a new political party could bring the values of environmentalism and environmental philosophy more directly to public policy and the political process. Green parties continue to stand as a "moral challenge" to beliefs about governing in the Western world.[21]

While there is now an international network of Green parties, the Global Greens established in 2001, they have been most successful in Western Europe, where those in radical politics and single-issue environmentalism joined together. Green parties are generally a coalition across environmental and social justice perspectives from green (ecological) to red (socialist), often proving difficult for ongoing unity. The rainbow includes a range from light green (eco) to dark green (agrarian romantics), from red (leftists, socialists, social democratic) to red-green (ecosocialists), and from Green anarchists/social ecologists to socialist/ecosocial feminists. One of the key divides in Green parties has been between fundamentalists who reject industrial capitalism and those who choose a practical approach by working within existing systems, using step-by-step legislative change.[22]

Rather than combative, uncivil political processes, Greens attempt to engage in politics from a much different frame. They promote a more participatory politics, authentic democracy, and consensus decision-making, especially avoiding ruthless partisanship and conflictual or attack-style processes. Greens have been most successful in voting systems based on "proportional representation" rather than "first-past-the-post" systems.[23] This shift in voting system has enabled Green parties to obtain representation and work effectively through negotiation within "coalition governments,"[24] which typically arise through proportional representation systems.

Starting in the 1980s, Green parties were simultaneously organized throughout Europe—including Germany, Switzerland, Belgium, Italy, France, and Sweden—as well as in Australia (Tasmania), New Zealand, and Canada. In fact, the first ecological party was the United Tasmania Group in Australia, established in 1972 against the building of a dam which would destroy significant rainforest, a precursor to their Green party. Similarly, was the Values Party in New Zealand and the PEOPLE later Ecology Party in the UK. These proto-Green parties, often bringing together support from the anti-nuclear, environmental, peace and New Left movements, led to full Green parties in many places.[25]

In West Germany, the Green party was established in 1980. In 1983, 28 members of the Green party were elected to the West German parliament.[26] In 1996, Ralph Nader was nominated by the US Green party as the first presidential nominee, finishing fourth, although the American two-party system is much less amenable to the long-term success of a Green party. One of the most successful Green parties in Africa has been the *Mazingira* Green Party in Kenya, a *Swahili* word for environment.[27] Wangari Maathai, of Green Belt and Nobel Peace Prize fame, ran as a presidential candidate for this party in 1997. In Africa, there are now 19 Green parties. In Latin America, Indigenous peoples have been important contributors to Green parties, although Green parties are relatively weak in North and South America. Nevertheless, individuals have achieved electoral success, such as Elizabeth May in Canada and Marina Silva in the Brazilian Green Party who served as Minister of the Environment.[28]

Green parties have become part of an international network of discourse and mutual assistance. At the Rio Earth Summit in 1992, a Charter of the Global Greens was approved, and a federation of Green parties with membership from every continent was established.[29] Their primary platform is based on four pillars: grassroots democracy, social justice and equal opportunity, ecological wisdom, and nonviolence.[30] Their ten values include the foregoing four pillars as well as decentralization, community-based economics and economic justice, feminism and gender equality, respect for diversity, personal and global responsibility, and future focus and sustainability.[31] Overall, Green parties are part of the global Green movement that encompasses party politics, the movement politics of EMOs, and the organizational power of ENGOs.

The Science of Ecology

During this burst of organizing, a new science discipline aided environmental research and theory. Departing from a mechanistic understanding of the natural world, a branch of modern biology called ecology emerged as a new theoretical science. From the late 1800s and early 1900s, German zoologist Ernst Haeckel, considered the "father" of ecology, created a holistic approach to biology,[32] discussed further in Chapter 4. It would be the 1980s when the terms ecology and ecosystems entered common parlance.

Briefly, *ecology* is the study of the interrelationships between organisms and their environments, particularly the processes by which they interact and relate to the physical and chemical components of their surrounding environment.[33] The term *ecosystem* refers to an interconnected web of species which share habitat, exchange nutrients, and recycle wastes,[34] called a "natural economy."[35] Ecosystems can be explored at various levels, including the planetary scale, bioregional scale, ecosystem scale, human scale, and microbial scale.[36] *Biodiversity*, the richness of life, results from variety within and between species, unique to geographic location, as well as from unique linkages within and between ecosystems.[37]

Environmentalists began to turn to ecological explanations of environmental issues, expanding science-informed policy-making and problem-solving. Environmental philosopher Paul Shepard considered ecology a "subversive science," a basis for a resistance movement that could undo human exceptionalism.[38]

In 1973, Norwegian philosopher Arne Naess posed the philosophical concept of *deep ecology*, going beyond ecological awareness to a deep, personal understanding of self as integrated in the natural world, an ecological self. Naess felt that reform environmentalism was anthropocentric, shallow, and piecemeal in approach, therefore unable to generate the necessary deep, long range changes in the human/nonhuman relation as a needed culture shift. Deep ecology, as an ecocentric philosophy, considers the preservation of ecosystems first, with humans as only one among many species. Humans are not exempt from, but dependent on, ecological laws and natural processes. Naess did not offer a critique of capitalist production, but he promoted the transformation of consciousness. This generated biocentric environmental groups like Earth First! in 1980, which do not necessarily assume the continuance of humans.

Third Wave Environmentalism

While First Wave environmentalism is considered to represent the conservation and preservation efforts in the late 19th and early 20th centuries, and Second Wave environmentalism as the representation of growing public awareness and single-issue campaigns in the mid-20th century, this growth as an organized movement in the 1970s and 1980s, constituted by professional organizations, comprised Third Wave environmentalism. The 1960s counterculture and New Social movements laid the groundwork for this maturing of environmentalism. It was continually fed by the outrage accompanying constantly emerging environmental issues.

In the 1980s, new environmental catastrophes and vigorous responses continued to occur, such as acid rain, the ozone hole, nuclear reactor meltdown, Amazon deforestation, and toxic waste sites. The New Ecological Paradigm (NEP) was designed in 1978 by environmental sociologist Riley Dunlap and colleagues to measure public levels of environmental concern and the various facets of an environmental world view.[39] With the rise of ecology and new environmental philosophies, the Third Wave became increasingly complex with distinct threads appearing, including environmental justice, ecofeminism, social ecology, spiritual ecology, and Traditional Ecological Knowledge (TEK), discussed further in the context of education in Chapter 4.

Acid rain became a pressing global issue during this time. Acid rain is created from fossil fuel emissions, metal ore smelting, and coal-driven electrical utilities. Acidic particles of sulfuric and nitric acids are deposited by rain, snow, and hail. These particles acidify natural water bodies causing eutrophication and lake die-off; a weakening of forests and forest die-back; contamination of soil, fish, and seafood, as well as the deterioration of human buildings and monuments. Through rigorous international clean air agreements struck from the 1970s to 1990s, emissions reductions were achieved, reducing sulphur oxide emissions by 88% and some nitrogen oxides by 50% in 2017, substantially reducing the problem across the globe.[40]

In 1985, three scientists[41] discovered a significant reduction in stratospheric ozone levels over the Antarctic stations, starting in the late 1970s.[42] They linked the ozone holes to the chlorofluorocarbons (CFCs) used in aerosol cans and in cooling systems such as air conditioners and refrigerators. A global agreement was signed increasing restrictions on their use. The 1987 Montreal Protocol, the only treaty signed by all nations, who constitute the 198 UN Member States, stipulated phase-outs by 2010.[43] As a result, the ozone hole closed in 2020 although full recovery is not expected until 2050, given the lifetime of these chemicals.[44]

In 1986, the Chernobyl disaster of a Russian nuclear plant meltdown and subsequent fire led to plumes of radioactive waste being released into the atmosphere for 9 days, particularly over the USSR and Western Europe. It was estimated to be 400 times more potent than the combined atomic bombing of Hiroshima and Nagasaki. Many staff and firefighters sustained acute radiation injuries and died soon thereafter. The local city of Pripyat, Russia was evacuated with a total of over 100,000 dislocated from the area. A 30-kilometre permanent Exclusion Zone was established around the reactor site, still in effect. It is estimated that at least 4,000 died from radiation-related causes, much higher if all of Europe is considered. This gave new impetus to the anti-nuclear movement.

In the 1960s, deforestation in the Amazon rainforest made way for soy production and cattle ranching as part of international development, peaking in the 1990s after eliminating approximately 20% of the Amazon rainforest. At that time, it was estimated that the Amazon, called the lungs of the planet, produced as much as 20% of atmospheric oxygen, although that has been reconsidered to be about 6%. Deforestation not only reduces oxygen production and removes a carbon sink, it also adds substantially to carbon dioxide emissions, reduces biodiversity, and displaces Indigenous peoples. Deforestation was eventually reduced through the following actions:

- consumer boycotts and media campaigns,
- significant international pressure on corporate exporters,
- a soy then beef moratorium,
- a land registry,
- international monies paid to Brazil for forest protection, including for active monitoring and enforcement, and
- lawsuits for those selling illegal products from newly deforested areas.[45]

These strategies reduced deforestation by as much as 70%. Now, however, despite a 2015 pledge to achieve an 80% reduction by 2020, the rate of deforestation reached record highs under the auspices of the Bolsonaro ultra-right-wing government in Brazil.[46]

Collectively, these issues led to increasing diversity in environmental thinking and collaboration. When Lois Gibbs, a White woman, led the Love Canal protests, she was joined by Black organizers, leading to a mass evacuation prompting the first federal funding to rectify an environmental issue. Similarly, in 1984, a gas leak at a Union Carbide pesticide plant in India killed almost 4,000 people and injured a half million, called the Bhopal disaster. It highlighted the importance of identifying the impact of corporate negligence on marginalized groups. Similarly, Indigenous tribes have been paid significant monies for toxic waste to be dumped on their land. These are just three examples of vulnerable low-income and minority groups experiencing higher risks of negative exposure to environmental dangers. The placement of highways, landfills, military research facilities, industrial production and waste sites, nuclear plants, and extractive industries tend towards areas with populations considered to have less political power.

These kinds of disasters led to the *environmental justice* movement which seeks an end to environmental racism and such corporate negligence. A push for the creation of a Bill of Citizen's Rights occurred in many countries. The bill includes the right of citizens to know, the right to be safe from harmful exposure, and the right to compensation.[47] Rather than solely utilizing public funds, corporations have become increasingly responsible for financing clean-up. Seventeen principles of environmental justice were crafted in 1991 by The First People of Color Environmental Leadership Summit.[48] The seminal 1994 book *Unequal Protection: Environmental Justice & Communities of Color* by Robert Bullard details these "sacrifice zones," early struggles, and coalition building.[49]

With ongoing disasters in the 1980s, environmentalism continued to grow, with ENGOs doubling in membership and growing in sophistication. There was a realization, however, that public awareness, cultural attitude shift, and legislation were not enough.[50] ENGOs realized that they needed to focus their targets and develop niche marketing, whether endangered species or wetland protection. However, this began to create siloed campaigns and competition for donors. By the mid-1980s, there was "list fatigue" with the overuse of mailing lists tapping the same individuals for funding across concerns.[51]

While ENGO endeavours are not often discussed as educational undertakings, ENGOs incorporate public communication and educational materials as part of their mandates to inform the public and generate support. Framing environmental information proved critical, but it was largely a didactic endeavour, using one-way, top-down information, communication, and education strategies. Also typical were the fear, shaming, and blaming approaches, part of apocalyptic predictions, which began to repel potential supporters.

EMOs began to establish offices in multiple nations for an international presence, such as Greenpeace and Earth First!. Disgruntled with reform environmentalism, they prioritized *direct action*, building from the foundations of deep ecology or political ecologism, which characterized the Third Wave. While they chose radical actions to garner media attention, they were not necessarily working for fundamental systemic change but rather "generating awareness of the environmental depredations of economic development" while demonstrating that the natural environment is of intrinsic value.[52] Often these actors were highly educated and working in social and educational sectors, distinct from older movements like labour. They worked through symbols and culture, often able to mobilize a wide range of social support. Some campaigns took on an insurgency flavour to illustrate dire circumstances.

As previously described, women began to take many leadership roles in local grassroots movements, fighting for issues "close to home" that addressed the basics of life. These small groups were rich in social learning, research learning, and served as an entry point for activism.[53] The intertwining domination of women and the natural world was recognized, particularly the use of violence for control as well as patriarchal decision-making. In 1980, Carolyn Merchant published *The Death of Nature*, detailing the "age-old association" of women and nature[54] and this domination structure. She understood this as a combined assault on women and children's bodies as well as the body of the Earth.[55]

In 1987, a conference on *ecofeminism* was held, calling "women to lead an ecological revolution to restore planetary ecology."[56] Women's leadership, though mostly local, was global in reach—from the *Chipko* "tree hugging" movement in India, to the Kenyan Green Belt movement planting millions of trees to reverse desertification, to Greenham Common in England protesting nuclear missiles, to the *Mirrar* in Australia led by Yvonne Margarula protesting uranium mining. A spectrum of ecofeminisms developed, from liberal ecofeminism to socialist and radical ecofeminism.

In 1987, Riane Eisler proposed a "partnership ethic that treats male, female, and the nonhuman species as partners, each needing sufficient room for life-generating processes."[57] In 1988, scientist Vandana Shiva wrote *Staying Alive: Women, Ecology and Development* where she charged that the development paradigm is the twin to the Industrial Revolution. She described how rural Indian women not only resist destruction but also how they regenerate the integrity of interdependent living systems.

Emerging out of the civil rights movement of the 1960s and the questioning of capitalism, Murray Bookchin published the *Ecology of Freedom* in 1984, which synthesized ecology, anthropology, and political theory into what is now called *social ecology*. He advocated an end to hierarchy and domination and a move towards freedom and democracy, as smaller decentralized, ecological, and self-governed societies. Rather than a shift in values and worldviews, he argued for a fundamental economic transformation that is sustainable, democratic, and just. Arguing the failure of both state socialism and industrial capitalism, social ecologists continue to call for ecological socialism as confederations of decentralized, independent communities.

In 1971, process theologian and transdisciplinarian John Cobb Jr wrote *An Ecology of Theology* and then in 1989, *For the Common Good: Redirecting the Economy Toward Community, Environment, and a Sustainable Future*. In 1988, cultural historian and ecotheologian Father Thomas Berry wrote *The Dream of the Earth*, offering a wholistic vision that challenged the narrowness, blindness, and deafness of the industrial era which has limited our perceptual capacity. He proposed that our current modern "story" is no longer adequate and that the renewal of human consciousness towards a mutually enhancing human–earth relationship is already under way. This generated *spiritual ecology*, where humans are considered to have a moral duty to maintain ecosphere integrity and live within natural world processes and ecosystem limits.

Spiritual ecology has helped to reclaim nature-based spiritualities that were lost or suppressed historically. Spiritual ecology would link Christian process theology,

Buddhism, and complexity science into an ecological postmodernism, also known as "restructive" or "constructive" postmodernism.[58] While mainstream religious institutions probed their sacred writings and traditions for environmental values, ancient goddess worship and "neopagan" rituals were revived, building on ecofeminist and gender partnership principles. Animism is also being reclaimed. Defined by religious studies scholar Graham Harvey,

> Animists are people who recognise that the world is full of persons, only some of whom are human, and that life is always lived in relationship to others. Animism is lived out in various ways that are all about learning to act respectfully (carefully and constructively) towards and among other persons.[59]

He differentiates between Old Animism, that was described by early anthropologists, and New Animism as a relational way of being based on reciprocity.

As Indigenous populations reclaimed their traditions, nature-based ethics, and healing rituals, they drew cross-cultural learners to their model of life-giving cultures. While careful to protect areas of their Indigenous Traditional knowledge, rituals, and inherent rights, some Indigenous people began to share their general worldview, knowledge, and teaching stories. As *Seneca* Chief Oren Lyons asserted, "We'll have peace with the earth only when we have peace with each other. And we'll have peace with each other only when we have justice,"[60] linking ecology, justice, and peace within diverse human communities.

As *Chippewa* Melissa Nelson says, "we have not vanished . . . we are 'back from extinction,'"[61] referring to the colonial goals of genocide. Nelson suggests that the operating principle in Indigenous cultures is to "embrace and practice cognitive and cultural pluralism . . . [to] respect and celebrate cultural diversity as an essential part of engendering peace."[62] Within diverse Indigenous Knowledge Systems (IKS) resides *Traditional Ecological Knowledge* (TEK) or "native science," which teaches the understandings and practices for following natural laws in a particular place.[63] As Gregory Cajete explains in his seminal book, *Native Science*, Traditional Ecological Knowledge is not segmented into categories or disciplines as colonial knowledges but rather is encompassed within the overall task of "seeking life . . . [as] a participant in the great web of life."[64]

With all the new forms of environmental devastation, strong environmental organizing, and this growing diversity of environmental thinking, new social analyses emerged mid-20th century, which more clearly linked environmentalism with development thinking. While at times these two social movements would find themselves on opposite sides of an issue, it would become evident by the 1990s that they needed to work together.

Looming Population Crisis

In 1968, Garrett Hardin published an essay called *The Tragedy of the Commons*, illustrating how self-interest pursued in relation to common land would inevitably lead to the degradation of common resources. This was a critique of a foundational principle of capitalism. Paul and Anne Ehrlich published the *Population Bomb* in 1968, discussing the social upheaval and possibilities of famine from human overpopulation. Following this, the Club of Rome's *Limits to Growth* in 1972 detailed the interactions between population, food production, industrialization, pollution, and consumption of nonrenewable natural resources. They posed the possibility of massive decline if limits to growth were not established.

While the population of the world had been 1 billion in 1804, it reached 2 billion in 1927 and 3 billion in 1960, within ever shorter intervals. While many environmentalists were adamant that restrictions to population growth were necessary, new social research found that population numbers stabilize and then decrease with rising personal income, as occurred over several hundred years among European nations. Put succinctly, "development is the best contraceptive."[65] The time lag, though, for "developing" nations to industrialize then experience decreased population growth was considered problematic. Overweening concerns about population growth led China to implement a one-child policy from 1980 to 2016. In 1975, India carried out the enforced sterilization of more than 8 million poor villagers, mostly men.[66] African countries called "population controls" racist and refused to participate.[67]

Disputing modern assumptions that "bigger is better" and "one size fits all," Fritz Schumacher advocated for *Small is Beautiful* in 1973, suggesting that technology, economics, and social organization are optimal and less vulnerable in smaller human-based, autonomous units. Nevertheless, he said, ideas should remain big and wide.[68] Edward Goldsmith crafted *A Blueprint for Survival* in 1972 recommending decentralization and deindustrialization. He suggested that ecologically sound principles are more attainable in smaller communities.[69] These two issues of population and size of social organization bring us to the emergence of international development, as part of the broader social justice movement, running parallel to the environmental movement at this time.

Waves of International Development

Wolfgang Sachs, German development theorist and Green activist, says that the 40 years between 1949 to 1989 have been the "Age of Development."[70] This Age of Development has been predicated upon a form of social Darwinism that saw nations on a linear continuum of development moving ever towards "Western rationalism and Western institutions."[71] The message was "if you want to develop, be like us (the West)."[72] European progress was seen as:

- partly biological in that different natural environments endowed societies differently,
- partly cultural in that some areas of the world were more innovative, rational, and predisposed to development, progress, and civilization, and
- partly sociological in that some societies are the "victors in struggle for existence," and the wealthy are the victors in the "industrious and striving behaviour" that is rewarded in industrial capitalism.[73]

Development was explicitly seen as a process of Westernization. As development theorists Richard Peet and Elaine Hartwick summarize,

> Development is a founding belief of modernity. . . . In development, all the modern advances in science and technology, in democracy and social organization, in rationalized ethics and values, fuse into a single humanitarian project of deliberately and cooperatively producing a far better world for all.[74]

Sachs says development became "the idea which oriented emerging nations in their journey through post-war history."[75] Development theory hit an impasse, though, in the 1990s, when postdevelopment theories began to gain traction.[76] Today, conventional development is highly suspect or rejected as a preferred societal path.

Yet, from the end of WWII, "development" became not only a promoted societal path but also a "mental structure" that many now say needs dismantling as part of decolonizing efforts.[77] Over the last half of the 20th century, international development became a key framework shaping international relations between the First (industrialized), Second (communist bloc), and Third (nonaligned and/or industrializing) worlds. Over these 40 years, from 1949 to 1989, two broad pathways were promoted—that of industrial modernization predicated on conventional economic growth theory, and development predicated on radical or Critical Theory.

Development as Industrialization

As described in Chapter 2, coming out of WWII and the dissolution of Pax Britannica and other European empires, the term "development" successfully dissociated colonialism from the "exercise of helping" through development assistance.[78] What had been considered "uncivilized and backward nations" were now "underdeveloped nations" in need of financial and technological assistance.[79] As American President Truman explained in his 1949 inauguration speech, it was the moral duty and humanitarian mission to help

> free peoples of the world, through their own efforts, to produce more food, more clothing, more materials for housing and more mechanical power to lighten their burdens . . . I believe that we should make available to peace-loving peoples the benefits of our stories of scientific and technical knowledge in order to help them realize their aspirations for a better life . . . based on the concepts of democratic fair dealing.[80]

The United States led this effort, supported by the new United Nations (1945), later joined by the European nations after WWII reconstruction was complete.

The UN had a broader mission than any one nation in providing a world environment of international peace and security, economic stability, friendly international relations, and social progress. International development was considered "international cooperation for an assault on worldwide hunger, disease, illiteracy, and all the economically and socially disruptive forces, which might lead to international turmoil."[81] The UN has maintained its work as expressed in Article 55 of the UN Charter, "to promote the solution of international economic problems, higher standards of living, full employment, and conditions of economic and social progress and development."[82]

International or "overseas aid" was designated for immediate humanitarian crises, and "development assistance" was provided for longer term projects which assisted developing nations in moving past the perceived indignity and insecurity of subsistence living and high mortality. It was believed that through a universal process of efficient economic improvement for individuals and nations, national security and prosperity could be ensured. The UN target of 1% of each member nation's GDP towards Official Development Assistance (ODA) was part of a "rearrangement of wealth between nations,"[83] based on the scale of assistance from the United States to Europe during the Marshall Plan. Since then, the normalized target for ODA has been .7% of each country's GDP.[84] The UN then set about establishing the multilateral funds and agencies through which international stability and global prosperity could be fostered.

In sum, during the 1950s, the lack of industrialization was seen as the cause of "poverty."[85] Development was *economic* development, towards "economic diversification, stable growth which reflects a balance between the agricultural and industrial sectors, and improved living standards."[86] The first task for developing countries was to build industrial capacity—from dams, roads, and factories—for tapping natural resources through mining, forestry, and plantation agriculture. Local governments and populations were forced to comply with national plans of industrialization, often under increasingly authoritarian governments, as development crusaders pointed to the road of progress. However, the "growing Cold War standoff" between the Western and Eastern blocs led to developing nations establishing a bloc of nonaligned nations in 1955, called the "Third World."[87] This included many countries newly independent of their colonial masters.

Development as Modernization

The first official UN Development Decade (1960–1970) was launched by American President John F. Kennedy, now considered the Golden Age of Development. The sense of philanthropy, optimism, and empathy filled scores of young people with moral idealism. They would take part in countless volunteer, professional, and exchange programmes, from the Peace Corps, Alliance for Progress, Food for Peace, to University Service Overseas. In the Peace Corps alone, Kennedy sent "16,000 workers to Latin America between 1962 and 1967, to conduct civic action projects, improve agricultural practices, and help educate the local populations."[88]

Conventional development thinking thus shifted from having a sole focus on industrial infrastructure to creating new social institutions—political to financial, health to education. A new "development industry" grew as governmental agencies, intergovernmental institutions, university programmes, overseas practitioners, and charitable programmes were involved in the practice of development. Their goals were "social progress, redistribution of wealth, efficient administration, political stability, and democratic participation,"[89] considered to be the common elements across industrialized nations.

Young people throughout the West were involved in a range of new opportunities, from Miles for Millions fundraising marches and Model UN assemblies to the explosion of nongovernmental organizations (NGOs), including International Volunteer Cooperation Organizations (IVCOs) participating in humanitarian aid and development assistance activities, such as Red Cross, Oxfam, and Voluntary Service Overseas (VSO). As Canadian Mitchell Sharp explained, "There is one good and sufficient reason for international aid and that is that there are less fortunate people in the world who need our help."[90] To ensure domestic support within Western nations for aid and development, development education was offered to citizens in tandem with overseas projects.

American economic historian Walt Rostow's linear continuum called *The Stages of Economic Growth* (1960), illustrated in Figure 3.5, was the model of modernization utilized by the development community.[91] It was meant to be a deliberate alternative to Karl Marx's theory of history, namely historical materialism. Rostow asserted that "the most profound economic changes are viewed as the consequence of non-economic human motives and aspirations,"[92] acknowledging the connection between the economic and the social spheres in promoting industrialization.

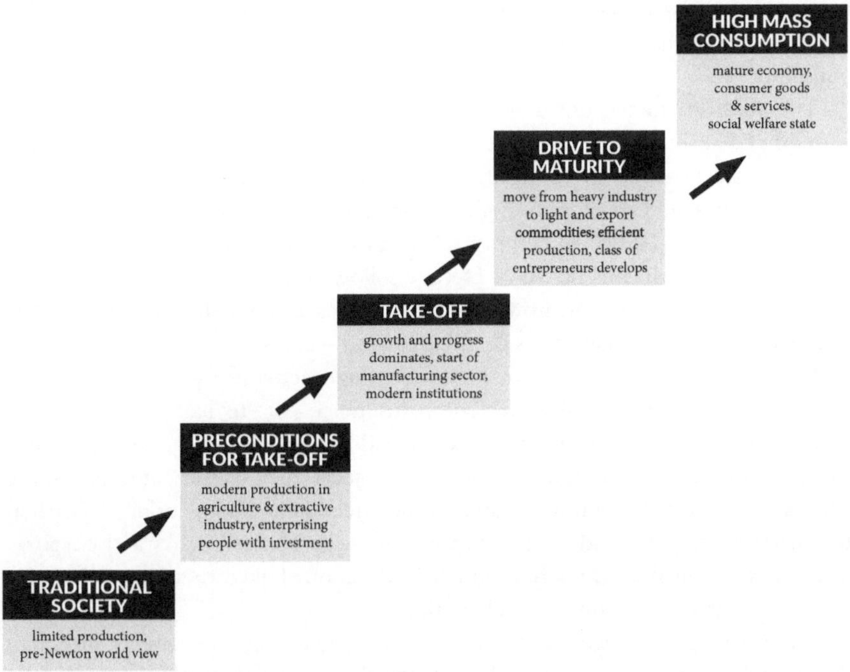

FIGURE 3.5 Walt Rostow's Stages of Modernization[93]

In Rostow's model,[94] Stage One was the stage of *traditional societies*, where there was limited production, often agricultural, accompanied by hierarchical kinship structures and a pre-Newtonian, fatalistic worldview.

Stage Two was establishing the *preconditions for take-off*, namely adopting the ideas of economic progress, modernity, nationalism, and modern science. New production techniques in agriculture would begin raising availability of foodstuffs, and enterprising people would invest in further modernizing agriculture and extractive industries.

Stage Three was *take-off*, a watershed when growth becomes "normalized." The resistance of traditionalism would be overcome through modernizing institutions. Belief in economic growth and social progress would dominate. A class of entrepreneurs would develop, investing in manufacturing and commercial endeavours.

Stage Four is one of fluctuating progress in the *drive to maturity*, towards becoming fully industrialized nations. Nations would shift from heavy industry to light industry, concentrating on export commodities, and taking their place in the international economy.

Finally, Stage Five is the age of material abundance and high mass consumption.[95] In Rostow's view, a *mature economy* concentrates on durable consumer goods and services, augmented by a rise in real income which creates the necessary consumers. Nations then shift towards a social welfare state and continual upgrading of modern technology. Modernization, as it was called, aimed for the transformation of societies into "efficient producers . . . and consumers of Western industrial products."[96] Swiss development historian Gilbert Rist cites Rostow's underlying assumption that "[c]ommunism is likely to wither in the age of high mass consumption."[97]

The ultimate international goal was building a prosperous world capitalist economy. The European and American experience would be condensed into a universal, accelerated process through aid and development. It was expected that each country would develop through an annual growth rate of approximately 5%, measured by their GDP.[98] In contrast to the 1950s and the policy of development as industrial growth, this first UN Decade followed the precepts of Rostow and Keynesian economics where growth needed to be balanced with equity of distribution, like the social welfare states in industrialized nations.

However, the conflicting aims of development activities were not disguised. Development aid was considered "cheap insurance . . . to halt communism in Asia."[99] "Pre-capitalist" states were encouraged not to fall under the communist sphere of influence. Through development projects, they would come to see the economic and social progress available in the "free, democratic world."[100] Further, the investment, loans, and market opportunities for Western banks and corporations were a bonanza, as many industrialized nations linked their aid to benefits accrued in their own countries, called "tied aid."

Large-scale projects were not necessarily producing expected results, although the growth rates were at expected levels. Envisaging quick results in the rise of gross national product (GNP), there was a growing realization that development might be a longer-term project, attainable perhaps by the end of the century. Naïve optimism gave way to assertions that aid itself could not guarantee stable and self-sustaining development, peace, or democracy.[101] Further, it was not just industrialization, new technology, and expanded agricultural output that was needed, but more focus on social transformation.

Nations and especially their citizens needed to speedily modernize. To create Western-styled political, social, and cultural institutions, people needed to abandon traditional ways of thinking, especially kinship ties, subsistence agriculture, communality and cooperative relations, and territoriality. They needed to accept the concepts of private property, contract law, capitalist markets, commodities, waged and mobile labour, medicalization of health, schoolization of education, industrialism, and liberal democratic nation-states. They needed to adopt modern values and beliefs, including concepts of freedom of choice, freedom of expression and free press, social and geographic mobility, occupations and mobility, nuclear families, achievement, profit, secularization, and efficiency,[102] all comprising what it meant to be modern.

However, many young people who worked in "developing" nations began to see contradictory, paternalistic, self-interested, and often ineffectual international development policies and programmes. Many charged their own Western nations with neocolonialism that perpetuated maldevelopment. This awakening informed their rising solidarity with emerging leaders fighting for self-determination—from Cuba to Chile, Namibia to Tanzania.

Developing countries themselves were increasingly disillusioned as they saw developed nations using international aid for their own narrow political and economic interests.[103] The wealth often created an internal elite in developing nations, often fed by corruption and mismanagement. The majority of "Third World" populations remained untouched by development. In fact, "more people were poor than before the mission began."[104]

This led to the organizing of a solidarity bloc of developing nations in 1964 called the Group of 77 or G77, still in existence, meeting as part of the UN Conference on Trade and Development (UNCTAD), also founded in 1964. The G77 membership quickly grew to 134, enhancing their negotiating power for an interdependent world economy, rather than a continuation of economic imperialism. The promised trickle-down of economic benefits from modernization was not reaching the poor. As Brazil stated:

> [we have] rejected the position of waiting until the economic growth by itself solves the problem of income distribution; or the theory of waiting for the cake to grow bigger! [. . . This] may take far longer than social conscience will permit . . . to improve rapidly the level of welfare of wide sectors of the population.[105]

World Bank President Robert McNamara then pledged bank resources for meeting the basic needs of people in the least developed countries.

Development as Basic Needs

The second UN Development decade, from 1971 to 1980, moved away from big industrial infrastructure and the social interventionism of modernization to concentrate on basic needs: food, clothing, shelter, and essential community services such as safe drinking water, sanitation, health, and education. This was considered *human* development rather than just economic development. Rather than top-down development, this was development *from below*. It was still expected that this would help nations transition towards self-reliance. With their most basic needs met, the most impoverished classes could then become part of the economy through new purchasing power.

However, the self-interested political economies of donor nations continued, given the need to placate citizens regarding tax dollars spent outside the country. In exchange for "gift-giving," benefits to donor countries included exchanging aid for diplomacy gains, new trade partners, disposal of surplus commodities, and tied aid, meaning aid products and aid expertise purchased in donor countries. Any real attempt to establish locally grown self-sufficiency measures were often politically, economically, or militarily prevented to maintain the dominant economic position of industrialized nations. Further, negative trade arrangements, debilitating food and agricultural policies, exploitative private foreign investment, and blockage of redistributive policies especially land redistribution, further undermined the development of industrializing nations.

Developing nations were becoming incensed at the injustices. The 1970s was a time of growing power and voice in the Third World, or "Third Worldism,"[106] a willingness to confront the self-interested means and neocolonial economic domination of industrialized nations. New alternatives were emerging, such as Julius Nyerere's philosophy of *ujamaa* or socialist self-reliance for Tanzania.[107] Many nations were interested in a blend of economic systems that would distribute wealth more fairly, prevent corporate or industrial parasitism as well as individual economic acquisitiveness and competitiveness.[108] For them, such a blend would preserve their unique traditional cultures, often with a more collectivist basis, especially regarding the land and the living world. *Ujamaa*, a *Swahili* word referring to the spirit of an extended family, in this context meant cooperative economics, which would prevent what was perceived as the "anti-social" orientation built into capitalism.[109]

In 1967, the Group of 77 adopted the Algiers Charter which "summed up all their grievances" against industrial countries.[110] In 1974, they proposed a New International Economic Order (NIEO) which called for equal terms in the international trading system stacked against them, including changes in trade barriers, financial resourcing, technology transfer, and market pricing. They also called for noninterference in their domestic affairs and an enabling of collective self-reliance.

Development as Partnership

Given these grievances, the focus of international development in 1975 became "partnership." The First World would work with the Third World in looking for constructive solutions that met unique needs, including smaller scale development. It was surmised that through North–South dialogue, comprehensive economic and social development could be jointly planned rather than continuing the top-down, universal models. Further, the Third World insisted that if they were to industrialize and become part of the world capitalist economy, they needed open markets and fair prices for their commodities.

In 1977, the Independent Commission for International Developmental Issues, or Brandt Commission, was established to address the "North–South Divide" in terms of global poverty and income disparity.[111] Importantly, this report introduced the notion of North and South (hemispheres) into development parlance, illustrated in Figure 3.6.

No longer accepting the imposition of one model of development onto all Third World nations, the Brandt Report acknowledged the issue of the "North's domination of the international economic system, its rules and regulations, and its international institutions of trade, money and finance."[112] They argued for a restructuring of the world economy to enable fairer terms of trade and assist in the economic growth and social development of the South, rather than just provide more dollars. To address basic needs first, they suggested "a large-scale transfer of resources from North to South in order to revive a failing world economy,"[113]

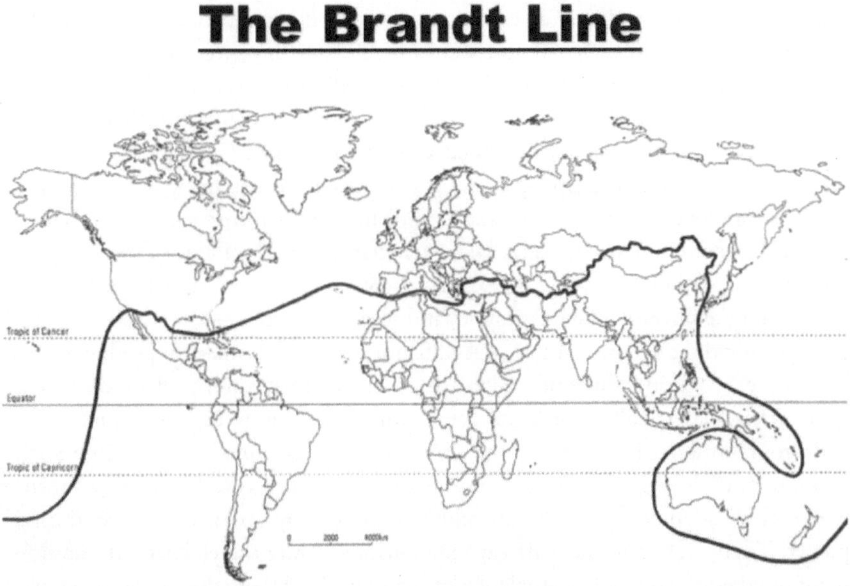

FIGURE 3.6 Map: The Brandt Line—North–South Divide

Credit: The Independent Commission on International Development Issues, and Willy Brandt, *North-South,* Map "North–South Divide" @1980 Massachusetts Institute of Technology, by permission of The MIT Press.

acknowledging that 40% of the population in Third World nations could not acquire the basic necessities of life, making this a moral argument. Brandt argued that the North and South had mutual interests, a way to bring them all to the negotiating table.

Numerous international conferences on structural economic reform were then held, but, unfortunately, the internationalism of the 1960s and 1970s as well as the attempts to address the grievances and proposals were derailed in two ways. First, by the impact of the OPEC oil crises in the 1970s and second, by the neoliberal economic backlash and structural adjustment policies of the 1980s and 1990s. Both events profoundly shifted the relations between the North and South. This would irrevocably change environmentalism and the pursuit of international development. One achievement, however, was the forgiveness of aid-related debt for the 12 least developed nations in 1977.[114]

OPEC Oil Crisis and Economic Disruption

From 1970, while environmentalism and developmentalism were gaining traction, several economic disruptions occurred. First, the United States walked away from the Bretton Woods Accord and the gold standard, leaving all currencies to float against the American dollar, a way to address their recession. This would lead, however, to constant instability in the foreign exchange system.

Second, in 1973, the new cartel of OPEC (Organization of Petroleum Exporting Countries) created an oil crisis. Arab oil-producing nations disputed their categorization as Third World and were unwilling to be extorted continually by industrialized nations. So, they used oil as their political weapon.[115] OPEC reduced production, quadrupled the price of oil, and threatened to nationalize the producers. While this momentarily added weight to the NIEO negotiations, it also divided the Third World, many countries of which permanently lost ground on the development trajectory. This was a hard hit for all North and South nations who had become reliant on external rather than internal sources of oil, including North America. This "first oil shock" along with a "second oil shock" in 1979 had far reaching environmental, economic, and social impacts.

Not only did the national incomes of Saudi Arabia and other oil producers soar, but Western nations implemented price controls and rationing of oil and gas as well. Their leaders called for lower speed limits and energy conservation research and programmes. There was a surge in environmental innovation, including the redesign of buildings using passive and active solar energy as well as energy alternatives including nuclear power, wind and wave power, hydropower, and geothermal power. There were also multiple car redesigns, including the electric car, development of ethanol, and higher efficiency gas engines. Had these designs been retained and implemented on a large scale, we would not be experiencing the extent of the climate crisis today. It was a moment of opportunity that was largely missed, although it did establish the research and innovation groundwork for the climate-related work of today.

The third "disruption" was the development of the silicon chip in the 1960s, leading to the call for globalization. Taking advantage of this new technology, Japan quickly shifted into electronics and away from heavy industry and intensive use of oil. Their prosperity meant they would assume the role of world banker from the United States. Similarly, industry in Western nations needed to retool to take advantage of digital technology to reduce overall costs, particularly labour costs. Rather than retool, though, it was easier and would lower operating costs more to move business "offshore" and simply rebuild factories. Some of the recipients would be the Newly Industrializing Nations (NICs),[116] including the Asian Tigers of Hong Kong, Singapore, South Korea, and Taiwan.[117]

The old industrial era in Western nations appeared to be ending. With slack consumer demand and falling profit margins, capital wanted the freedom to move, or rather "globalize." This meant breaking their social contract, or "class compromise"[118] as British geographer David Harvey calls it, of providing consumer goods and services as well as good paying jobs, usually unionized, in their home nations. Corporate business pressured for the establishment of international free trade zones so they could rationalize and streamline production in new factories, escape union contracts and environmental laws, and thus raise profit margins. Capital also wanted a "leaner" production structure—a central firm with integrated vertical and horizontal subcontractors globally. A new definition of the "service" sector emerged, called just-in-time production,[119] producing goods as they are needed, reducing the need for inventory. Further, rather than just making commodities, many companies found they could make more money in currency speculation, aided by computer trading. This led to the casino economy, where private corporations including banks risked large amounts of money on options trading and other derivatives, futures trading, and currency trading,[120] sometimes at their own peril.

The fourth disruption came from the excess of petrodollars held by the Middle East oil nations. While many nations were already indebted from large industrial projects, they were cash hungry for further development.[121] So, they now took the petrodollar loans being offered through Western investment banks. Simultaneously, however, Northern nations were experiencing stagflation, the rise of both inflation or escalating prices, as well as economic stagnation. Starting in the United States, nations decided to raise interest rates substantially, in some cases up to 20%, to quickly stop inflation.[122] The debt load of all nations, especially Third World nations, was suddenly staggering. Highly indebted nations were thrown into precarious financial situations, including "most of the Latin American nations, many African nations, and some Asian nations."[123]

The Lost Decade of the 1980s: Debt and Dependency

The rising interest rates on international loans created a massive wealth transfer where now the South was paying millions more in dollars than it received from the North, a financial haemorrhaging. In 1970, total outstanding debt in Latin America was $29 billion, but by 1982, the debt reached $327 billion.[124] In 1980–1981,

Latin America received $13 billion in assistance, but they paid out $25 billion per year for interest in a reverse flow.[125]

As early as 1982, Mexico defaulted on their foreign debt, which had escalated to $80 billion, throwing some American banks into crisis as well as threatening the international financial system.[126] This became known as the debt crisis which created the "Lost Decade" as Argentina, Brazil, and other countries either defaulted or threatened to default. Northern nations, feeling the strain themselves, pulled back their aid and development funding, further decreasing the ability of the South to manage economically.

Opposition to Modernization Theory had been taking shape in the South as early as the 1950s in the form of Underdevelopment Theory and Dependency Theory within World Systems theory.[127] Many Third World intellectuals and communities did not accept the capitalist trajectory. They asserted that capitalist development had a necessary corollary, that of underdevelopment.[128] Amiya Bagchi, from India, argued that once there had been contact between countries "further along the capitalist path, the lagging societies could not possibly develop in an autonomous fashion . . . contaminating"[129] their systems. "Western European capitalism retarded the development of anything resembling capitalism in Third world countries."[130] In other words, the development of the advanced capitalist countries was enabled through the deliberate underdevelopment of former colonies, expropriating their wealth for the benefit of the North. Underdevelopment theorists argued that Modernization Theory had simply ignored the mercantilist phase of colonialism that made industrial capitalism possible.

Dependency theorists from Latin America saw all nations locked into a metropole and periphery relation, part of a world system and global division of wealth. Some argued that unequal exchange relations in the international market could be avoided by some form of protectionism, particularly of their fledgling industries, and by import substitution to kickstart new industry. Others argued that Southern nations were caught in unequal production relations, as capitalism is at different stages of development in different places, taking advantage of lesser developed economies and thwarting their self-sustaining growth. External assistance thus prevents endogenous industrial development, exaggerating Southern dependency. Further, it is often enforced by authoritarian, repressive regimes and military dictatorships which were, in some cases, covertly installed by Northern nations.[131] Some theorists advocated a radical break from this system, while others advocated hard bargaining to gain relative advantages. In sum, Samir Amin argued that unequal exploitation in production leads to unequal exchange which leads to unequal development.[132]

Other opposition emerged as well. Ivan Illich was concerned over the institutionalization and technocracy of social services, including the expansion of schooling, which robs people of their creativity, know-how, and organic community relations. In 1971, he advocated for *Deschooling Society* and, in 1973, for a return to *Tools for Conviviality* where communities re-establish effective local and independent means for human flourishing.

Picking up from Illich were Gustavo Esteva from Mexico and Suri Madhu Prakash from India. Gustavo offers this cogent (excerpted) story that traces his childhood before the international development agenda appeared in the 1940s to its subsequent impact upon his identity and community.

> As a child who was as yet not underdeveloped, I had a tradition and the dream of continuing and enriching it. I had a place in the world and concrete customs. These gave me precise orientation in my daily life. . . . I was only 13 years old when I got underdevelopment, that illusion, that kind of sickness. . . . My father died a little after I got underdevelopment. This era was not for him. . . .
>
> I fully immersed myself into making the new promise a reality. Bravely, I became the first generation of students with a profession that was something like business administration. . . . I would be a pioneer of my country's struggle to develop itself. . . . My dreams became tinted by fantastic images of prosperity I saw at the movies made in Hollywood. I still remember clearly my fascination with the first mall, created by Sears and Roebuck a few blocks from my middle-class home. The old department stores, established by the French a century before, were left immediately obsolete. Our *tianguis* and traditional markets became dirty, incoherent, chaotic. The market of Oaxaca suddenly became a cherished, folkloric curiosity.
>
> I still respected and loved the world of my grandmother. But clearly, I did not want it for myself or my offspring; nor for my country. My nation and people desperately needed to be "developed"—and I wanted to be part of that epic venture. For that purpose, I could not appeal to my customs, to my experiences, to my traditions. Only expert's advice, education, and professional handbooks could help. I came under a system of permanent education. Everything I was learning in the school or in my jobs was rapidly becoming obsolete. If I wanted to keep the pace of development, run and win, to fulfill my duty to myself, family, and country, I should learn as fast as possible. . . .
>
> The culturally diverse imaginations and dreams of different peoples, full of energy only a few years before, were progressively dismantled and reduced to the illusion of soon possessing, with development aid, the American way of life. . . . By accepting the economic definition of learning, we embraced scarcities of teachers and schools. [. . . The] story is precisely a story of loss, the story of real, living men and women who became, through the experience of development, uprooted living dead.[133]

Brazilian educator Paulo Freire critiqued the oppression implicit in both feudal landholding arrangements and industrial modernization, reinforced by Catholic church dogma and conventional education systems. He asserted that conventional schooling is typically a form "banking education"[134] that silences. Schooling reproduces these relations of domination, therefore reinforcing the oppression of the poor. "The teacher talks about reality as it if it were motionless, static, compartmentalized, and predictable . . . to "fill" the students . . . [with] contents that are detached from reality."[135]

In his 1970 *Pedagogy of the Oppressed*, he developed a pedagogy for literacy learning that enabled the poor to problematize their poverty while becoming literate, "reading the world while reading the word."[136] His literacy process was called *conscientization* as adult learners became conscious of the root causes of their oppression and then took collective action to improve their lives. For this, he would be exiled. Liberation theologians and religious orders helped establish base Christian communities throughout Latin America as their response to the Church's complicity in oppressive national circumstances, risking murder and disappearance.

Neoliberal Globalization as Backlash

Into this economic crisis of the 1980s came the ideas of the Chicago School of Economics, particularly those of Milton Friedman and F. A. Hayek. It represented a stark turn away from the Keynesian social welfare state and a turn towards market fundamentalism. David Harvey explains that there were several economic and political threats to the global ruling elites and classes, namely, the expansion of democracy which was clearly illustrated by the New Social Movements as well as the expansion of social democracy parties, particularly throughout Europe.[137] Therefore, neoliberalism became a backlash project for the "restoration or reconstruction of the power of economic elites."[138]

The international gathering that crafted these ideas into a global agenda in the 1970s was the Trilateral Commission, including Europeans, Japanese, and North Americans from business, politics, and new economic thinktanks funded by the wealthy.[139] They felt that the expansion of the welfare state was threatening to undo capitalism. Therefore, their explicit goal was dismantling the welfare state. Their prescriptions for economic restructuring would be called the Washington Consensus.

Their rhetoric, however, built upon specific ideas of freedom, undifferentiated from other ideas of freedom. As David Harvey explains, democracy was originally built upon the freedoms of conscience, religion, and expression; of the right to vote; of communication, speech, and the press; of assembly, movement, and association; and of life, liberty, and security, all protected by the rule of law and equality under the law. Yet, the neoliberal use of the term freedom is restricted to the idea of freedom of private property, the freedom of capital in the market, and individual liberty and choice within the market. Freedom in "[t]he neoliberal project is to disembed capital from constraints . . . liberating corporate and business power and re-establishing market freedoms."[140] As Karl Polanyi described in 1944, this was

> [T]he freedom to exploit one's fellows, or the freedom to make inordinate gains without commensurable service to the community, the freedom to keep technological inventions from being used for public benefit, or the freedom to profit from public calamities secretly engineered for private advantage.[141]

While neoliberals advocated for freedom in the marketplace, each person was accountable for their own well-being, including losses. "Individual success or failure are interpreted in terms of entrepreneurial virtues or personal failings . . . rather than attributed to any systemic property."[142] This directly undercut the assumptions of the social welfare state which attempted to balance individual and collective welfare and to protect societal members from the excesses of capitalism.

Neoliberalism was also committed to "attacking all forms of social solidarity" such as unions protecting workers and social movements advocating for the rights of marginalized groups or goals in the common interest. Hence, new derogatory rhetoric dismissed "demands of special interest groups," including the environmental movement. Any attempt to regulate private ownership and corporate activity was "attacked as a denial of freedom."[143] David Harvey cited Polanyi's view that neoliberalism always carries within it the turn to authoritarianism and fascism, where the "good freedoms are lost, and the bad ones take over."[144] In other words, excessive self-interest and individual liberty can undermine core social and political freedoms.

Politically, neoliberalism was ushered in through the rise of the New Right, led by Ronald Reagan in the United States, Margaret Thatcher in the United Kingdom, Pinochet in Chile, and here in Canada, Brian Mulroney, who set about zealously implementing these new economic prescriptions in the 1980s. In addition to more freedoms for globalizing business, neoliberal economics proposed much smaller government by deregulating and privatizing public services into profit-making companies. The rhetoric was that there was too much "government interventionism" or "nanny state" as Thatcher called it. Market dynamics would no longer be kept out of specific sectors, including health and education. They argued against full employment and worked to vigorously weaken unions, maintaining there is a natural rate of unemployment which keeps wages and salaries lower, enhancing profit making. Finally, they asserted, there was "too much" democracy[145] with marginalized groups around the world gaining too much voice. While these ideas had previously been heresies, they now gained traction as a way out of the global economic crisis.

New Right philosophy was embraced in three, often contradictory, ways. First, it was the overhauling of "progressive" conservatism in addition to vigorous opposition to both liberalism and social democracy. The values of libertarianism were promoted through schools, media, and the law, including the rights of "the individual; freedom of choice; market security; laissez-faire; and minimal government."[146] Second, as neo-conservatism, it advocated for strong government, social authoritarianism, a society that disciplines opposing voices, and re-establishing the hierarchy of elites and subordinated groups.[147] A third strand was fundamentalist religious beliefs, which comprised a moral crusade against liberalism. Groups like the Moral Majority contested the separation of church and state and sought to bring religion back into politics, hearkening back to older eras. The New Christian Right in the United States sought to reverse many social policies, such as reincluding Christian prayer in schools and eliminating access to abortion and same sex marriage. In the extreme libertarian position, there is a fascist element that advocates for the necessity of strong control over the public, strict obedience to government and the market, sharp suppression of opposition to these prescriptions, and ultranationalism.[148]

Structural Adjustment Plans as Third Development Decade

The rally cries in the 1980s were about "going global," establishing "free trade zones," "privatizing" the public sector, "downsizing" government, and "deregulating" oversight of industry including the rise of monopolies.[149] In response, the European Union, North American Free Trade zone, and the Pacific Rim free trade zones were established. Factories, often in free trade zones close to the border, such as the *maquiladoras* in Mexico, enabled foreign firms to operate duty- and tariff-free.

Austerity fiscal policies were implemented in many nations, eroding their economic and social infrastructure. Internationally, neoliberalism led to Structural Adjustment Plans (SAPs) where every country—South, North, East, and West—needed to restructure their economies according to these principles to access IMF international loans. Development took a back seat to economic adjustment throughout the third Development Decade (1981–1990). Ronald Reagan declared that the New International Economic Order was dead.[150] The goal of a new economic order predicated on justice and equity was replaced by "short term pain for long term gain."

Countries were forced into major cuts to government programmes, freezing wages, privatizing public enterprises, promoting exports, reducing domestic consumption (particularly imports), reducing tariffs and other trade barriers, devaluing currency to encourage foreign investment, and removing subsidies on food and other staples so prices would rise to curb domestic consumption while increasing foreign export income to pay down debt. Southern nations were often required to sell their assets, such as companies, land, or mineral rights, at "fire-sale prices."[151] There was a decline in real wages globally, as middle classes shrank. Massive tax cuts for corporations and the wealthy were implemented ostensibly to create investment capital. As economist Joseph Stiglitz commented, the "poor countries are in effect subsidizing the richest."[152]

By the end of the 1990s, these policies had predictably led to the massive enrichment and expansion of the global wealthy class (the 1% as coined by the 2011 Occupy movement). In the UK, particularly, it was a move against the aristocratic classes in favour of a new class of entrepreneurs, changing the composition of the upper class.[153] By 1996, the combined income of the world's 358 richest people was more than the income of 45% of the poorest of the world's population, 2.3 billion people.[154]

It was also a profound deepening of poverty within, not only between, countries. Real wages fell by 20% or more in many countries at the same time that social programmes were also being slashed.[155] Globally, "[t]he poverty rate climbed sharply, from 40.5% to 48.3% in 1990, and would return to 1980 levels only in

2005."[156] This means that Southern nations lost not only a decade but also 25 years of development. In sum, the standard of living of millions of people came second to the needs of banks, financial institutions, and the wealthy global elite.[157]

Global hostility led to protests and food riots throughout the South. In some development forums, policies of "adjustment with a human face" were advocated to compensate for the degradation in living conditions, health, nutrition, and education, especially of the most vulnerable.[158] The International Monetary Fund and World Bank themselves determined that their policies led to a doubling of absolute poverty globally, while increasing economic output five-fold. Wealth was not trickling down but being held by the wealthy. A few leaders in both the IMF and World Bank walked away disgusted, in the face of these failed policies, persistently implemented.[159] One IMF research director said that neoliberal economic policy, with the spreading and deepening of poverty, was the greatest economic failure of the 20th century.[160]

By 1989, the Berlin Wall fell, signalling the end of the Cold War. Nelson Mandela was released from prison in South Africa in 1990, eventually ending the formal system of apartheid. Capitalism was considered victorious. Many advocated for a "peace dividend" where relations within the UN could take a friendlier turn, and military spending could be redirected into international aid to overcome the setback in development. This would never materialize.

Somewhat ironically, the fourth Development Decade (1991–2000) was focussed on good governance and an enabling environment for development.[161] The UN tried to regain leadership of the development agenda away from the IMF and World Bank by promoting people-centred and rights-based development. New UN summits related to women, children, the Earth, and education tried to achieve some consensus on global development goals. Instead, the New Right "boasted" of their ability to shift global public opinion in just a decade and create a backlash against the social welfare state and social justice goals, with neoliberal economics remaining ascendant today. In particular, the New Right created a vicious backlash against claims of an environmental, especially climate, crisis.[162]

Sustainable Development: Merger of Environmentalism and Developmentalism

Prior to late 1980s and early 1990s, the international development and environmental constituencies were largely separated, both organizationally, programmatically, and in their social analyses which were based on different disciplines - science and social science. Until this time, Southern leaders considered "the environment" largely a First World concern, as it was Northern nations who were polluting and ravaging the natural world. For them, the responsibility lay with Northern nations to rectify their own actions. Poverty and hunger were Third World priority issues, addressed through development.

However, there was a growing awareness that environmental issues could not be addressed without considering the model of human and societal development.

Further, alternative development models other than industrialism and capitalism kept emerging, as the failures of the conventional model were exposed.

Sustainable Development as Fourth Development Decade

Into the tumultuous context of the late 1980s and early 1990s came the concept of sustainability and sustainable development. The terms "sustainability" and "sustainable society" were first used by the World Council of Churches in 1974 to describe the need to balance environmental protection with development issues.[163] The term "sustainable development" was first used by the *World Conservation Strategy*[164] in 1980 to mean the maintenance of essential ecological processes and life-support systems on which human survival and development depend. Lester Brown then used the term in his 1981 book *Building a Sustainable Society*, advocating for reduced population growth, resource conservation, and renewable energies. In 1984, Norman Meyer published *Gaia: An Atlas of Planet Management* where he also used the term sustainability to describe current issues, events, and statistics in an atlas format.[165]

The term gained most prominence from the 1987 Brundtland Commission report, *Our Common Future*, which commissioned 22 people from developing and developed nations who were both development *and* environmental specialists. Their task was to collectively identify strategies to address simultaneously environment and development concerns,[166] explicitly merging environmentalism and developmentalism. The Brundtland Commission inserted itself into the contentious North–South relationship to find a productive way forward, considering *both* the environmental issues of the North and the poverty and hunger issues of the South. In other words, "how could respect for nature be married with a concern for justice?"[167]

They worked around highly polarized debates, especially the suspicions of the South who considered environmental concerns as another cover for blocking their development and reducing aid.[168] The Commission tried to find a middle way, balancing environmental protection with economic growth. The term "sustainable development" was then defined as "development that meets the needs of the present without compromising the ability of future generations to meet their needs."[169] This meant that the South would continue their growth but within sustainable means, and the North would become sustainable by attaining ecoefficiencies within their existing level of development.[170] Sustainable development then became the focus of the fourth Development Decade in the latter 1990s and beyond.

The significance of the term sustainability is that it represents a synthesis of principles: ecological integrity, economic well-being, and social equity. Equity is sought *within* generations (intragenerational) situated in both the North and South, as well as equity *between* generations (intergenerational) in considering the needs of future generations.[171] Further, the North was coming to the realization that if the South followed the same development trajectory, it would require two planets or more of resources.[172] Nevertheless, the foundational assumptions of perennial growth and economic competition were not readily questioned, yet.

The term sustainable development meant to convey that poverty reduction and environmental protection should go hand-in-hand. As the development community argued, environmental protection could not be achieved on the backs of the world's poor. For them, affluence was more responsible for environmental damage than poverty. They asserted that the environmental community was offering social prescriptions that were ahistorical and apolitical by not accounting for the history and political consequences of colonialism and the promises of international development.[173] The environmental community argued in response that development among the poor could not be achieved by weakening the natural systems which are the very basis for human existence. Southern nations argued that *they* were the postmaterialists who lived on far less and that their environmental movements were as advanced in blockading, demonstrating, and achieving wins, especially community control, which saved livelihoods as well as forests, oceans, soil, and biodiversity on which they depend.[174]

The Brundtland Report highlighted these intersections: "A world in which poverty is endemic will always be prone to ecological and other catastrophes."[175] The concept of sustainable development not only addressed the needs of the impoverished but also "implied limits . . . based on the ability of the biosphere to absorb the effects of human activities."[176] Thus, the report identified that needs and limits were both necessary—addressing poverty and the needs of the South as well as placing limits on human endeavours in the North for environmental protection. It was "the first time, the Commission had considered environmental concerns arising from an economic, social, and political perspective rather than solely from a science base, as in previous studies."[177] Finding this balance has been the ongoing challenge of sustainable development.

Despite many good intentions, the concept of sustainable development has proven to be ambivalent in definition and not explicitly operational.[178] Some would say that this elasticity is the strength of the concept, offering possibilities for remaking it differently in each context. It was this "constructive ambiguity' and "strategic flexibility"[179] that explained the broad uptake. Others suggested that sustainable development ought not to be seen as an end but as a process, requiring ongoing debate and dialogue appropriate to each context.

Yet, sustainable development was commonly seen as the product of too much compromise between the two lobbies.[180] Sustainable development has often been called an oxymoron, where development as growth, anywhere, is not compatible with conservation and environmental limits. It enabled the "greening" of growth in the North, seen as just a way to protect affluent lifestyles and camouflage existing corporate activities. Further, if the conventional model of development for the Third World was questioned, it would invalidate the past 50 years of development prescriptions around growth, trade, and "catching up."[181]

Therefore, the promise of sustainable development hinged on market solutions, or "market environmentalism,"[182] as well as on technological fixes which could address injustice and develop substitutions for depleted natural resources. Sustainable development would remain within this conventional "technocratic approach predicated on global-scale management by a centralized international elite."[183] Because of this, sustainable development has been considered reformist, characterized as "weak sustainability," further described in Chapter 5.

Nevertheless, many saw the "sustainability" discourse as having tremendous potential. It could generate a transdisciplinary field of study, with the breadth and depth necessary to take on complex global issues.[184] It could help to redefine core concepts, operating principles, and dominant designs of colonial mercantilism and capitalist industrialism. It could point the way into a fundamental transformation of human habitation that could find its place in the web of life, a form of "strong sustainability."[185] Figure 3.7 illustrates the elements accepted as common across various concepts of sustainability.[186]

Since then, much effort has been expended to find integrated solutions that recognize both needs and limits and do not unfairly discriminate against the North or the South, despite deep controversies and disparate visions. While the Brundtland definition still dominates, particularly in international dialogue, new conceptual language is emerging—from vibrant, resilient, thriveable, and flourishing societies based on regenerative practices.[187] Over the past 40 years, sustainability as a concept has diffused globally, gaining remarkable global currency. It has proliferated into every discipline and sector along with a host of declarations, indicators, and measurement processes for accountability.[188]

The concept of sustainability is:

- a challenge to conventional thinking and practice
- about long-term and short-term well-being
- comprehensive, covering all the core issues of decision-making
- recognition of the links and interdependencies, especially between humans and the biophysical foundations for life
- embedded in a world of complexity and surprise, in which precautionary approaches are necessary
- recognition of both inviolable limits and endless opportunities for creative innovation
- about an open-ended process, not a state
- about intertwined means and ends—culture and governance as well as ecology, society, and economy
- both universal and context dependent

FIGURE 3.7 The Shared Essentials of the Concept of Sustainability

Credit: From *An Introduction to Sustainable Development* by Jennifer A Elliott, p. 25. Copyright @ 2013 Fourth Edition by Imprint. Reproduced by permission of Taylor & Francis Group.

Derailing Environmentalism and Sustainable Development?

Indian physicist Vandana Shiva suggests, "the more successful a social change movement is, the more aggressive the backlash movement."[189] Given the transition to neoliberalism, the 1980s and the 1990s were a time of significant defunding and delegitimation of environmental movements and the prevention of meaningful funding for sustainable development, particularly after the Rio Earth Summit in 1992. It is fair to say that the momentum of sustainability and sustainable development was derailed by the rise of neoliberal global economics and right-wing politics, eclipsing the liberal democratic consensus and internationalism that originally stood behind these initiatives. Sustainable development became a contest between UN-led initiatives and commitments, and neoliberal market-driven initiatives.

Open hostility to the voice of environmental social movements increased as neoliberal regimes pursued a "right-wing agenda of unrestricted exploitation of natural resources"[190] and the gutting of existing environmental laws. The sources of funding for ENGOs were reduced significantly starting in 1985. The push for "environmental regulation [was] now seen as government oppression not protection."[191] With the growing professionalization of the environmental movement, neoliberals portrayed them as rich, elitist lobbyists who were insensitive to social justice and "working man's issues."[192] They rallied those working in the natural resource industries or on the land (ranchers, miners, loggers, farmers, fisherman, trappers, and hunters) by labelling environmental activists "extremists, fanatics, communists, cultish, alarmists, ecoterrorists, anti-family, anti-human, anti-civilization, anti-private property, anti-trade, and anti-Christian."[193] The Right wing deliberately polarized the debate, especially between jobs and the environment as well as instilled fear, charging that environmentalists were trying to destroy their jobs and their lives. They called those in the resource industries "freedom fighters and heroes."[194]

This would eventually grow into the current culture wars based on resentment, especially regarding race, but also marginalized groups from LGBTQ+ to women's rights. Environmental journalist Andrew Rowell suggests that with the Soviet regime gone, environmentalists became the new scapegoats.[195] The New Right would utilize both threats by labelling environmentalists "watermelon Marxists,"[196] green on the outside and red on the inside.

The growth in membership of environmental groups began to stall, and by the mid-1990s, some of the largest environmental groups, such as Greenpeace USA, would cut their staff by 40%. American polls found that public support for environmental protection dropped from 80% to 51% between 1989 and 1992. Restrictions in funding from an economy in stress firmly suppressed the gains being made in terms of social justice and environmental integrity, while greatly magnifying corporate influence. ENGOs were often forced to turn away from shrinking governmental funding and toward corporate donors and sponsors, sometimes the very corporations that they were challenging. This became a deliberate strategy of many corporations, funding environmental organizations while explicitly working against them. This also led to a form of tied funding, with either indirect or direct control over messaging and public engagement, particularly in the school system, addressed further in Chapters 5 and 7.

Climate Change Gains Visibility

One of the key factors neoliberal proponents did not foresee in the 1990s was climate change. In 1979, the first international conference on world climate was held by the World Meteorological Organization. By 1985, climate modelling predicted a doubling of $CO_{2,}$ leading to a call for international cooperation. In 1988, respected climatologist, James Hansen, gave a speech to Washington on the risks.[197] James Hansen and his research team predicted that there would be catastrophic and irreversible impacts if the world was to reach 2 degrees of warming (about 450 parts per million or ppm),[198] including mass starvation, water famine, land loss, violent conflict, and significant population displacement.[199] Their conclusions were if all fossil fuels were fully exploited, life itself would be endangered, including humanity and millions of other species. This led directly to the establishment of the UN Intergovernmental Panel on Climate Change (IPCC) in 1988, involving 2,000 to 2,500 global scientists in ongoing climate change assessment.[200]

Particularly since the 1970s, Earth's temperature has been rising. Since 2000, we have had the 10 warmest years on record. By 2011, earth's surface temperature had risen a half degree since 1880. This is due to an increasing accumulation of greenhouse gases, primarily CO_2, trapping more heat in the atmosphere and raising the overall temperature. A warmer atmosphere holds more water vapour and therefore increases the intensity of storms and extreme weather. In 2000, the Worldwatch Institute reported that the Earth's ice cover was melting at a more rapid rate than originally predicted, having lost 40% in 30 years.[201] With a 3-foot ocean level rise, most of the world's coastal cities with hundreds of millions of people would be displaced or perish, 40 inhabited islands would disappear, and 30% of global cropland would be lost.

In 1880, CO_2 was 285 ppm. The goal was to keep greenhouse gases under 350 ppm, otherwise positive feedback systems would amplify the warming trend, called global warming. Author Bill McKibben who wrote *The End of Nature* (1989), one of the first books on climate change, founded 350.org in 2008 to build a global climate movement with the goal of keeping CO_2 under 350 ppm.

It was clear in the 1990s that environmental issues and particularly climate change were no longer national issues but cross-border issues. A globalized "market makes it possible to take resources (oil, wood, water, etc.) from one region, to consume them in another region, and to dispose of the waste in yet another."[202] Climate change, especially, impacts everyone. Thus, the dynamics of a global economy resulting in climate change would need to be addressed through a sustainable development "spanning science, economics, and international relations."[203]

Rio and the 1992 Earth Summit

Importantly, the Brundtland Report laid the groundwork for a pivotal 1992 international summit in Rio de Janeiro, called the Earth Summit (called The United Nations Conference on Environment and Development or UNCED). In 1987, work had begun immediately on an Earth Charter,[204] a document building from the Brundtland Report that could guide the transition to sustainable development. It was here that the concept of sustainable development would be worked out, attentive to the political arena of international development.[205] In the end, the charter was an idea before its time.

In 1990, the IPCC released its first report in the lead up to a second World Climate Conference and world treaty negotiation. They developed a climate framework with 150 nations,[206] completed for signature at the Rio Earth Summit. Going into the summit, the United States was only 4% of the world's population but responsible for 36% of the carbon emissions, so their participation in any solution was critical.[207] Yet, the South resented the permanent inequality and poverty that the North-driven agenda would likely create, particularly the expensive environmental regulatory regimes which they could not afford.

The Earth Summit was the largest international conference that had ever been held, involving over 8,000 delegates, 100 heads of state, and 3,000 NGO delegates in a parallel conference called the Global Forum. In the end, the Rio Declaration became the achievable consensus, with Agenda 21 as the "bible of sustainable development,"[208] building from ecological, climate, earth systems, and conservation sciences. Importantly, it laid out 27 key principles that informed an agenda for sustainable development action, recognizing "the complex interdependences of environmental, social, and economic development."[209] Key themes were integrated: social and economic dimensions, resource conservation and management, a strengthened role for grassroots organizations, as well as the means of implementation to achieve the goals.

Other important conventions were also signed, such as the Convention on Biodiversity, Declaration on the Forest, and, particularly, the Convention on Climate Change. This was signed by 160 nations, establishing a path towards emissions reductions. It also dedicated $125 billion to the South to assist them in reaching their climate change goals as part of Agenda 21.[210] This was a watershed moment. It generated much optimism and would structure ongoing dialogue into the early 21st century.

The surprise at Rio was the strong activist voice of NGOs, as citizen groups from a full range of sectors and interests, North and South, gathered. At the citizen level, poverty and environment were no longer antagonistic issues but a chance to build "the world we want." This was the beginning of a far "more environmentally conscious age"[211] built on ecological science and a burst of visionary and innovative projects integrating social and environmental dimensions. Nongovernmental practitioners across all social movements, not just in the environmental movement, became the vanguard for sustainable solutions, from the ground up.[212] They worked on the Earth Charter in various clusters, including a Business Council on Sustainable Development and the Planet Earth Council of scientists.

Agenda 21 came into force in 1994, with annual follow-up meetings called COPs (or Conventions of the Parties), the first occurring in Berlin in 1995. While funds were available for emissions reductions, not enough development funding was provided.[213] In 1995, the second assessment of the IPCC gave authoritative evidence that the global mean surface temperature had increased between 0.3°C and 0.6°C and could be attributed largely to the result of human activity, primarily fossil fuel use, land-use change, and agriculture.[214]

Originally developed in 1987, Canadian Maurice Strong, who chaired Rio, and Russian Mikhail Gorbachev decided in 1997 to make the Earth Charter a declaration for civil society instead. It comprises four pillars: the interdependence of all life, love and responsibility, democracy and freedom, and justice across generations, with 16 subsidiary principles. Since 2000, millions of individuals and 50,000 organizations across the spectrum of civil society signed the document. In 1992, an Indigenous Peoples Earth Charter[215] had been endorsed by thousands of Indigenous groups. Both documents are important as teaching and organizing tools.

In 1997, another financial crisis hit, this time in East and Southeast Asia, despite their previous economic progress. Thailand needed to devalue its currency which triggered financial withdrawals and threatened another global financial crisis. The volatility of the global economy was increasingly evident.

In Kyoto, 1997, the discussions around the Kyoto Protocol included legally binding targets and timetables for industrialized nations, as part of the world's first greenhouse gas emissions reduction treaty. The Americans stipulated they would not sign anything if there were no meaningful emissions cuts by "developing" countries, an evasive tactic even though the majority of Americans supported emissions cuts. In the end, the agreement was signed by over 100 states with the goal of reducing emissions by 5.2% below 1990 levels by 2010. While 50–70% reductions were recommended by the IPCC, the United States made a commitment to 7% and the EU 8%, though the treaty would not be ratified until 2005. In 1997, Greenpeace put forward the idea of a carbon budget where countries would be limited to a total allowable amount of emissions, adopted in just a few countries.

In the United States, George Bush succeeded Bill Clinton as president in 2000, walking away from the Kyoto Protocol. Rather than making the hard decisions domestically, the United States committed to their own unilateral response. Primarily, they wanted to establish a "carbon market" where emissions trading could occur, another source for corporate profit-making. Other industrialized nations utilized loopholes in the treaty, such as relying on emissions trading and lucrative contracts for the alternative energy industry as part of aid packages. This funding of a Clean Development Mechanism for Southern nations would be offered in exchange for carbon offsets in developed nations.[216]

Zapatista Revolution: Building Endogenous Sustainability

In 1994, the Zapatistas were an armed movement of Indigenous leaders from the state of Chiapas in Mexico, protesting the enslavement, exploitation, and government persecution of the Mayan population over centuries. They continue to be the largest population in the state and the most impoverished.

On the day the North American Free Trade Agreement was to be signed, the Zapatistas released their Declaration for Life.[217] It was their protest against neoliberal globalization policies, particularly the privatization of community-owned lands. They were asserting their right to develop on their own terms, outside of the global economy, a truly endogenous sustainability.

While they were routed by the Mexican army, they did sign a peace accord in 1996. While not legally implemented, the Zapatistas declared their autonomy, redistributed land among peasants, and created complex democratic organizations to govern themselves. Most importantly, they have been building community sustainability upon their ancient Indigenous knowledge, languages, and culture. Their model of sustainability will be discussed further in Chapter 8.

This is one example that the conventional model of development was losing legitimacy in the South. What became known as the "Fourth World," largely Indigenous people committed to subsistence economies, were excluded permanently from the global economy. As Dana Stuchul, Madhu Suri Prakash, and Gustavo Esteva argue,

> [B]eginning in the 1990s, an entirely new social class was explicitly created: disposable people. In capitalist societies, there has always been the unemployed, the 'surplus population.' . . . Today, the situation is very different. Vast numbers of people now serve no use for capital. They are disposable. And today, the disposable are being disposed of—a new iteration of war.[218]

The Zapatistas call this the Fourth World War, a perpetual war that keeps people hungry and dispossessed.

Battle of Seattle: Fourth Wave Environmentalism

With a growing desperation in the "South" regarding neoliberal economics, social movements began to coalesce. Similarly, the volatility in Southern Europe related to debt loads and loss of government programmes led to massive national protests, becoming more violent as the suffering increased.[219] After Rio, with growing awareness of interconnected issues and NGO connectivity, there was the beginning of a transnational social movement called the Global Justice movement.[220] In an attempt to slow down the process of globalization and mobilize people particularly across European countries into a broad net of solidarity, concerns across organizations and issues fused.

The first intense face of a transnational, transsectoral movement was the Battle of Seattle in 1999. The World Trade Organization (WTO) was meeting in Seattle to lay out trade parameters for the coming millennium. Over 50,000 activists from around the globe converged—trade unionists, environmentalists, students, urban community leaders, peasant and small farmers, women's groups, church groups, and countless other citizen groups related to peace, human rights, and social justice. They protested the rapacious global financial system, including the deregulation of labour, which pitted unions against each other globally, and the elimination of environmental regulations. There was outrage at further concentrations of power and wealth and at the right of 1,000 World Bank and IMF economists to dictate the life conditions for the majority of people and the planet.[221] It was "the closest thing the world had so far seen to a popular front against global capitalism,"[222] though their strategies and goals were diverse. Thus, fourth-wave environmentalism was characterized by the growth of this transnational civil society movement that builds coalitions across all previous social movements.

More specifically, it was the first visible sign of global civil society resisting the neoliberal revolution and calling for the dismantling of the WTO. Banners such as "Teamsters and Turtles" spoke about new common ground between labour and environmentalists. Environmentalists dressed up as butterflies and polar bears and unionists dressed in hard hats and high visibility vests illustrated the common damages, while creating a media spectacle. Mostly peaceful, and through no central logic but extensive on-the-ground networking, they closed off parts of the city. As best they could, holding hands, they created a circle around the building and WTO participants. Seattle police were overwhelmed, resorting to pepper spraying, jackbooting, beating, and arresting protesters, raising the emotions and the stakes. This was the beginning of the counter-globalization movement.

The End of Development?

After 50 years, an impasse in international development theorizing had been reached, which many were calling "the end of development." In *The Development Dictionary*, 1993, Wolfgang Sachs and many intellectuals questioned the Age of Development, critiquing the constellation of concepts that comprised development discourse. They concluded that "all the hopes and desires which made the idea fly, are now exhausted: development has grown obsolete."[223]

It was evident that the idea of development was based on competition, where the developed world would always ensure that they rank on top. The redistribution of wealth was rhetoric only. Southern theorists charged that Northern development theory is part of ongoing "empire." Only with the development of Southern theory and practices, could alternatives be found.[224]

Further, the end point of a "completely developed world" based on continual growth was a "misconceived enterprise" that would annihilate the biophysical foundation of the planet.[225] Further, it was creating a cultural monoculture, and, thus, for all these reasons was a "self-defeating discourse."[226] Development enacted structural violence against the bodies and labour of women, mothers, and children, rather than centring gender relations and women's experiences.[227] As Rist argues, in the end, globalization replaced development, even though the convictions behind the enterprise of development, such as poverty eradication, remained.[228]

Poststructural, postcolonial, and Indigenous discourse contested all the foundational concepts and theories. They argued that notions of development from below, endogenous development, and participatory development were still rooted in the old paradigm. In all its forms, development was a social construction that enabled continued domination through the economic and cultural logics of the West. Despite good intentions, the professionalization and institutionalization of development and the assumptions of "problems" and "abnormalities"[229] were themselves parts of this homogenization of Westernism. Development was an "imaginary" that was "sick, dying, gone."[230]

Wolfgang Sachs concluded, "Delusion and disappointment, failures and crimes have been the steady companions of development and they tell a common story: it did not work."[231] John Isbister, involved in the development enterprise for decades, came to the conclusion that though development was founded on the promise that the "rich could join in a partnership with the poor, that they could adopt the struggle against world poverty as a common endeavour, [it] was betrayed."[232] "[T]he promises made to [the Third World] have not been kept."[233]

A postdevelopment era was proposed and a new agenda imagined.[234] The ways forward, such as those identified in *The Post-Development Reader*,[235] are varied. They include *radical pluralism* drawing from the ideas of Wendell Berry, Gandhi, Illich, Schumacher, Esteva, and Prakash. These ideas assert human-scale institutions and technologies as well as protecting local cosmo-visions. They promote voluntary simplicity or minimalism involving both ecological efficiency and material sufficiency. They reimagine Earth-based spiritualities in promoting the development of the soul, the moral and ethical, and the life of the mind. Through a reappraisal of noncapitalist, nonmodern, and nonwestern societies, and consideration of *Indigenous Traditional knowledge* with modes of thinking and being that are ecological and wise, new pathways become visible.[236] One alternative economic model is an economy of gift and reciprocity.[237] Another is a move towards *postmaterialism*, primarily postmaterial values that reconstrue growth as growth towards the highest human values and community strength, rather than materialism and material wealth, at both the individual and economic system levels.[238] *Zero growth and degrowth* have been proposed in light of the instability of the global economy which has been unable to establish equity and justice or protect and restore ecological integrity.

Conventional development has also demonstrated that capitalism and democracy are at odds.[239] Much like the Zapatistas, there has been advocacy for direct, deep, participatory, or radical democracy, outside current ideological polarities. Carolyn Merchant, in *Radical Ecology*, advocates for a new ontological paradigm of relationality. For her, this includes "new patterns of production, reproduction, and consciousness that will improve the quality of human life and the natural environment."[240] She proposes resolving the contradictions between production and ecology, between production and reproduction, and between homocentric and ecocentric approaches. This can only be done standing outside the dominant world order and rethinking the current worldview. It is turning towards a social transformation through the processes of relationality.[241] To these ideas, we will turn in Chapter 8.

The Millennial Turn and Fourth Wave Environmentalism

Millennium Development Goals 2000–2015

In 2000, at the Millennium Summit in New York, a new opportunity existed to re-establish a sense of partnership and commitment to sustainable development. The UN launched the Millennium Development Goals or MDGs for the next 15 years (listed in Box 3.1), considered a departure from the sole focus on economic growth, particularly structural adjustment, to include human development.[242] Given the uneven results of globalization, there was a recognition that economic development did not necessarily translate into human development. So, there was another call for a nondiscriminatory trade system.[243] The MDGs would assist nations in developing sustainable development plans, with a concentrated focus on reducing extreme poverty by 2015. Other commitments during this time frame was economic diversification to reduce vulnerabilities in times of downturn as well as increasing Official Development Assistance, from .3% up to the .7% longstanding commitment.[244]

BOX 3.1 MILLENNIUM DEVELOPMENT GOALS

The eight Millennium Development Goals from 2000 to 2015 were:

1. eradicate extreme poverty and hunger,
2. achieve universal primary education,
3. promote gender equality and empower women,
4. reduce child mortality,
5. improve maternal health,
6. combat HIV/AIDS, malaria, and other diseases,
7. ensure environmental sustainability, and
8. develop a global partnership for development.[245]

The MDGs were "relatively successful,"[246] as extreme hunger and poverty (those living on less than $1/day) were reduced by 50% between 1990 and 2010, 5 years ahead of schedule. The number of those living on less than $1.25/day fell from 36% to 12%. The number of those undernourished fell by 10% in 2011. However, regional trends tell more of the reality, with sub-Saharan Africa poverty reduced by only 16%.[247] Yet, economic realities improved in Brazil, Russia, India, and China, or what are called BRIC nations. In advance of the millennial turn, the Jubilee 2000 international coalition of 70 NGOs, based on the biblical notion of jubilee as a debt forgiveness year, lobbied for the cancellation of third-world debt over the latter part of the 1990s.[248] As a result, $100 billion dollars of debt was cancelled for the poorest 35 nations.[249]

Then, in January 2001, growing out of the Battle of Seattle, Porto Alegre in Brazil hosted the first World Social Forum, as the first and largest global gathering of civil society. Through democratic, participatory, and deliberative processes with tens of thousands of people, they drew up a Charter of Principles that envisioned a planetary society that lives in peaceful relations with the Earth. Their assertion was "another world is possible."[250] They vigorously opposed neoliberal globalization and the domination of capital in all its imperialistic forms.

The World Social Forum continues to meet almost every year, as the counter-globalization movement to the World Economic Forum, posing alternative policies and innovations. ENGOs and EMOs are now part of this transnational activist community which enables sustained dialogue across borders.[251] Protest has been organized to every global economic meeting, whether the G7, G20, WTO, or World Economic Forum in Davos, forcing these meetings into more and more "fortress-like settings."[252]

In 2002, Rio+10 was in Johannesburg, as the Summit on Sustainable Development. It was considered one of the most inclusive summits, including a range of sectors from business to Southern NGOs. Yet, the widening disparities and ongoing ecological degradation continued to be the largest challenges, now overtly considered the result of globalization, adding significant threats to global security and democracy.[253]

9/11 and Disaster Capitalism

On 11 September 2001, two hijacked commercial airliners were flown into the twin towers of the World Trade Center in New York as a suicide attack by 19 members of the militant Islamic extremist network Al-Qaeda. One other plane was flown into the Pentagon and another aimed for a federal building but crashed in a field near Shanksville, Pennsylvania. In total, almost 3,000 people were killed in these coordinated attacks, the most deadliest terrorist attack in history. The response to the attack was termed the War on Terror, as a global counterterrorism campaign to root out armed, militant groups who oppose the industrial, modern West and its policies. The West cut off diplomatic ties to the governing Taliban in Afghanistan and then, as part of the War in Afghanistan, pursued members of and bombed Taliban, Al-Qaeda, and other terrorist camps. In December 2001, the US-led coalition invaded Afghanistan and ousted the Taliban, remaining until 2014, under this operation.

What neoliberal strategists had learned over the previous 20 years was that disaster can create optimal conditions for the implementation of radical capitalism. Naomi Klein calls this "disaster capitalism:"

> That is how the shock doctrine works: the original disaster—the coup, the terrorist attack, the market meltdown, the war, the tsunami, the hurricane—puts the entire population into a state of collective shock. The falling bombs, the bursts of terror, the pounding winds serve to soften up whole societies much as the blaring music and blows in the torture cells soften up prisoners. Like the terrorized prisoner who gives up the names of comrades and renounces his faith, shocked societies often give up things they would otherwise fiercely protect.[254]

Thus, the War on Terror enabled a broader agenda of conflict, an expansion of neoliberalism, and tightening of civil liberties, with many nations passing anti-terror legislation while redirecting attention away from these economic agendas. Klein explains that, rather than the stable markets required previously by capitalism, the possibility for superprofits and public opinion shifts are much higher during times of disaster, when fear and disorder reign.[255] After 9/11, "the fear of terror was greater than the fear of living in a surveillance society."[256]

In the United States, the fight on terror facilitated "the creation of the disaster capitalism complex—a full-fledged new economy in homeland security, privatized war, and disaster reconstruction tasked with nothing less than building and running a privatized security state, both at home and abroad."[257] The goal was a "hollow government,"[258] where governments are not about governing or providing services but outsourcing tasks through subcontracts to "the generally superior private sector."[259] For instance, in the push to apply market logic to the US military, the private company "Halliburton was responsible for creating the entire infrastructure of a US military operation overseas. All that was required of the army was to provide the soldiers and weapons—they were, in a way, content providers, while Halliburton ran the show."[260] In these years, contracts were let for homeland security, intelligence, surveillance, airport security, welfare payments, data management, communications, prisons, aspects of education and health care, as well as disaster response, to name a few.

The War on Terror provided far-reaching opportunities for profitable contracts, underwritten with public funds, embedded in the overall shift towards privatizing government and security as a permanent economy.[261] Further, instead of waiting for disasters to exploit them, they could also be engineered. For instance, the shock and awe doctrine in Iraq gave American capitalists access to oil fields and significant private security and reconstruction contracts. Not only could one site of terrorists be eliminated, but Iraq could become a model of a free-market democracy.

The end result of this shift has been "corporatism: big business and big government combining their formidable power to regulate and control the citizenry."[262] As Canadian political philosopher John Ralston Saul explains, "the corporatist movement was born in the twentieth century as an alternative to democracy. It proposed the legitimacy of groups over that of the individual citizen"[263] better known as fascism. Neoconservatives have aligned with neocorporatists in denigrating concepts as equality and justice with the "overall effects on the individual [as] passivity and conformity in those areas which matter and non-conformism in those which don't."[264]

Corporate Camouflage and Climate Denialism: More Backlash

The corporate world, particularly in the United States, was feeling increasingly embattled by the environmental movement and the high-emotion level of support. Right-wing thinkers explained,

> Industry cannot win purely with a public relations drive and therefore needs to initiate a pro-industry activist movement not only to win the public's hearts and minds but to fight the environmental movement. . . . It takes a movement to fight a movement.[265]

Following this rationale, they turned to several strategies. The first was "greenwashing," which involves camouflaging destructive practices through labelling and misleading advertising. For instance "General Motors, the largest multinational on the planet, the largest producer of vehicles and a major defence contractor, suddenly announced the company's '20 years of environmental progress.'"[266] Rowell calls this "greenbabble" which involves "deception and consumer confusion."[267] Companies like Chevron created public relations campaigns that are a "selective presentation of facts with little context."[268] "Greening" became so extensive that Friends of the Earth created the "Green Con Awards" to identify for the public which companies were offering misleading or disingenuous advertising.[269]

The second was to develop "front groups" which are NGOs with environmental sounding names that undermine science-based environmental knowledge.[270] For instance the Global Climate Coalition comprises some of the most powerful corporations who deliberately contest the credibility of climate science and reduce public support. Such groups exist in each sector, from consumer safety to forestry. Despite their names, they are overtly pro-capitalist and anti-environmentalist. Industry also developed their own independent law firms, calling them "public interest law firms,"[271] paralleling Ralph Nader's Public Interest Research Groups.

The fight was also *within* the environmental movement. Called the Wise Use movement, building on the original name by foresters, it was a coalition between those in the raw resource and agricultural industries who wanted public land designated for multiple uses rather than protected. As one Wise Use architect suggested, "We don't even care what version of Wise Use people believe in as long as it protects private property, free markets, and limits government."[272]

The third strategy was "counter-science" or "pseudoscience" particularly to delegitimate climate change science.[273] Experts and scientists were hired, often those who were unpublished in peer-reviewed and legitimate journals. They were paid to discount any legitimate research despite broad scientific consensus. As more climate science was released, corporations, industry, the New Right, and other neoliberal vested interests such as right wing talk show hosts began to assert that there was a "climate debate." The media, pursing their ethic of balanced reporting and most often corporately owned, sought out "both sides" to illustrate they were representing both sides of this so-called "scientific debate."

Fourth, to prevent the Not-In-My-Back-Yard or NIMBY phenomenon, corporations hired public communications and marketing specialists to craft messages. Then, they used grassroots, or what Rod Bantjes calls "astroturf" organizers given their synthetic nature, to find potential supporters and give the appearance of a mobilized public in their favour.[274] In other words, corporations used these supporters, who were in fact not self-organizing, to speak on their corporate behalf, protecting their vested interests.

Fifth, they used a binary discourse to limit public discussion within a "for/against" or "demonization/victimization" stance, rather than illustrate a spectrum of alternatives. Through their own media, and particularly now through social media, they shape public opinion, even against their supporters' own interests.[275]

In all these ways, disinformation has been on the rise, distinct from misinformation, in being crafted to deliberately deceive. Even though there was a significant scientific consensus on climate science, outright climate denialism was trumpeted throughout the media via corporate funded organizations and thinktanks, such as the Global Climate Coalition and Global Climate Council.[276] Then, in IPCC meetings, after significant lobbying, scientists on the payroll of the fossil fuel industry were given a spot at the table. These scientists, allegedly producing no recent research and discredited in their fields, became mouthpieces for climate denialism. The media provided space for this climate denialism and the idea that climate changes were natural and "no danger."[277]

Numerous nations with ultraright governments, who chose to adopt the convenience of climate denialism, were called "the Carbon Club." Their goal was to offload responsibility at international meetings. One tactic was complaining about the carbon emissions of China and India. While not insignificant, 90% of emissions are from industrialized nations.[278] In 1997, while the corporate world was stalling the inevitable energy transition in public, behind closed doors BP assigned $1 billion to solar energy development and Shell assigned $500 million to renewable energy options, readying for the inevitable pivot.[279] In 2021, a Dutch court ruled that Shell must reduce its carbon emissions by 45% in 2030.[280]

To counter this climate denialism, Al Gore, formerly a journalist, wrote *Earth in the Balance* in 1992, just prior to his election as US Vice President, generating public support for authentic climate science. He even proposed a Global Marshall Plan towards ecological ends. After his vice-presidency, he would create a slideshow entitled *An Inconvenient Truth* which became a 2006 award-winning documentary, accompanied by a book of the same name.[281] American polls have indicated that it did much to raise public awareness and change minds about the science of climate change. It also enhanced the climate movement through activist training of thousands as climate reality leaders and educators via The Climate Reality Project and Inconvenient Youth.[282]

Climate Accords and Reality

In 2000, rather than cutting emissions by 5% as part of the Kyoto Protocol goals, American emissions were up by 30%, Canadian emissions were up by 130%, Australian emissions were up by 17%, and the Netherlands emissions were up by 17%.[283] Citizen and activist anger rose as many nations experienced increased flooding, droughts, and storms. In the EU, fuel protestors were not so much protesting the increasing price of fuel as the lack of forethought in providing better options for public uptake, such as better public transit, more fuel-efficient vehicles, more aggressive development of renewable energies, and better scaling towards the goals.

In 2001, the Millennium Ecosystem Assessments were established under the auspices of the UN, an integrated assessment of ecosystem health and human well-being.[284] Through the work of almost 1,400 scholars, the goal was to "establish the scientific basis for actions needed to enhance the conservation and sustainable use of ecosystems and their contributions to human well-being."[285] In their first report of 2005, they found that humans have changed ecosystems more rapidly than any other historical period, with irreversible losses; the changes have benefited human well-being but degraded ecosystems placing future generations in peril, a barrier to achieving the Millennium Development Goals. They recommended significant changes in policies, institutions, and practices to reduce these "negative trade-offs."[286]

In 2001, James Hansen was asked by President Bush to be a part of his Climate Task Force. However, contrarians were also called into the task force with Hansen, presenting conflicting evidence that did not follow the scientific method. Hansen was forced to conclude, "I believe the biggest obstacle to solving global warming is the role of money in politics."[287]

In 2005, the Kyoto Protocol was ratified by 191 nations and went into force until 2012, with 36 nations meeting the binding targets. The developing world, including India and China, were not expected to reach these targets. Russia finally committed to the Protocol, and emissions trading began in the EU. When it was clear that some nations could not meet the goals, a flexibility mechanism was granted, allowing for emissions trading or write-offs from funding emission reductions in other countries via the Clean Development Mechanism. Emissions, however, still increased by 32% over the 20 years between 1990 and 2010. Yet, in the Stern Review, 2006, the costs of inaction regarding climate change were shown to have far more costs and devastating impacts than the costs of immediate action.[288]

Frustrated while watching Climate Summit negotiations, the mayors of major cities gathered at their own summits, establishing the C40 cities in 2005. The primary issue was that megacities (over three million people) produce 70% of the world's emissions.[289] They consume two/thirds of the global energy supply, including the 25% that is already being generated by renewable sources. Since 2005, megacities can sign a Global Covenant to become a net-zero carbon city. Originally, C40 cities agreed to peak their emissions by 2020, reduce them by 50% by 2030, and achieve carbon neutrality by 2050. By 2020, 34 cities were on target for peaking their emissions.

C40 cities have succeeded by retrofitting building infrastructure and stipulating all new buildings be net zero/carbon neutral, among many other actions from active transportation[290] to reducing solid waste. A source of pride for their residents, Oslo has a carbon budget, resulting in a 50% reduction in emissions by 2022. Superblocks that are car-free, such as in Barcelona, use car-shares and bike-shares to reach carbon neutrality. C40 cities now have significant funding partners, research expertise, measurement inventories, and peer teaching to develop replicable change.[291] Other city-to-city networks and communities of practice, such as 100 Resilient Cities as well as Strong Cities, have also been taking on the Sustainable Development Goals (SDGs), since 2015, using them as planning tools.[292]

At COP13, 2007, the Bali Road Map was a comprehensive action plan which included mitigation strategies and adaptation strategies.[293] While mitigation measures—reducing emissions and enhancing carbon sinks like forest protection—are crucial,[294] so are adaptation measures—adjustments in ecological, social, or economic systems that respond to expected or actual climate impacts.[295] In 2008, an adaptation fund for developing nations was established.

In Copenhagen, 167 nations signed the 2009 Copenhagen Accord, pledging a deep cut in global emissions to hold the global temperature below 2°C, although small island states argued for 1.5°C. Different responsibilities were placed on developed and developing nations, though the accord was not legally binding.[296] Developed nations pledged $30 billion in fast-start funding for Southern nations between 2010 and 2012.

2008 Financial Crash

Amidst this international climate wrangling, the 2008 Financial Crash occurred, called the sub-prime mortgage crisis. In both the United States and Europe, cheap mortgages had been offered with very low interest rates, enabling people to purchase without a significant downpayment or adequate financial means. In the short term, this stimulated house construction and expanded supply, driving growth. Banks and mortgage companies, however, were gambling by bundling cheap mortgages together, then selling and reselling them on financial markets.

In 2007, the United States needed to raise interest rates. Suddenly, many owners could no longer afford their mortgages, often owing more money than their home was worth. Millions would default, and many mortgage companies went bankrupt, threatening the stability of the banks. Given the tightly linked financial system, the global economy went into shock. Unemployment soared, especially youth unemployment, leading to what has been described as a "lost generation."[297]

The Arab Spring in 2010 unleashed a revolutionary fervour of protests against autocratic rule, economic decline, extreme poverty, high food prices, corruption, and human rights violations. In Fall 2011, the Occupy Movement gathered in a major park on Wall Street, New York. The Occupy Movement quickly went global, protesting economic and political injustice, particularly the predatory lending practices that triggered the global financial crisis and the economic divide between the superrich and the "rest of us." The key slogan was "We are the 99%," referring to the economic expansion between 2002 and 2007 where the income of the top 1% of Americans grew 10 times faster than the income of the bottom 99%. In 2011, there were more people living in poverty globally than the whole world population in 1900.

In 2009, evidence of climate change was accelerating. The UK Environmental Justice Foundation found that 26 million people were already environmental refugees.[298] Just before the meeting in Doha, 2012, the World Bank released a report indicating that present activity puts the globe on course for an increase of 4°C in global temperature. It warned about tipping points in which change will increase dramatically, including melting permafrost and glaciers and massive forest die-off in the Amazon.

In Doha, a second Kyoto commitment period from 2012 to 2020 was discussed. Under the Doha Amendment, nations agreed to an 18% cut in emissions between 2013 and 2020.[299] Canada and the United States, however, withdrew from continued participation in the Kyoto Protocol, two of the largest emitters, but 147 nations accepted.

As part of the constant volatility of capitalism and "vulnerability of the global financial architecture,"[300] such as the 2008 crash, the first developed nations began to experience debt and deficit problems in 2010, including Greece, Portugal, Ireland, Spain, and Cyprus. As Eurozone members, they could not devalue their currency, and so they applied severe austerity measures. Numerous rounds of tax increases were implemented along with government programme cuts, resulting in 25% unemployment. As part of the resulting humanitarian crisis, there were riots and protests as well as a brain drain from these countries, as people emigrated.

2015 Paris Agreement, Agenda 2030, and The Future We Want

The year 2015 was "a landmark year for multilateralism and international policy shaping."[301] At Rio+20 in Brazil 2012, under the theme of "The Future We Want," the commitment to sustainable development was re-evaluated and solidified, as well as an identification of the challenges in moving to a green economy. Climate change was posing profound challenges for sustainable development, but the achievements of some MDGs proved that focussed international efforts can make a difference.[302]

Neoliberal financial policies were increasingly questioned. As Stiglitz once again asserted, "those countries that are the most closely integrated into the global economy, having followed the advice of the international economic institutions and opened themselves up the most, have been among the worst hit"[303] by financial instability. Government intervention and national protectionism were again seen as being necessary in certain circumstances. The Addis Ababa Action Agenda for Financing Development was signed in 2015.

The IPCC reports in 2001 and 2007 brought "the development agenda into the climate change arena,"[304] stimulating the call for climate justice amidst growing humanitarian crises. As part of climate adaptation, the idea of climate resilience took hold, ensuring communities develop the capacity to prepare for, recover from, and adapt to increasing climate impacts.[305] By 2013, the IPCC Fifth Assessment noted that CO_2 is 40% higher than in pre-industrial times, with most of the human impact between 1951 and 2010. The Earth's surface had already warmed by .85°, with Arctic Sea ice declining by almost 4% a decade. They advocated for "aggressive mitigation"[306] to avoid the worst consequences. In 2015, the Paris Agreement forged common cause among 195 nations who agreed to combat climate change by unleashing significant actions and investment towards a low-carbon, resilient, and sustainable future. The Sendai Framework for Disaster Risk Reduction was also signed in 2015.

In 2013, the United Nations General Assembly proposed 17 Sustainable Development Goals (SDGs) for 2015–2030, to replace the Millennium Development Goals (see Box 3.2).[307] Adopted in 2015, this became known as Agenda 2030, including 169 targets across the goals.[308] The scale of ambition and level of comprehensiveness surpassed all other previous documents.[309] The key recognition was that goals cannot be siloed if policy coherence was to be achieved. Rather, the SDGs needed to be interdependent across economic, social, and environmental dimensions.[310] Further, the goals would now apply to both developed and developing nations, each defining their own priorities for effective progress towards Agenda 2030. A pledge of US $500 billion by business leaders, NGOs, and other groups was given to promote green growth.

BOX 3.2 SUSTAINABLE DEVELOPMENT GOALS

The 17 Sustainable Development Goals (SDGs) from 2015 to 2030 are:

1. No poverty
2. Zero hunger
3. Good health and well-being
4. Quality education
5. Gender equality
6. Clean water and sanitation
7. Affordable and clean energy
8. Decent work and economic growth

9. Industry, innovation, and infrastructure
10. Reduced inequalities
11. Sustainable cities and communities
12. Responsible consumption and production
13. Climate action
14. Life below water
15. Life on land
16. Peace, justice, and strong institutions
17. Partnerships for the goals

In 2017, with slow economic growth and a decline in world trade, 1.4 billion people around the globe faced vulnerable employment conditions, with increasing income inequality.[311] That said, there was growing international recognition that traditional subsistence sectors should not be disturbed, and that conventional development should not necessarily be followed, but be appropriate to unique circumstances.

The Climate Justice Movement

As part of the transnational counter-globalization movement, the climate justice movement was now taking hold, with linkages across the North and the South. Large demonstrations, particularly of young people, were common, including groups such as Rising Tide, Sunrise Movement, and Extinction Rebellion. They called for "radical, alternative measures to ensure the wellbeing of all humanity in harmony with nature built on the principles of solidarity, justice and Mother Earth's rights."[312]

In 2018, 16-year-old Greta Thunberg began her protests in front of the Swedish parliament, as a School Strike for Climate. She captured the anger, fear, and despair that millennials were feeling, that their "future was being stolen" through limited progress on emissions cuts. This grew into the global *Fridays for Future* movement. While youth had been playing a more prominent role in the United Nations conferences over recent years, Greta was a lightning rod that galvanized them. She went on to lead rallies throughout the world in 2019, culminating in a Global Climate Strike in Montreal. She spoke to the United Nations Climate Change Conference in 2018 and then the UN Climate Action Summit in 2019, with 15 other young people. In 2020, she spoke to the World Economic Forum and EU Environment Committee, reiterating her statement that "our house is on fire." What has become "the Greta effect" has had leaders admitting their guilt of inaction and promising more, but it remains to be seen what tangible actions will result. One impact has been "flight shaming" as people become reluctant to use air travel, on the basis of emissions.

In 2018, the IPCC released a special report announcing that global warming could reach 1.5°C as early as 2030. They reiterated that human activities have caused at least 1°C of global warming above pre-industrial levels, as a call for strengthening global responses to climate change, sustainable development, and eradication of poverty.[313] Without significant action, they say, there is high confidence that the result will be profound changes in natural systems and therefore human systems. In comparing the differential impacts of global warming at 1.5°C and at 2°C, they demonstrated that the Paris Agreement commitments would still exceed 1.5°C of warming. If sustainable development, poverty eradication, and reducing other inequalities are to succeed, they were adamant that global emissions "must decline well before 2030."[314] The IPCC advocated that sustainable development is very supportive of the needed transformation for limiting global warming, added support for achieving Agenda 2030. They stressed that the need for global coordination "across national, sub-nation, civil society, private sector, Indigenous peoples, and local communities" is unavoidable and requires significant capacity-building.

UN Secretary-General António Guterres' called the 2021 IPCC special report a "code red" for humanity, requiring decisive action.[315] The full IPCC report released in February 2022[316] states that "Human-induced climate change is causing dangerous and widespread disruption in nature and affecting the lives of billions of people around the world, despite efforts to reduce the risks. People and ecosystems least able to cope are being hardest hit."[317] They suggest we are facing severe, unavoidable, cascading, and irreversible "multiple climate hazards" in the next 20 years as global warming inches towards 1.5°C (2.7°F).[318] Accelerated action making deep cuts in emissions is required to address increasing loss of human life, loss of biodiversity, and loss of human infrastructure, including low-lying coastal habitations.

Another Cultural Shift: Rage and the Silence Breakers

In early 2016, construction began on an oil pipeline under two rivers, one lake, and the Standing Rock reserve in the United States, without any meaningful consultation with Indigenous people, a glaring example of environmental racism. Young people and tribal leaders from the *Dakota/Lakota* Nation reserve established a water protectors camp to defend the waters, several hundred sites of cultural significance, and burial grounds, drawing thousands of others in solidarity.[319]

Seven months later, as sacred ground and a camp were being bulldozed, protestors defending the land were attacked by riot police and their dogs and sprayed with water cannons in freezing weather. Many were arrested and strip-searched, the first of several violent attacks also involving a private security firm and the National Guard.[320] Capturing media attention, the *Dakota* took their issue before both the UN Human Rights Commission and the UN Forum on Indigenous Issues. Under the Obama Presidency, the Dakota Access Pipeline partners were eventually denied an easement and construction stopped. This was reversed by President Donald

Trump. Bypassing any environmental assessment, the pipeline was completed in 2017. The pipeline continues to operate during ongoing court challenges. There was a temporary stop work order, but that was overturned despite an incomplete environmental assessment.

In Canada, 2021, mass graves of school children were being unearthed on former residential school grounds. The first discovery, using ground penetrating radar technology, was at *Tk'emlúps te Secwépemc* First Nation in British Columbia, which uncovered 215 unmarked graves at the former Kamloops Indian Residential School. In the late 1800s up to 1990, First Nations, Inuit, and Metís children were forcibly taken to residential schools. The children were not allowed to speak their languages, wear their traditional clothing or long hair, practice their culture, or speak to siblings. They endured psychological, physical, and sexual abuse largely at the hands of Church-run and state-run institutions, recalled by many courageous survivors.[321] More than 1,700 graves of little children have been found so far,[322] but the estimates are as high as 7,000, in the "Every Child Matters" effort.[323]

This is the latest discovery as Canada engages a Truth and Reconciliation process between Indigenous and settler people regarding attempted cultural genocide, land theft, and systemic impoverishment.[324] The Truth and Reconciliation Commission was established as part of a settlement agreement.[325] Painful truth telling has also included the Missing and Murdered Indigenous Women as well as Children Commission[326] and the routine harassment and brutality experienced by Indigenous people on a regular basis in their defence of their land, against the predations of forestry, mining, and pipeline companies. This reality is repeated globally, from Ecuador to Nigeria.

In 2008, Tarana Burke used the term "Me Too" to encourage solidarity among women who were survivors of violence, particularly women and girls of colour.[327] In 2017, Ashley Judd accused Hollywood director Harvey Weinstein of sexual assault. Later in the year, Alyssa Milano responded with "#MeToo," asking those who have been sexually harassed or assaulted to use the hashtag. The #MeToo movement quickly proliferated around the globe in an avalanche of mass disclosures regarding violence against women, particularly sexual violence. As a result, violence against women was pushed to the top of the global human rights agenda, asserting that it is "everywhere, hidden in plain sight, normalized so we didn't even see it."[328]

The #MeToo movement has identified serial perpetrators from France to Finland to Fiji in entertainment, music, sports, media, medicine, churches, the judiciary, military, and governments. More accurately, it has been many connected movements across the globe: #NiUnaMás in Latin America; #lifeinleggings in the Caribbean; #WoYeShi in China, the #BalanceTonPorc movement in France, #EverydaySexism in the UK; #YoTambien in Spain, #QuellaVoltaChe in Italy, and #stilleforopptak in Norway. Similar hashtags exist in Palestine, India, Israel, Japan, Kenya, Nigeria, Pakistan, Vietnam, and South Korea, to name a few. The "silence breakers" were honoured by *Time Magazine* as the 2017 People of the

Year, including Tarana and Alyssa. Movement leaders state that the ultimate goal is "disrupting all systems that allow sexual violence to flourish."[329]

In May 2020, African American George Floyd died from injuries sustained from police kneeling on his neck. Captured live by bystanders, enraged citizens of Minneapolis took to the streets. Initially peaceful, police attacked protesters to break them up, leading to looting, fires, and vandalism. One in a long series of Black American deaths at the hands of police, most often men, tens of thousands took to the streets in 140 cities as part of the Black Lives Matter movement.[330] Property and lives were lost over the next few months as well as hundreds of arrests happened, while white supremacists and other resentment groups took to the streets. In many cities, the National Guard was called in, raising tensions further. Floyd's death was ruled a homicide, and one police officer was charged with second-degree murder. Three other officers were charged with aiding and abetting the killing.

Black Lives Matter has become a global movement, as protests spread against racially motivated violence, particularly by police. In 2015, the UN announced the International Decade for People of African Descent, remembering the 15+ million men, women, and children who died as part of the 400 years of the transatlantic slave trade and the ongoing racism and injustice that is the legacy of slavery.[331] In the United States, there is a call for reparations for 250 years of slavery and Jim Crow brutality.[332]

Together, these movements bring the history of Western civilization full circle in terms of accountability. These movements are asserting a restoration of nonmodern cultures, ways of knowing, knowledge systems, and ways of being. This attempted erasure, through violent means, of ethnodiversity and biodiversity (living beings, cultures, and ecosystems) have been profoundly linked.

Global Pandemic and Nuclear Threat

The COVID-19 or coronavirus attacking respiratory systems, was first identified in Wuhan, China, in late 2019. In March 2020, one nation after another was locked down on the advice of the World Health Organization. In the attempt to stop the spread, the world was brought to a complete standstill. Whole cities and skies were quiet as people hunkered down in their homes. As of late 2022, more than 6.5 million people have died, and almost 500 million have been infected, as pharmaceutical scientists rushed to develop a vaccine.[333] By 2021, several vaccines became available, but debates about efficacy and harms began to circulate, sometimes through misinformation on social media or by leaders hoping to profit from the chaos. The realities of racial discrimination and health equity manifested as did the struggle between public health recommendations/requirements and those advocating for full individual rights and liberty. The economic disruption, including massive layoffs, supply chain disruption, and large institution closures, created a global recession and spiral of inflation. However, the temporary reduction in emissions, evidence of global resolve, and the move towards workplace flexibility are part of a significant shift, enabling many to see beyond current structures.

In 2022, Russia carried out an unprovoked invasion of Ukraine. Since the Ukrainian 2014 uprising which deposed Russian-friendly leadership and Ukraine's migration towards membership in the North Atlantic Treaty Organization (NATO), Russia has questioned their existence as a separate state. In 2014, Russian annexed Crimea, and then, on the pretext of liberating two regions which were controlled by pro-Russian separatists, Luhansk and Donbas, they invaded several locations in eastern Ukraine. The determined bravery of the Ukraine army, paramilitary, reservists, and foreign volunteers have held off a Russian conquest, but numerous cities and towns have been substantially damaged by bombing, including civilian targets. Evidence of war crimes is mounting.[334] At the end of 2020, 7.8 million of almost 44 million Ukrainians are now refugees, fleeing the country, while millions are internally displaced.[335] President Volodymyr Zelensky has spoken to the leadership of many nations and international bodies seeking assistance to secure their airspace, alongside material assistance and economic sanctions. Most concerning has been the nuclear "sabre rattling." Once again, the world has been brought to the brink of potential nuclear war, with additional threats of chemical and biological weapons. Much is at stake in terms of governance among a global network of nations.

So, How Did We Get Here?

> When [people] lack a critical understanding of their reality, apprehending it in fragments which they do not perceive as interacting constituent elements of the whole, they cannot truly know that reality. To truly know it, they would have to reverse their starting point: they would need to have a total vision of the context in order to subsequently separate and isolate its constituent elements and by means of this analysis achieve a clearer perception of the whole.[336]
>
> Those who cannot remember the past are condemned to repeat it.[337]

Too often, existing generations are not aware of all the predecessor movements and achievements that have addressed ongoing environmental and social justice issues across centuries. Transformative educators need to be critical students of history, including root causes of issues, social movements, and ongoing social change. As philosopher educator Paulo Freire says in the aforementioned first quote, through a broad historical and critical analysis, a vision with a more complete context can emerge, from which we identify both root causes of environmental and social injustice as well as possibilities. We are able to identify propitious moments for action. In the second quote, Spanish philosopher George Santayana reminds us that it is only through awareness and reflexivity that we can avoid repeating what has been attempted unsuccessfully in the past. However, the corollary is also true, in which we can try that which was aborted or ignored, as reservoirs of potential beyond the current progress trap of Western civilization.

As Chapter 2 illustrated, we exist in a much larger cosmological and geological history, constantly eclipsed through the proclivities of rationalism and modernism. Premodern civilizations have attributes and social organizations worth examining, as do nonmodern cultures, including Indigenous and Southern ways of knowing and being, discussed further in Chapter 8.

The Western worldview started to take shape in the Classical periods of Greece and Rome, providing the conceptual ground—from rationalism, dualism, humanism, to democracy. From Early Modernism in the late Middle Ages and early Renaissance, additional concepts emerged, from a linear sense of history, the importance of individual conscience, to the early beginnings of capitalism. The Scientific Revolution gave us an inert and mathematically ordered universe as well as a scientific method that led to the industrial technologies we now use to "subdue" the earth and each other, as Francis Bacon originally promoted. The Enlightenment elaborated further ideas—liberty and freedom, equality, progress, private property, secularism, and fully fledged individualism.

The Colonial empires exported these ideas and their "rightful dominance" throughout the world, creating the conditions for continual war, conflict, racism, sexism, speciesism, and all other forms of division that we know today. It was the spread of death for whole peoples and the living world. The Age of Revolutions—Scientific, French, American, and Industrial revolutions—led to the nation states, representative democracies, and industrialism we know today.

The Late Modern era of the 20th century saw the formal end of colonialism but the rise of economic neocolonialism, partly through international aid and development. The 20th century was one of the most violent periods in history, given two world wars, the ideological extremes of the Cold War with its ongoing regional proxy wars, callous financial systems causing significant suffering, and the sheer power of the capacity for mass death.[338]

Most importantly, says historian Eric Hobsbawn, "the historical memory was no longer alive,"[339] as "most young men and women at the century's end grow up in a permanent present," without knowledge of "basic facts" and how they "hang together."[340] Addressing this lapse as well as enabling learners to name the significance of their present within a broad historical context is a vital task facing educators today. Understanding the push and pull of ideologies, the impact of changemakers within tragedies and catastrophes, and the public power of critical analyses and innovative ideas can help us connect the disparate dots of how we got here. Suspending our investments in these systems as we carefully and honestly examine them is another important way forward, particularly as educators and learners.

Tracing the threads backward, there have been many variants of resistance to industrialism and its range of material practices and consequences, especially since the Industrial Revolution. While new technology has enabled the industrial model to shift various energy systems (whale oil and coal to oil and gas and now solar, geothermal, and wind) as well as enabled resource substitution for waning resources, from beaver pelt fashions to polypropylene fleece and gortex, industrialism still threatens the carrying capacity of the natural world and the ability to regenerate. Even the so-called postindustrial system is reliant on scarce resources for digital devices, such as the copper, tellurium, lithium, cobalt, manganese, and tungsten in cell phones.[341] Further, while large-scale utility providers, such as energy grids or regional water systems, are useful in terms of efficiencies, they are highly vulnerable in an era of climate events. The story of international development helps demonstrate the fallacies of large-scale development, the levels of corruption, mismanagement, ideological zeal, and economic self-interest that accompany the conventional model of development, which have yielded dramatic losses, suffering, and other negative consequences for millions of people around the globe. Nevertheless, the humanitarianism and the call to international fairness, solidarity, and accountability, especially in the face of an existential threat, remains vital.

The protest against the most exploitative and rapacious practices of capitalism and the remarkable flexibility that capitalism has shown as a system has continued over several centuries. Capitalism has been critiqued or adjusted in multiple ways, but truly effective action for change has been imprisoned within ideological binaries or silenced. Capitalism, as an economic system predicated on growthism[342] and profit-making, is not a given. Our current economic system is founded on the basest of human values, from greed, competitive antagonism, to egocentrism. These motivators feed people's desire for power over others, to gain wealth at the expense of others, and to amass consumer goods that constantly degrade the life system upon which we depend—from minerals to mammals, forests to fish, soil to sea, and all the ecosystems in which they relate.

There is a call now for developing systems and structures that build upon the highest of human values, for a moral economy that honours all of Life. Many innovations over several centuries, often at small local levels, have maintained a market system but without capitalist assumptions, yielding a closer connection between buyers and sellers in terms of scale and for undertaking appropriate planning in times of stress. This is taken up in more detail in subsequent chapters.

Globally, there also is much frustration with existing representative democratic systems, not necessarily the most effective form of democracy. Many innovations with direct and deep democracy are being implemented globally. They seek to undo large, impersonal, bureaucratic systems based on accepting certain levels of unemployment, poverty, devastation, and death. Existing systems accept a loss of democracy and rise in poverty as the normal consequences of international competitiveness. These systems will continue to squeeze nations and individuals to work harder but "get ahead less," losing the conviviality of life. This has been the history of Empire over human history.

The next chapters are for educators and all others working for social change. A recounting of the journey of education, its constitution and reconstitution within larger systems, is offered to prod your historical and sociological imagination. Out of this, a pluriverse of possibilities is proposed that can help us in this momentous epochal shift to break through to another way of thinking and being.

Notes and References

Chapter line drawing by Canadian artist Erin Marie Lange Christensen. www.erinchristensen.com. Reproduced with permission.

1 J. R. McNeill, *Something New Under the Sun: An Environmental History of the Twentieth-Century World* (New York: W. W. Norton & Company, 2000), p. 362.
2 The Latin root of "radical" is radix or "root," as in the "root of a problem." www.merriam-webster.com/words-at-play/radical-word-history.
3 Excerpt from full page ad in the *New York Times*, January 18, 1970.

4 The Earth Day organizing committee included Denis Hayes, Andrew Garling, Arturo Sandoval, Stephen Cotton, Barbara Reid, and Bryce Hamilton along with Senator Gaylord Nelson.
5 Charles Reich, *The Greening of America* (New York: Bantam Books, 1970).
6 "The History of Earth Day." www.earthday.org/history/. Accessed February 10, 2021.
7 Manuel Castells, *The Power of Identity* (Oxford: Blackwell, 2009).
8 Christopher Rootes, "Environmental Movements," in David Snow, Sarah Soule and Hanspeter Kriesi (Eds.), *The Blackwell Companion to Social Movements* (Malden, MA: Blackwell Publishing, 2007), p. 608.
9 Donatella Della Porta and Mario Diani, *Social Movements: An Introduction* (Oxford: Blackwell Publishing, 2006), pp. 178–179.
10 Direct action is understood as action that works "to achieve an end directly and by the most immediately effective means (such as a boycott or strike)." From www.merriam-webster.com/dictionary/direct%20action.
11 For statistics, see Carolyn Merchant, *Radical Ecology: The Search for a Livable World* (New York: Routledge, 2005), p. 167.
12 For more history, see https://buffalonews.com/news/local/history/a-history-of-the-love-canal-disaster-1893-to-1998/article_5df93af9-e5fe-5ae4-be74-efed7dbf43ed.html.
13 Carolyn Merchant, *Radical Ecology: The Search for a Livable World* (New York: Routledge, 2005), p. 168.
14 Bernard Goldstein, "Environmental Movement," Encyclopedia.com. www.encyclopedia.com/earth-and-environment/ecology-and-environmentalism/environmental-studies/environmental-movement. Accessed February 11, 2021. For more discussion, see also, Merchant, *Radical Ecology*, pp. 166–169.
15 Merchant, *Radical Ecology*, p. 165.
16 Rootes, "Environmental Movement," p. 614.
17 From www.environmentandsociety.org/.
18 From www.unep.org/about-un-environment.
19 From www.unep.org/about-un-environment.
20 Merchant, *Radical Ecology*, p. 166.
21 Ramachandra Guha, *Environmentalism: A Global History* (New York: Longman, 2000).
22 Guha, *Environmentalism*, p. 94.
23 For key differences between proportional representation and first-past-the-post voting systems, see https://keydifferences.com/difference-between-first-past-the-post-and-proportional-representation.html.
24 When no one party gains a majority of votes, several parties may form a temporary coalition to rule over the period of the mandate, particularly common in proportional representation systems. See www.britannica.com/topic/coalition-government.
25 Derek Wall, *The No-Nonsense Guide to Green Politics* (Ottawa and Toronto: New Internationalist Publications and Between the Lines Publishing, 2010).
26 Wall, *The No-Nonsense Guide to Green Politics*, p. 10.
27 Wall, *The No-Nonsense Guide to Green Politics*, p. 17.
28 Wall, *The No-Nonsense Guide to Green Politics*, p. 18.
29 Merchant, *Radical Ecology*, pp. 176–181.
30 Merchant, *Radical Ecology*, p. 176.
31 Merchant, *Radical Ecology*, p. 176.
32 From www.encyclopedia.com/science/encyclopedias-almanacs-transcripts-and-maps/developments-ecology-1900-1949.
33 Oswald Schmitz, *Ecology and Ecosystem Conservation* (Washington: Island Press, 2007), p. 7.
34 Ernest Callenbach, *Ecology: A Pocket Guide* (Berkeley: University of California Press, 1998).
35 Schmitz, *Ecology and Ecosystem Conservation*, p. 7.
36 Callenbach, *Ecology*, n.p.
37 Schmitz, *Ecology and Ecosystem Conservation*, p. 7.

38 Merchant, *Radical Ecology*, p. 94.
39 Mark Anderson, "New Ecological Paradigm (NEP) Scale," in *The Berkshire Encyclopedia of Sustainability: Measurements, Indicators, and Research Methods for Sustainability* (Great Barrington, MA: Berkshire Publishing Group, 2012). See https://umaine.edu/soe/wp-content/uploads/sites/199/2013/01/NewEcologicalParadigmNEPScale1.pdf.
40 See Gene E. Likens. www.britannica.com/science/acid-rain and Thomas Butler and Gene Likens at www.britannica.com/explore/savingearth/acid-rain.
41 Joe Farman, Brian Gardiner, and Jonathan Shanklin
42 Susan Solomon, "The Discovery of the Antarctic Ozone Hole," *Nature*, Vol. 575 (2019), pp. 46–47. https://doi.org/10.1038/d41586-019-02837-5.
43 See https://www.unep.org/ozonaction/who-we-are/about-montreal-protocol.
44 Solomon, "The Discovery of the Antarctic Ozone Hole."
45 For more information, see the World Wildlife Fund's 2015 *Living Forests Report* at https://wwf.panda.org/discover/our_focus/forests_practice/forest_publications_news_and_reports/living_forests_report/. See also canopyplanet.org.
46 See Chris Mooney, "Destruction of the Amazon is Speeding Up – Just When the Planet Can Least Afford It," *Washington Post*, December 2, 2016, and Jeff Tollefson, "Battle for the Amazon," *Nature*, Vol. 520 (2015), pp. 20–23.
47 See Merchant, *Radical Ecology*, pp. 169–176.
48 Merchant, *Radical Ecology*, pp. 174–175. For a list of the principles, see Elliott, *An Introduction to Sustainable Development*, p. 52 or www.ejnet.org/ej/.
49 Robert Bullard, *Unequal Protection: Environmental Justice & Communities of Color* (San Francisco: Sierra Club Books, 1994).
50 Peter Martin, "A WWF View of Education and the Role of NGOs," in *Education for Sustainability*, John Huckle and Stephen Sterling (Eds.) (London: Earthscan, 1996), pp. 40–54.
51 Lorraine Elliot, Britannica Online Encyclopedia, s.v., "Environmentalism." www.britannica.com/topic/environmentalism. Accessed February 13, 2021.
52 Rootes, "Environmental Movements," p. 614.
53 Rootes, "Environmental Movements," p. 617.
54 Carolyn Merchant, *The Death of Nature: Women, Ecology and the Scientific Revolution* (San Francisco: Harper & Row Publishers, 1980), xix.
55 Merchant, *The Death of Nature*, p. xvi.
56 Merchant, *The Death of Nature*, p. xv.
57 Cited from Elizabeth A. Lange, "Environmental Adult Education: A Many-Voiced Landscape," in Carole Kasworm, Amy Rose, and Jovita Ross-Gordon (Eds.), *Handbook of Adult and Continuing Education* (Thousand Oaks, CA: Sage Publications, 2010), pp. 305–315. Based on Riane Eisler, *The Chalice and the Blade: Our History, Our Future* (New York: HarperSanFrancisco, 1987) and *Tomorrow's Children: A Blueprint for Partnership Education in the 21st Century* (Boulder, CO: Westview Press, 2000).
58 See both John Cobb, *Process Theory as Political Theology* (Philadelphia: Westminster Press, 1982) and Thomas Berry, *The Dream of the Earth* (San Francisco: Sierra Club Books, 1988).
59 Graham Harvey, *Animism: Respecting the Living World* (New York: Columbia University Press, 2005), p. xi.
60 Chief Oren Lyons is quoted by Kenny Ausubel, "Preface: Remembering the Original Instructions," in Melissa Nelson (Ed.), *Original Instructions: Indigenous Teachings for a Sustainable Future* (Rochester, VT: Bear & Company, 2008), p. xxi.
61 Melissa Nelson (Ed.), Original Instructions: Indigenous Teachings for a Sustainable Future (Rochester, VT: Bear & Company, 2008).
62 Nelson, *Original Instructions*, p. 4.
63 Nelson, *Original Instructions*, p. 12.
64 Gregory Cajete, *Native Science: Natural Laws of Interdependence* (Santa Fe, NM: Clear Light Publishers, 2000), p. 2.

65 Helen Kopnina and Eleanor Shoreman-Ouiment, *Sustainability: Key Issues* (London: Routledge Earthscan), p. 10.
66 Ellwood, *The No-Nonsense Guide to Degrowth and Sustainability* (Oxford: New Internationalist, 2014), p. 174.
67 See Maggie Black, *The No-Nonsense Guide to International Development* (Toronto: New Internationalist and Between the Lines, 2002), pp. 91–93.
68 Fritz Schumacher, *Small is Beautiful* (London: Sphere Books, 1974).
69 Edward Goldsmith and Robert Allen, *A Blueprint for Survival* (New York: Penguin, 1972).
70 Wolfgang Sachs (Ed.), *The Development Dictionary: A Guide to Knowledge as Power* (Johannesburg and London: Witwatersrand University Press and Zed Books, 1992).
71 Richard Peet and Elaine Hartwick, *Theories of Development: Contentions, Arguments, Alternatives* (New York: The Guilford Press, 2009), p. 104.
72 Peet and Hartwick, *Theories of Development*, p. 104.
73 Paraphrased from Peet and Hartwick, *Theories of Development*, pp. 104–106, 111.
74 Peet and Hartwick, *Theories of Development*, pp. 1–2.
75 Sachs, *The Development Dictionary*, p. 1.
76 See Majid Rahnema and Victoria Bawtree, *The Post-Development Reader* (London, Dhaka and Halifax: Zed Books, University Press Ltd, and Fernwood Publishing, 1997).
77 Sachs, *The Development Dictionary*, p. 1.
78 Marianne Gronemeyer, "Helping," in Sachs, *The Development Dictionary*, pp. 53–60.
79 Sachs, *The Development Dictionary*, p. 2.
80 Found in Arturo Escobar, "Planning," in Sachs, *The Development Dictionary*, pp. 132–145.
81 Black, *The No-Nonsense Guide to International Development*, p. 17.
82 United Nations, *World Economic and Social Survey 2017* (*WESS 2017*) (New York: Department of Economic and Social Affairs, 2017). From file:///C:/Users/EAL/Desktop/WESS_2017-FullReport.pdf, p. 3.
83 Black, *The No-Nonsense Guide to International Development*, p. 19.
84 This would eventually change to GNI or Gross National Income. UN, *WESS 2017*.
85 Black, *The No-Nonsense Guide to International Development*, p. 16.
86 UN, *WESS 2017*, p. 18.
87 Black, *The No-Nonsense Guide to International Development*, pp. 15–16.
88 Thomas Wright, *Latin America in the Era of the Cuban Revolution* (Westport, CT: Praeger), p. 72.
89 In Black, *The No-Nonsense Guide to International Development*, p. 20.
90 Mitchell Sharp, "Canada's Stake in International Programmes," *Dialogue 1961-New Dimensions in International Relations* (Ottawa: Canadian National Commission for UNESCO, 1961), p. 47.
91 Walt Rostow, *The Stages of Economic Growth: A Non-Communist Manifest*o (ebooks.com: Barakaldo Books, 1960/2020).
92 Rostow, *The Stages of Economic Growth*, p. 2.
93 The following descriptions are paraphrased from Rostow, *The Stages of Economic Growth*, pp. 16–28.
94 The following descriptions are paraphrased from Rostow, *The Stages of Economic Growth*, pp. 16–28.
95 Citations from Walt Rostow in Gilbert Rist, *The History of Development: From Western Origins to Global Faith* (London: Zed Books, 1997), pp. 94–99.
96 Ankie Hoogvelt, *The Third World in Global Development* (London: Macmillan Publishers, 1982), p. 116.
97 In Rist, *The History of Development*, p. 99.
98 UN, *WESS 2017*, p. 67.
99 Keith Spicer, *A Samaritan State? External Aid in Canada's Foreign* Policy (Toronto: University of Toronto Press, 1966), p. 15.
100 Lester Pearson, *Pearson Report: A New Strategy for Global Development* (Paris: The UNESCO Courier, February 1969). https://unesdoc.unesco.org/ark:/48223/

pf0000056743. Quoted in Keith Spicer, *A Samaritan State? External Aid in Canada's Foreign* Policy (Toronto: University of Toronto Press, 1966), p. 252.

101 Pearson, *Pearson Report.*

102 Hoogvelt, *The Third World in Global Development*, p. 118.

103 Pearson, *Pearson Report.*

104 Black, *The No-Nonsense Guide to International Development*, p. 20.

105 Claes Brundenius, *Revolutionary Cuba: The Challenge of Economic Growth with Equity* (Boulder, CO: Westview Press, 1975/1984), p. 2.

106 Rist, *The History of Development*, p. 140.

107 See Julius Nyerere, *Ujamaa: Essays on Socialism* (Oxford: Oxford University Press, 1968).

108 Julius Kambarage Nyerere, *Ujamaa: The Basis of African Socialism*, April 1962, p. 2. https://ethics.utoronto.ca/wp-content/uploads/2017/11/Nyerere-UjamaaThe-Basis-of-African-Socialism-1962.pdf.

109 Julius Kambarage Nyerere, *Ujamaa: The Basis of African Socialism*, April 1962, p. 3. https://ethics.utoronto.ca/wp-content/uploads/2017/11/Nyerere-UjamaaThe-Basis-of-African-Socialism-1962.pdf.

110 Rist, *The History of Development*, p. 143.

111 Willy Brandt, *North-South: A Programme for Survival* (Cambridge: MIT Press, 1980).

112 STWR, *The Brandt Report: A Summary* (London: Share the World's Resources, 2006), p. 4. https://sharing.org/information-centre/reports/brandt-report-summary.

113 STWR, *The Brandt Report*, p. 5.

114 See Amiya Kumar Bagchi, *The Political Economy of Underdevelopment* (Cambridge: Cambridge University Press, 1982); Hoogvelt, *The Third World in Global Development*; and historical documents at the North-South Institute.

115 Black, *The No-Nonsense Guide to International Development*, p. 23.

116 Seehttps://worldpopulationreview.com/country-rankings/newly-industrialized-countries.

117 Black, *The No-Nonsense Guide to International Development*, pp. 23–24.

118 David Harvey, *A Brief History of Neoliberalism* (Oxford: Oxford University Press, 2005), p. 10.

119 Discussed at length in Patricia Marchak, *The Integrated Circus: The New Right and the Restructuring of Global Markets* (Montreal: McGill-Queen's University Press, 1991).

120 www.oxfordlearnersdictionaries.com/definition/english/casino-capitalism See also Harvey, *A Brief History of Neoliberalism*, p. 33.

121 Harvey, *A Brief History of Neoliberalism*, p. 27.

122 The second rise in interest rates was called the Volcker shock, after the Chairman of the US Federal Reserve Bank. See Harvey, *A Brief History of Neoliberalism*, p. 23.

123 UN, *WESS 2017*, Chapter 3.

124 www.federalreservehistory.org/essays/latin-american-debt-crisis.

125 UN, *WESS 2017*, p. 62.

126 Rist, *The History of Development*, p. 172.

127 Black, *The No-nonsense Guide to International Development*, p. 23. Dependency theorists on different continents include Samir Amin, Andre Gunder Frank, Osvaldo Sunkel, Immanuel Wallerstein, and Walter Rodney.

128 A key point by Andre Gunder Frank. See *Capitalism and Underdevelopment in Latin America* (New York: Monthly Review Press, 1967).

129 Amiya Bagchi, *The Political Economy of Underdevelopment* (Cambridge: Cambridge University Press, 1982), p. 13.

130 Bagchi, *The Political Economy of Underdevelopment*, p. 13.

131 Harvey, *A Brief History of Neoliberalism*, p. 28.

132 Samir Amin, *Unequal Development* (New York: Monthly Review Press, 1976).

133 Gustavo Esteva, *Unpublished Text from Culture and Development Conference* (Edmonton, Canada: Providence Renewal Centre, 1994), pp. 2–4. Used with permission.

134 Banking education refers to the

> act of depositing, in which the students are the depositories and the teacher is the depositor. Instead of communicating, the teacher issues communiqués and makes

> deposits which the students patiently receive, memorize, and repeat . . . In the banking concept of education, knowledge is a gift bestowed by those who consider themselves knowledgeable upon those whom they consider to know nothing. Projecting an absolute ignorance onto others, a characteristic of the ideology of oppression, negates education and knowledge as processes of inquiry.
>
> Paulo Freire, *Pedagogy of the Oppressed* (New York: Seabury Press, 1970), p. 58

135 Freire, *Pedagogy of the Oppressed,* p. 57.
136 Freire, *Pedagogy of the Oppressed*, pp. 75–78.
137 Harvey, *A Brief History of Neoliberalism*, p. 15.
138 Harvey, *A Brief History of Neoliberalism*, p. 19.
139 Discussed at length in Marchak, *The Integrated Circus.*
140 Harvey, *A Brief History of Neoliberalism*, pp. 11, 13.
141 Polanyi is cited in Harvey, *A Brief History of Neoliberalism*, p. 36.
142 Harvey, *A Brief History of Neoliberalism*, p. 67.
143 Harvey, *A Brief History of Neoliberalism*, p. 36.
144 Harvey, *A Brief History of Neoliberalism,* p. 37.
145 Discussed at length in Marchak, *The Integrated Circus.*
146 Marchak, *The Integrated Circus*, p. 95.
147 Marchak, *The Integrated Circus*, p. 94.
148 Marchak, *The Integrated Circus*, p. 94.
149 Marchak, *The Integrated Circus.*
150 UN, *WESS 2017*, p. 68.
151 Harvey, *A Brief History of Neoliberalism*, p. 20.
152 Stiglitz is cited in Harvey, *A Brief History of Neoliberalism*, p. 74.
153 Harvey, *A Brief History of Neoliberalism*, p. 31.
154 Harvey, *A Brief History of Neoliberalism*, pp. 34–35.
155 UN, *WESS 2017*, pp. 65–66.
156 United Nations, *WESS* 2017, p. 61. www.un.org/development/desa/publications/graphic/wess2017-global-economy.
157 Paraphrased from Harvey, *A Brief History of Neoliberalism*, p. 73.
158 UN, *WESS 2017*, p. 64.
159 This includes Harvard economist and IMF advisor Jeffrey Sachs and Joseph Stiglitz who was a chief economist with the World Bank. Both have said that it was a bad prescription, too narrow in economic strategies and too secretive and unaccountable in operations. The IMF itself in a 2000 report indicated that the policies spread economic insecurity while output grew, where one-fifth of the world's population now suffer more poverty and a lower quality of life. In Wayne Ellwood, *The No-Nonsense Guide to Globalization* (Oxford: New Internationalist, 2001), pp. 96–97.
160 Ellwood, *The No-Nonsense Guide to Globalization*, p. 97.
161 Black, *The No-Nonsense Guide to International Development*, pp. 122–124.
162 Marchak, *The Integrated Circus.*
163 Simon Dresner, *The Principles of Sustainability* (London: Earthscan, 2002).
164 Written by the International Union for Conservation of Nature and Natural Resources. www.environmentandsociety.org/mml/iucn-ed-world-conservation-strategy-living-resource-conservation-sustainable-development#:~:text=The%20World%20Conservation%20Strategy%20of,conservation%20rather%20than%20hinder%20it.
165 Some of this history was found in Jacobus Du Pisani, "Sustainable Development – Historical Roots of the Concept," *Environmental Sciences*, Vol. 3, No. 2 (2006), pp. 83–96. https://doi.org/10.1080/15693430600688831.
166 Called the World Commission on Environment and Development (WCED, 1987).
167 Rist, *The History of Development*, p. 180.
168 Carl Mitcham, "The Concept of Sustainable Development: Its Origins and Ambivalence," *Technology and Society*, Vol. 17 (1995), pp. 311–326.

169 Brundtland Commission, *Our Common Future* (Oslo: WCED, 1987), p. 8. https://sustainabledevelopment.un.org/content/documents/5987our-common-future.pdf.
170 Ecoefficiency here means doing more with less or, in this case, creating more value with less material used. See Kopnina and Shoreman-Ouiment, *Sustainability*, pp. 13–14.
171 Simon Dresner, *The Principles of Sustainability* (London: Earthscan, 2002).
172 Black, *The No-nonsense Guide to International Development*, p. 94.
173 Elliott, *An Introduction to Sustainable Development*, p. 42.
174 Guha, *Environmentalism*, pp. 98–124.
175 Rist, *The History of Development*, p. 181.
176 Rist, *The History of Development*, p. 181.
177 Elliott, *An Introduction to Sustainable Development*, p. 9.
178 Rist, *The History of Development*, pp. 182–183.
179 Elliott, *An Introduction to Sustainable Development*, p. 19.
180 David Selby and Fumiyo Kagawa, *Sustainability Frontiers: Critical and Transformative Voices from the Borderlands of Sustainability Education* (Opladen, Berlin & Toronto: Barbara Budrich, 2015).
181 Rist, *The History of Development*, pp. 184–185.
182 Elliott, *An Introduction to Sustainable Development*, pp. 48–49.
183 Elizabeth A. Lange, "Transformative Sustainability Education: From Sustainababble to a Civilization Leap," in Marcella Milana, Sue Webb, John Holford, Richard Waller, and Peter Jarvis (Eds.), *The Palgrave International Handbook on Adult and Lifelong Education and Learning* (London: Palgrave Macmillan, 2018). Paraphrased from David Orr, *Ecological Literacy* (Albany, NY: SUNY, 1992).
184 David Orr, *Ecological Literacy*. See more discussion of strong sustainability in Chapter 5.
185 Orr, *Ecological Literacy*.
186 From Elliott, *An Introduction to Sustainable Development*, p. 25. Adapated from Robert Gibson, *Sustainability Assessment* (London: Routledge Earthscan, 2005).
187 Lange, *Transformative Sustainability Education*, p. 403.
188 Andre Edwards, *The Sustainability Revolution* (Gabriola Island, BC: New Society Publishers, 2005).
189 Andrew Rowell, *Green Backlash: Global Subversion of the Environmental Movement* (London: Routledge, 1996), pp. 22, 374.
190 Rowell, *Green Backlash*.
191 Rowell, *Green Backlash*, p. 6.
192 Rowell, *Green Backlash*, p. 23.
193 Rowell, *Green Backlash*.
194 Rowell, *Green Backlash*, pp. 12, 21, 28.
195 Rowell, *Green Backlash*, p. 44.
196 Rowell, *Green Backlash*, p. 45.
197 For a description, see James Hansen, *Storms of my Grandchildren: The Truth about the Coming Climate Catastrophe and Our Last Chance to Save Humanity* (New York: Bloomsbury, 2009).
198 Simon Dresner, *The Principles of Sustainability* (London: Earthscan, 2002).
199 Hansen, *Storms of My Grandchildren*.
200 From Elliott, *An Introduction to Sustainable Development*, p. 104.
201 Dinyar Godrej, *The No-Nonsense Guide to Climate Change* (Toronto: New Internationalist and Between the Lines, 2001 and 2006), p. 94.
202 Rist, *The History of Development*, p. 186.
203 UNDP, *Human Development Report 2007/8: Fighting Climate Change: Human Solidarity in a Divided World* (New York: United Nations Development Programme, 2007), p. 4.
204 Find at https://earthcharter.org/.
205 Elliott, *An Introduction to Sustainable Development*, p. 8.
206 https://unfccc.int/timeline/.
207 Godrej, *The No-Nonsense Guide to Climate Change*.

208 Rist, *The History of Development*, p. 190.
209 Elliott, *An Introduction to Sustainable Development*, p. 10.
210 Black, *The No-nonsense Guide to International Development*, p. 96.
211 Black, *The No-nonsense Guide to International Development*, p. 108.
212 Black, *The No-nonsense Guide to International Development*, p. 108.
213 Elliott, *An Introduction to Sustainable Development*.
214 See https://archive.ipcc.ch/pdf/climate-changes-1995/ipcc-2nd-assessment/2nd-assessment-en.pdf.
215 Find at www.ienearth.org/kari-oca-2-declaration/.
216 Godrej, *The No-Nonsense Guide to Climate Change*, p. 109.
217 You can find their *Declaration for Life* at https://enlacezapatista.ezln.org.mx/2021/01/01/part-one-a-declaration-for-life/.
218 Dana Stuchul, Madhu Suri Prakash, and Gustavo Esteva, "From Fear to Hope: Learning from BIPOC in Hard Time—COVID 19, Climate Collapse & Racial Violence," *International Journal of Lifelong Education*, Vol. 40, No. 4 (2021), p. 416.
219 Rootes, "Environmental Movements," p. 622.
220 Donatella Della Porta, *The Global Justice Movement: Crossnational and Transnational Perspectives* (Boulder, CO: Paradigm Publishers, 2007).
221 Ellwood, *The No-nonsense Guide to Globalization*.
222 In Rod Bantjes, *Social Movements in a Global Context: Global Perspectives* (Toronto: Canadian Scholar's Press, 2007), p. 320. Also see film: "This is what democracy looks like." www.thisisdemocracy.org.
223 Sachs, *The Development Dictionary*, p. 1.
224 Raewyn Connell, *Southern Theory: The Global Dynamics of Knowledge in Social Science* (Cambridge: Polity Press, 2007).
225 Sachs, *The Development Dictionary*, p. 3.
226 Sachs, *The Development Dictionary*, p. 4.
227 Peet and Hartwick, *Theories of Development*, pp. 240–274.
228 Rist, *The History of Development*, p. 249.
229 Arturo Escobar in Richard Peet, *Theories of Development* (New York: The Guilford Press, 1999), p. 147.
230 Richard Peet, *Theories of Development* (New York: The Guilford Press, 1999), p. 149.
231 Sachs, *The Development Dictionary*, p. 1.
232 John Isbister, *Promises Not Kept: Poverty and the Betrayal of Third World Development* (Bloomfield, CT: Kumarian Press, 2006), p. 234.
233 Isbister, *Promises Not Kept*, p. x.
234 Frans Schuurman (Ed.), Beyond the Impasse: New Directions in Development Theory (London: Zed Books, 1993).
235 Majid Rahnema, *The Post-Development Reader* (London: Zed Books, 1997).
236 Categorizations by Peet, *Theories of Development*, 151–153, based on Rahnema, *The Post-Development Reader*.
237 Rist, *The History of Development*, p. 233.
238 Roots, "Environmental Movements," p. 618.
239 Merchant, *Radical Ecology*, p. 225.
240 Merchant, *Radical Ecology*, p. 8.
241 Merchant, *Radical Ecology*, p. 254.
242 UN, *WESS 2017*, p. 73.
243 See also Heather Farley and Zachary Smith, *Sustainability: If It's Everything, Is It Nothing?* (London: Routledge, 2020), p. 84.
244 UN, *WESS 2017*, pp. 94.
245 www.un.org/en/conferences/environment/newyork2000.
246 UN, *WESS* 2017.
247 All the statistics are from UN, *WESS 2017*, pp. 91, 93
248 Della Porta, *The Global Justice Movement*, pp. 132–133.

249 From www.advocacyinternational.co.uk/featured-project/jubilee-2000#:~:text=Jubilee%202000%20was%20a%20global,and%20mobilisation%20across%2062%20countries.
250 See https://fsm2016.org/en/sinformer/a-propos-du-forum-social-mondial/.
251 William Fisher and Thomas Ponniah, *Another World is Possible: Popular Alternatives to Globalization at the World Social Forum* (Halifax: Fernwood Books, 2003).
252 Black, *The No-Nonsense Guide to International Development*, p. 127.
253 Elliott, *An Introduction to Sustainable Development*, p. 11.
254 Naomi Klein, *The Shock Doctrine: The Rise of Disaster Capitalism* (New York: Metropolitan Books, 2007), p. 17.
255 See Naomi Klein, *The Shock Doctrine.*
256 Klein, *The Shock Doctrine*, p. 303.
257 Klein, *The Shock Doctrine*, p. 299.
258 Klein, *The Shock Doctrine*, p. 294.
259 Klein, *The Shock Doctrine*, p. 288, 295.
260 Klein, *The Shock Doctrine*, p. 292.
261 Klein, *The Shock Doctrine*, p. 12, 306.
262 Klein, *The Shock Doctrine*, p. 307.
263 John Ralston Saul, *The Unconscious Civilization* (Concord, ON: House of Anansi Press, 1995), p. 16.
264 Saul, *The Unconscious Civilization,* p. 27, 2.
265 Rowell, *Green Backlash*, p. 13.
266 Rowell, *Green Backlash*, p. 101.
267 Rowell, *Green Backlash*, p. 101.
268 Rowell, *Green Backlash*, p. 102.
269 Rowell, *Green Backlash*, p. 103.
270 Bantjes, *Social Movements in a Global Context*, pp. 258–259.
271 Rowell, *Green Backlash*, p. 7.
272 Rowell, *Green Backlash*, p. 14.
273 Rowell, *Green Backlash*, pp. 135–150.
274 Bantjes, *Social Movements in a Global Context*, pp. 258–264.
275 For more discussion, see Bantjes, *Social Movements in a Global Context.*
276 Godrej, *The No-Nonsense Guide to Climate Change*, pp. 87–91.
277 Godrej, *The No-Nonsense Guide to Climate Change*, p. 91.
278 Godrej, *The No-Nonsense Guide to Climate Change*, p. 95.
279 Godrej, *The No-Nonsense Guide to Climate Change*, p. 100.
280 Greenpeace, *Greenpeace* at 50, *Greenpeace Magazine*, Vol. 22, No. 2, p. 6.
281 Al Gore, *An Inconvenient Truth: The Planetary Emergency of Global Warming and What We Can Do about It* (Emmaus, PN: Rodale Press, 2006).
282 See https://climaterealityproject.org/blog/inconvenient-truth-then-and-now and https://climaterealityproject.org/blog/very-inconvenient-youth.
283 Godrej, *The No-Nonsense Guide to Climate Change,* p. 111.
284 www.millenniumassessment.org/documents/document.356.aspx.pdf.
285 www.millenniumassessment.org/documents/document.356.aspx.pdf, p. 9.
286 www.millenniumassessment.org/documents/document.356.aspx.pdf, p. 1 of Synthesis.
287 James Hansen, *Storms of My Grandchildren* (London: Bloomsbury, 2009), p. x.
288 Elliott, *An Introduction to Sustainable Development*, p. 109.
289 David Miller, *Solved: How the World's Great Cities Are Fixing the Climate Crisis* (Toronto: University of Toronto Press, 2020).
290 Active transportation refers to human-powered transportation.
291 See www.c40.org/ and Miller, *Solved.*
292 See www.brookings.edu/blog/future-development/2018/11/29/can-us-cities-help-the-world-achieve-the-sustainable-development-goals/.
293 https://unfccc.int/process/conferences/the-big-picture/milestones/bali-road-map.
294 https://unfccc.int/topics/mitigation/the-big-picture/introduction-to-mitigation.

295 https://unfccc.int/topics/adaptation-and-resilience/the-big-picture/what-do-adaptation-to-climate-change-and-climate-resilience-mean.
296 Elliott, *An Introduction to Sustainable Development*, p. 102.
297 Ellwood, *The No-Nonsense Guide to Degrowth and Sustainability*, pp. 9–103.
298 Ellwood, *The No-Nonsense Guide to Degrowth and Sustainability*, p. 63.
299 https://unfccc.int/kyoto_protocol.
300 UN, *WESS 2017*, p. 106.
301 https://sdgs.un.org/goals.
302 Kopnina and Shoreman-Ouimet, *Sustainability*, p. 156.
303 Elliott, *An Introduction to Sustainable Development*, p. 39.
304 Elliott, *An Introduction to Sustainable Development*, p. 107.
305 www.c2es.org/document/what-is-climate-resilience-and-why-does-it-matter/.
306 www.ipcc.ch/report/ar5/syr/.
307 UN, Department of Economic and Social Affairs, "Sustainable Development." https://sdgs.un.org/goals. Used by permission.
308 See https://sdgs.un.org/goals and https://sustainabledevelopment.un.org/globalsdreport/.
309 UN, *WESS 2017*, p. 12.
310 Farley and Smith, *Sustainability*, pp. 86–87 and Kamarulazizi Ibrahim, Kanayathu Koshy, and Walter Leal Filho, *The Future We Want: Post Rio+20 Sustainable Development and the Role of Education for Sustainable Development*. In Kopnina and Shoreman-Ouimet, *Sustainability*, p. 156.
311 UN, *WESS 2017*, pp. 17–22
312 Elliott, *An Introduction to Sustainable Development*, p. 111.
313 www.ipcc.ch/sr15/chapter/spm/.
314 www.ipcc.ch/sr15/chapter/spm/.
315 https://press.un.org/en/2021/sgsm20847.doc.htm#:~:text=Today's%20IPCC%20Working%20Group%201,of%20people%20at%20immediate%20risk.
316 The report can be found at: https://www.ipcc.ch/report/ar6/wg2/downloads/report/IPCC_AR6_WGII_SummaryForPolicymakers.pdf.
317 Quote is from https://www.ipcc.ch/report/ar6/wg2/downloads/press/IPCC_AR6_WGII_PressRelease-English.pdf.
318 Quote is from https://www.ipcc.ch/report/ar6/wg2/downloads/press/IPCC_AR6_WGII_PressRelease-English.pdf.
319 www.npr.org/sections/thetwo-way/2017/02/22/514988040/key-moments-in-the-dakota-access-pipeline-fight.
320 www.npr.org/sections/thetwo-way/2017/02/22/514988040/key-moments-in-the-dakota-access-pipeline-fight.
321 For instance, see Chief Robert Joseph, *NAMWAYUT: A Pathway to Reconciliation* (Vancouver: Page Two, 2022).
322 See https://storymaps.arcgis.com/stories/cfe29bee35c54a70b9621349f19a3db2.
323 www.cbc.ca/news/indigenous/residential-school-children-deaths-numbers-1.6182456.
324 See the final report at https://nctr.ca/records/reports/. For reconciliation education, see https://www.reconciliationeducation.ca/.
325 The Calls to Action can be found here: https://www2.gov.bc.ca/assets/gov/british-columbians-our-governments/indigenous-people/aboriginal-peoples-documents/calls_to_action_english2.pdf.
326 www.mmiwg-ffada.ca/final-report/.
327 www.chicagotribune.com/lifestyles/ct-me-too-timeline-20171208-htmlstory.html.
328 www.unwomen.org/en/news/in-focus/end-violence-against-women/2014/rights; United Nations General Assembly, *Declaration on the Elimination of Violence against Women*. 85 Plenary Meeting (1993). Downloaded October 5, 2017 from www.un.org/documents/ga/res/48/a48r104.htm.
329 Associated Press. (2017). "Silence Breakers" named *Time* Magazine's Person of the Year. New York, December 6.

330 www.nytimes.com/article/george-floyd-protests-timeline.html.
331 www.un.org/en/observances/decade-people-african-descent/slave-trade.
332 Ta-Hehisi Coates, "The Case for Reparations," *The Atlantic*, June 2014. From www.theatlantic.com/magazine/archive/2014/06/the-case-for-reparations/361631/.
333 https://covid19.who.int/. See also https://www.worldometers.info/coronavirus/.
334 See https://www.ohchr.org/en/hr-bodies/hrc/iicihr-ukraine/index.
335 See https://data.unhcr.org/en/situations/ukraine.
336 Freire, *Pedagogy of the Oppressed*, p. 95.
337 Quote from George Santayana, *The Life of Reason*, 1905.
338 Eric Hobsbawn, *The Age of Extremes: A History of the World 1914–1991* (New York: Vintage Books, 1996).
339 Hobsbawn, *The Age of Extremes*, p. 3.
340 Hobsbawn, *The Age of Extremes*, p. 3.
341 www.nhm.ac.uk/discover/your-mobile-phone-is-powered-by-precious-metals-and-minerals.html.
342 Kopnina and Shoreman-Ouimet, *Sustainability*, p. 22.

4

ENVIRONMENTAL EDUCATION

Situating Environmental Education

> Too much environmental and global education has been outer-directed (looking out on the world) and has denied interiority (inner journeying). The hidden agenda of this tendency is to collude with mechanism by implying that our inner self is outside the universe. . . . It is unlikely that environmental and global education can ever impact our culture unless we embrace a radical interconnectedness that revives mystery, a sense of the ineffable, the unknowable.[1]

Both environmental education (EE) and sustainability education (SE) sit at the intersection of multiple fields—education, philosophy, science, biology, chemistry,

DOI: 10.4324/9781003159643-4

physics, earth science, atmospheric science, ecology, geography, environmental science/studies, social science, history, sociology, peace studies, anthropology, international development, human psychology, Indigenous studies and cultural preservation, regenerative agriculture, technology, social change—and all their related professions. Environmental and sustainability education (ESE) also sits at the intersection of educational policy, research, theory, and practice. Further, Alan Reid argues, there is a "broad and contested range of accounts of the origins and traditions of environmental education around the world."[2] And as Theodore Roszak expressed earlier, there is much critique of the existing forms of environmental education. As such, a comprehensive summary of environmental and sustainability education is problematic, especially at the global level, tending to dehistoricize and decontextualize. However, a modest historical recounting is possible, despite the risks of generalization and a limiting of geographic coverage, building on the historical context of the previous two chapters. While this chapter focusses specifically on the North American context, the United States and Canada primarily, it illustrates international connections[3] and hopefully has some resonance with the development of EE and SE in other places.

Chapter 2 offered a big historical context—some origins of the Western worldview—as well as a brief history of colonialism, industrialism, and the First and Second Waves of environmentalism. Chapter 3 described the rise of environmental movements (EMOs) from 1970, green politics, and the proliferation of environmental nongovernmental organizations (ENGOs) as part of Second and Third Wave environmentalism. It also discussed waves within the international development enterprise, the backlash to progressive movements with the rise of neoliberal economics, and then Fourth Wave environmentalism with the emergence of multisector transnational movements.

This chapter, Chapter 4, discusses in broad strokes only, the early development of environmental education, which mirrored historical events, while navigating shifting ground within the field of education. The intention of this chapter is to convey the origins, early theorists, broad strands, and persistent debates in environmental education, then to offer a contextual assessment and critique of the field. The field is broad, so this review is not comprehensive of the wide range of voices, especially current voices. The goal, in part, is to tell a somewhat different historical story. The next chapter discusses the early development of sustainability education and then the growing integration of environmental and sustainability education, for some.

As shown earlier, it is evident that environmental concern existed for centuries and that the antecedents of environmental education were originally located in nonformal settings, such as organized, episodic adult and child learning in community settings as well as informal learning through literature and print media, before it entered formal settings such as schooling and higher education.[4] Further, when school-based environmental education becomes cut off from sister programmes like adult environmental education or community programming for young people, it is isolated from important sites of support for creating transformative change.

Antecedents of Environmental Education

> In a time characterized by so much desertification of alternatives, it is as difficult to imagine the end of capitalism, colonialism, and patriarchy as to imagine that they will have no end. The imagination of the end is being corrupted by the end of imagination.[5]

There are three commonly accepted antecedents to environmental education—conservation education, nature studies, and outdoor education—now coexisting with/in environmental education. This brief reprise will discuss additional facets of the antecedents to environmental education, including the workings of capitalism, colonialism, and patriarchy.

First Wave environmentalism from the 1800s contested the spread of death from colonial mercantile[6] policies that threatened many species with extinction. It recorded the degradation of land and forests through plantations, spread of agriculture, importation of non-native species, and rapacious harvest of "resources" through hunting, mining, and forestry. The death of human and nonhuman species occurred simultaneously through slavery and brutal colonial rule.

For instance, as British anthropologist Hugh Brody explains, if Indigenous peoples were not understood as being human, civilized, Christian, or with any viable economic system, especially as hunter gatherers, then they had no rights.[7] Europeans felt fully justified in the taking of land, plundering of resources, and genocidal elimination of people, except when a contrary action benefitted them. Indigenous people often felt that they simply needed "to yield or perish."[8] Further, it was considered the European duty to share their own "spiritual and material achievements"[9] with the "savages, primitives, pagans, unwhipped, untamed, degenerate, and innocent children of nature."[10] The thinking was that if you did not kill them, they would kill you first, so this was deemed a security issue as well as an economic and moral issue.[11]

The original moral debate about Indigenous people occurred in 1550 between Spanish priest Bartolomé de Las Casas, who was an early settler in Mexico, and Sepulveda, who was a Spanish Renaissance philosopher who never left Europe. From his experience, Las Casas argued that Indigenous people had a social and economic system that should be respected, although his thinking was part of the Noble Savage myth. He argued they should have the option of Christianity *through education*. He went on to record the savagery of colonization. Sepulveda disagreed, saying Indigenous people were less than human and required European Christian control, including enslavement.

The lineage of Las Casas came to include Rousseau, Montesquieu, Marx, as well as the Romantic and social reform movements of the 18th century. Together, they decried the oppression, exploitation, environmental degradation, and alienating qualities of colonization and industrial capitalism. However, as Brody emphasizes, these debates "had very little to do with a resolve that any particular Noble Savages be left their lands and economic system."[12] Unfortunately, these debates still reverberate as settler nations contend with this heritage and persistent beliefs, even as educational systems attempt decolonization. The primary point here is that the fortunes of Indigenous people and people of colour as well as the nonhuman and more-than-human worlds have always been woven together. Yet, this integration has not always been perceived by the environmental movement and the contradictions are little discussed by environmental education.

The usual history of environmental education revolves around the natural sciences and the development of schooling. Malcolm Swan, professor of outdoor teacher education, explains that as early as the 1700s in North America, some of the earliest concerns by both Thomas Jefferson and George Washington were the damaging soil practices of the tobacco industry.[13] At this time, education typically was only offered to the upper classes through private tutors, the churches, and later through the establishment of universities. It was Jefferson who promoted the need for a more widely available "republican education" to create the foundations for a sovereign state and representative form of government. In particular, he promoted the use of natural science and observation of the natural world for building a new kind of society.[14]

In the 1800s, James Audubon and other proponents advocated for environmental protection, particularly against the massive losses of hardwood forests and bird species from the Eastern seaboard westward.[15] His was an "epic quest to document all of the continent's avian life."[16] As a naturalist, his powers of observation and artwork were key aspects of natural science at this time, in his case ornithology. As there were no cameras, this was one way to preserve a visual image of a species, which, in Audubon's case, delighted European audiences with their novelty. In some cases, it was a record of a species before their extinction, as in Figure 4.1.

Yet, Audubon, born on a sugar plantation in Haiti, was involved in the slave trade. Further, he utilized Indigenous knowledge in *The Birds of America* with no attribution. To render his drawings, he killed the birds first and then articulated them, which was conventional at that time, yet such collections certainly contributed to the erasure of species.

Similarly, John Muir, revered as a key forest preservationist and promoter of national parks, advocated that Indigenous people be removed from planned parks. He considered them a detraction from pristine wilderness as well as "lazy," in that they only work enough for subsistence. In Yosemite, the seven tribes of the *Ahwahnechee* had lived in the Yosemite Valley for thousands of years in harmony with the natural world. Yet they were evicted.[17] Thus, the preservation movement was actually a protection against colonials themselves and later capitalists. Yet, it was Indigenous peoples who paid the price.

This process has been repeated around the world where tribal people are no longer given access to the land of their livelihoods. The land has been repeatedly set aside and preserved for the pleasure of well-heeled tourists, white middle class and often well educated urban folk, and, in some places, big game hunters.

Traditionally in North America, the only parks that people of colour and newcomers have had access to are urban parks, in terms of affordability and

FIGURE 4.1 Audubon's Drawing of Passenger Pigeons

Credit: Courtesy of the John James Audubon Center at Mill Grove, Montgomery County Audubon Collection, and Zebra Publishing.

accessibility. This remains true.[18] As well, given the history of slavery and illegal farm labour, African Americans, Latino Americans, and others can have negative relations with soil. Their interrelated experiences of racism, exploitative labour, food injustice, and loss of land, all require healing and reparations in order to return to a sense of sacredness about the Land.[19] Thus, environmental sociologist Dorceta Taylor argues that wildlife protection, forest conservation, and wilderness preservation has had a distinct race, class, and gender face. "People of color [have been] driven off their land, corralled onto reservations, enslaved, and used as low-wage labourers."[20]

Over the years, people of colour have found alternate pathways into the environmental movement and environmental education through issues such as civil rights, workplace hazards, farmwork injustice, denial of land, housing conditions, health and sanitation issues, toxic waste management, protection of water and air, and laws governing mining and other raw resource industries. "Economic elites, however, used racial fear and hatred to prevent a coalition between poor Whites and African Americans from forming,"[21] especially around environmental issues.

In terms of gender, women often made uncelebrated mountain ascents and wilderness trips even before the likes of Thoreau and Muir. White middle- and upper-class women tied their mountaineering to calls for suffrage as statements about freedom—of movement, voting rights, and clothing choices. Taylor explains, "female outdoor enthusiasts . . . defying stereotypes of white womanhood . . . expressed sympathy for the plight of Indians and were instrumental in founding or preserving parks and historic sites"[22] to assist them. In other words, "environmentalists . . . made little effort to incorporate social justice concerns into their environmental agendas, although the women were more inclined to do so than the men."[23] This history of intersectional realities illustrates the systemic connectedness of all humans and life forms, which had been ignored but now fuels the environmental and climate justice movements of today.

In the late 1800s and early 1900s, many conservationists supported scientific racism and eugenics, extrapolating the principles of park conservationism, such as "culling the weak," to human populations.[24] This thinking is still alive in fascist and White supremacist movements. These are just a few examples of the multiple forms of violence undergirding colonial and now neocolonial knowledges.[25]

At times, lone colonial voices critiqued not only the loss of species but also the attempted genocide of peoples through slavery and Indigenous "clearances,"[26] but they were few. As discussed in Chapter 2, Indigenous peoples themselves assessed the spread of Western civilization, finding it wanting of admirable virtues, particularly deeper forms of freedom and democracy. Their voices were not acknowledged.[27] Nevertheless, Indigenous methods of governance deeply informed the development of American republicanism as well as Canadian proclivities, specifically notions of inclusivity, egalitarianism, negotiated compromise, and acceptance of constant tension between diverse groups, only now being acknowledged.[28]

As part of the social reform movement, in the early 1800s, American Horace Mann led the Common School movement, subject to heated debate among the elite as to whether the "common" people *should* have access to education. Mann advocated for public schooling to build a sense of national identity, morality, and to "fulfill the new nation's political, social, and economic needs."[29] However, a public school system had many requirements: including well-prepared teachers who eventually would require 2-year normal school training, state funding to build schools, and curriculum to standardize learning. Edgerton Ryerson advocated similarly in Canada, yet by the early 20th century in Canada, only half of the nation's children attended school with any regularity.[30] Thus, it was the next hundred years which saw a dramatic expansion of the state apparatus for mass schooling, including mandatory attendance policies and full provision of Grades 1 to 12 by the 1940s and 1950s.

The earliest curriculum in the 1800s comprised reading, writing, spelling, and "speaking," but later included mathematics, history, geography, and natural history.[31] After 1850, the American science curriculum had elements of astronomy, physiology, geography, geology, natural philosophy, physics, natural history, zoology, chemistry, and botany.[32] By the late 1800s, many organizations, from the *Audubon Society* to *National Wildlife Federation*, advocated for curricular inclusion and development of educational materials related to nature studies.[33]

After 1900, nature studies would enter US elementary schools, general science would enter junior high, and biology would enter the high schools.[34] Rural school leaflets encouraged nature study alongside new available textbooks by Wilbur Jackman (1891 and 1904) and Anna Botsford Comstock (1911).[35] Teachers began to promote first-hand nature observation in and out of the school classroom—from cloud formations to animal behaviour. With the flood of immigrants into western North America, this met the need for nation-building as well as familiarization with new environs. It was not until after the 1920s that specialized science courses and laboratory work based on the scientific method even entered higher education.

Teaching at this time was "encyclopaedic" based on rote memorization of facts, drills, and literature recitation, accompanied by the threat of physical punishment. Cleanliness, punctuality, and obedience comprised cultural socialization. As Canadian educational historian Kas Mazurek argues, "In reality the creation of free, public schooling in Canada witnessed the arrival of a rather heavy-handed, authoritarian, and coercive institution."[36] An even more dire face of this phenomenon was the children of Indigenous peoples were forcibly taken to residential schools. Their language and culture were demeaned and violently suppressed and many experienced physical, psychological, and sexual assault. Many Indigenous children would die in residential schools, with their deaths and location of bodies unknown to their parents.[37]

In the late 1800s and early 1900s, given the rapid privatization of land and sharp decline in flora and fauna, there was a more pressing call for establishing national parks aimed at preserving areas of pristine beauty for general public access, tourism, and recreation. Parks would serve as an intact natural legacy for future generations. Preservation was advocated by influential naturalists, like John Muir, through newspaper essays, magazine articles, books, and in Congress.[38] Eventually, organizations like the *Sierra Club* and *American Nature Study Society* encouraged nature study in the new parks, becoming nature or park interpretation in the 20th century, not only for school groups but also for families and "adventure" groups.

Legislation in different American states would eventually mandate the study of forests, plant life, fish, and "game" within schools.[39] By the 1930s, the prairie "Dust Bowl" necessitated a change in farming practices to conserve water and agricultural soil. This was promoted through associations, such as the *Soil and Water Conservation Society* and *Civilian Conservation Corps* in the United States or the *Farmer's Institutes* and *Ducks Unlimited* in Canada, as well as through extension agents in departments of agriculture. They often used radio broadcasts, local meetings,[40] and printed materials to reach and educate adults in new agricultural practices.

In the early 1900s, John Dewey's philosophy of pragmatism evolved from the transcendentalism of Emerson and Thoreau, as described in the previous chapter. Dewey's progressive education movement utilized experiential learning or "learning by doing," as well as change through innovation thinking, more suited to developing an industrial workforce and generating "social progress."[41] As a more scientific approach to schooling, progressivist ideas derived from psychology and child development theory promoted "artistic and interpersonal development" through a "more democratic and humanistic learning environment."[42]

From the 1930s to 1950s, Dewey's promotion of learning outside the physical classroom and in the natural world helped inspire outdoor education.[43] Outdoor education was also particularly pursued as part of the school camping and summer camping movement. Outdoor education can be seen as part of the back-to-nature movement, emerging as society increasingly urbanized. Outdoor education can also be seen as a part of the social reform movement that sought to develop a moral and disciplinary framework for children and youth, especially those living in difficult conditions such as poverty.[44] These programmes were inspired either by paramilitary training camps, like Scouts and Cubs, or religious character training as with the origins of the YMCA. For instance the Scout Law is

> A Scout is helpful and trustworthy,
> Kind and cheerful,
> Considerate and clean,
> and wise in the use of all resources.[45]

Professor Del Giorno explains that science only entered schooling once the public was "cognizant of its importance to the industrial and economic growth of the country."[46] Generally, she says, new developments in science took 30 years to be incorporated in curriculum, especially until textbooks became more abundant and changed regularly. It was not until the 1950s that the application of scientific methodology was understood as a route for solving daily problems.[47] Environmental educator John Disinger asserted that nature study "dwindled and died" as science education grew more robust by the mid-1950s,[48] eventually shaping the approaches to environmental education.

It would be the convergence of multiple factors after WWII that created a propitious moment for the expansion of science education and then the rise of environmental education. The collective shock from Russians launching the first satellite into space, Sputnik in 1957, led to the fear that the United States "lagged far behind in technological capacity,"[49] impacting national security. This was the single most significant event that accelerated science education in the United States as well as Canada.

In the United States, the National Defense Education Act created $1 billion in funding for "high quality teaching and learning in science and mathematics."[50] In Canada, new education spending led to programme expansion in mathematics and the natural sciences as well as new research facilities, technical institutes, and community colleges.[51] Wartime spawned many technological innovations and production systems, which were applied postwar to the industrialization of agriculture, air and vehicle mobility, and development of new household as well as workplace machines, necessitating technical education and training. A powerful consumer ideology was promoted after the war, creating a continual growth economy and generating the prosperity of the 1960s.

As populations became increasingly urbanized during the 1950s and 1960s, outdoor settings for education were further promoted to build "social, spiritual, and recreational opportunities"[52] for city and town children and young adults. Schooling began to include courses in outdoor living, physical education, and recreation skills—from swimming and paddling to fire starting and tenting. It represented a "reintroduction of nature into children's lives"[53] and an example of education that balanced both indoor and outdoor settings. Schooling and higher education also expanded, in part, as an equal opportunity effort that was part of the developing social welfare state as well as for preparing young adults for professions in the expanding social and human services.

However, within the optimism and prosperity of the 1950s and 1960s, the postwar baby boom, and expanded scientific research, concerns began to escalate about pollution, overpopulation, nuclear war, and food toxins.[54] As home TVs increased in number, images of industrial waste and destruction were now seen alongside images of international poverty and famine. Environmental damage and social justice issues slowly penetrated public awareness. There was a growing mass realization that technological industrial society was leading to irreversible changes, many of them international in scope. It also appeared that natural resources might be limited.[55] Nevertheless, there was still an unfettered belief in technology to solve the consequences of human behaviour.

Emergence of Environmental Education

Second and Third Wave environmentalism would beget a dramatically different entity called environmental education.[56] Environmentalism created a seismic cultural shift that is now acknowledged, despite periods of backlash, denial, and political paralysis. Arguably, the environmental movement has made a "compelling achievement as an educational social movement."[57] Thus, environmental education can be seen as one facet of the larger environmental movement that tilled the ground for its emergence.

During the 1950s and 1960s, Second Wave environmentalism focussed on single issues related to advanced industrialism—from smog to toxins and atomic testing to large-scale dams. The New Social Movements, including the peace/antiwar, anti-poverty, women's rights, civil rights, Indigenous rights, gay rights, and consumer rights, laid the groundwork for a counterculture that rejected the excesses of industrial capitalism while promoting an expansion of human rights, social justice, and environmental protection.

In historical moments of societal self-critique, there is typically a turn towards public education to "modify or improve social [and environmental] conditions" or to keep up with "ongoing economic, labour market, and policy transformations."[58] In this way, schooling can at once be seen as a conservative institution and also a social change mechanism. Professor of environmental politics Andrew Dobson highlights that it is often influential organizations and individuals who take their concerns into "politically charged spheres," advocating for policies including educational policies, who help create change.[59] Thus, as the founder of Earth Day, it was American Senator Gaylord Nelson (see Figure 4.2) who was a key influencer and championed numerous key pieces of legislation from 1970 to 1980, including the National Environmental Education Act in October 1970. The American Congress passed the first environmental education act almost without opposition.[60]

The period from the 1970s through to the early 1980s is now considered the heyday of the environmental education field in North America through its rapid expansion and optimism.[61] In the Act, the purpose of environmental education was to "promote among citizens the awareness and understanding of the environment, our relationship to it, and the concern and responsible action necessary to assure our survival and to improve the quality of life."[62] Environmental education was defined as the "study of the factors influencing ecosystems, mental and physical growth, living and working conditions, decaying communities and population pressures,"[63] acknowledging linkages between the human and natural worlds.

That said, the Environmental Education Act received little priority, housed in the US Department of Health, Education, and Welfare. Despite promises, funding

FIGURE 4.2 Senator Gaylord Nelson Speaking

Credit: Wisconsin Historical Society, WHI-48015. Reprinted with permission.

was erratic. When the Act was renewed in 1990, the responsibility for EE was shifted to the US Environmental Protection Agency which prioritized environmental education[64] and led to a renewal of focus.

Educational administrators debated at length how best to implement environmental education.[65] Was environmental education a discrete unit of study, a topic, a new discipline, a way of teaching, a place to teach, or a perspective that permeated all education?[66] The attributes of this new entity, Environmental Education, challenged current notions of how education should be organized, including disciplinary compartmentalization and topic comprehensiveness.

In the midst of surging interest in the "environment" in the late 1960s, John Disinger explains that it was a small group of American community college educators who first began developing and sharing instructional materials which formed the early beginnings of environmental education and the National Association for Environmental Education (NAEE), in 1970.[67] However, much to their surprise, they were quickly joined by teachers and teacher educators who were also in need of resources and professional development.[68] Environmentalists from nongovernment organizations (ENGOs) as well as civil servants, including park naturalists and conservation educators, were also interested in environmental education, in learning to foster an "environmentally knowledgeable public."[69]

The NAEE expanded into the North American Association for Environmental Education (NAAEE) the next year, 1971,[70] with the inclusion of Canadians and, later, Mexicans. In 1973, the Alliance for Environmental Education was formed in the United States supporting educators across the American states. In Canada, education is a provincial responsibility, so teacher organizations across the country fostered professional development opportunities related to environmental education. Educational publishing houses quickly generated a flood of curricular materials for educator use.[71]

Parallel EE organizations would develop across multiple nations at this time, including England, Scotland, and Australia. For instance in England, the antecedent was the National Rural Studies Association which become the UK National Association for Environmental Education in 1971.[72] Within 20 years, there were innumerable environmental education organizations around the globe as well as international conferences hosted by the UN.

In 1972, it was the Stockholm UN Conference on the Human Environment which recommended that environmental education (EE) be recognized and promoted in all countries. In 1975, the UNESCO International Workshop on Environmental Education crafted the *Belgrade Charter for Environmental Education*, which provided the first international goal statement:

> [T]o develop a world population that is aware of, and concerned about, the environment and its associated problems and which has the knowledge, skills, attitudes, motivations, and commitment to work individually and collectively toward solutions of current problems and the prevention of new ones.[73]

Two years later, in 1977, a seminal document, the *Tbilisi Declaration*, was signed by over 300 UNESCO and United Nations Environment Programme (UNEP) representatives and delegates, as a unanimous accord. *Tbilisi* stipulated that it was imperative to defend and improve the environment for current and future generations. It expanded on the goals and objectives for EE and provided momentum for the uptake of environmental education globally.[74]

BOX 4.1 GOALS OF ENVIRONMENTAL EDUCATION, 1977

The goals of environmental education are:

1. to foster clear awareness of, and concern about, economic, social, political, and ecological interdependence in urban and rural areas;
2. to provide every person with opportunities to acquire the knowledge, values, attitudes, commitment, and skills needed to protect and improve the environment;
3. to create new patterns of behaviour of individuals, groups, and society as a whole towards the environment.[75]

The ultimate goal was behaviour change, shifting the human–environment relationship. *Tbilisi* approached EE as comprehensive lifelong education[76] for all ages and all levels, within both formal and nonformal education, the approach taken in this book. They listed five objectives for EE.

BOX 4.2 OBJECTIVES OF ENVIRONMENTAL EDUCATION

The categories of environmental education objectives are:

Awareness—to help social groups and individuals acquire an awareness and sensitivity to the total environment and its allied problems.

Knowledge—to help social groups and individuals gain a variety of experience in, and acquire a basic understanding of, the environment and its associated problems.

Attitudes—to help social groups and individuals acquire a set of values and feelings of concern for the environment and the motivation for actively participating in environmental improvement and protection.

Skills—to help social groups and individuals acquire the skills for identifying and solving environmental problems.

Participation—to provide social groups and individuals with an opportunity to be actively involved at all levels in working towards resolution of environmental problems.[77]

These five elements of awareness, knowledge, attitudes/values, skills, and participation/action have figured into most approaches to environmental education, yet their emphasis and relationship continue to be debated and researched. While UN documents are generally aspirational rather than foundational theoretical pieces, these documents pointed the way forward.[78]

In 1980, the International Union for Conservation of Nature (IUCN) published the *World Conservation Strategy* in partnership with the World Wildlife Fund (WWF) and the UN Environment Programme (UNEP). The IUCN, established in 1948 as the first global environmental union, is composed of both governmental bodies as well as civil society organizations who work together on global conservation matters. Their role has been developing key international conventions. The *World Conservation Strategy* was a groundbreaking document which identified three conservation priorities: that of maintenance of essential ecological processes and life support systems, preservation of genetic diversity, and sustainable utilization of species and ecosystems, as seen in their symbol in Figure 4.3.[79]

The IUCN ascertained the central and urgent issue to be the "almost limitless capacity of human beings for building and creation, matched by equally great powers of destruction and annihilation . . . underlin[ing] the need for conservation."[80] They also noted the global interrelatedness of environmental issues, requiring globally coordinated action. They identified action priorities to be undertaken as part

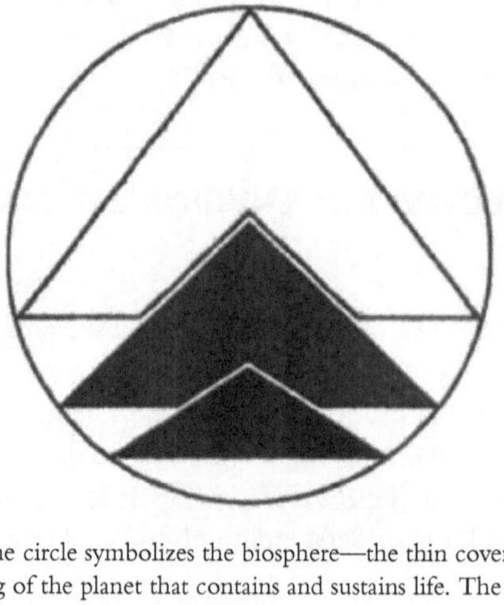

The circle symbolizes the biosphere—the thin covering of the planet that contains and sustains life. The three interlocking, overlapping arrows symbolize the three objectives of conservation:

—maintenance of essential ecological processes and life-support systems,
—preservation of genetic diversity,
—sustainable utilization of species and ecosystems.

FIGURE 4.3 IUCN—The Symbol of Conservation Priorities, 1980

Credit: Used with the permission of the IUCN.

of the third UN development decade, calling it sustainable development, discussed further in the next chapter. Their conclusion was that "[u]ltimately the behaviour of entire societies towards the biosphere must be transformed if the achievement of conservation objectives is to be assured."[81] As British environmental education professor Joy Palmer summarizes, all these various initiatives "emphasized the urgent need for world-wide environmental education."[82]

Early Definitional and Boundary Debates

In the United States, a variety of terms for this new entity were initially used—environmental management education, resource use education, and outdoor education. Yet, it was the term "environmental education" used by Clay Schoenfield in 1968 and later by Robert Roth which persisted in North America. In 1969, the *Journal of Environmental Education* was launched with a Fall issue. In it, William Stapp set forth the concept of environmental education and its purposes (see next page).[83] Also in 1969, James Swan challenged outdoor and conservation educators in an issue of *Phi Delta Kappan* asserting that "environmental education" was now the most appropriate concept for the historical moment. As a result, many outdoor and conservation education organizations changed their name and orientation or merged their orientations, such as "environmental *and* outdoor education."

Sheer numbers dictated that the NAAEE would become school-centric, shifting the focus from college educators to K/P-12 education. That said, the broad tent of environmental education includes not only primary and secondary schooling, but also:

- higher, vocational, and continuing education;
- workplace, union, government, and professional development education;
- civil society including environmental non-governmental organizations (ENGOs) and nongovernmental organizations (NGOs) who include EE programming from churches and museums to zoos and parks; and
- social movement actors (EMOs).

It has been the cross-fertilization among these sectors engaged in education which has created important momentum. For instance ENGOs were instrumental in assisting in curricular development in their areas of expertise as well as in offering environmental education programming. Professors in higher education also contributed significantly to curriculum and teacher professional development.

When environmental issues impact communities, citizens voluntarily undertake research to advocate for changes in policy, funding, and industry practices.[84] With the expansion of ENGOs and EMOs, environmental education became increasingly paid and professionalized in these community contexts. Thus, the NAAEE was quickly joined by community educators and citizen volunteers outside of schools—from naturalists, conservation educators, and outdoor educators to environmental managers and science educators working in government, public institutions, nonprofits, and business sectors. This range of members and hence a spectrum of views and vested interests have created ongoing tensions within the organization. Nevertheless, the NAAEE became "the world's flagship environmental education organization" that legitimized the field, won public support for EE, and is now linked to over 50 nations either through its individual members or affiliate organizations.[85]

As expected, some of the earliest scholarly debates in the nascent field were about defining and distinguishing EE from other learning endeavours. Debates included how much a universal, descriptive, or precise definition of EE was needed. Some advocated for a conceptual framework of thinking about environmental education based on philosophy and theory, while others simply wished to get the job done by developing curricular and pedagogical materials. The first definition of EE was provided by seminal theorist William Stapp in 1969:

> Environmental education is aimed at producing a citizenry that is *knowledgeable* concerning the biophysical environment and its associated problems, *aware* of how to help solve these problems, and *motivated to work toward their solution*.[86]

It is fair to say that the definition of EE has remained highly contested and rife with tensions across considerably different perspectives.[87]

In 1972, upon reviewing the emerging literature, Arthur Lucas classified EE into three approaches:

> Education *about* the environment . . . is concerned with cognitive understanding including the development of skills necessary to obtain this understanding, education *for* the environment, which is directed environmental preservation . . . [and] education *in* the environment is characterized by a technique of instruction . . . usually . . . the world outside the classroom.[88]

Education about includes nature studies, ecology, and conservation. *Education in* facilitates learning in the natural world as a site of learning about natural and built environments. *Education for* was understood as the application of knowledge for social or political action related to environmental issues.[89]

Some now add *education with* where humans consider themselves part of the natural world, working positively within natural processes. Later, *education as* would be added, discussed in the next chapter. Work was further done to distinguish environmental education from environmental communication and environmental interpretation.[90]

Environmental Communication

Environmental communication, closely linked to science communication, communication theory, and media studies, addresses any type of communication concerned with environmental issues. As Canadian philosopher Marshall McLuhan originally asserted in 1964, "the medium is the message."[91] In other words, how we communicate environmental issues either perpetuates existing understandings or shifts them.

Environmental communication is now a field of inquiry that is founded on two assumptions that "the ways we communicate powerfully shape our ecological perceptions; and these perceptions inform how we act with/in the more-than-human world."[92] Environmental communication is now used by journalists, activists, scientists, lobbyists, politicians, civil servants, popular culture influencers, marketers, and public relations folks. It uses traditional media as well as social media in communicating environmental research. It studies communication processes, elements, audiences, and the impact of messaging. Environmental communication is aimed either towards citizens, specific segments of the public, professional sectors, or national and international policy makers. Environmental communication *pedagogy,* then, assists in the analysis of contexts, assumptions, and principles of environmental communication.

Closely related is *climate change communication*, arising since the turn of the 21st century with increasing urgency. Related to risk communication, it also needed to contend with the rise and influence of social media and its role as an "echo chamber" for climate denialism, often ideologically motivated.[93] "Early climate communication was largely focussed on awareness raising and explaining the science of climate change (assuming that knowledge alone would suffice to move people to action)."[94] Like educators, climate communicators found that this movement to action could not be assumed and that multidisciplinary and multisectoral approaches were necessary to help shape messaging. Climate communicators examine climate news, specific audiences, and "the role of values, beliefs, worldviews, identity, and meaning-making" in crafting messaging.[95] Over time, storytelling or narrative formats have proven to be most effective in conveying value frames for specific audiences, in relation to the realities of climate change.[96] It has also been recognized that hopeful emotions, specific and viable actions, as well as practical support must be offered[97] to empower citizens. While environmental communication is vital for mass consumption, it is environmental education which has a longer-term role to play among specific groups of learners, for more sustained and deeper understanding and engagement.

Environmental Interpretation

Environmental interpretation, whether in parks, museums, zoos, nature centres, historic sites, or aquaria, has been defined as communication that is both an informational and inspirational process to create intellectual and emotional connections between natural sites or species and the learning audience.[98] For some, it is about interpreting the technicalities of the natural sciences in a way that can be understood by a lay audience to help them develop their own informed perspectives. For others, it is about bringing an audience into a powerful experience that enhances not only an understanding but also an appreciation for, and desire to protect, the natural and human cultural legacy bound up in any natural landscape.[99] Freeman Tilden[100] coined "interpretation" in the 1950s and derived the first six principles of interpretation.

BOX 4.3 SIX PRINCIPLES OF ENVIRONMENTAL INTERPRETATION

1. Any interpretation that does not somehow relate to what is being displayed or described to something within the personality or experience of the visitor will be sterile.
2. Information, as such, is not Interpretation. Interpretation is revelation based upon information. But they are entirely different things. However, all Interpretation includes information.
3. Interpretation is an art, which combines many arts, whether the materials presented are scientific, historical, or architectural. Any art is in some degree teachable.
4. The chief aim of Interpretation is not instruction, but provocation.
5. Interpretation should aim to present a whole rather than a part and must address itself to the whole [human] rather than any phase.
6. Interpretation addressed to children (say, up to the age of 12) should not be a dilution of the presentation to adults, but should follow a fundamentally different approach. To be at its best, it will require a separate programme.[101]

William Lewis wrote the seminal *Interpreting for Park Visitors* (1980), stressing the interactivity between the interpreter, the visitor, and unique landscapes.[102] In this sense, interpretation is about helping an audience to "see" a hidden story behind the landscape and visible cultural history.[103] Since then, Larry Beck and Ted Cable have identified 15 guiding principles for interpreting nature and culture in the 21st century.[104]

Ecology and EE

As discussed in the previous chapter, the science of ecology and the language of "eco" and "ecological" burst into popular consciousness in the 1980s.[105] Yet, the concepts of ecology, environmental science, and environmental studies are still confused. Further, the rhetoric of "ecology" is not necessarily rooted in the science of ecology, adding to misunderstanding.

For clarification, *environmental science* is the branch of science that studies the larger context of life and Earth ecosystems, originally focussing on the atmosphere, lithosphere, hydrosphere, and biosphere, but now investigating questions such as human population, resources, and damage.[106] It studies the *science* facet of environmental issues. However, given "mission creep," there is now much overlap with environmental studies (see Figure 4.4).[107]

Arguably, *environmental studies* is a larger multidisciplinary field concerned with the *environ*, a French word meaning to encircle or surround, first used in the early 20th century.[108] Therefore, environmental studies concerns itself with not only the natural world, but also the complexity of physical, geographical, and biological conditions alongside social, cultural, and political elements that surround individual organisms. It utilizes the *humanities and social sciences* facet in addition to the sciences to study the socio-cultural-technological origin of environmental issues and how the natural world and human world shape each other.

Ecology, originally defined by Ernst Haeckel in the mid-1800s, is "the body of knowledge concerning the economy of nature."[109] Now, ecology is generally considered "the study of the relationships between organisms and their environment."[110] Ecology studies relationships, particularly the vital connections between plants, animals, and the physical world around them. It also provides information

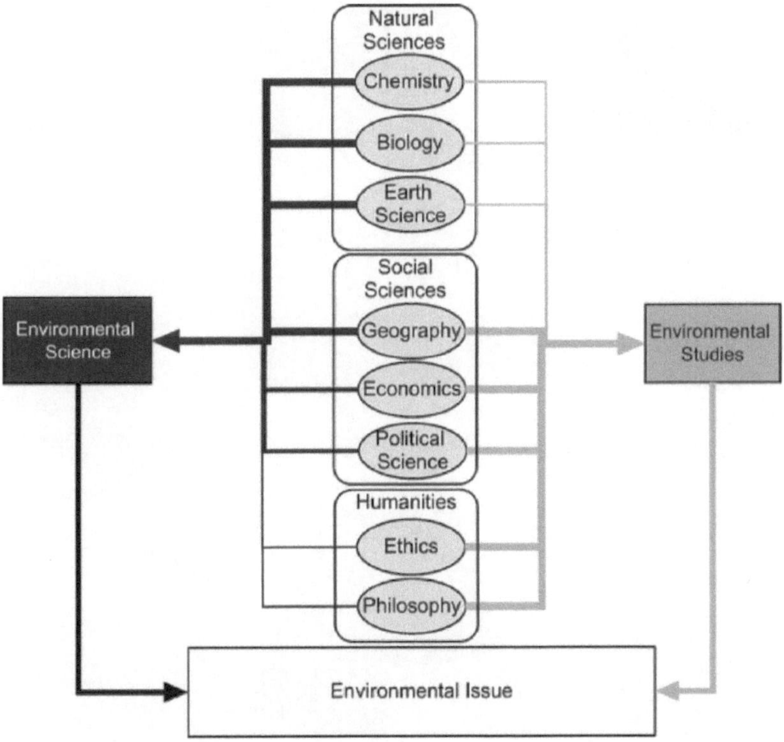

FIGURE 4.4 Differentiating Environmental Science and Environmental Studies

Credit: Differentiating Environmental Science and Environmental Studies by Steven Cooke. Reprinted by permission of Springer Nature.

about the functions and benefits of ecosystems and how humans can interact and utilize Earth's elements in ways that leave natural systems vibrant and healthy, rather than depleted, for future generations.

Ecology has brought about a way of seeing systems, rather than individual organisms in hierarchical relation. Systems in this usage are not referring to mechanistic systems predicated on clocks or computers. Systems thinking can still be based on Cartesian thinking and bureaucratic and industrial forms of precision and efficiency, as traditionally utilized in organizational and management theory.[111] Rather, ecology builds on the emergence of systems thinking from biology pioneers who first saw "living organisms as integrated wholes," soon to be called living systems theory.

Ecology studies "all the relationships that interlink all members of the Earth Household."[112] The three living systems include "organisms, parts of organisms, and communities of organisms."[113] Over time, it was understood that living systems

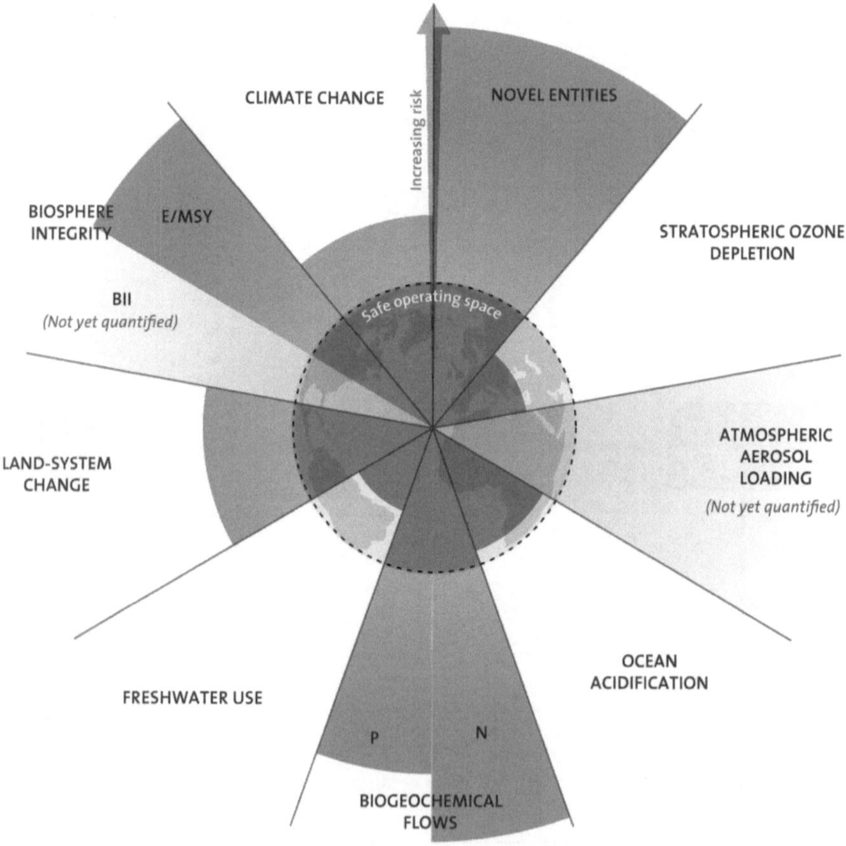

FIGURE 4.5 Planetary Boundaries, 2022

Credit: Designed by Azote for Stockholm Resilience Centre, based on analysis in Persson et al. 2022 and Steffen et al. 2015. CC by 4.0.

are the web of life, interacting as networks which are nested within other networks, nonhierarchically.[114] In this living systems perspective, there are properties that a system exhibits, which do not arise in smaller elements of the system, whether an organism or ecosystem.

Austrian American physicist and living systems educator Fritjof Capra advocates for a *holistic worldview*. The principles of living systems or "language of nature"[115] are key to a perceptual shift, explored in the next chapter. Together with Luigi Luisi, they propose systems thinking in education, healthcare, organizational life, food production, energy, and the economy through "designing for life . . . informed by basic ecological awareness . . . solutions [that] are ecologically literate . . . part of an ecodesign revolution."[116]

Another example is the ongoing research from the Stockholm Resilience Centre at Stockholm University on the ecological thresholds or tipping points of the nine planetary systems. In 2009, Johan Rockström and his colleagues identified the boundaries within these planetary systems, which maintain a safe operating space for humanity's activities.[117] They have been continually revising the impact of human activity in relation to these boundaries, first in 2015 and again in 2022, as seen later in the chapter. The nine planetary systems are climate change, stratospheric ozone depletion, atmospheric aerosol loading, ocean acidification, biogeochemical flows, freshwater use, land system change, biosphere integrity, and novel entities. In Figure 4.5, the inner circle illustrates the safe operating zone, and the dark grey indicates the four areas where humanity has already exceeded the boundaries, particularly: biodiversity loss as part of biosphere integrity, nitrogen cycles as part of biogeochemical flows, climate change, and plastics as part of novel entities.

An Evolving Ecology of EE Approaches

In 2005, Lucie Sauvé[118] offered a typology that maps 15 strands in three general categories, representing early and later developments within the EE field. Her analysis compared each strand in its conception of the environment, primary aim, and dominant approach, including pedagogical strategies.[119] One can see the field of environmental education as an evolving ecology, a web of interconnecting approaches, where each strand is playing a different but mutually enriching role in the environmental education ecosystem, thereby creating diversity, nourishing new ideas, offering multiple pathways to change, and fostering resilience.

With slight reorganizing and broader groupings than the Sauvé typology, the following EE typology with four primary strands emerged (see Figure 4.6). Often, educators draw from multiple strands, so these strands are not mutually exclusive, but show the general assumptions and tendencies within each strand. Such a typology is a heuristic only, to assist educators in more clearly understanding, comparing, and critiquing their educational practices. The descriptions are *highly* generalized, whereas, in practice, they are often nuanced, hybridized, and often a matter of emphasis. Each has a role to play, often related to context.

Strand of EE	*Strand One: Naturalist/ Conservationist/ Environmental Science*	*Strand Two: Environmental Literacy/ Environmental Citizenship*	*Strand Three: Deep Ecology/ Ecoliteracy/ Ecoconsciousness/ Ecojustice/ Place-Based Education*	*Strand Four: Critical EE/ Environmental Justice and Critical Multicultural EE/Ecopedagogy/Land-Based and Indigenous EE/ Commons-Based EE/ Ecofeminist Education*
Focus	Human relationship to "nature"	Managing human behaviour in relation to environment	Ecological science; deep ecology; cultural crisis; ecological consciousness that humans are part of natural world	Root causes and intersecting patterns of power and domination
Conception of Environment	Naturalist and environmental science where "nature" is object of study, positivist; Conservationist where "nature" is resource to conserve	Natural world as "problem;" "nature" as science lab; "nature" as living milieu for humans	Natural world as living systems; natural world as worthy of moral consideration; ecology as matrix of design	Natural world as dominated and assaulted by technocapitalism/ patriarchy/racism; natural world as alive
Primary Aim	Reconnecting to nature; change towards conservation behaviours; derive scientific knowledge	Environmental literacy and citizenship through responsible behaviour; build environmental knowledge, sensitivity, and action; outcomes based	Harmonize human and natural systems; duty to preserve and protect; ecoliteracy; cultural shift towards biophilia; systems thinking; ecological identity; reinhabitation of place	Structural critique; economic transformation; transform all structures of Western society, deracialization and decolonization; social and environmental justice; alternative global project; re-emergence of nonwestern knowledges, ways of knowing and being

Pedagogy	Cognitive, experiential, sensory, immersion; science methodology; empirical, transmissive learning and communication	Project based using issue analysis, values clarification, and social change towards environmental problem solving; discovery and inquiry learning; empirical scientific method rooted in ecology	Science of ecology; multiple intelligences; experiential, holistic, spiritual education; contemplative towards a planetary or ecological consciousness; inscape and landscape	Socio-politico-economic analysis as ecopedagogy; ethic of care; embodied and intuitive knowing; land-based pedagogy; regeneration of autonomous cultures rooted in place; community and commons based; Indigenous pedagogy;
Primary Critique and Ethics	Accommodationist; education "about" and "for;" first-order learning as knowledge acquisition and behaviour change; shallow, strong anthropocentrism as environment is "out there;" and humans have prime moral consideration	Reformist to change decisions; technocratic; education "in" and "for;" second-order learning; weak anthropocentric ethics with static view of "the environment;" individual-focussed	Psychology based; cultural transformation only; education "in, with and about;" second- and some third-order learning; ecocentric or biocentric ethics; some attention to power and social injustice	Socio-politico-economic transformation; education "in" and "for;" second- and some third-order learning; welfarist ethics

FIGURE 4.6 Evolving Ecology of EE—Broad Strokes

Strand One: Naturalist/Conservationist/Environmental Science

The earliest strands in the field were the naturalist, conservationist, and environmental science approaches to EE.

The *naturalist* focusses on nature as the object of study. The conception of the environment is "nature" or "the natural world." The primary aim has been reconnecting learners with the natural world through the communication of scientific knowledge, natural history, or sensory immersion through hikes, discovery learning, sensory exploration, and nature centres. It draws across the range of cognitive, sensory, affective, and experiential learning.

The *conservationist* considers environment as "resource" to be conserved. The goal is changing behaviours towards conservation. The approach is the transmission of empirical knowledge and/or skills-building for conservation with the aim of contributing to successful environmental management of natural spaces and species. It could involve a hands-on conservation project, environmental audit, or learning certain codes of behaviour.

The *environmental science* approach teaches science fundamentals along with the scientific method, including lab work and observation skills, for studying and acquiring knowledge about the natural world. In this case, the environment is conceived as an "object of study." It is a form of "education about" largely relying on cognitive and empirical understanding.

As Sterling[120] suggests, these areas in the field are accommodationist to the status quo in society and education. They tend to utilize first-order learning in Bateson's typology[121] where the learning is acquisitional and performative, as in changing the learner's response. Generally, in Lucas' typology, it is "education about" the natural world but includes "education for" the purpose of preservation and conservation.

One critique of this strand, says Palmer, is that "the scientific research paradigm has tended to dominate the relatively young and evolving body of environmental education research."[122] It builds upon the positivist tradition in science, accepting existing views of nature, "wilderness," and use of the scientific method. It is often strongly anthropocentric in considering "the environment" as being "out there." It gives humans prime moral consideration by enabling learners to develop reasoned choices and use their free will to choose behaviours that are "good for the environment."

This strand of environmental education tends to be shallower in responding to environmental issues. It does not necessarily challenge the human origins of the environmental crisis, mechanistic science, or siloed disciplines but works within existing parameters. Overall, this strand uses a more transmissive mode of teaching and learning, even though this may be participatory or involve a variety of modalities of learning.[123] Often, these methods of learning are most appropriate for one-time contact such as field trips or campground visits, for awareness building, and for highly structured classroom learning requiring knowledge acquisition.

Strand Two: Environmental Literacy/Environmental Citizenship

The second strand is primarily concerned with managing ourselves as humans in relation to the environment, focussing on behaviour change. In the early days, William Stapp with Dorothy Cox, Harold Hungerford with John Ramsay and Trudi Volk, John Disinger, Charles Roth, and Milt McClaren all spearheaded ideas around environmental citizenship and environmental literacy, the two frameworks which dominated North American theory and practice. The starting point for both is finding the connections with existing curriculum for infusing environmental education.

Environmental citizenship has been conceptualized as learning the capacity for citizenship behaviour related to the environment. In other words, it combines citizenship ideals with scientific practices and environmental knowledge.[124] Originally, Hungerford, Ramsay, and Volk defined environmental citizenship as "building environmental knowledge and sensitivity as well as responsive action"[125] to the impacts of industrial society. Positivist research on environmental citizenship has been used for determining the entry-level, ownership, and empowerment variables that play a synergistic role in developing pro-environmental behaviour change.[126]

Finn Jørgensen and Dolly Jørgensen now define environmental citizenship as the

> [R]esponsible pro-environmental behaviour of citizens who act and participate in society as agents of change in the private and public sphere, on a local, national, and global scale, through individual and collective actions, in the direction of solving contemporary environmental problems, preventing the creation of new environmental problems, achieving sustainability, as well as developing a healthy relationship with nature.[127]

As an example, citizens become involved in citizen science projects using collaborative and interdisciplinary scientific practices. The goal of such projects is not just creating scientific knowledge but, in the process, changing citizen attitudes to, and relationships with, the natural world.[128] As part of learning, learners may examine underlying causes of degradation as well as learn about avenues for creating change within a democratic system.[129] Projects can range from salmon and stream habitat reclamation, to bird counts and bird protection strategies, to monitoring and mitigating ocean shore erosion, to eradication of invasive species and reintroduction of native species. It can employ a range of pedagogy involving critical reflection, play, and socio-emotional learning as part of the process, dependent on age and context.

Environmental literacy, a term initially used by the American Nixon administration in the late 1960s, has been outcomes-based in terms of creating "an environmentally literate citizenry,"[130] evident through observable behaviours. Environmental literacy has referred to the "capacity to perceive and interpret the relative health of environmental systems and to take appropriate action to maintain, restore, or improve the health of those systems."[131] Roth's original 1968 model comprised four elements, namely knowledge, skills, affect (attitudes/values), and behaviour, which were all needed for understanding and assessing environmental systems. He later expanded the elements so that the affect category included environmental sensitivity, and the behaviour category included personal investment and responsibility.

Environmental literacy has tended to involve issues investigation by asking questions and defining issues, conducting investigations, interpreting data, and generating explanations, following the scientific method. From there, participants are encouraged to identify solutions and a plan of action that can address an issue. Environmental literacy is considered distinct from scientific literacy in that it is rooted in an ecological and not mechanistic science paradigm.[132]

Three levels of literacy are considered—*nominal* in recognizing basic concepts, *functional* with a broader knowledge base regarding interactions between natural and social systems, and *operational* in both breadth and depth of understanding and ability to act knowledgeably.

With regard to environmental literacy, debate has also addressed the concept of literacy. Considering literacy as limited to textuality limits conceptions of environmental literacy. Andrew Stables and Keith Bishop advocate a much broader view of literacy, from which strong environmental literacy can be derived.[133] For instance, learning to "read" the environment is considered a stronger conception of environmental literacy, perhaps even broader than environmental education.

Both environmental literacy and environmental citizenship are generally inquiry-based and use a learner-based discovery model. They variously include an issues analysis component examining the human-environment relation, a problem-solving component that draws from empirical science, and a values clarification component related to the building of environmental responsibility and citizenship skills. In terms of learner readiness, the first stage generally emphasizes "environmental awareness," then moves into behaviour change through the stages of "environmental literacy, responsibility, competence and citizenship."[134]

In both approaches, the concept of nature is either as "a problem or issue" or as a "living milieu" and "learning lab." The educational goal is for learners to be knowledgeable as well as feel a sense of belonging and responsibility for one's habitat. It focusses on acquiring skills for environmental problem-solving "in and for the environment." In terms of a replicable learning process, it is second-order learning as "learning to learn."

The pre-eminent example of environmental literacy in North America has been the Zoo School at the School of Environmental Studies, an optional 2-year high school on zoo property in Minnesota, started in 1995.[135] As one example, students research an issue to solve a real problem, and then, in a case of invasive species, construct a technical report for municipal officials and a guide of best practices for park gardeners.[136] The architecture of the school was also designed for best practices, with welcoming work environments and flexible workstations.

Sauvé and Sterling both suggest these approaches, at least in the early years of their development, have been reform-based in that they problematize human activity in relation to the environment. They aim to modify or reform existing decision-making through issue resolution processes and behaviour change. Research from within this strand tends to take an interpretivist or constructivist approach. Practitioners usually take education outside the four walls into new and interesting learning sites. However, problem solving and inquiry approaches, while useful, can still be emblematic of Western ways of knowing, particularly mechanist ways of knowing. As John Fien suggests, much of environmental education has "been based on a technocratic approach to environmentalism which favours initiating young people into the concepts and skills needed for finding scientific and technological solutions to environmental problems without addressing their root social, political, and economic causes."[137] Both approaches are considered weak anthropocentrism as they still foreground human activity and can hold a more static, objective view of "the environment." It concentrates on individual responses rather than examining societal systems and collective responses.[138] Newer practices are addressing these critiques in important ways.

Strand Three: Deep Ecology/Ecoliteracy/Ecoconsciousness/ Ecojustice/Place-Based Education

By the 1990s, as the understanding of environmental issues grew more complex, it was clear that a deepening cultural analysis should be added to the traditional science approach using a participatory pedagogy, largely from a deep ecology and the living systems view.[139] The primary inspiration for *deep ecology* was the natural philosophy or *ecosophy* of Arne Naess,[140] which challenged the dominant mechanistic paradigm of science and culture. As Naess explained, he found mainstream environmentalism piecemeal and reformist in approach, neither offering a deep enough critique of Western society nor a strong ethical basis for a needed culture transformation.[141]

In 1986, William Devall and George Sessions[142] published their book *Deep Ecology* which systematized a range of ecology writing and portrayed a way of living for bringing humans back into harmony with the natural world. They called for a new ontology or way of being, a new epistemology or way of knowing, a new psychology or idea of self, as well as a new metaphysics or nature of reality.[143] The focus was on transforming cultural worldviews towards seeing humans as belonging within the living world. Core principles of deep ecology include planetary consciousness, reinhabitation of place, respect for carrying capacity, and ecocentric ethics with a duty to preserve the integrity of the living world.[144]

For environmental education and ecodesign professor David Orr, the ecological crisis is "in large measure, a failure of education."[145] Both physicist Fritjof Capra and Orr envision a necessary but "long, difficult, and risky transition to a world we can barely see."[146] To address the failings of education, Orr differentiates between schooling and learning, suggesting that postmodern education must be the integration of schooling with active learning, currently too atypical.

Ecological literacy then, teaches how the living world is the "matrix for all design" and can help us derive the practices that can be alternatives to the "old project."[147] For this reason, Orr says "all education should be environmental education."[148]

He sketches out an ecological literacy that moves past the "old project of dominating nature" towards the ability of understanding the principles of natural systems which form the substrate for the web of life. He also calls for the "cultivation of political wisdom, moral virtue, and self-knowledge" as part of visionary leadership and active citizenry.[149] In this way, ecological literacy has both a biological and social imperative.

Both David Orr and Michael Stone, previously the Senior Editor for the *Center for Ecoliteracy*, suggest that ecological literacy includes a "basic comprehension of ecology, human ecology, and the concepts of sustainability, as well as the wherewithal to solve problems."[150] Ecological literacy is understanding how the physical world works, how humans have become destructive within it, and how humanity can harmonize their systems within natural systems. They assert that the drivers of environmental issues will not be addressed without political reform and collective social action to restore damaged ecosystems. Ecological literacy also encourages attention to monitoring the health of the natural world and what is needed to sustain life. It calls for "biophilia," a term from biologist E.O. Wilson, which describes that appropriate motivations spring from our affections and sense of kinship with all life.[151]

Psychologist Daniel Goleman, who has been writing on the multiple intelligences, defined ecoliteracy as the integration of emotional, social, and ecological intelligence.[152] He was building on Howard Gardner's 1983 work that broadened the narrow notion of IQ into multiple intelligences: bodily-kinesthetic, interpersonal, verbal-linguistic, logical-mathematical, intrapersonal, visual-spatial, musical, and, later, naturalistic. For Goleman, like E.O. Wilson, humans are wired to connect, so ecological literacy is our knowledge of, empathy for, and cooperation within all living systems. The critical task then is "genuinely prepar[ing] young people for the ecological challenges presented by this entirely unprecedented time in human history."[153]

By the 1990s, "ecological consciousness" was a concept reflecting an "emerging synthesis of the psychological . . . and the ecological," said history professor Theodore Roszak.[154] The goal of catalysing an *ecological consciousness* was to move beyond anthropocentrism, individual egocentrism, and human exceptionalism, towards eco or biocentrism. In an ecological consciousness, one sees oneself in an equal relationship with the "other than human" world, where "nature and natural entities (trees, species, ecosystems) are given moral consideration."[155] The environment is no longer an "out there" or landscape. It is also not separate from 'in here" or inscape.[156] Rather, "the environment" is a living system in which we are embedded, worthy of moral consideration. The goal is for learners to "rediscover their ties to the natural world in order to experience full mental health."[157] Thus, both the biophysical and the sociopsychological become integrated within the living world, aligning with ecological principles in multiple ways.

A further development was Mitchell Thomashow's concept of an "ecological identity"[158] meaning "the ways people construe themselves in relation to the earth."[159] Naess had also described becoming an "ecological self,"[160] seeing oneself in relation to the living world. Overall, this strand is considered second-order learning as in learning new ways of learning. In terms of developing an ecological consciousness and ecological identity, it can sometimes be third-order learning, in reflecting deeply on structures of learning and worldviews. Learners coming to see themselves as part of nature, not above or apart from it, entails a profound transformation in worldview and philosophy.

Ecojustice education includes the work of C.A. Bowers as well as Rebecca Martusewicz and colleagues. The central premise of ecojustice education is that the "ecological crisis is really a cultural crisis."[161] Through examining underlying belief structures, it attempts to understand assumptions about our human place in the world.[162] A deep cultural analysis can shift both the human regard for the natural world as well as change patterns of domination, thereby ameliorating negative impact.[163] Thus, ecojustice education builds on biocentric ethics where all living beings are given moral standing, as they possess a good of their own.[164]

Ecojustice education demands holistic education honouring a multiplicity of ways of knowing beyond the cognitive, empirical, and rational, including the metaphoric and performative. While drawing from the science of ecology, practices typically include interactive and sensory engagement with the Earth, such as in the work from Stephen Van Matre.[165] For Martusewicz and colleagues, it also examines hierarchized thinking, dualism, and the structures of knowledge and language. Further, it explores the cultural dynamics of androcentrism, racism, heteronormativity, and socioeconomic class. There is a willingness to explore Indigenous and nonwestern cultures whose ways of thinking and practices have long acknowledged an interdependent relationship with the Earth. As part of cultural transformation, learning may draw from cultural myths and stories, as in the work of Michael Caduto.[166]

These approaches concern education "in, with, and about." However, ecocentrism *sometimes* foregrounds the processes of ecological systems while backgrounding human injustice, rather than seeking a balance of moral interests. It also may not problematize the political economic system which in large part generates much of "culture," including the dominator relations of the Western world.[167] Some approaches do cross over into the critical approaches to environmental education. As critical theorists assert, we cannot think away structures of domination. Nevertheless, ecojustice learning is a vital piece of learning and reflection, deeply probing our way of living and relating.

Bioregionalism is related to place-based education as a facet of ecological education that rethinks political boundaries according to bioregions.[168] It advocates for economies of self-sufficiency within a natural region. As Kirkpatrick Sale[169] elaborates, rather than a global economy, decentralized economies are more diverse and therefore offer more stability, local cooperation, and self-sufficiency for communities. Bioregionalism also offers a critique of the mobility of industrial societies which alienate people from place.

Place-based education[170] is based on "deep local knowledge of place . . . [that is] the restoration of the essential links between a person and her place,"[171] says environmental educator David Sobel. By learning to dwell in place, humans can re-enact the ancient moral responsibility of maintaining ecosphere integrity. These approaches go beyond curricular reform to more deeply reinhabiting geographically unique places, restoring filial and responsible relationships with all living beings.[172] As Sobel asserts, "place-based education takes us back to basics, but in a broader and more inclusive fashion . . . teach[ing] about both the natural and built environments."[173]

David Orr emphasizes that place-based learning counteracts the rootless mobility fostered by modern society through relearning "the arts of inhabitation."[174] Such learning would "reorder" our educational priorities by teaching how to build prosperity locally without ruining our own place, crafting a "place-focussed politics," building an "ecological concept of citizenship" literally from the ground up, and teaching "public spiritedness and genuine service."[175]

There are a range of critiques to this deep ecology strand. The enthusiastic uptake of ecology has led to the charge that some ecological education does not necessarily stay true to the science of ecology. It may wander into "ecological thinking"[176] as understood within the conventional mechanistic paradigm, or it may extrapolate beyond what science supports.

Social ecologists Murray Bookchin and George Bradford[177] charge that some deep ecologists are "anti-human" in their biocentrism, in which all life forms are "morally interchangeable," or that the health of the Earth should come before human health. Instead, social ecology critiques capitalism's inherent rapaciousness, profit drive, and global wealth accumulation as well as the destructive colonization of so many diverse cultures for the purposes of money-making. It connects social and economic democracy with ecological principles, as a deeper way to redesign society and transform business. Bookchin is also critical of "mystical ecologists" who take up mythopoetic language, moving away from rationalism to spirituality.[178] He promotes an ecology that is not "antitechnological, antirational, or antiscientific."[179] Rather, he says, it is capitalism that is "systemically anti-ecological" and is the root cause of biosphere degradation,[180] which can no longer be ignored.

By way of contrast, *spiritual ecology,* according to Joanna Macy and others, defends the need to go beyond current ways of knowing and being towards rebuilding systems and worldviews modelled on ecological processes.[181] She uses living systems theory and deep ecology to design educational experiences that reconnect humanity to the natural world, as a pathway to revolutionizing human perceptions. She deliberately addresses the emotional, psychological, and spiritual by suggesting gratitude practices (affirmation work) as well as owning fear and despair (despair work), as despair and fear can mask a "deadened mind and heart."[182] She highlights that

> The greatest destruction in our world is not being inflicted by psychopathic tyrants or terrorists. It's being done by ordinary people—law-abiding, churchgoing, family-loving "moral" people—who are enjoying their sport utility vehicles, their vacation cruises, and their burgers, and are oblivious to where those pleasures come from and what they really cost. Oblivious not

to what those things cost at the store, but to what they cost when all the uncounted effects of their production and use are added up.[183]

Drawing from Buddhism in an overtly spiritual approach, her pedagogy fosters "compassionate action as an antidote to numbness and paralysis"[184] enabling the transition towards an ecologically sustainable world.

Only a few of these approaches also address the structures of the English language related to mechanism, as one part of the domination structures of Western society, although this is addressed in the ecojustice approach. This third strand, however, puts learners up against the constraints of a society operating on opposing, nonecological assumptions and principles. They may feel powerless to address the overall design of human habitats centred on cars, fossil fuels, chemicals, urban policies and bylaws, engineered landscapes, rational bureaucracies, and other attributes of industrial society, all fuelled by an unending race for profit and perennial growth.

Strand Four: Critical EE/Environmental Justice and Critical Multicultural EE/Ecopedagogy/Land-Based and Indigenous EE/Commons-Based EE/Ecofeminist Education

In a potent critique, Anna Gahl Cole suggested that environmental education needs cultural diversity to overcome its Western-centrism and the white, middle class, male-dominated structures that parallel the environmental movement. Further, she called for a rethinking of the ways in which power, race, class, gender, and politics shape human interactions with land.[185] The critical theory orientations in strand four assert that a shift in values and culture itself will not address the fundamental economic transformation that is needed, particularly given the urgency of climate change.

Critical environmental education, then, examines socio-political-economic realities and the intersecting power dynamics of domination by class, race, gender, sexuality, and ability. The *environmental justice* movement addressed how people of colour as well as low-income communities and nations disproportionality bear the brunt of environmental consequences from toxic manufacturing to landfills and dump sites to extractive industries. As Christina Marouli explains, "the worldviews of marginalized people have not been represented in environmental education as the environmental justice movement has shown."[186]

A critique of environmental education emerged out of the first National People of Color Environmental Leadership Summit in 1991, charging that the majority of practitioners were Euro-American and financially stable within the middle and upper socio-economic classes.[187] As Dorceta Taylor, professor of environmental sociology explained,

> The advent of the environmental justice movement marks a radical departure from the traditional, reformist ways of perceiving, defining, organizing

> around, fighting, and discussing environmental issues; it challenges some of the most fundamental tenets of environmentalism that have been around since the 1800s.

Similarly, in 1995, Running Grass says, "[a]s the Environmental Justice Movement challenges the definitions, concepts, and philosophies of mainstream environmentalism, it is also challenging—and changing—those of environmental education."[188] In response, the NAAEE adopted actions for diversifying the field, including culturally relevant education resources by 2002. Julian Agyeman asserts that much more "discussion of power, race, class, gender, sexuality, ability/disability" as social justice issues, needs to be taken up within "Western/conventional models of environmental education."[189] Without this, environmental education is "reproducing an exclusionary and exclusive environmental ideology."[190]

These perspectives have informed the development of *multicultural environmental education (MEE) or critical multicultural environmental education*, starting in the early 1990s. It blends EE with multicultural education to increase the access of culturally diverse groups to environmental education as well as nondominant worldviews.[191] As Blanchet-Cohen and Reilly found in their research, when working in contexts of significant linguistic, ethnic, and racial diversity, teachers need to develop their own awareness as well as be supported in fostering interactive dialogue across a plurality of values and beliefs, uncommon lived experiences, learning styles, and other paradoxes.[192]

As Marouli summarized, "multicultural environmental education envisions a multicultural society at peace with the natural world and itself; and, environmental education curricula and programs should be based on the direct and significant involvement of families and communities."[193] More than content diversity and cultural diversity, it addresses process, how teaching unfolds including processes of respect, learner empowerment, and valuing cultural lenses on the environment—in a "culturing of the environment,"[194] which represents a "paradigm shift."

Ecopedagogy[195] draws from the heritage of critical pedagogy including the work of Paulo Freire. It offers not only a cultural analysis but a structural and power analysis of modern society as well. Social change cannot occur without examining the capitalist system with its rapacious profit motive and extractive industries, vast polarization of wealth, and the patriarchal and racist structures that drive hierarchies dividing the human world and discounting elements and entities of the natural world. The dominant culture is suffused with the assumptions of industrial capitalism. Ecopedagogues assert culture cannot be understood or changed without addressing this dominant political economic form.

Rather than emerging from the global North, this approach emerged from the Global South, primarily Latin America. The goal of ecopedagogy, says Richard Kahn, is to critique this whole "matrix of domination" that is part of "global technocapitalist" structures, including the social constructs and reductionism of conventional science and the marginalizing of resistance movements often discounted in mainstream environmentalism as being "too radical."[196]

Ecopedagogues assert that change cannot evolve painlessly without power struggles, as evolutionary science supposes.[197] For instance the alternatives to the climate change crisis, which are offered by billionaires, simply leave the social and economic structures that enriched them fully intact. The power structures that enable this enrichment and leave impoverishment intact must be deconstructed. They advocate for the development of competencies for socio-politico-economic analysis and ecologically based community.

Brazilians Angela Antunes and Moacir Gadotti, both related to the Paulo Freire Institute, assert that ecopedagogy refers to an

> [a]lternative global project concerned with nature preservation" (Natural Ecology) and the impact made by human societies on the natural environmental (Social Ecology), but also as a new model for sustainable civilization from the ecological point of view (Integral Ecology), which implies making changes on economic, social, and cultural structures.[198]

The concept of ecopedagogy emerged during the 1992 Rio Summit and was a way to connect environmental education with grassroots political organizing and alternative social institutions that could catalyse social transformation. In 2004, an Ecopedagogy Charter was signed and has been promoted in multiple international fora. It aims at transformation on different fronts: cosmological, technological, scientific, a organizational. In sum, ecopedagogy works with learners to understand "how political economy and ideology produce the domination of nature."[199]

Related to ecopedagogy is a strand of ecological education built upon Indigenous knowledge or *land-based pedagogy*. For Indigenous peoples, place-based learning requires decolonization. Land pedagogy is an act of repatriation of land, a dismantling of colonialism, and a reclaiming of nonwestern ways of knowing and being, no longer reproducing the modern Western form of consciousness.[200] As *Michi Saagiig Nishnaabeg* scholar Leanne Simpson suggests, *Indigenous Environmental Education* addresses "the degradation of Indigenous ways of life, knowledge systems, traditional governance systems, foods and cultures, alongside the destruction of the places in which they lived."[201] She explains,

> [O]ur spiritualities, identities, languages, and systems of governance come from the land. The sustenance of our wisdom, worldviews, philosophies, and values come from the land. The source of our knowledge and our teachers themselves come from the land and the spirit-world it encompasses.[202]

She suggests that Indigenous EE requires a necessary critique of Western knowledge and science as well as a healing from colonizing processes. This needs to involve the incorporation of Elders as experts and keepers of knowledge, Elders as guardians of tradition and culture, pedagogies connecting to the Land, Indigenous instructors, the use of Indigenous language, and using Indigenous ways of teaching and learning.

In *commons-based environmental education*, living exemplars like the Zapatistas in Chiapas, Mexico, aim at regenerative education, which regenerates the ancient soil of cultures. Small local cultures like the *Mayan* Zapatista are concretely and consciously rerooting themselves in their particular soil, rebuilding the arts of living and dying within their ecosystem, which they consider animate as a force that speaks to and teaches them. The goal of the 1994 Zapatista Revolution was Indigenous autonomy and resistance against neoliberalism.

Gustavo Esteva and Madhu Prakash, building from their childhood experiences in Mexico and India respectively, describe how in these small places, they are reclaiming what it is to be real women and men with real culture, outside of economic assumptions and outside formal governance structures such as the nation-state. They illustrate how people can once again walk with dignity and wisdom, "not [in] the shadows of the upper classes"[203] or "like stones, like weeds growing by the side of the road . . . [with] lives [of] less value than their machines or animals."[204]

In South Africa, environmental educator Heila Lotz-Sisitka also embraces decolonization learning, not only for Indigenous peoples and those in the Global South, but also as a route beyond "existing forms of being in order to reimagine new becomings that are oriented to the common good (i.e. processes of emergence)."[205] She argues that decolonization not only needs to occur in the previous colonies but also in "modern minds shaped by market forces and global capital in all places of the world . . . taming minds and people."[206]

Some of the approaches cross over into a newly emerging strand based on Relationality, further discussed in Chapter 6. Overall, in Strand Four, educators and learners tend to come up against disciplinary boundaries and institutional constraints. Architectural spaces "box" in thinking as as well as the bodies living in them, predicated upon industrialism. Change-oriented education can be viewed as ideologically driven and therefore explicitly political in nature, which continues to trigger major backlash from the New Right and educators who question advocacy-oriented education.

Persistent Conundrums and Growing Critique

Conceptual struggles have characterized the history of environmental education. There have been multiple debates over theoretical foundations, theoretical frameworks, pedagogical interpretation, curricular/programme context and location, and a growing critique on the effectiveness of EE.

Science and Social Science

Early debates focussed on how to institutionalize EE, including the most effective ways it could become part of schooling curricula and higher education content.[207] Debates centred on whether a new school subject was required or if EE should be rooted in an existing discipline, such as a specific science. Until the 1990s, EE has remained rooted in the science disciplines, relying on the scientific method for teaching about environmental issues. Yet, the science base was often considered a deterrent for teachers in other disciplines in offering too narrow of a scope. As early as 1980, Arthur Lucas questioned the 'disciplinary chauvinism' in an EE rooted in science education, considered a lack of regard for other disciplinary engagements and approaches.

As other disciplines engaged environmental education, a new charge emerged that EE had wandered away from the natural and social sciences. It was suggested that some non-science practitioners were using outdated or misrepresented scientific frameworks.[208] A deeper concern was that mechanistic science itself was part of the root problem, constraining the scope of, and societal critique offered by, environmental education.[209]

Mainstreaming and Margins

A related discussion has been about "mainstreaming" EE in existing curricula or if EE would be better positioned in "alternative" curriculum orientations. If mainstreamed, some have asked if environmental education imperils itself with its own extinction by becoming tokenistic or unrecognisable, structured within mainstream education.[210] If it becomes an alternative curriculum, then it risks marginalization.

By 1990, after two decades, Hungerford and Volk concluded that EE was not yet mainstream, and therefore "only a fraction of our young learners are being exposed to logically developed, well-articulated EE programmes."[211] Further, it was charged that too many programmes stayed at the awareness level. While some developed higher conceptual awareness of environmental issues, many did not go far enough to involve investigation, evaluation, and action skills.[212]

Another debate has been if EE should be an "add-on" topic in a subject or if EE concepts and processes should be infused across subjects. In North America, with a well-developed school system, environment education has generally been infused into existing subjects rather than one interdisciplinary subject.[213] Yet, there have been worries that this approach risks identity dispersal and that it leads to sporadic rather than sequential knowledge building. By the 1990s, Deborah Simmons found that this full integration of EE was still not occurring.[214]

EE is now understood as an interdisciplinary even transdisciplinary field that demands integration of knowledge across many fields. Yet, this continues to challenge disciplinary divisions in both schooling and higher education. Minnesota was one of the first states to promote EE as a "non-disciplinary curriculum," given that environmental education "encompasses all subjects usually found in the school curriculum." However, this approach has not been common.[215]

Depth and incremental knowledge building has occurred more naturally in higher education. Higher education, however, is still organized by disciplines making interdisciplinary study difficult unless positioned in an interdisciplinary programme or an arms-length institute. Nevertheless, since the 1970s, there has been rapid expansion of higher education programmes, including environmental science, environmental studies, environmental sociology, and environmental psychology, from which educators draw. Environmental education has been present both in teacher preparation and teacher professional development. However, 30 years later, the number of teacher preparation institutions offering environmental education courses in Canada has remained low.[216]

Knowledge-to-Action Assumption

An underlying assumption in early EE was the linear model of teaching for behaviour change. Scholars began to realize "a wide gap between a persuasive vision and current practice."[217] The model of knowledge→attitudes→behaviour change seemed to be having little impact on learners. Research consistently demonstrated that the link between environmental knowledge and behaviour change has *not* been a close one, just as public environmental concern does not necessarily translate into pro-environmental action, a central conundrum taken up in later chapters.[218]

Much of the early EE research focussed on outcomes.[219] Generally, it used experimental (with control groups) and semi-experiential methods to assess the development of pro-environmental knowledge, skills, and attitudes (KSAs). Further, using social psychology, researchers attempted to establish the link between educational practice and knowledge, attitude, and behaviour changes.[220]

In general, these studies found that the longer the exposure and the more systematic the instruction, the more the knowledge, attitudes, and behaviour shifted and persisted over time. Thus, residential experiences such as camps or multi-day workshops were shown to be more effective than brief field trips. Knowledge levels were found to be easier to change than attitudes and behaviours. Hands-on involvement proved to be more effective than passive viewing of TV or films and using worksheets. In-class instruction in support of outdoor experiences was also very effective. It was also found that children as young as kindergarten age are able to form environmental responsibility values and act as effective agents of change. For adults, involvement in a change behaviour within a learning situation increases the likelihood of persistence outside the learning situation, even with a short intervention. However, attitude shift becomes increasingly difficult the older the participant gets.[221]

Environmental Advocacy and Education

What education and thus environmental "education is, is not, should be, and should not be"[222] has been vigorously discussed. One persistent question has been about the relation between education and advocacy: whether it should be either/or or both/and? Theorists such as Bob Jickling have consistently suggested that EE should be about "good" education first. It should teach for solid inquiry and critical thought, not skills training or advocacy. In a variation, Ian Robottom and Paul Hart advocated going beyond science to explore the social and political implications of environmental decisions as part of social and civic responsibility. For them, educators rely too heavily on behaviour modification as a primary aim, rather than fostering independent critical thinking.[223]

Three early approaches to advocacy were identified:

- including "hands-off" which is value-free or value-neutral and does not promote any specific outcome from EE but the acquisition of natural science knowledge;
- the "soapbox" (perjorative term) or "instrumental" approach (partially pejorative term) describing EE aimed towards pro-environmental worldviews and responsible environmental behaviours; and
- "balanced exposure or bias-balance" which presents a range of viewpoints to learners.[224]

In 1996, as a response to such critiques and debates, the NAAEE developed six criteria for EE: fairness and accuracy, depth, emphasis on skills building, action orientation, instructional soundness, and usability.[225]

An underlying concern has been that advocacy can too easily migrate towards indoctrination, especially with children.[226] Yet, David Orr and David Greenwood suggest that environmental education will be ineffective in advancing its own goal of creating an environmentally or ecologically literate citizenry as long as it continues to discipline itself within the norms of general education.[227] Peter Martin agrees that this mutes the potential of EE as a transformative educational practice.[228]

Another position is that education is always ideologically invested, whether explicit or not, with its dual role of transmitting the existing knowledge and values of society as well as teaching for social transformation to meet future societal challenges.[229] Some believe that not acknowledging the ideologies informing directions in education is disingenuous. Following Paulo Freire,[230] critical pedagogues assert that education can never be neutral but always has an ideological root. As corporate and other actors increasingly seek to influence or provide environmental education materials, teachers and students must have a well-developed critical capacity not only regarding environmental issues but also regarding the motivations and values of environmental actors as well.

Instrumental Approach or Not?

As Alan Reid identifies, another debate has been the place and critique of "instrumentalism" in curriculum policy and practice. Jickling critiques teaching for skills and advocacy as utilitarian and instrumental thinking, still part of the dominant rationality. For decades, others have advocated for a more holistic way of knowing beyond analytic and rational approaches through sensory experience, emotions, intuition, and creativity.[231] Others, however, argue *for* an instrumental education as necessary to promote and help shape responsible environmental behaviours. It has been widely accepted that behaviour-related instruction is necessary, although the links between knowledge, attitudes, and ethical behaviours are not well known and highly contextual.[232]

Catastrophe Education

In 2005, David Sobel charged that environmental education had become too narrow and carried too much baggage.[233] He suggested that while EE has moved beyond nature study and natural sciences to broader issues of education, now it seems to be "catastrophe education—learning about rainforest destruction, ozone depletion, toxic waste, and endangered species."[234] Hart called it a "pedagogy of despair."[235]

The Failure of EE?

After decades, the severity of environmental degradation has only been deepening. Kahn explains that a "crisis of environmental education" had been reached.[236] Studies continued to illustrate a persistent lack of public understanding of basic environmental ideas.[237] Environmental education did not appear to be making an appreciable difference.[238] Some believed that EE lost its way, or perhaps that it never started from the right place.[239] Others were concerned about its deeply anthropocentric approach and isolation from the community.[240] As Roger Hart concluded,

> Environmental education must be radically reconceived in order to be seen as fundamental to the residents of communities from all social classes in all countries. We need programmes based on the identification and investigation of problems by residents themselves, with "action research" as the dominant methodology. There is of course a central place for the teaching of environmental science—"ecology" as it is often called—but this should at first be directly related to the local environment. Environmental education from this perspective is intrinsically tied to community development in general.[241]

Environmental nonprofit director Charles Saylan and scientist/educator Daniel Blumstein charged that

> [E]nvironmental education has failed to bring about the changes in attitude and behaviour necessary to stave off the detrimental effects of climate change, biodiversity loss, and environmental degradation that our planet is experiencing at an alarmingly accelerating rate. . . . This collective inability to act is brought about in part by educational institutions that generally do not provide tools necessary for critical thinking and for understanding the modern world. Nor do they teach individual responsibility and social engagement, two fundamental tenets of free and democratic societies.[242]

The research agenda has broadened as it became clearer that "environmental education was a movement for fundamental education reform,"[243] taken up in Chapter 7. Education professor Stephen Gough argued, however, that "education cannot be expected to 'save the planet' through, or mostly through, its own direct or indirect effects."[244] It is only one factor in the constellation of factors impacting planetary health. Saylan and Blumstein assert that environmental education is an essential civic responsibility which requires environmental education to permeate the entire education system in flexible and adaptive ways. They also charge that politicians have prevented this curricular integration, keeping it marginalized due to the impact of highly paid and influential lobbyists. This charge and more recent developments in EE will be discussed over the next few chapters.

Adult Environmental Education

Adult environmental education has been variously taken up in environmental movement organizations (EMOs) as social movement learning,[245] in environmental nongovernment organizations (ENGOs) as community-based learning,[246] and in nonformal adult education as environmental adult education (EAE)[247] also known as adult environmental education (AEE),[248] including in government, churches, unions, and business and corporate sectors.[249] Each of these settings are outside formal educational organizations and are largely for adults but can include young adults and children, dependent on context. Overall, there has been more pedagogical freedom and fewer institutional constraints on adult environmental education activities. Australian environmental sustainability educator Karen Malone explains that as awareness of the environmental crisis has grown, environmental issues were no longer "seen as 'scientific' problems (to which science could invent scientific solutions), but that all sections of the community had to begin to play a role."[250]

There had been lone calls in the 1970s[251] "to rally the field of adult education to take up an environmental discourse."[252] For instance, out of the teach-ins that were part of the first Earth Day, research illustrated that people attended because of the speaker profiles, the speakers' role as opinion leaders offering a counter analysis, and for suggestions on specific actions they could take.[253] It was found that most did take some consumer-based actions, but few engaged in lifestyle change.[254] Other researchers found that environmental actions like sit-ins have limited educational value, given the lack of participatory issue analysis,[255] prompting a call for adult environmental education that might ensure "continued citizen involvement" over time.[256]

That said, environmental education for adults remained underdeveloped until the 1990s, thereafter becoming a unique body of knowledge.[257] During most of the 1980s, adult environmental education was largely occurring in individual programming with little attention to theory development, collective attention, or international focus.[258] The initial lack of uptake in the adult education field has been ascribed to several reasons: the well-developed affiliation of EE with schooling, an individualist focus which runs against the social purposes on which the field was founded, the locatedness within science education and perceived need for specialist knowledge, the focus on behaviour change as a scientific endeavour, the lack of understanding of the philosophical and scientific controversies related to environment and ecology, environmentalist polemics and polarization, as well as lack of a theoretical critique on industrial capitalism, or, in later years, an intersectional analysis examining multiple forms of domination.[259]

In 1972, the International Council of Adult Education (ICAE) was established as an international non-governmental body for the adult education movement. Adult education associations and university-based adult educators from Canada, the United States, Britain, India, Tanzania, Ghana, Nigeria, Jamaica, and Venezuela were some of the early members. Now ICAE serves 800 NGOs and 75 nations, promoting the right to learn, literacy, youth and adult education, lifelong learning, as well as addressing critical global issues through education.[260] It is considered the "most influential and comprehensive international organization in the field of youth and adult education."[261] ICAE is associated with UNESCO and has observer status with several UN bodies such as UNEP. ICAE defines itself now as a "strategic network that promotes adult learning and education as a tool for active citizenship and informed participation of people."[262]

Nonformal adult education has its roots in social movements, considered a social movement itself. It emerged as part of the social reform movement, a collective and organized movement seeking social change in terms of giving adults access to learning if they could not access schooling or to continuing learning, especially for those in groups experiencing isolation, poverty, or exploitation. Nonformal adult education has been considered liberatory in that it often works explicitly with/for marginalized populations, seeking not only "personal enlightenment" but also social justice and active citizenship.[263] Thus, ICAE has worked to maintain linkages with other social movements.

> By the early 1980s the women's movement, the peace movement, trade union movements, movement of indigenous people, and other movements were gaining momentum, and the Council, among many other activities, worked to strengthen the links between the adult education movement and other movements. Social movements were identified as primary means of transformation, where adult educators, could demonstrate the potential of their field in the service of the movements.[264]

In 1989, ICAE established the *Learning for the Environmental Programme* or LEAP to "articulate the theory and practice of environmental adult education,"[265] with the task of distinguishing a different role for nonformal environmental education.[266] Given its social movement roots, adult environmental education was distinct from school-based EE in explicitly promoting environmental advocacy and supporting social action in communities. As UNESCO stated in 1986,

> The strength of nonformal environmental education lies in the fact it does not operate within a given set of rules with strict structure, curriculum, and examination procedures. Nonformal environmental education is more capable of responding to local environmental issues, which have more social meaning and usefulness to the community and less dominated by academic requirements.[267]

At Rio in 1992, ICAE designed the participatory process that led to the NGO treaty on *Environmental Education for Sustainable Societies and Global Responsibility*.[268] After this, environmental adult education was explicitly put on the 1997 agenda for ICAE. It was here that the "environment" was conceptualized as "an integrated human and nonhuman context" with discussions about the opportunity to "develop new ecological frameworks for social and cultural activity" and learning "new knowledge, new attitudes, and new forms of local and planetary citizenship participation."[269]

At Rio, environmental adult educator Darlene Clover explained that it was further recognized that environmental adult education has already been practiced across generations and centuries by Indigenous people and people of the South. But "there has been little documentation on their practices or attempts to theorize what this might mean in terms of environmental change."[270] Summarizing the spirit of Rio, Marta Benavides asserted that adult education overall needs to become "environmental and just in its essence, spirit, and practice."[271]

One of the key tenets of adult education is that a democracy requires informed citizens, and another is that life itself and all its facets should be the fundamental subject matter. Thus, EAE/AEE generally takes a participatory and transformative approach that is learner-driven. Nonformal adult education also assumes that learners desire self-directed education or to participate in their own educational decision-making. It further considers that adult learners have rich experiential resources that they bring to learning.

Building on the original work of Malcolm Knowles,[272] nonformal adult education is predicated on: relevance to learner needs, attention to adult readiness and internal motivation to learn, learning aimed at a specific audience to help resolve a need, learning with clear purposes and goals, and flexible curriculum and structure.[273] These general principles for nonformal adult education also inform the theory and practice of adult environmental education. Most notably, EAE/AEE starts where adults are at and with what motivates them. From here, educational content and process is formulated.[274]

Another distinctive feature is that EAE/AEE strives to "move beyond the 'behaviour change' agenda"[275] and information transmission approaches, viewed as rooted in positivist science. Adult environmental educators generally take a transformative approach to "develop a critical consciousness that challenges systemic and structure problems [. . . that are] embedded in destructive ideological frameworks of capitalism and globalization."[276] In these lifelong and lifewide settings, Clover explains, environmental adult education globally is seen to address:

1. oppression based on race, gender, sexual orientation, socioeconomic class, ethnicity, religion, and national origin;
2. citizen participation, democracy, and civil society. and
3. social action.[277]

As she asserts, "environmental and social injustices are intertwined within a complex and intimate, interdependent relationship, and neither form of injustice can be adequately addressed or acted upon in the absence of the other."[278] "To be most effective, environmental adult education must be linked to local, national, or global activism and be based on a discourse of democracy, accountability, equity, and hope."[279] Coming from a critical theory perspective, Clover asserts that "environmental adult education is an activist-based political pedagogy"[280] as part of a "strengthened participatory democracy."[281]

At the 1997 UNESCO conference on adult education, the resulting *Hamburg Declaration and Agenda for the Future* and accompanying handbook, *Environmental Adult Education: Awareness to Action*, represented an international watershed for EAE. A working definition of environmental adult education resulted:

> A permanent process in which individuals gain awareness of their environment and acquire the knowledge, values, skills, experiences, and also the determination which will enable them to act individually and collectively to solve present and future environmental problems . . . as well as to meet their needs without compromising those of future generations.[282]

The purpose for adult environmental education was to "play an important role in sensitizing and mobilizing communities and decision-makers towards sustained environmental action."[283] Together, they declared:

BOX 4.4 HAMBURG DECLARATION AND AGENDA FOR THE FUTURE, 1997

We commit ourselves to:

35. Promoting the competence and involvement of civil society in dealing with environmental and development problems:

a. by making use of adult education activities in order to increase the capacity of citizens from different sectors of society to take innovative initiatives and to develop programmes based on ecologically and socially sustainable development;
b. by supporting and implementing adult education programmes designed to give people the chance to learn and interact with decision-makers on environmental and development issues, in particular on the need for changes in production and consumption patterns;
c. by integrating Indigenous and traditional knowledge of the interaction between human beings and nature into adult learning programmes, and by recognizing that minority and Indigenous communities have special authority and competence in protecting their own environment;

d. by ensuring the accountability of decision-makers in the context of policies relating to the environment, population, and development;
e. by integrating environmental and development issues into all sectors of adult learning and developing an ecological approach to lifelong learning.[284]

To further establish a theoretical foundation, adult educator Pierre Walter extrapolated the philosophical approaches of adult education to EAE/AEE.[285] For instance he explains that the *liberal arts* approach is didactic but often includes "critical analysis of underlying structural roots of environmental problems" using science, logic, and rational debate, although it "does not advocate for change."[286] Such approaches are often used in libraries, zoos, parks, and museums.

The *progressive* approach is more experience- and learner-based with embodied and problem-solving learning, such as wilderness leadership and master gardening programmes. The *behaviourist* approach uses positive and negative reinforcement (incentives and disincentives) to effect behaviour change, such as programmes aimed at promoting recycling and active (human-powered) transportation, eliminating plastic waste, or reducing carbon footprints. The *humanist* approach promotes personal and spiritual development through connectedness with the natural world, creative and arts-based approaches, and catalysing social involvement.

The *critical* approach involves consciousness-raising and a critical, intersectional analysis of the links between social oppression, racial injustice, and environmental degradation. It aims at social transformation by assisting adults in thinking critically about positions and mechanisms of power, exposing implicit as well as explicit values, and identifying who does and who does not benefit from specific beliefs and practices, including the natural world.[287]

As Walter describes, individuals can clearly undergo a profound transformative learning process. Whether it is social movement learning in the context of an environmental protest defending the natural world, perhaps being jailed,[288] or scientists such as Rachel Carson and Aldo Leopold who experienced dramatic, life-changing events from witnessing the death of another species or facing racist and sexist attacks for their research work, they begin to question their assumptions. They begin asking different questions leading to an "inner reworking of individual identity and consciousness"[289] with emotional, spiritual, embodied, cosmological as well as cognitive impacts. In such cases, they go on to "create counternarratives and counterknowledges" that have public impact as well as an impact on corporate and state responses.[290] Yet, as Stephen Gough *et al.* caution, assuming or insisting on a "specific destination for lifelong learning in environmental education is self-defeating precisely because it closes a door on . . . "imaginative engagement."[291] An educator can only establish learning conditions that create an opportunity for transformation.[292]

After Rio and the turn towards ecology, Paul Belanger proposed another moniker, Environmental Lifelong Learning (ELL),[293] picking up the turn to lifelong learning and education by UNESCO in the 1970s and 1980s. UNESCO conceptualizes adult education as a part of lifelong learning, which

> [G]oes beyond the traditional distinctions between initial and continuing education . . . [toward a] learning society, in which everything affords an opportunity for learning and fulfilling one's potential, [through] learning to know, learning to do, learning to live together, and learning to be.[294]

Belanger recognized the role that adult education could play in the ecological transformation of *all* educational practices, formal and nonformal. Belanger calls this the possibility of creating an "ecology of learning" with

> synergy between formal, nonformal, and informal learning . . . catalyzing the learning potentials of each environment so as to create a cultural dynamic in which the learner can live in creative tension and reflexive relationship in his or her environment . . . as a pedagogy of place.[295]

Belanger identified the four most effective elements for environmental lifelong learning as: educating in the natural environment, pursuing content rooted in daily life contexts, synergy between lifelong and lifewide education across sectors, and interactive communities who have feedback cycles associated with their sustainability initiatives.[296] Picking up the concept of environmental literacy, Ralf St. Clair integrates deep ecology with critical literacy, considering adult environmental literacy as being necessary to functional literacy.[297]

Then in the 2000s, a definitive shift towards transformative learning as the foundation for EAE/AEE occurred. As Edmund O'Sullivan and Marilyn Taylor define, it is "the transformation of our fundamental assumptions and beliefs about ourselves and our relationship to our environment" that is necessary, shifting away from modern ways of being as "achievers and users."[298]

Together with Shirley Follen and Budd Hall, Darlene Clover identified the key principles for transformative environmental adult and popular education that builds a basis for citizen action.

BOX 4.5 KEY PRINCIPLES OF ENVIRONMENTAL ADULT AND POPULAR EDUCATION

- We need to passionately reconnect with the rest of nature through all our senses and emotions;
- we need to critically examine the unjust power relations behind contemporary social and environmental trends;

- we learn best by beginning with our daily lived experiences and by understanding our own locations, contexts, and histories;
- we are responsible for taking personal and collection action to recreate a healthy planet for all species both now and in the future;
- we can all transcend the bonds of traditional education; and
- we are all artists, poets, storytellers, songwriters, dreamers, and more.[299]

Robert Hill,[300] Deborah Barndt,[301] Tuere Bowles,[302] Sandra Tan,[303] Dip Kapoor,[304] and Benjamin Feinstein[305] are just a few who have been engaging EAE/AEE from an environmental justice lens that examines the intersections of white privilege, anti-racism, social activism, Traditional Ecological Knowledge (TEK), immigrant/newcomer learning, and popular education. Since the late 1990s and early 2000s, this has been continued in newer currents within EAE/AEE. For instance drawing from the work of Chet Bower and Rebecca Martusewicz *et al.* on schooling, Audrey Dentith, Onah Thompson, and Wendy Griswold elaborate ecojustice adult education as "opposing the dominance of one group over another, of humans over nonhumans, or humans over nature."[306]

While EE is not normally distinguished within the integrated approach of Indigenous teaching, *Pueblo* Gregory Cajete and *Metís* Gregory Lowan-Trudeau describe Indigenous environmental education as land-based. In other words, the natural world is neither a "wilderness" or "isolated refuge" that "wanderers, idealists" or urbanites flee to for nourishment nor is it a recreational or commodity resource to be conserved for long-term enjoyment or extraction.[307] For Indigenous people, the land is home,[308] needing protection only from the Western political economy. For them, the natural world is not wild or forbidding, something to *museumize* by leaving undisturbed. They have lived in harmony through an intimate relationship with the land in some places "for 30,000 years or more."[309] As a Pueblo Elder says to Cajete, "It is this place that holds our memories and the bones of our people . . . This is the place that made us."[310] Land is a place of consciousness, identity, and sacred orientation in space.[311] Thus, decolonial approaches to AEE and sustainability involve disrupting normalized settler and colonizer stories as well as normalizing Indigenous epistemologies and ontologies.[312]

Knowledge creation has previously been the province of scholars, but Participatory Action Research (PAR) has sought to democratize research. PAR has been a community-based approach to knowledge creation, which is then utilized for the political, economic, and social interests of the community, generally marginalized communities.[313] From the experiences at the Highlander Center in Appalachia, it was found that,

> People may discover for themselves dominant knowledge or interpretations of reality which do not conform to their own experience . . . Or the process

> of popular investigation may reveal previously hidden information that does confirm through "official" knowledge what the people have suspected from their own experience.[314]

Using environmental popular education, Joaquin Esteva and Javier Reyes from Mexico have worked with remote Indigenous people addressing the soil erosion, sewage, and lake death that accompanied modernization. They affirmed that there is "an indissoluble relationship between development and environment that encompasses, among other things, the possibility that every individual and community can live satisfactorily when ecological balance is understood and respected."[315] For them, then, "the task for environmental popular education should be *conscientization*, that is, drawing out an embedded environmental consciousness through the collective development and acquisition of new knowledge for the development of a more sustainable society."[316]

Land-based AEE has also been taken up in Africa and in Mongolia for the reclamation of pastoralist practices of previously nomadic groups, who had "balance[d] the co-existence of humans and animals."[317] Rosa Muraguri-Mwololo describes how socio-environmental degradation has occurred in Kenya first through colonialism then modernism. Typically, ecological wisdom was handed down orally from generation to generation, but this knowledge has been "abused" with schooling nearly extinguishing it. By using environmental adult education principles, they were able to facilitate community action learning to mitigate these impacts. Salvatore Engel-Di Mauro and Karanga Keita Carroll also describe a land-based environmental education blended with critical pedagogy and African-centred perspectives.[318] The goal is "facilitating people's reconnection to places and the ecosystems they inhabit in ways that do not reduce them, among other things, to objects of conquest, or things to be exploited for profitability and individual gain."[319] Similarly in Mongolia, Batchuluun Yembuu describes how modernization has encouraged young adults to move away from Mongolia and think negatively about Traditional Knowledge. Through nonformal community learning centres that bring together knowledge holders with young adults, the knowledge will persist and promote "environmental protection and preservation of sustainable and minimalist lifestyles, which are increasingly important in a climate challenged world."[320]

Moacir Gadotti explains that PAR has been extrapolated into the ecopedagogy movement as a way to empower movements for environmental justice.[321] Ecopedagogues link environmental, social, and cultural literacy and more robustly address the root drivers of the environmental crisis.[322] In the South, ecopedagogy assists in the development of alternative social institutions which can form the foundation of a post-anthropocentric society that no longer "negates life,"[323] while addressing social injustices that drive environmental degradation. It builds on the aesthetic and spiritual understandings of the natural world of those who see themselves as forgotten people.

By 2010, this author had identified that many AEE practitioners were drawing from multiple theoretical currents, integrating them in unique ways. For instance Edmund O'Sullivan combined deep ecology, ecopsychology, ecospirituality, and ecofeminism for transformative learning towards ecological consciousness. Merging deep ecology, ecofeminism, and critical Freirean popular education, Darlene Clover has developed a transformative pedagogical approach that embraced learning "with, for, through, and about nature."[324] In ecopedagogy, Richard Kahn blends critical theory with an ecological ethic to offer both a powerful analysis of the economic and political forces generating environmental catastrophe as well as an emancipatory vision. Ecopedagogy seeks to "bridge the politics of the academy with forms of grassroots political organizing capable of achieving social and ecological seeking transformation."[325] David Greenwood embraces critical and place-based EE as well as radical multiculturalism, thereby acknowledging the intersections of race, power, and place.[326] Frederique Apffel-Marglin[327] and Patricia Palulis[328] combine Indigenous ways of knowing through story, subversive spiritualities that bring back a connected and relational way of knowing through ritual and ceremony. They engage biocultural regenerative agriculture in the Peruvian High Amazon, to decolonize and indigenize teacher education, while teaching for ecoliteracy.

In sum, environmental concerns have crossed the centuries; yet, one of the key issues has been the balancing of justice in the natural world with justice in the human world, particularly between the North and the South, and between relatively privileged and marginalized groups. Over time, a more complex and layered analysis with multiple approaches has evolved. While school-based EE has been constrained by structure, curriculum, and contested purposes, adult EE has been able to stay connected to the community and the larger environmental movement, rejuvenating its energy and refreshing its focus. After the Bruntland and Rio conferences, education discourse would shift towards sustainability and "achieving sustainable development in environmental adult education."[329] To this dialogue we now turn, in Chapter 5.

Notes and References

Chapter line drawing by Canadian artist Erin Marie Lange Christensen. www.erinchristensen.com. Reproduced with permission.

1 Theodore Roszak, *The Voice of the Earth* (New York: Simon and Schuster, 1992), p. 47.
2 Alan Reid, "A Non-Technical Introduction to Curriculum Challenges for and from Environmental Education," in Alan Reid (Ed.), *Curriculum and Environmental Education: Perspectives, Priorities and Challenges*, Kindle Edition (London: Routledge, 2018), p. 28.
3 For instance, UNESCO connections, UNESCO as the United Nations Educational, Scientific and Cultural Organization. See www.unesco.org/en/brief.
4 Formal education carries credentials, has a set curriculum, and is usually provided by an educational institution. Nonformal education is organized by educational and other institutions or groups. It is often noncredentialled or offers nondegree credentials, part-time, with a variety of delivery methods. Informal education is learning that occurs daily whether individually or in groups. While there may be goals for learning, it is not learning that is formally organized. Bruce Spencer and Elizabeth A. Lange, *The Purposes of Adult Education: An Introduction* (Toronto: Thompson Educational Publishing, 2014), p. 12.
5 Boaventura de Sousa Santos, *The End of Cognitive Empire: The Coming of Age of Epistemologies of the South* (Durham: Duke University Press, 2018), p. ix.
6 Mercantilism is the economic system "by which governments used their economies to augment state power at the expense of other countries. Governments sought to ensure

that exports exceeded imports and to accumulate wealth in the form of bullion (mostly gold and silver)." www.britannica.com/topic/mercantilism.

7 Hugh Brody, *Maps and Dreams: Indians and the British Columbia Frontier* (Vancouver: Douglas & McIntyre, 1981), p. 49.

8 Brody, *Maps and Dreams*, p. 58.

9 Brody, *Maps and Dreams*, p. 51.

10 Brody, *Maps and Dreams*, p. 59.

11 Brody, *Maps and Dreams*, p. 58.

12 Brody, *Maps and Dreams*, p. 50.

13 Malcolm Swan, "Forerunners of Environmental Education," in Noel McInnis and Don Albrecht (Eds.), *What Makes Education Environmental?* (Washington DC and Louisville, KT: Environmental Educators Inc and Data Courier Inc, 1975), p. 5.

14 Gerald Gutek, "Thomas Jefferson: Advocate of Republican Education," in *Historical and Philosophical Foundations of Education* (Columbus, OH: Merrill Prentice Hall, 2001), pp. 149–167.

15 Swan, "Forerunners of Environmental Education," p. 5.

16 Gregory Nobles, *The Myth of John James Audubon*. www.audubon.org/news/the-myth-john-james-audubon.

17 Stephen Corry, "The Colonial Origins of Conservation: The Disturbing History Behind US National Parks," *TruthOut*, August 25, 2015. https://truthout.org/articles/the-colonial-origins-of-conservation-the-disturbing-history-behind-us-national-parks/.

18 Elizabeth A. Lange, Peter Vogels, and Zenobia Jamal, *Learning a Language, Learning the Land: Newcomers, Parks, and Language Learning* (Edmonton: Research and Evaluation Report for Alberta Tourism, Parks, and Recreation, Edmonton Community Adult Learning Association, Edmonton Mennonite Centre for Newcomers, and the City of Edmonton, 2011).

19 Leah Penniman, "By Reconnecting to Soil, We Heal the Planet and Ourselves," *Yes! Magazine*, Spring 2019. https://www.yesmagazine.org/issue/dirt/2019/02/14/by-reconnecting-with-soil-we-heal-the-planet-and-ourselves.

20 Dorceta Taylor, *Race, Class, Gender, and American Environmentalism: General Technical Report* (Corvallis, OR: United States Department of Agriculture, Pacific Northwest Research Station, April 2002), p. i.

21 Taylor, *Race, Class, Gender, and American Environmentalism*, p. 24.

22 Taylor, *Race, Class, Gender, and American Environmentalism*, p. 7.

23 Taylor, *Race, Class, Gender, and American Environmentalism*, p. 7.

24 Corry, "The Colonial Origins of Conservation."

25 Nobles, *The Myth of John James Audubon*.

26 Clearing land to turn it into private property, called enclosure, meant clearing it of people and large mammals. On the North American plains, this was the buffalo and Indigenous people. See James Daschuk, *Clearing the Plains: Disease, Politics of Starvation, and the Loss of Aboriginal Life* (Regina: University of Regina Press, 2013).

27 See Chapter 2 in this book and also David Graeber and David Wengrow, *The Dawn of Everything: A New History of Humanity* (Toronto: Signal McClelland & Stewart, 2021) as well as Daniel Paul, *We Were Not the Savages: Collision between Europe and Native American Civilizations* (Halifax: Fernwood Publishing, 2006).

28 John Ralston Saul, *A Fair Country: Telling Truths about Canada* (Toronto: Viking Canada, 2008).

29 Gerald Gutek, "Horace Mann: Leader of the Common School Movement," in *Historical and Philosophical Foundations of Education* (Columbus, OH: Merrill Prentice Hall, 2001), pp. 188–210.

30 Wotherspoon, *The Sociology of Education in Canada*, p. 62.

31 Gutek, *Historical and Philosophical Foundations of Education*.

32 Bette Del Giorno, "The Impact of Changing Scientific Knowledge on Science Education in the United States since 1850," *Science Education*, Vol. 53, No. 3 (1969), pp. 191–195. https://doi.org/10.1002/sce.3730530304.

33 Swan, "Forerunners of Environmental Education," p. 11.
34 Del Giorno, *The Impact of Changing Scientific Knowledge on Science Education in the United States since 1850*, p. 192.
35 Swan, "Forerunners of Environmental Education," p. 7.
36 Kas Mazurek, "Interpreting Educational History – With a Commentary on the Social Context of Early Public Schooling," in Nick Kach, Kas Mazurek, Robert S. Patteron, and Ivan DeFaveri (Eds.), *Essays on Canadian Education* (Calgary: Detselig Enterprises Ltd, 1986), p. 31.
37 See the Truth and Reconciliation Commission of Canada at https://www.rcaanc-cirnac.gc.ca/eng/1450124405592/1529106060525.
38 https://vault.sierraclub.org/john_muir_exhibit/life/muir_biography.aspx.
39 Swan, "Forerunners of Environmental Education," p. 8.
40 Gordon Selman and Paul Dampier, *The Foundations of Adult Education in Canada* (Toronto: Thompson Educational Publishing, 1991), pp. 40, 64–69.
41 Terry Wotherspoon, *The Sociology of Education in Canada* (Don Mills, ON: Oxford University Press, 2009), p. 66.
42 Wotherspoon, *The Sociology of Education in Canada*, p. 66.
43 Swan, "Forerunners of Environmental Education," p. 10.
44 Downloaded from www.pgpedia.com/o/organized-camping-movement, May 18, 2021.
45 Scouts Canada. www.scouts.ca/about/about.html.
46 Del Giorno, *The Impact of Changing Scientific Knowledge on Science Education in the United States since 1850*, p. 192.
47 Del Giorno, *The Impact of Changing Scientific Knowledge on Science Education in the United States since 1850*, p. 195.
48 John Disinger, "Environmental Education's Definition Problem," in Harold Hungerford, William Bluhm, Trudi Volk, and John Ramsey (Eds.), *Essential Readings in Environmental Education* (Champaign, IL: Stipes Published, 1998/2001/2005), p. 17.
49 Office of the Historian, "Milestones: 1953–1960." https://history.state.gov/milestones/1953-1960/sputnik#:~:text=On%20October%204%2C%201957%2C%20the,accomplish%20this%20scientific%20advancement%20first.
50 David Roos, "How the Cold War Space Race Led to U.S. Students Doing Tons of Homework," *History Stories*, August 13, 2019. www.history.com/news/homework-cold-war-sputnik#:~:text=How%20the%20Cold%20War%20Space,U.S.%20focus%20on%20schoolchildren's%20workloads.
51 Wotherspoon, *The Sociology of Education in Canada*.
52 Kenneth V. Pike, "School Camping Has Come to Stay," *Journal of the American Association for Health, Physical Education, and Recreation*, Vol. 22, No. 6 (1951), pp. 23–24. https://doi.org/10.1080/23267232.1951.10629939.
53 Downloaded from www.pgpedia.com/o/organized-camping-movement, 18 May 2021.
54 Swan, "Forerunners of Environmental Education," p. 12.
55 James Aldrich and Anne Blackburn, "What Makes Environmental Education International?" in Noel McInnis and Don Albrecht (Eds.), *What Makes Education Environmental?* (Washington DC and Louisville, KT: Environmental Educators Inc and Data Courier Inc, 1975), p. 353.
56 John Kirk, "The Quantum Theory of Environmental Education" cited in Disinger, "Environmental Education's Definitional Problem," pp. 20–21.
57 Richard Kahn, *Critical Pedagogy, Ecoliteracy, & Planetary Crisis: The Ecopedagogy Movement* (New York: Peter Lang, 2010), p. 1. See also p. 5.
58 Wotherspoon, *The Sociology of Education in Canada*, pp. 57, 74.
59 Andrew Dobson, *Environmental Politics: A Very Short Introduction* (Oxford: Oxford University Press, 2016, cited in Reid, *Curriculum and Environmental Education*, p. 13.
60 Swan, "Forerunners of Environmental Education," p. 16.
61 Reid, *Curriculum and Environmental Education*, p. 15.

62 Cited in Alan Reid, "A Non-Technical Introduction to Curriculum Challenges for and from Environmental Education," (London: Routledge), pp. 35–36.
63 Cited in Alan Reid, "A Non-Technical Introduction to Curriculum Challenges for and from Environmental Education," (London: Routledge), pp. 36–37.
64 John Disinger and Charles Roth, *Environmental Literacy, ERIC/CSMEE Digest November 1992* (Columbus, OH: ERIC Clearinghouse for Science, Mathematics, and Environmental Education, ED351201).
65 Swan, "Forerunners of Environmental Education," pp. 16–17.
66 Swan, "Forerunners of Environmental Education," p. 18.
67 John Disinger, "Tensions in Environmental Education: Yesterday, Today and Tomorrow," in Hungerford, Bluhm, Volk, and Ramsey (Eds.), *Essential Readings in Environmental Education*, pp. 1–2.
68 Disinger, "Tensions in Environmental Education," p. 1.
69 Disinger, "Tensions in Environmental Education," p. 2.
70 Disinger, "Tensions in Environmental Education."
71 Swan, "Forerunners of Environmental Education," p. 16.
72 https://naee.org.uk/about-naee/.
73 UNESCO, "The Belgrade Charter: A Global Framework for Environmental Education," *UNESCO-UNEP Environmental Education Newsletter*, Vol. 1, No. 1 (1976), p. 104.
74 UNESCO/UNEP, "The Tbilisi Declaration," *Connect, the UNESCO-UNEP Environmental Education Newsletter*, Vol. 3, No. 1 (1978). Also in Hungerford, Bluhm, Volk, and Ramsey (Eds.), *Essential Readings in Environmental Education*, p. 13.
75 https://unesdoc.unesco.org/ark:/48223/pf0000032763.
76 UNESCO/UNEP, "The Tbilisi Declaration," in Hungerford, Bluhm, Volk, and Ramsey (Eds.), *Essential Readings in Environmental Education*, p. 13.
77 https://unesdoc.unesco.org/ark:/48223/pf0000032763.
78 Lucie Sauvé, "Environmental Education between Modernity and Postmodernity: Searching for an Integrating Educational Framework," *Canadian Journal of Environmental Education*, Vol. 4 (1999), pp. 9–25.
79 IUCN, *World Conservation Strategy* (Gland, Switzerland: IUCN-UNEP-WWF, 1980).
80 IUCN, *World Conservation Strategy*, p. I.
81 IUCN, *World Conservation Strategy*.
82 Joy Palmer, *Environmental Education in the 21st Century: Theory, Practice, Progress, and Promise* (London: Routledge Falmer, 1998), p. 77.
83 Swan, "Forerunners of Environmental Education," p. 16.
84 Dean Bennett, "Community and Adult Environmental Education," in Noel McInnis and Don Albrecht (Eds.), *What Makes Education Environmental?* (Washington, DC and Louisville, KT: Environmental Educators Inc and Data Courier Inc, 1975), p. 133.
85 Kahn, *Critical Pedagogy, Ecoliteracy, & Planetary Crisis*, p. 5. See also https://naaee.org/.
86 William Stapp, "The Concept of Environmental Education," *Environmental Education*, Vol. 1, No. 1 (1969), pp. 30–31 (emphases added).
87 Joe Heimlich, *Environmental Education* (Bloomington: Phi Delta Kappa Educational Foundation, 2002).
88 Arthur Lucas, *Environment and Environmental Education: Conceptual Issues and Curriculum Implications*. PhD Dissertation, Ohio State University cited in Disinger, "Environmental Education's Definition Problem," p. 25 (emphases added).
89 Michael Mappin and Edward Johnson (Eds.), *Environmental Education and Advocacy: Changing Perspectives of Ecology and Education* (Cambridge: Cambridge University Press, 2005), p. 3.
90 John Disinger, Ed McCrea, and David Wicks, *The NAAEE Third Years of History 1971–2000* (Washington, DC: NAAEE, 1995/2001). www.naaee.org.
91 Marshall McLuhan, *Understanding Media: The Extensions of Man* (New York: Mentor, 1964).
92 Tema Milstein, Mairi Pileggi, and Eric Morgan (Eds.), *Environmental Communication Pedagogy and Practice* (London: Routledge Earthscan, 2017).

93 Susanne Moser, "Reflections on Climate Change Communication Research and Practice in the Second Decade of the 21st Century: What More is There to Say?" *WIREs Clim Change*, Vol. 7 (2016), p. 351. https://doi.org/10.1002/wcc.403.
94 Moser, *Reflections on Climate Change Communication Research and Practice in the Second Decade of the 21st Century*, pp. 345–369. https://doi.org/10.1002/wcc.403.
95 Moser, *Reflections on Climate Change Communication Research and Practice in the Second Decade of the 21st Century*, p. 350.
96 Moser, *Reflections on Climate Change Communication Research and Practice in the Second Decade of the 21st Century*, p. 350.
97 Moser, *Reflections on Climate Change Communication Research and Practice in the Second Decade of the 21st Century*, p. 352.
98 Larry Beck and Ted Cable, *Interpretation for the 21st Century* (Champaign, IL: Sagamore Publishing, 2002).
99 The National Association for Interpretation in Lisa Brochu and Tim Merriam, *Personal Interpretation: Connecting Your Audience to Heritage Resources* (Fort Collins, CO: Interp-Press, 2002).
100 Freeman Tilden, *Interpreting our Heritage* (Chapel Hill, NC: The University of North Carolina Press, 1957/1967/1977).
101 Tilden. *Interpreting our Heritage,* p. 9.
102 Brochu and Merriam, *Personal Interpretation.*
103 Beck and Cable, "Interpretation for the 21st Century."
104 These principles are: 1. Lighting a Spark, 2. Interpreting in the Information Age, 3. Importance of the Story, 4. Provocation, 5. Holistic Interpretation, 6. Interpretation throughout the Lifespan, 7. Bringing the Past Alive, 8. Modern Tools of Interpretation, 9. Enough is Enough, 10. Technique before Art, 11. Interpretive Writing, 12. Attracting Support and Making Friends, 13. Interpreting Beauty, 14. Promoting Optimal Experiences, and 15. Passion. From Beck and Cable, *Interpretation for the 21st Century*.
105 Mappin and Johnson (Eds.), *Environmental Education and Advocacy*, p. 9.
106 Bill Freedman, *Environmental Science: A Canadian Perspective* (Scarborough: Prentice Hall Canada, 1998), p. 513.
107 Steven Cooke and Jesse Vermaire, "Environmental Studies and Environmental Science Today: Inevitable Mission Creep and Integration in Action-Oriented Transdisciplinary Areas of Inquiry, Training, and Practice," *Journal of Environmental Studies and Sciences*, Vol. 5 (2015), pp. 70–78.
108 Fritjof Capra and Pier Luigi Luisi, *The Systems View of Life: A Unifying Vision* (Cambridge: Cambridge University Press, 2014), p. 66.
109 Robert L. Smith and Thomas M. Smith, *Ecology and Field Biology* (San Franciso: Benjamin Cummings, 2001), p. 3.
110 Freedman, *Environmental Science*, p. 2.
111 Capra and Luisi, *The Systems View of Life*, p. 59.
112 Capra and Luisi, *The Systems View of Life*, p. 66.
113 Capra and Luisi, *The Systems View of Life*, p. 67.
114 Capra and Luisi, *The Systems View of Life*, p. 68.
115 Fritjof Capra, "Speaking Nature's Language: Principles for Sustainability," in Michael Stone and Zenobia Barlow (Eds.), *Ecological Literacy: Educating our Children for a Sustainable World* (San Francisco: Sierra Club Books, 2005), pp. 18–29.
116 Capra and Luisi, *The Systems View of Life*, p. 442.
117 Johan Rockström, Will Steffen, et al., "Feature: A Safe Operating Space for Humanity," *Nature*, Vol. 461, No. 24 (2009).
118 Lucie Sauvé, "Currents in Environmental Education: Mapping a Complex and Evolving Pedagogical Field," *Canadian Journal of Environmental Education*, Vol. 10 (2005), pp. 11–37. See also Sauvé, *Environmental Education between Modernity and Postmodernity*, p. 13.
119 Sauvé, "Currents in Environmental Education," pp. 11–37.

120 Stephen Sterling, *Sustainable Education: Re-visioning Learning and Change* (Cambridge: Green Books, 2001), p. 11.
121 Bateson's typology is first order learning as adaptive within an existing set of alternatives, second order learning as reflective on the process of learning, and third order learning as transformative as changing the available alternatives, with the possibility of a fourth level. See Gregory Bateson, *Steps to an Ecology of Mind* (Chicago: University of Chicago Press, 1972/2000), p. 293.
122 Palmer, *Environmental Education in the 21st Century*, p. 102.
123 Anna Gahl Cole, "Expanding the Field: Revisiting Environmental Education Principles through Multidisciplinary Frameworks," *Journal of Environmental Education*, Vol. 38, No. 2 (2007), pp. 35–44.
124 Finn Arne Jørgensen and Dolly Jørgensen, "Citizen Science for Environmental Citizenship," *Conservation Biology*, Vol. 35, No. 4 (2021), p. 1344.
125 Harold Hungerford and Trudi Volk, "Changing Learner Behaviour through Environmental Education," *Journal of Environmental Education*, Vol. 21, No. 3 (1990), pp. 4–6.
126 Hungerford and Volk, "Changing Learner Behaviour through Environmental Education," p. 11.
127 ENEC (European Network for Environmental Citizenship). Defining environmental citizenship. ENEC, Lemesos, Cyprus (2011). http://enec-cost.eu/our-approach/enec-environmental-citizenship/. [Google Scholar]
128 Andrew Dodson, "Environmental Citizenship: Towards Sustainable Development," *Sustainable Development*, Vol. 15 (2007), pp. 276–285.
129 ENEC (European Network for Environmental Citizenship), Defining environmental citizenship.
130 Disinger and Roth, "Environmental Literacy," *ERIC/CSMEE Digest* (Columbus, OH: ERIC Clearinghouse for Science, Mathematics, and Environmental Education, ED351201 (1992).
131 Disinger and Roth, "Environmental Literacy."
132 Disinger and Roth, "Environmental Literacy."
133 Andrew Stables and Keith Bishop, "Weak and Strong Conceptions of Environmental Literacy," *Environmental Education Research*, Vol. 7, No. 1 (2001), pp. 89–97. https://doi.org/10.1080/13504620125643.
134 William Scott, "25 Years On: Looking Back at Environmental Education Research," *Environmental Education Research*, Vol. 26, No. 12 (2020), pp. 1681–1689.
135 Kahn, *Critical Pedagogy, Ecoliteracy, & Planetary Crisis*, p. 7. See also www.edutopia.org/school-environmental-studies-project-learning.
136 Diane Curtis, *It's All Happening at the Zoo School: Innovative Education with Practical Applications* (George Lucas Educational Foundation, September 26, 2002). www.edutopia.org/school-environmental-studies-project-learning.
137 John Fien, "Teaching for a Sustainable World: The Environmental and Development Education Project for Teacher Education," *Environmental Education Research*, Vol. 1, No. 1 (1995), pp. 21–33.
138 Rebecca Martusewicz, Jeff Edmundson, and John Lupinacci, *Ecojustice Education: Toward Diverse, Democratic and Sustainable Communities* (New York: Routledge, 2011).
139 Fritjof Capra, *The Web of Life: A New Scientific Understanding of Living Systems* (New York: Anchor Books, 1996). See also Capra and Luisi, *The Systems View of Life*.
140 Arne Naess, "The Shallow and Deep, Long Range Ecology Movements," *Inquiry*, Vol. 16 (1973), pp. 95–100. Also see Arne Naess, *Ecology, Community and Lifestyle* in D. Rothenberg (Ed.) (Cambridge: Cambridge University Press, 1989).
141 Carolyn Merchant, *Radical Ecology: The Search for a Livable World* (New York: Routledge, 2005), p. 91.
142 William Devall and George Sessions, *Deep Ecology: Living as if Nature Matters* (Salt Lake City: Peregrine Smith, 1985).
143 Merchant, *Radical Ecology*, p. 92.

144 Merchant, *Radical Ecology*, p. 93.
145 David Orr, *Ecological Literacy: Education and the Transition to a Postmodern World* (New York: SUNY, 1992), p. x.
146 Orr, *Ecological Literacy,* p. x.
147 Orr, *Ecological Literacy*, p. 33.
148 Orr, *Earth in Mind: On Education, Environmental, and the Human Prospect* (Washington, DC: Island Press, 1994), p. 12.
149 Orr, *Ecological Literacy*, p. xi.
150 David Orr, "Foreword," in Michael Stone and Zenobia Barlow (Eds.), *Ecological Literacy: Educating our Children for a Sustainable World* (San Francisco: Sierra Club Books, 2005), p. xi.
151 David Orr, "Foreword in Stone and Barlow," *Ecological Literacy*. See also Orr, *Ecological Literacy*, p. 87.
152 Daniel Goleman, Lisa Bennett, and Zenobia Barlow, *Ecoliterate: How Educators Are Cultivating Emotional, Social and Ecological Intelligence* (San Francisco: Jossey-Bass, 2012), p. 2.
153 Goleman, Bennett, and Barlow, *Ecoliterate*, p. 2
154 See Theordore Roszak, Mary Gomes, and Allen Kanner, *Ecopsychology: Restoring the Earth, Healing the Mind* (San Francisco: Sierra Club Books, 1995), p. 4.
155 Susan Clayton and Susan Opotow, *Identity and the Natural Environment: The Psychological Significance of Nature* (Cambridge: The MIT Press, 2003), p. 8.
156 Pierre Dansereau, *Inscape and Landscape* (Toronto: Canadian Broadcasting Corporation, 1973).
157 Clayton and Opotow, *Identity and the Natural Environment*, p. 7.
158 Mitchell Thomashow, *Ecological Identity: Becoming a Reflective Environmentalist* (Cambridge: MIT Press, 1995).
159 Thomashow, *Ecological Identity*, p. 3.
160 Arne Naess, *Ecology, Community, and Lifestyle* (New York: Cambridge University Press, 1989) cited in Clayton and Opotow, *Identity and the Natural Environment*, p. 8.
161 See C. A. Bowers, *Education, Cultural Myths, and the Ecological Crisis: Toward Deep Changes* (New York: SUNY, 1993) and *Educating for an Ecologically Sustainable Culture: Rethinking Moral Education, Creativity, Intelligence and Other Modern Orthodoxies* (New York: SUNY, 1995).
162 Martusewicz et al., *Ecojustice Education*, p. 10.
163 Rebecca Martusewicz, Jeff Edmundson, and John Lupinacci, *Ecojustice Education: Toward Diverse, Democratic and Sustainable Communities* (New York: Routledge, 2011), p. 11.
164 Sinclair Macrae, *An Introduction to Ethics: Theories, Perspectives, Issues* (Toronto: Pearson Education Canada, 2003), p. 178.
165 Steve Van Matre, *The Earth Speaks* (Warrenville, IL: The Institute for Earth Education, 1983).
166 Michael Caduto and Joseph Bruchac, *Keepers of the Earth: Native Stories and Environmental Activities for Children* (Saskatoon: Fifth House Publishers, 1989).
167 For more of my critique, see Elizabeth A. Lange, "Is Freirean Transformative Learning the Trojan Horse of Globalization and Enemy of Sustainability Education? A Response to C.A. Bowers," *Journal of Transformative Education*, Vol. 10, No. 3 (2012), pp. 3–21. https://doi.org/10.1177/1541344612453880.
168 See Van Andruss, Christopher Plant, Judith Plant, and Eleanor Wright, *Home! A Bioregional Reader* (Gabriola Island: New Society Publishers, 1990).
169 Kirkpatrick Sale, "Principles of Bioregionalism," in Jerry Mander and Edward Goldsmith, *The Case Against the Global Economy: And for a Turn Toward the Local* (San Francisco: Sierra Club Books, 1996), pp. 471–484.
170 David Sobel, *Placed-Based Education: Connecting Classrooms & Communities* (Barrington, MA: The Orion Society, 2004/2005).
171 Laurie Lane-Zucker in Sobel, *Placed-Based Education,* p. ii.

172 David Gruenewald and Gregory Smith, *Place-Based Education in the Global Age* (New York: Routledge, 2008).
173 Sobel, *Place-Based Education*, p. 9.
174 Orr, *Earth in Mind*, p. 170.
175 Orr, *Earth in Mind*, pp. 163–170.
176 Mappin and Johnson, *Environmental Education and Advocacy*, p. 9.
177 George Bradford, *How Deep is Deep Ecology?* (Hadley, MA: Times Change Press, 1989).
178 Murray Bookchin, *Which Way for the Ecology Movement?* (Edinburgh: AK Press, 1994), pp. 21–22.
179 Bookchin, *Which Way for the Ecology Movement?* p. 23.
180 Bookchin, *Which Way for the Ecology Movement?* p. 23.
181 Joanna Macy and Molly Young Brown, *Coming Back to Life: Practices to Reconnect our Lives, Our World* (Gabriola Island, BC: New Society Publishers, 1998).
182 Macy and Brown, *Coming Back to Life*, p. 25.
183 Macy and Brown, *Coming Back to Life*, p. 12.
184 Macy and Brown, *Coming Back to Life*, p. ix.
185 Anna Gahl Cole, "Expanding the Field: Revisiting Environmental Education Principles through Multidisciplinary Frameworks," *Journal of Environmental Education*, Vol. 38, No. 2 (2007), pp. 34–44.
186 Christina Marouli, "Multicultural Environmental Education: Theory and Practice," *Canadian Journal of Environmental Education*, Vol. 7, No. 1 (2002), pp. 26–42.
187 Dorceta Taylor, *Race, Class, Gender, and American Environmentalism*, General Technical Report PNW-GTR-534 (USDA Forest Service, April 2002).
188 Running Grass, "Environmental Education and Environmental Justice: A Three Circles Perspective," *NYSOEA Pathways* (Spring 1995), pp. 9–13.
189 Running Grass and Julian Agyeman, *Reorienting Environmental Education for Environmental Justice* (Second National People of Color Environmental Leadership Summit II Resource Paper Series, October 23, 2002), pp. 2–3.
190 Grass and Agyeman, *Reorienting Environmental Education for Environmental Justice*, p. 3.
191 Marouli, "Multicultural Environmental Education," p. 28.
192 Natasha Blancet-Cohen and Rosemary Reilly, "Teachers' Perspectives on Environmental Education in Multicultural Contexts: Towards Culturally-Responsive Environmental Education," *Teaching and Teacher Education*, Vol. 36 (2013), pp. 12–22.
193 Marouli, "Multicultural Environmental Education," p. 33.
194 See the "Editorial" by Julian Agyeman, Cultural Environmental Education: From First Nation to Frustration, *Canadian Journal of Environmental Education*, Vol. 7, No. 1 (2002), pp. 5–12.
195 Kahn, *Critical Pedagogy, Ecoliteracy, & Planetary Crisis*.
196 The "matrix of domination" is from Patricia Hill Collins (1993) used in Kahn, *Critical Pedagogy, Ecoliteracy, & Planetary Crisis*, p. 9.
197 Merchant, *Radical Ecology*, p. 109. See George Bradford, *How Deep is Deep Ecology?* (Hadley, MA: Times Change Press, 1989).
198 Angela Antunes and Moacir Gadotti, "Ecopedagogy as the Appropriate Pedagogy to the Earth Charter Process," in Peter Blaze Corcoran (Ed.), *The Earth Charter in Action: Toward a Sustainable World* (Amsterdam: KIT Publishers, 2005) cited in Kahn, *Critical Pedagogy, Ecoliteracy, & Planetary Crisis*, p. 18.
199 Douglas Kellner, "Afterword," in Kahn, *Critical Pedagogy, Ecoliteracy, & Planetary Crisis*, p. 152.
200 See the Special Issue of *International Journal of Lifelong Education* entitled "Indigenous Voices and Decolonizing Lifelong Education" edited by Tiffany Prete and Elizabeth A. Lange, Vol. 40, No. 4 (July/August 2021).
201 Leanne Simpson, "Indigenous Environmental Education for Cultural Survival," *Canadian Journal of Environmental Education*, Vol. 7, No. 1 (2002), pp. 13–25.
202 Simpson, "Indigenous Environmental Education for Cultural Survival," p. 15.

203 Esteva and Prakash, *Grassroots Post-Modernism*, p. 197.
204 Mayor Ana María's speech to the *Intercontinental Encounter for Humanity and Against Neoliberalism* held in Chiapas cited in Esteva and Prakash, *Grassroots Post-Modernism*, pp. 173–4.
205 Heila Lotz-Sisitka, "Decolonisation as Future Frame for Environmental and Sustainability Education: Embracing the Commons with Absence and Emergence," in Peter Blaze Corcoran, Joseph Weakland, and Arjen Wals (Eds.), *Envisioning Futures for Environmental and Sustainability Education* (Wageningen, The Netherlands: Wageningen Academic Publishers, 2017), p. 45.
206 Lotz-Sisitka, "Decolonisation as Future Frame for Environmental and Sustainability Education," p. 46.
207 John Disinger, *Trends and Issues in Environmental Education: EE in School Curricula*. ERIC Clearinghouse for Science, Mathematics, and Environmental Education (Columbus, OH: NAAEE, 1987).
208 Mappin and Johnson, *Environmental Education and Advocacy*, p. 1.
209 Arthur Lucas, "Science and Environmental Education: Pious Hopes, Self Praise, and Disciplinary Chauvinism," *Studies in Science Education*, Vol. 7 (1980), pp. 1–26.
210 Alan Reid (Ed.), *Curriculum and Environmental Education* (London: Routledge, 2018).
211 Hungerford and Volk, "Changing Learner Behaviour through Environmental Education," pp. 8–22.
212 Elaine Andrews, *Assessing Water Quality Education Needs for the Nonformal Youth Audience* (US Department of Agriculture: Extension Service, 1992).
213 Deborah Simmons, "More Infusion Confusion: A Look at Environmental Education Curriculum Materials," *Journal of Environmental Education*, Vol. 20, no. 4 (1989), pp. 15–18. https://doi.org/10.1080/00958964.1989.9943034.
214 Simmons, "More Infusion Confusion," pp. 15–18.
215 Alan Reid (Ed.), "A Non-Technical Introduction to Curriculum Challenges for and From Environmental Education," in *Curriculum and Environmental Education*, Kindle Edition (London: Routledge, 2018), p. 41.
216 Emily Low, "Trend of Environmental Education in Canadian Pre-Service Teacher Education Programs from 1979–1996," *Canadian Journal of Environmental Education*, Vol. 7, No. 1 (2002), pp. 199–215.
217 Scott, *25 Years On*, p. 1688.
218 John Disinger, *Survey Report on Current Practice: Environmental Education in U.S. School Curricula*. ERIC Clearinghouse for Science, Mathematics and Environmental Education (Columbus, OH: NAAEE, 1987). See also Mappin and Johnson, *Environmental Education and Advocacy*, p. 14.
219 Frank Leeming, William Dwyer, Bryan Porter, and Melissa Cobern, "Outcomes Research in Environmental Education: A Critical Review," *Journal of Environmental Education*, vol. 24, no. 4 (1993), pp. 8–21.
220 Leeming, Dwyer, Porter, and Cobern, "Outcomes Research in Environmental Education," pp. 8–21.
221 Leeming, Dwyer, Porter, and Cobern, "Outcomes Research in Environmental Education," pp. 8–21.
222 Disinger, McCrea, and Wicks, *The NAAEE Thirty Years of History 1971–2001*.
223 Ian Robottom and Paul Hart, "Behaviourist EE Research: Environmentalism as Individualism," *Journal of Environmental Education*, Vol. 26, No. 2 (1995), pp. 5–9.
224 Thomas Tanner, *Ecology, Environment, and Education* (Lincoln, NE: Professional Educators Publications, 1974), p. 91.
225 North American Association for Environmental Education (NAAEE), *Environmental Education Materials: Guidelines for Excellence* (Troy, OH: NAAEE, 1996).
226 Bob Jickling, "Environmental Education and Environmental Advocacy: The Need for a Proper Distinction," in Rowland Lorimer and Michael M'Gonigle (Eds.), *To See Ourselves/To Save Ourselves: Ecology and Culture in Canada* (Montreal: Association for Canadian Studies, 1991), pp. 169–176.

227 Orr, *Ecological Literacy*.
228 Peter Martin, "A WWF View of Education and the Role of NGOs," in John Huckle and Stephen Sterling (Eds.), *Education for Sustainability* (London: Earthscan, 1996/2001).
229 Lucie Sauvé, "Currents in Environmental Education: Mapping a Complex and Evolving Pedagogical Field," *Canadian Journal of Environmental Education*, Vol. 10 (Spring 2005), pp. 11–37.
230 Paulo Freire, *Pedagogy of the Oppressed*, New York: Seabury Press, 1970.
231 Sauvé, "Currents in Environmental Education," p. 20.
232 Gerald Culen, "The Status of Environmental Education with Respect to the Goal of Responsible Citizenship Behaviour," in Hungerford, Bluhm, Volk, and Ramsey, *Essential Readings in Environmental Education*, p. 37.
233 Sobel, *Place-Based Education*, p. 8.
234 Sobel, *Place-Based Education*, pp. 8–9.
235 Paul Hart, *Teachers' Thinking in Environmental Education: Consciousness and Responsibility* (New York: Peter Lang, 2003) cited in Reid, *Curriculum and Environmental Education*, p. 49.
236 Kahn, *Critical Pedagogy, Ecoliteracy, & Planetary Crisis*, p. 11.
237 Kahn, *Critical Pedagogy, Ecoliteracy, & Planetary Crisis*, p. 6.
238 Hungerford and Volk, "Changing Learner Behaviour through Environmental Education," p. 15.
239 Reid, *Curriculum and Environmental Education*, Kindle Edition.
240 Michael Bonnett, "Environmental Education and the Issue of Nature," *Journal of Curriculum Studies*, Vol. 39, No. 6 (2009), pp. 707–721.
241 Roger Hart, *Children's Participation: The Theory and Practice of Involving Young Children in Community Development and Environmental Care* (New York: UNICEF, 1997) cited in Sobel, *Place-Based Education*, p. 9.
242 Charles Saylan and Daniel Blumstein, *The Failure of Environmental Education [And How We Can Fix It]* (Berkeley: University of California Press, 2011), pp. 1–2.
243 Scott, *25 Years On*, p. 1681.
244 Stephen Gough, "Lifelong Learning: Towards a Theory of Practice for Formal and Nonformal Environmental Education and Training," *Canadian Journal for Environmental Education*, Vol. 6 (2001), pp. 178–196.
245 For examples see D. Chovanec, E. Lange, and L. Ellis, "Social Movement Learning: A Catalyst for Action," in Marie Hammond-Callaghan and Matthew Hayday (Eds.), *Mobilizations, Protests, and Engagements: Canadian Perspectives on Social Movements* (Halifax, NS: Fernwood Publishing, 2008), pp. 186–201; Pierre Walter, "Dead Wolves, Dead Birds, and Dead Trees: Catalysts for Transformative Learning in the Making of Scientist-Environmentalists," *Adult Education Quarterly*, Vol. 63 (2011), pp. 24–42; Jessica Kovan and John Dirkx, "Being Called Awake: The Role of Transformative Learning in the Lives of Environmental Activists," *Adult Education Quarterly*, Vol. 53 (2003), pp. 99–118. https://doi.org/10.1177/0741713602238906; and Elizabeth A. Lange and Aaron Chubb, "Critical Environmental Adult Education in Canada: Student Environmental Activism," in Patricia Cranton and Leona English (Eds.), Reaching Out Across the Border: Canadian Perspectives in Adult Education, *New Directions for Adult and Continuing Education*, No. 124 (2010), pp. 61–72. https://doi.org/10.1002/ace.356.
246 For example see Edmund O'Sullivan and Marilyn Taylor, *Learning toward an Ecological Consciousness: Selected Transformative Practices* (New York: Palgrave Macmillan, 2004).
247 EAE is the term initially used by Darlene Clover.
248 AEE is the term initially used by Pierre Walter. I use these terms jointly and interchangeably.
249 For example see Lilian Hill and Darlene Clover, "Environmental Adult Education: Ecological Learning, Theory, and Practice for Socioenviromental Change," *New Directions for Adult and Continuing Education* (Fall 2003), No. 99; Darlene Clover (Ed.), *Global Perspectives in Environmental Adult Education* (New York: Peter Lang Publishing, 2004);

and Elizabeth A. Lange, "Environmental Adult Education: A Many-Voiced Landscape," in Carol Kasworm, Amy Rose, and Jovita Ross-Gordon (Eds.), *Handbook of Adult and Continuing Education* (Thousand Oaks, CA: Sage Publications, 2010), pp. 305–331.

250 Karen Malone, "Critical Environmental Adult Education: Taking on the Environmental Challenge in Clover," in Clover, *Global Perspectives in Environmental Adult Education*, p. 141.

251 Edward Kormondy, *Environmental Education: The Adult Public* (Washington, DC: American Institute of Biological Sciences, 1970); W. Jeske, "Should Adult Education Programs Include Courses Relating to Environmental Concerns?" *Adult Leadership*, Vol. 21 (1973), pp. 283–284, 307–308; Lars Emmelin, "The Need for Environmental Education for Adults," *Convergence*, Vol. 9, No. 1 (1976), pp. 45–53.

252 Robert Hill, "Environmental Adult Education: Producing Polychromatic Spaces for a Sustainable World," in Sharan Merriam, Bradley Courtney, and Ronald Cervero (Eds.), *Global Issues and Adult Education: Perspectives from Latin America, Southern Africa, and the United States* (San Francisco: Jossey-Bass, 2006), pp. 265–266 cited in Lange, *Environmental Adult Education*, p. 307.

253 David Lingwood, "Environmental Education Through Information-Seeking: The Case of an "Environmental Teach-In," *Environment and Behavior*, Vol. 3 (1971), pp. 230–261. https://doi.org/10.1177/001391657100300302.

254 Lingwood, "Environmental Education Through Information-Seeking."

255 Emmelin, "The Need for Environmental Education for Adults," pp. 45–53.

256 Caitlin Secret Haugen, "Adult Learners and the Environment in the Last Century: An Historical Analysis of Environmental Adult Education," *Electronic Green Journal*, No. 29 (2009), ISSN: 1076-7975.

257 Haugen, "Adult Learners and the Environment in the Last Century."

258 Haugen, "Adult Learners and the Environment in the Last Century."

259 An intersectional analysis comes out of feminist theory. Intersectionality refers to an analytical framework for understanding how different aspects of a person's social and political identities together create different forms of discrimination and privilege. For more on the lack of uptake, see Elizabeth A. Lange, "Environmental Adult Education: A Many-Voiced Landscape," in C. Kasworm, A. Rose, and J. Ross-Gordon (Eds.), *Handbook of Adult and Continuing Education* (Thousand Oaks, CA: Sage Publications, 2010), pp. 305–315 and Elizabeth A. Lange, "Transformative Sustainability Education: From Sustainababble to a Civilization Leap," in Marcella Milana, Sue Webb, John Holford, Richard Waller, and Peter Jarvis (Eds.), *The Palgrave International Handbook on Adult and Lifelong Education and Learning* (London: Palgrave Macmillan, 2018), pp. 397–420, for more explanation. See also Pierre Walter, "Adult Environmental Education," in Tonette Rocco, Cecil Smith, Robert Mizzi, Lisa Merriweather, and Joshua Hawley (Eds.), *The Handbook of Adult and Continuing Education 2020 Edition* (Sterling, VI: Stylus and AAACE), pp. 314–321.

260 International Council of Adult Education, *History of ICAE* (Toronto: ICAE, 2022). http://icae.global/.

261 International Council of Adult Education, *History of ICAE* (Toronto: ICAE, 2022). http://icae.global/.

262 http://icae.global/.

263 Bruce Spencer and Elizabeth A. Lange, *The Purposes of Adult Education: An Introduction* (Toronto: Thompson Educational Publishers, 2014), p. 13. For more discussion on social movements, see Donatella Della Porta and Mario Diani, *Social Movements: An Introduction* (Malden, MA: Blackwell Publishing, 1999/2006), pp. 20–22.

264 International Council of Adult Education, "History of ICAE" (Toronto: ICAE, 2022). http://icae.global/.

265 Clover, *Global Perspectives in Environmental Adult Education*, p. vii.

266 Malone, "Critical Environmental Adult Education," p. 141.

267 UNESCO, *Guidelines of the Development of Non-Formal Environmental Education*, 1986, p. 20 cited Malone, "Critical Environmental Adult Education," p. 143.

268 Moema Viezzer, "Environment, Education and Sustainable Development," *Convergence*, Vol. 39, No. 4 (2006), pp. 45–50.
269 Clover, *Global Perspectives in Environmental Adult Education*, p. ix. See also Viezzer, "Environment, Education and Sustainable Development," p. 48.
270 Clover, *Global Perspectives in Environmental Adult Education*, pp. x–xi.
271 Marta Benavides, "Lessons from 500 Years of a 'New World Order' – Towards the 21st Century: Education for Quality of Life," *Convergence*, Vol. XXV, No. 2 (1992), pp. 37–45 cited in Clover, *Global Perspectives in Environmental Adult Education*, p. x.
272 Malcolm Knowles, *The Adult Learner: A Neglected Species* (Houston: Gulf, 1973/1984).
273 Mark K. Smith, "What Is Non-formal Education?" *The Encyclopedia of Pedagogy and Informal Education (Pascal International Observatory, 2001).* https://infed.org/mobi/what-is-non-formal-education/.
274 Edmund O'Sullivan and Marilyn Taylor (Eds.), *Learning toward an Ecological Consciousness: Selected Transformative Practices* (New York: Palgrave Macmillan, 2004), pp. 3–4.
275 Clover, *Global Perspectives in Environmental Adult Education*, p. xiv.
276 Clover, *Global Perspectives in Environmental Adult Education*, p. xiv.
277 Lilian Hill and Darlene Clover, "Editors' Notes," in Environmental Adult Education: Ecological Learning, Theory, and Practice for Socioenviromental Change, *New Directions for Adult and Continuing Education*, No. 99 (Fall 2003).
278 Darlene Clover, "Environmental Adult Education: Critique and Creativity in a Globalizing World," in Hill and Clover, *Environmental Adult Education,* pp. 5–13.
279 Clover, "Environmental Adult Education." See also Elizabeth A. Lange, "Transformative and Restorative Learning: A Vital Dialectic for Sustainable Societies," *Adult Education Quarterly*, Vol. 54, No. 2 (2004), pp. 121–139. https://doi.org/10.1177/071713603260276
280 Clover, *Global Perspectives in Environmental Adult Education*, p. xvi.
281 Kent Mikkelson, "Environmental and Adult Education—Towards a Danish Dimension," *Convergence*, Vol. XXV, No. 2 (1992), pp. 71–74 cited in Clover, *Global Perspectives in Environmental Adult Education*, p. xvi.
282 UNESCO, "Environmental Adult Education: Awareness and Environmental Action," In *Adult Learning and the Challenges of the 21st Century*. Booklet 6a (Hamburg: CONFINTEA V, 1999).
283 CONFINTEA V Adult Education, *The Hamburg Declaration: The Agenda for the Future, International Conference on Adult Education* (Hamburg: UNESCO-Institute for Education, 1997).
284 CONFINTEA V, *The Hamburg Declaration.*
285 Walter, "Adult Environmental Education," pp. 315–317.
286 Walter, "Adult Environmental Education," p. 315. See also Pierre Walter, "Philosophies of Adult Environmental Education," *Adult Education Quarterly*, Vol. 60, No. 3 (2009), pp. 3–25. https://doi.org/10.1177/0741713609336109.
287 Sauvé, "Currents in Environmental Education," p. 23.
288 Pierre Walter, "Adult Learning in New Social Movements: Environmental Protest and the Struggle for the Clayoquot Sound Rainforest," *Adult Education Quarterly*, Vol. 57, No. 3 (2007), pp. 248–263. https://doi.org/10.1177/0741713606297444.
289 See Pierre Walter, "Dead Wolves, Dead Birds, and Dead Trees: Catalysts for Transformative Learning in the Making of Scientist-Environmentalists," *Adult Education Quarterly*, Vol. 63, No. 1 (2011), pp. 24–42.
290 Walter, "Adult Learning in New Social Movements," p. 260.
291 Stephen Gough, Kim Walker, and William Scott, "Lifelong Learning: Towards a Theory of Practice for Formal and Non-formal Environmental Education and Training," *Canadian Journal of Environmental Education*, Vol. 6 (2001), pp. 178–196.
292 Elizabeth A. Lange, "Interrogating Transformative Learning: Canadian Contributions," in Tom Nesbit, Susan Brigham, Nancy Taber, and Tara Gibb (Eds.), *Building on Critical Traditions: Adult Education and Learning in Canada* (Toronto: Thompson Educational Publishing, 2013), pp. 107–118.

293 Paul Belanger, "Learning Environments and Environmental Education," in Hill and Darlene Clover, *Environmental Adult Education*, pp. 79–88.
294 CONFINTEA V, *The Hamburg Declaration*, p. 10.
295 Belanger, "Learning Environments and Environmental Education," pp. 79.
296 Belanger, "Learning Environments and Environmental Education."
297 Ralf St. Claire, "Words for the World: Creating Critical Environmental Literacy" for Adults in Hill and Clover, *Environmental Adult Education*, pp. 69–78.
298 Edmund O'Sullivan and Marilyn Taylor (Eds.), *Learning toward an Ecological Consciousness: Selected Transformative Practices* (New York: Palgrave Macmillan, 2004), p. 2.
299 Darlene Clover, Shirley Follen, and Budd Hall, *The Nature of Transformation: Environmental, Adult and Popular Education* (Toronto: Transformative Learning Center, University of Toronto, 1998).
300 Robert Hill, "Environmental Justice: Environmental Adult Education at the Confluence of Oppressions," in Hill and Clover, *Environmental Adult Education*, pp. 27–38.
301 Deborah Barndt, "Blessings on the Food, Blessings on the Workers: Arts-based Education for Migrant Worker Justice," *Canadian Journal of Environmental Education*, Vol. 18 (2014), pp. 59–79.
302 Tuere Bowles, "The Dimensions of Black Women's Learning in the Environmental Justice Movement in the Southeastern US," in Laura Servage and Tara Fenwick (Eds.), *Proceedings of the Joint International Conference of the Adult Education Research Conference (AERC) and the Canadian Association for the Study of Adult Education (CASAE)* (Halifax: Canadian Association for the Study of Adult Education, 2007), pp. 55–60.
303 Sandra Tan, "Anti-racist Environmental Adult Education in a Trans-Global Community," in Clover, *Global Perspectives in Environmental Adult Education*, pp. 3–22.
304 Dip Kapoor, "Indigenous Struggles for Forests, Land, and Cultural Identity in India: Environmental Popular Education and the Democratization of Power," in Clover, *Global Perspectives in Environmental Adult Education*, pp. 41–58.
305 Benjamin Feinstein, "Learning and Transformation in the Context of Hawaiian Traditional Ecological Knowledge," *Adult Education Quarterly*, Vol. 54, No. 2 (2004), pp. 25–32.
306 Audrey Dentith and Onah Thompson, "Teaching Adult Ecojustice Education," in Audrey Dentith and Wendy Griswold (Eds.), Ecojustice in Adult Education: Theory and Practice in the Cultivation of the Cultural Commons in *New Directions for Adult and Continuing Education*, No. 153 (2017), pp. 65–75.
307 Gregory Lowan-Trudeau, *From Bricolage to Métissage: Rethinking Intercultural Approaches to Indigenous Environmental Education and Research* (New York: Peter Lang, 2015), p. 12.
308 Gregory Cajete, *Look to the Mountain: An Ecology of Indigenous Education* (Skyland, NC: Kivaki Press, 1994) cited in Lowan-Trudeau, *From Bricolage to Métissage*, p. 16.
309 Gregory Cajete, "'Look to the Mountain:' Reflections on Indigenous Ecology," in Gregory Cajete (Ed.), *A People's Ecology: Explorations in Sustainable Living* (Santa Fe: Clearlight Publishers, 1999), p. 3.
310 Cajete, "Look to the Mountain," p. 3.
311 Cajete, "Look to the Mountain," p. 3.
312 Adult sustainability education will be discussed in the next chapter, especially the work of *Maori* Lewis Williams. See also Marie Battiste on settler *vis a vis* Indigenous epistemologies.
313 Elizabeth A. Lange, "Reconceptualizing Participatory Action Research for Sustainability Education," in Dip Kapoor and Steven Jordan (Eds.), *Education, Participatory Action Research and Social Change* (New York: Palgrave Macmillan Publishing, 2009), pp. 123–136.
314 John Gaventa, "Participatory Research in North America," *Convergence*, Vol. 21, No. 2/3 (1988), pp. 19–28.
315 Joaquín Esteva and Javier Reyes, "Environmental Popular Education Strategies for Creating Sustainable Communities in the Patzcuaro Lake Basin Region of Mexico," in Clover, *Global Perspectives in Environmental Adult Education*, p. 103.

316 Esteva and Reyes, "Environmental Popular Education Strategies for Creating Sustainable Communities in the Patzcuaro Lake Basin Region of Mexico," p. 104.

317 Rosa Muraguri-Mwololo, "Re-Invigorating Pastoralist Environmental Practices through Collective Learning: A Case of Nomadic Ethnic Groups of Northern Kenya in Clover, *Global Perspectives in Environmental Adult Education*, pp. 23–40.

318 Salvatore Engel-Di Mauro and Karanja Keita Carroll, "An African-centered Approach to Land Education," *Environmental Education Research*, Vol. 20, No. 1 (2014), pp. 70–81.

319 Engel-Di Mauro and Carroll, "An African-centered Approach to Land Education," pp. 70–71.

320 Batchuluun Yembuu, "Intergenerational Learning of Traditional Knowledge through Informal Education: The Mongolian Context," *International Journal of Lifelong Education*, Vol. 40, No. 4 (2021), pp. 355.

321 Moacir Gadotti, *Pedagogy of the Earth and Culture of Sustainability* (Sao Paulo, Brazil: Instituo Paulo Freire, 2000).

322 Kahn, *Critical Pedagogy, Ecoliteracy, & Planetary Crisis*, p. 11

323 Paulo Freire, *Pedagogy of Indignation* (Boulder: Paradigm Publishers, 2004), p. 47.

324 Darlene Clover, Shirley Follen, and Budd Hall, *The Nature of Transformation: Environmental, Adult and Popular Education* (Toronto: Transformative Learning Center, University of Toronto, 1998), p. 14.

325 Richard Kahn, *Critical Pedagogy, Ecoliteracy, & Planetary Crisis*, p. 18.

326 David Greenwood, "Culture, Environmental, and Education in the Anthropocene," in Michael Mueller, Deborah Tippins, and Arthur Stewart (Eds.), *Assessing Schools for Generation R (Responsibility)* (New York: Springer, 2013), pp. 279–292.

327 Frédérique Apffel-Marglin, *Subversive Spiritualities: How Rituals Enact the World* (Oxford: Oxford University Press, 2011).

328 Patricia Palulis, "Black Earth Green Moon Mama Allpa: Polyphonic Moments from Temple to Tambo," in Giuliano Reis and Jeff Scott (Eds.), *International Perspectives on the Theory and Practice of Environmental Education: A Reader* (Cham, Switzerland: Springer Nature, 2018), pp. 221–232.

329 Haugen, "Adult Learners and the Environment in the Last Century."

5

SUSTAINABILITY EDUCATION

Situating Sustainability Education

> The plain fact is that the planet does not need more successful people. But it does desperately need more peacemakers, healers, restorers, storytellers, and lovers of every kind. It needs people who live well in their places. It needs people of moral courage willing to join in the fight to make the world habitable and humane. And these qualities have little to do with success as our culture has defined it.[1]

Emergence of Sustainable Development

As recounted in the last chapter, environmental education consolidated its growth during the 1980s, paralleling the evolution of the environmental movement.[2]

DOI: 10.4324/9781003159643-5

In Rio 1997, the UN concept of sustainable development became the offspring of the coalescence between the environmental and international development endeavours. However, the late 1980s and 1990s were a time of significant economic upheaval globally, impacting educational efforts. It featured emerging tensions between at least two divergent societal pathways.

One pathway was the international dialogue around sustainable development, started by the International Union for the Conservation of Nature (1980) and the Brundtland Commission (1987). This dialogue aimed at balancing environmental protection with development needs, which shaped the historic dialogues at the Rio Earth Summit in 1992 and the resulting Agenda 21 action plan. As British environmental education researcher Joy Palmer suggests, one of the most significant outcomes of Rio was the recommendation that environmental and development education "be incorporated as an essential part of learning, within both formal and nonformal education sectors."[3] The other outcome was normalizing NGO involvement in international summits. As British education researcher John Huckle summarized, "sustainability provides the mediating bridge between the development and environment lobbies."[4]

The other pathway was the rise of the New Right and economic globalization from the 1980s. Proponents perpetuated environmental disinformation and climate denialism to protect corporate business from both declining profit margins after its "long boom" from 1945 to 1970[5] and from the continual expansion of democratic and human rights, including labour rights. In the international application of structural adjustment plans with austerity measures, the funding for, and national commitment to, international agreements weakened, particularly in North America. There was also significant defunding and delegitimation of environmental movements, slowing their growth. It was a retreat from environmental protection policies and resulted in less funding for the sustainable development initiatives agreed to in Rio. In subsequent international summits, market-driven solutions to sustainability were being promoted. This is an important part of the context in which sustainability developed, discussed in the next few sections.

The Crisis in *and* of *Education*

As part of this contextual landscape, environmental education and what would become sustainability education were situated within a growing crisis of education itself. Stephen Sterling[6] suggested that by 1992 and the Rio Summit, it was clear that there were two crises related to education: one a crisis *of* education and another a crisis *in* education.

The crisis *in* education was evident in questioning the purposes for and achievements of public education. North American education professors Karen Mundy and Lynn Murphy explain that while there were demands for greater expenditure on education internationally, there was also a "critical effort to rethink the models and purposes of schooling and its relationship to other means of ensuring social security, welfare, and solidarity on both a global and a national scale."[7] As American public intellectual and environmental education scholar David Orr clarifies in the opening quote, from an environmental perspective, more of the same education was not going to address the compounding crises created by Western society. This assertion was held across the ideological divide in education. However, each generated contrasting solutions.

The crisis *in* education referred to intransigent social issues not yet solved by access to education. In other words, the purpose of schooling for achieving redistributive justice for minority groups and building cultural solidarities for a more civil society was stalled, ameliorative at best. There was a shrinking capacity to embrace humanist and democratic goals as learning issues among students multiplied and as demands on schools to address societal issues multiplied. Schools appeared to be reaching their limits in contributing to a more just, equitable, and ecological respectful world, without more fundamental structural change. Further, the world of education was considered "closed," suspicious of noneducator critique, which can be a natural part of the specialization and professionalization processes. Yet, some internal education critics suggested that this crisis of purpose and process *required* outside thinking to refresh dialogue.[8]

The crisis *of* education refers to the relation between education and society. For some in the corporate and political sectors, the expansion of democracy and the "politics of inclusion"[9] had gone too far with ever more groups demanding recognition and equity.[10] The rationale of the neoliberal juggernaut was if people were kept busy scrambling to make a living in a competitive milieu, there would be less citizen involvement in democratic systems and more ability to (re)concentrate power. Neoliberal policies decreased the role of the State as regulator and arbiter of social goods, including education and health care.[11] Further, the equity and justice goals of schooling and related remedial programmes were on the chopping block, given the belief in natural hierarchies by the Right.

Austerity measures—harsh measures implemented on the pretext of reducing public debt—increased public pain and turmoil in many nations, including North America. This entailed removing subsidies from food and other staples; reducing the availability of, and now charging for, public health and social services including aspects of education programmes; freezing or cutting public sector wages and reducing union power including for teachers; and freezing and cutting pension funds, again including educators. In terms of broader government programmes, services like national parks or environmental protections were rolled back; public services like utilities and banks were deregulated or privatized; and corporate taxes were cut, benefiting owner profits while public workers endured wage rollbacks.[12] Stress, fear, pain, and anger predominated as unemployment rose, and the ranks of the poor and homeless expanded throughout the 1990s. Like Canada, many nations experienced homeless encampments and panhandling for the first time since the Great Depression.

In sum, Ortega y Gasset[13] has called this the "artificial intensification of practices" that led to decay in the first place. Since the 1990s, the result has been an escalation in environmental destruction, climate emissions, and social inequity. These policies also comprised an assault on schooling and higher education.

The Assault on Schooling and Higher Education

These two purposes of education—either for democratic citizenship or for economic reorganization—were openly at war within school systems around the globe. As part of dismantling the Keynesian welfare state, neoliberalism was attempting to privatize as many school systems as possible for purposes of downsizing government, pursuing efficiencies, and expanding profit-making. As Stephen Ball explains, economic globalization overall was "a threat to traditional forms of production and accumulation, and the opportunity for new forms."[14] Education was no longer about the public welfare[15] but now about "the education industry"[16] as one of these new forms of production and profit-making.

Yet, as Barlow and Robertson[17] claimed, the assault on Canadian schools, as elsewhere, was thinly veiled class warfare. "Under the guise of 'reforming' or even 'supporting' schools, there has been a great deal of activity advancing the competitive corporate ideology of the right wing."[18] New public–private partnership (PPP) funding and building schools as well as generous "contributions" from corporations towards science labs and curricular materials were intended to redesign education in concert with free trade and free enterprise goals.

To convince the public of the neoliberal argument, schools were painted as failing the public and needing renewal and far more accountability. One goal was to ramp up competitive culture through schooling,[19] preparing students for an increasingly global but hostile economy created by *laissez-faire* economics and free trade policies. As part of this business agenda, New Managerialism and the accountability movement began infiltrating school administration, curriculum, teacher training, and school evaluation, deliberately overriding the decision-making powers of professional educators. "Professional executives" were considered interchangeable across workplace sectors, with the goal of running all organizations, public or private, as if they were "for profit." Strategies included instituting performance measurements and bonuses for workers and institutions, including public institutions such as schools. Accountability strategies like testing regimes were implemented globally, and results were compared school by school, nation by nation. Peter Sacks argues that the "entitlement not to starve or freeze" guaranteed from the Great Depression by Keynesian economic policies became "entitlement to a middle-class life" for Baby Boomers, then morphed into "entitlement to success" for Generation X. Nevertheless, Generation X also experienced a growing fear and anxiety for their "diminishing economic prospects"[20] and even more stark prospects for the future, given climate change realities for millennials/Gen Y and Z.[21]

The public rhetoric from neoliberal leaders was that these new policies enhanced public choice. The real intention was to bring the market principle into education as a form of "consumer choice," rather than "citizen right to access" as had existed. As the privatization and market provision of formal education were promoted, many nonformal sites of adult and lifelong education, originally established for equalizing educational access, were defunded alongside development and environmental NGOs.

Adult and nonformal education were explicitly reoriented to job and entrepreneurial training and individual marketability. Schooling and higher education were no longer about elevating human potential and facilitating democratic citizenship. Workplace competencies were stipulated by business-related think-tanks, such as the Conference Board of Canada[22] who called for workers who were adaptable, flexible, and problem solvers with constantly positive attitudes and an entrepreneurial orientation.[23] All education was to now concentrate on what was called "learning for earning" in North America and "learning pays" in the European Union. Maltese critical education theorist Peter Mayo calls this a profound dessication of the purpose for, and integrity of, what it means to be educated.[24]

In higher education, a liberal arts university education and related scholarly research were painted as being hopelessly out of sync with societal needs. Rather, applied research for purposes of commercialization was promoted and funded simultaneously as state funding was withdrawn. Liberal arts programmes, ostensibly not offering financial value to the university, were painfully defunded, certainly in North America. University leaders rigorously applied business and leadership principles, fostering an increasingly competitive ethos with multiple ranking systems, pitting professor against professor, department against department, faculty against faculty, and university against university.[25] "Production" in universities and colleges was measured by newly instituted metrics—numbers of tuition-paying students, student satisfaction rankings, professor rankings, number of scholarly publications, and research funds received. Formerly collaborative professors and those dedicated to community work were now scrabbling on and over their colleagues for "bonus points" that might result in modest pay or status increases.

Taking Back Education?

The field of education was being forced to rethink its origins, purposes, and structures from a third perspective, contrasting with the social welfare or neoliberal perspective. The pattern of thinking inherited from the Scientific Revolution—*ala* Galileo, Newton, and Descartes—was being eclipsed by the revolutionary ideas of the New Science. Ecology, living systems theory, and quantum physics were shifting perceptions of reality, a Reality Revolution discussed further in Chapter 6. They were pointing to ways of thinking, which were transdisciplinary, connected, and integrated. The "modern drive to dominate nature"[26] was increasingly becoming suspect. No longer were scholars assuming a disembodied observer and utilizing reductionist methods of conventional research.

French philosopher Edgar Morin and others cast the traditional knowledge-production methods in universities as generating "mutilated knowledge."[27] Dialogues among scholars were concluding that "[t]ransdisciplinarity aims for a different way of thinking, and a different way of organizing knowledge,"[28] particularly needed for resolving pressing and complex societal issues, called "wicked public problems."[29] Disciplinary boundaries were slowly loosening, yet were diametrically opposed to the neoliberal constriction of perception and priorities into economics. In response, David Orr asserted that "we must take education back from those who intend it to be centralized, homogenized, standardized, technologized, and industrialized."[30]

Thus, there was a call for a "deeper transformation of the substance, process, and scope of education at all levels."[31] As one example, David Orr insisted that "all of education is environmental education."[32] For him, the very structure of schooling, design of buildings, and the insular and human-centric approach of the disciplines "reinforces passivity, monologue, domination, and artificiality."[33] This way of educating is no longer relevant or healthy, as it reinforces the dynamics that created current public crises.

In terms of environmental education, British educator Stephen Sterling echoed Orr: "We are educated by and large to 'compete and consume' rather than to 'care and conserve.'[34] Sterling asserted that more education, even more environmental education, was "not sufficient."[35] While science in the past had eschewed ethics, ethics were now re-entering the conversation. With the New Science, scientists and other scholars could no longer separate themselves from the outcomes and consequences of their work. Further, Orr advocated that mastery of one's person should be a vital focus of education, not just disciplinary subjects and their content. In contrast to neoliberal norms, he asserted that integrity, care, and thoughtfulness are just as important as cognitive knowledge. In this vein, knowledge and knowing are ethical practices. Understanding the impacts of knowledge frameworks and taking responsibility for enacting knowledge for the public good were beginning to be regarded as indivisible aspects of knowing and being in the world.[36]

In this 1992 context, Orr laid out six basic principles for rethinking education, anticipating future discourse over several decades:

1. recognition that all education is environmental education;
2. environmental issues are complex and cannot be understood through one discipline;
3. education occurs as a dialogue with "place," a relation which then defines us;

4. the process of education is as important as its content, changing the way we live;
5. experience in the natural world is essential for understanding the environment and to good thinking; and
6. education must be relevant to building a sustainable society, offering practical competence.[37]

Orr's vision was that education ought to be holistic, placed-based, relational, process-oriented, and connected to the natural world. Education should offer sustainability competences and ethics as well as cognitive knowledge. Thus, the discussions of sustainability and sustainable development were situated within this discourse of "taking back education." An additional context relevant to the development of sustainability education was the dialogue around the perceived failures of environmental education.

Perceived Failures of Environmental Education

Not only was there a crisis of education, but Richard Kahn also asserted there was a "crisis of environmental education."[38] Alongside Orr's critique and call for reform of the educational system, many long-time environmental educators began to examine more deeply the practice of environmental education (EE), introduced in Chapter 4.

Stephen Sterling[39] charged that environmental education had not fulfilled its mission: it had not "permeated all of education."[40] The increase in ecoliteracy among the general population was not as evident as expected. After 20 years, research on the impact of EE through schooling indicated that students had "high levels of awareness" but "poor understanding" of global trends, including environmental trends.[41] They could not connect the dots of current events, recognize links between issues and root causes, or distinguish between ideologies and their environmental impact. In sum, EE was not yet creating the foundation needed for environmentally respectful societies.

From a theoretical perspective, Lucie Sauvé suggested that environmental education was still operating in the modernist paradigm with the goal of reform primarily.[42] Perceiving the environment "as outside oneself" and "something to manage" is an example of the modernist paradigm in EE. It appeared that environmental education was not meaningfully addressing the human–environment link, especially the connection between the economy, environmental health, and human well-being.

Prior to the 1990s, the EE field had been charged with disinterest in social justice issues. While ecological science could now explain complex environmental dynamics, it was not necessarily offering solutions to pressing ecological issues while considering social impact, a disjuncture between science and society.

Other Antecedents to Sustainability Education: Adjectival Educations

A final feature of the 1990's context was the presence of what has been called "adjectival educations" or "education *for something*."[43] The field of education continuously renews and updates itself by developing new curricular and pedagogical foci that address emerging public issues. This has sometimes been considered pejoratively as the "bandwagon" syndrome. Yet, through such adjectival educations, the field of education can change fluidly and quickly in response to a changing social, cultural, and geophysical context.

Largely since the 1960s, new strands of education developed independently from each other, using adjectives such as: peace education, development education, multicultural education, gender education, LGBTQ+ equity education, and global education, briefly described later in the chapter. Environmental education was often considered to be one of these adjectival educations.

Together, these adjectival educations comprised a tapestry long considered a part of the liberal and progressive belief in education for fostering social improvement and addressing educational inequality.[44] Eventually, sustainability education would be considered by some to incorporate all these adjectival educations.

Peace Education

Emerging after two 20th-century world wars, peace education promoted the end of war through cultures of peace to maintain world order.[45] It started with the work of Maria Montessori who in the early 1900s reformed early childhood education through child-centred, self-directed learning that fostered the natural desire to learn. Through collaborative learning, students could learn the principles of co-operation and egalitarianism. After the second world war, she added a peace education aspect. In North America, her model waned for several decades, until the 1950s and 1960s, when didactic forms of education were questioned.[46] Her approach was reinforced by the Gandhian philosophy of nonviolence from the 1940s.

The Institute for World Order began in 1963 and, through the leadership of Betty Reardon, prepared teachers and curriculum materials to "study peace issues as well as international peacekeeping and dispute-settlement procedures."[47] It promoted an explicit values approach for preventing war, maximizing economic welfare, increasing social justice, broadening democratic participation, and restoring ecological balance.[48]

Johan Galtung, a Norwegian sociologist of peace and conflict, established some conceptual foundations for the field. He distinguished violence that is direct (behaviour acts), structural (patterns of inequity which can be the roots of conflict), or cultural (elements that justify violence).[49] He also defined peace in two ways: *negative peace* as the absence of war achieved by the reduction in armed conflict and *positive peace* which reduces the conditions that create violence and conflict.[50] For example, peace education addressed the spread of militarization as well as the dangers of nuclear armament within the arms race. For East Asian peace educators Toh Swee Hin and Virginia Floresca-Cawagas, hunger and poverty were considered a key manifestation of peacelessness,[51] an example of structural violence.

Peace education and conflict resolution were woven throughout school curriculum in the 1970s and 1980s, often integrated into social studies, civics, or history. Peace education focussed on these root causes of conflict, the capacities for peacemaking, as well as the values that support cultures of peace. Toh and Floresca-Cawagas identified six curricular components: militarization, structural violence, human rights, cultural solidarity, environmental care, and personal peace.[52] The UN University of Peace in Costa Rica, for example, is one higher education institution dedicated solely to the study of peace and peace education since 1980.[53]

The next generation of scholar-practitioners, such as Canadian Catherine Baillie Abidi, describes that understanding violence is another key part of peacebuilding.[54] Working for the Red Cross and now the Dallaire Institute, she explains that peace education is used in many active conflict situations. It addresses root causes of conflict as well as transforms ways of thinking, particularly the construction of "enemy" through dehumanization tactics.[55] Peace education is naturally tied to other adjectival educations, including development education, human rights education, and, later, global education.[56]

International Development and Human Rights Education

Another adjectival education has been international development education and human rights education. After the dissolution of colonial empires, notably *Pax Britannica*, and the establishment of the United Nations, "international education" was promoted to build support for a new international order based on human rights. Alongside the effort to address world poverty and "underdevelopment," development education was undertaken to build a sense of moral internationalism by educating people as "world citizens" able to function in and contribute to an increasingly international world.

While development NGOs in the United States tapped philanthropic funding for their public education work, Canada relied on federal funding, offered in support of its foreign policy, its development agency, and its view of itself as an honest international broker. Thus, development NGOs relied heavily on federal funding. Paralleling the changes in international development thinking, development education was state funded and offered through local community organizations across Canada in the late 1960s, with many faces reflecting the diversity of development theory.[57]

The dominant face was modernization, meant to augment the Western development agenda and foster public engagement. However, some forms of development education contested the modernization model and aimed to create an informed critique of aid and development policies and a conduit for voices from the South.[58] Often, these two purposes were held in tension.

In the 1970s, state funding helped establish "learner centres" across the country as resource centres from which to carry out adult education about world realities as well as foster ongoing support for international aid and development. Development education was also supported by the Canadian Council for International Cooperation, populated by the larger development NGOs.[59] Learner centres hosted international film festivals, profiled celebrations like World Food Day or International Week, and brought in high-profile guest speakers for community adult education and schooling. Many community educators used the principles of popular education,[60] replicating Latin American processes for building a "critical consciousness" and commitment to social justice, utilizing the pedagogy of Paulo Freire.[61]

> Popular education is called "popular" because its priority is to work among the many rural and urban poor who form the vast majority of people in most Third World countries. It is a collective or group process of education, where the teacher and students learn together, beginning with the concrete experience of the participants, leading to reflection on that experience in order to effect positive change.[62]

In 1988, significant funding was given to teacher's associations across Canada to foster development education in schools, eventually called global education in schools. Popular education principles were used to engage students while teaching social analysis as part of developing an informed citizenry. It was also hoped that development education would interest volunteers and young professionals to work in "developing" nations as part of Canada's international commitment. However, both development education and global education in Canada would largely meet their demise with the neoliberal withdrawal, or redirection, of federal funding in 1995.

Multicultural, Intercultural, and Anti-Racist Education

From the colonial then decolonization periods and resulting regional conflicts, population diversity was increasing in many Western nations through immigration and refugee protection. Nations such as Canada were implementing national policies on multiculturalism as their rates of immigration increased. Schools experienced demographic shifts, more "racially, culturally, and linguistically diverse."[63] Multicultural and, later, anti-racist education emerged to help shape values towards accepting cultural differences while reducing stereotyping and discrimination in the interests of "national unity."[64] However, the typical approach as the Business Council for Sustainable Development in 1992, to assist business in changing course.[103] Another was the Sustainable Development Solutions Network which pursues research and policy work, with regional hubs.[104] The International Institute of Sustainable Development was instituted in 1990, as a thinktank and Canada's voice to help accelerate sustainable solutions and shape sustainable development governance.[105]

Multiple metrics, management tools, benchmarking, and accountability mechanisms were designed to measure organizational progress towards a sustainability transition, whether corporations, small business, national and subnational governments, NGOs, or colleges and universities.[106] With various metrics quickly proliferating and designed for scalability, sustainability principles permeated, for instance, the operations of global cities through Local Governments for Sustainability (ICLEI)[107] and the Resilient Cities Network.[108]

Many organizations also took up the challenge of sustainability, adopting principle statements such as The Houston Principles, CERES Principles, or the International Chamber of Commerce Business Charter (ICC). They were guided by frameworks such as the Global Reporting Initiative, Dow Jones Sustainability Index, The Natural Step, UN Global Compact, and ISO family of international management standards.[109] New guides for ecological design and sustainability resulted in green building standards, from LEED (Leadership in Energy and Environmental Design) and Passivhaus to the World Green Building Council. As Tony Hodge described, the approach to implementing sustainability required a systemic approach; seeing human systems as ecosystems with patterned interactions; using a narrative approach to the organization; and then establishing assessments with baselines, targets, indicators, and measurements.[110]

Given this vast uptake, Paul Hawken has argued that sustainability was not solely a top down initiative. Aside from all the UN, workplace/corporate/professional, and governmental initiatives, he claimed that all the diverse activities in small nonprofit and nongovernmental organizations, rooted in their own global locales, *were* part of a vast movement, initially with no name. In the late 1990s and early 2000s, Hawken estimated there to be anywhere from 30,000 to 100,000 civil society groups globally who were addressing social justice, human rights, and environmental issues.[111] While this unrecognized movement did not meet the standard definition of a social movement, it was clear that this "global humanitarian movement arising from the bottom up,"[112] *was* a sustainability movement.

> [This movement] is dispersed, inchoate, and fiercely independent. It is has no manifesto or doctrine, no overriding authority to check with. It is taking shape in schoolrooms, farms, jungles, villages, companies, deserts, fisheries, slums—and yes, even fancy New York hotels. . . . Historically social movements have arisen primarily in response to injustice, inequities, and corruption. Those woes remain legion, joined by a new condition that has no precedent: the planet has a life-threatening disease, marked by massive ecological degradation and rapid climate change.[113]

was the "foods and festivals" approach, leaving intact an unjust and exclusionary social structure.

American education scholar Christine Sleeter explains that four helping models appeared in both education and social services. The *moral model* blames the victim for their problems. The *medical model* blames the social environment for problems, and the *enlightenment model* blames societal structures for the problems. Sleeter says that the latter two are benevolent helping models enacted by experts which can "disable" victims further. The fourth model is the *empowerment model* which views "persons as victims of problems created by society but as potentially active solvers of their own problems."[65] Rather than learning to "comply passively with the demands of a public institution,"[66] minority groups can learn to identify disempowering aspects as well as advocate for the inclusion of their cultures, languages, submerged histories, and lived realities in education and other institutions. Multicultural education aimed to empower students from marginalized groups by offering education that bridged unique cultural knowledge with mainstream education,[67] ultimately reforming mainstream education to some extent.

In cases where political entities have been concerned about their own cultural preservation, such as the province of Quebec in Canada, an intercultural education approach was taken. "Interculturalism involves establishing a dialogue among diverse cultural groups and integrating newcomers to the host society, the intent being to define together a new francophone social order where every culture finds its place."[68] Through educating for greater communicative capacity and positive attitudes towards cultural difference, the expectation was that cultural inclusion would be achieved. Debates remain, questioning the difference between cultural inclusion and structural inclusion, as an ongoing national dialogue.

Sleeter insists that critically oriented multicultural education has a specific "social change mission" to link race, social class, and gender issues with school reform, addressing those left behind or "at risk."[69] It was expected that peaceful coexistence was possible by rejecting nationalist patriotism, cultural ethnocentrism, and intolerance of difference. Peaceful co-existence was possible by fostering personal qualities of self-awareness, empathy, humility, nonjudgement, and trust.[70] At the social level, the overall goal was to "transform the social order in the interests of social justice."[71]

Antiracism education emerged out of the 1970s and 1980s, given concerns that "visible minority adults and children, despite the popularization of multicultural education"[72] still lacked social inclusion. The role of antiracism education was to examine more deeply how power relations are reproduced and how "democratic racism" and "colour blindness"[73] do not account for the lived experiences of racialization and systemic exclusion. While educators may celebrate difference, they can also: "suppress difference" via implicit assimilation; "insist on difference" through accommodation and sometimes segregation; "deny difference" by allowing meritocratic dynamics to determine success; "invite difference" emphasizing tolerance rather than structural change; and "critique difference" which questions power relations and how prejudice and discrimination work at both the individual and social levels.[74] These conversations have led to contemporary discussions around Africentric pedagogy, White fragility, postcolonial education, and critical race theory.

Gender and LGBTQ+ Equity Education

With the second wave of feminism in the 1960s and 1970s and the AIDS crisis in the 1980s, gender equity and LGBTQ+ education addressed the pervasive culture of sexism and homophobia in schools. It challenged the history of sexual and gender stereotyping and harassment in culture as well as the perpetuation of stereotypes and queer silence and silencing in textbooks, school curricula, and in educational research and leadership.

One goal was to end the deleterious impacts on girls and women in terms of lowered educational achievement and workforce access, in addition to other aspects of marginalization. By the late 1980s, it was clear that encouraging girls and young women in the fields of science and math, promoting women into leadership, and changing gender stereotypes in curricula were not enough.[75] The research of Mary Belenky and colleagues as well as that of Carol Gillian and Nel Noddings pointed to a "women's way of knowing"[76] and a more prominent use of "connected and care ethics."[77]

Other pedagogical strategies included shifting the focus to collaboration instead of competition and respectful dialogue instead of dominating dialogue. Studies from Myra and David Sadker illustrated the complexity of gender relations in learning spaces, from amount of time speaking to levels of attention and praise, to fostering male activity and female passivity, from kindergarten to higher education.[78] Kathleen Weiler also identified how social forces shaped women entering teaching and their pedagogical engagement with issues of power and oppression in both their classrooms and staffrooms.[79] By the 1990s, further research, particularly by June Larkin and Pat Stanton on "hallway pedagogy," illustrated the dynamics and levels of sexual harassment and aggression, from intimidation to groping.[80] Intersectional studies portrayed the layers of racism, heterosexism, and classism that accompanied sexual harassment. This theory and practice preceded antibullying education and the #MeToo movement, further aiming to end the psychological harassment and physical assault that have been part of the traumatizing experience of education, particularly for women and girls.

For members of the LGBTQ (now LGBTQIA2S+)[81] community, antihomophobia and queer education aimed to redress the educational exclusion of non-gender-conforming young people in both schooling and higher education. Canadian educationist André Grace considers this "cultural work" and "public pedagogy" in decreasing the marginalization, silencing, and high levels of symbolic and physical violence perpetuated in "homeplace, learning place, and workplace."[82] Overcoming stigmatization and shame as well as fostering dialogue and resilience among queer young adults, an invisible minority, can "counter their traditional status as fugitive learners"[83] whose "mere existence is often denied."[84] Changes to human rights legislation and education policies as well as changes in counselling, curriculum, pedagogy, discipline, availability of support groups, and school–parent–community relations have been required to ensure that anti-homophobia education was system-wide, as part of equity education. Further, educating non-queer adults helps improve civil entitlements, access to health services, personal security, and normalizes a spectrum of human sexuality.[85]

Global Education

With the fall of the Berlin Wall in 1989, technology and trade tightened the webs of international connection. The call to "go global" permeated national vocabularies and had an air of urgency, possibility, and inevitability. Modern industrialism appeared to be waning, while the postmodern and postindustrial era was rising. Globalizing the capitalist economic system and moving towards postindustrialism generated the necessity for educational systems to promote global awareness and competencies for global citizenship and knowledge work.

The underside of this optimism was a more sobering global reality. There was a decrease in national sovereignty through the neoliberal globalization of trade, the increasing miseries of poverty and hunger alongside growing wealth accumulation, mindless overconsumption of fast food to fast fashion, with looming environmental crises. No longer considered separate issues, the interweaving of peace, development, environmental, multicultural/antiracist, human rights, and women/gender/sexual diverse equity educations was considered integral to global education as an educational reform movement.[86]

American educators Barbara and Kenneth Tye argued that global education was qualitatively different from core disciplines such as history or geography as well as from the adjectival educations such as international/development education.

> Global education is the name . . . of a reform movement within contemporary American education that seeks to alter schools, universities, and nonformal educative institutions in ways that provide children and adults with the basic intellectual competencies needed to deal effectively and responsibly with . . . an increasingly globalized society.[87]

Thus, it would be global education that became the successor of this lineage of adjectival educations. Tye and Tye considered global education a holistic study of humankind, our ecological and cosmic home, and the global social and economic structure.

Yet, there were competing interests and understandings of global education, as it too was driven by divergent vested interests. One dominant understanding was the global competitiveness approach, an extension of modernization, which accepted the ideological assumptions of neoliberalism. An explicitly self-interested approach to global education, it aimed to preserve the economic supremacy and competitiveness of Northern nations by building global knowledge as well as language and intercultural skills for global social relations particularly business transactions. Higher education institutions often pursued "globalization" to expand their student population while charging increased tuition rates to international students. This necessitated services to meet the unique needs of international students, considered a facet of global education. Another concern within this approach was national security, as many people now carried multiple citizen identities and thus loyalties in an internationally mobile world.

A second and dominant approach to global education grew out of the long liberal tradition of societal improvement and cultural pluralism. It focussed on the effort of building intercultural appreciation and responsible global citizenship. Global education in this approach fostered a sense of common humanity within the context of growing global interdependence and increased migration. It focussed on "perspective consciousness" that one's world view will not be universally shared.[88] The American President's Commission on Foreign Language and International Studies[89] identified that American students were woefully inadequate in their understandings of complex issues like international relations, environmental degradation, and food security. Educational leaders called for an education in interdependence which would assist students in understanding perspective-taking, global responsibilities, and making responsive decisions in a changing world.

A third and contrasting perspective on global education was offering a systems (cultural) analysis that nurtured not only global awareness but also social action to address complex, interconnected global issues. Global educators Graham Pike and David Selby relied on General Systems Theory for developing their five dimensions of global education: spatial, temporal, global issues, global cultural crises, and human potential.[90] They used a systems analysis approach to address global issues as well as teaching analytic skills and issue resolution. Further, it involved learning about behaviour that could create a future that was based on "sane, human, and ecological (SHE)" patterns rather than "hyper-expansion (HE)" patterns. Global education taught for global interdependence, which Graham Pike and David Selby described as a "network of links, interactions, and relationships that encircle the planet like a giant, intricate spider's web so that the wider world is a pervasive and ubiquitous element in the routines of everyday life."[91]

A fourth perspective in global education was offering a structural (root cause) analysis by examining the changing shape of power relations within neoliberal globalization as well as the social and environmental consequences. Transformative global education endeavoured to shift inherited meanings and mental barriers that now threaten human survival in multiple ways, including a critique of more mobile industrial capitalism and its accelerated destruction around the globe. It aimed for transformative learning about injustice but also empowerment to act as knowledgeable global citizens. It also questioned the deep structure of schooling that reproduced learner passivity and dominant ideologies.

Tara Goldstein and David Selby argue that under the new rubric of global education, educators endeavoured to weave the adjectival strands together into a holistic and systemic approach. Thus, global education discourse had begun to highlight the pressing need for integrated thinking and holistic approaches, thus crossing disciplinary boundaries in education. In this way, global education became a terrain of educational politics and public debate on what shape individual nations and global society should take and how educational reform should address evolving global justice and environmental issues. It was into this complex and crowded educational landscape of the 1990s and early 2000s that the concept of sustainability entered educational discourse. These educational precedents and dialogue about education reform converged into the fertile ground of sustainability discourse.

Proliferation of Sustainability

There was much optimism as the new concept of sustainability initially appeared to move beyond existing limiting frameworks. In addition to the classic definitions described in Chapter 3, environmental scholar Julian Agyeman developed a definition of a "just sustainability" which places "great emphasis upon the need to: ensure a better quality of life for all, in a just and equitable manner, whilst living within the limits of supporting ecosystems."[92] Sustainability education in this conception holds potential to integrate the adjectival educations as well as contribute to educational reform.[93]

Chapter 3 discussed the origins of sustainability discourse, the conceptual debates about the term sustainable development, the common elements of sustainability, and the proliferation of the sustainability movement. In a quick review, it was generally understood as a synthesis of ecological integrity, economic well-being, and social equity, a pathway forward for global society. Sustainability was meant to be an equal valuing of three intersecting priorities, known as the "three Es" of environment, economy, and equity, as seen in Figure 5.1, also known as the "triple bottom

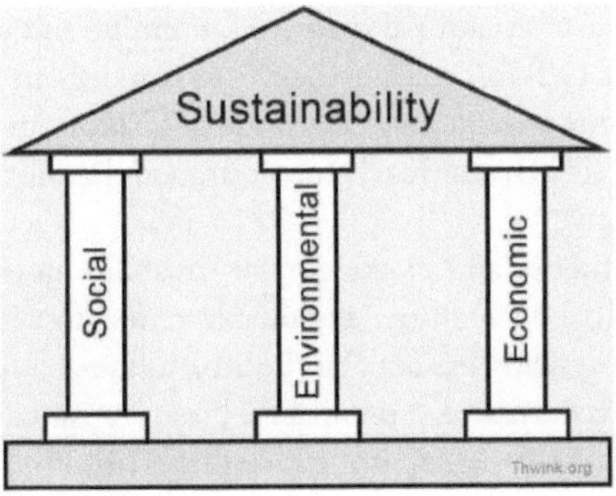

FIGURE 5.1 Three Pillars of Sustainability

Credit: Permission granted by Jack Harich of Thwink.org. Reproduced with permission of Thwink.org.

line" or "three Ps" of people, planet, and profits. The three Es became part of the dialogues at the World Summit on Sustainable Development in 2002.[94]

During UN international conferences, "sustainable development" was considered the term that best bridged the concerns of the North and South hemispheres. It was expected that a balance could be found "between the needs of the environment and those of humankind."[95] Sustainable *development* enabled nations with high levels of poverty to attain a quality standard of living for larger numbers of their people, while *sustainable* development would help curb the excesses of wealthier nations and generate less material-intensive activities and destructive growth that deprives future generations. This was captured in the classic 1987 Brundtland definition of sustainable development: *development that meets the needs of the present without compromising the ability of future generations to meet their own needs*,[96] striking a balance between the *needs* of the world's poor and material *limitations* for developed nations.[97] The Brundtland report explains: "Making globalization work better for the poor is seen as integral to sustainable development—as a human rights issue, as a moral concern, for peace and security and economic development in the future."[98]

Thus, it was a common conclusion that environmental, social, and economic crises could no longer be addressed independently, but needed a coherent, multifaceted, integrated decision-making approach.[99] It was also a valuing of not just *intra*generational equity—across nations—but also *inter*generational equity—between generations.[100]

Some understood sustainable development as a "shrewd slogan that made it possible to start an initial dialogue . . . between the worlds of business and politics and the world of the environment,"[101] given traditional enmity and suspicion. Despite ambivalence about the elasticity of the sustainability concept, including co-optation, greenwashing, or creating smokescreens, there was a massive uptake of sustainability.

Driven by the UN's adoption of sustainable development, many nations instituted sustainability roundtables or other mechanisms to develop national sustainable development strategies, examining how purposes and functions could be aligned with UN goals of sustainability development. This was fostered and supported by the UN with ongoing international dialogue, summits, and periodic assessments. For instance Timothy O'Riordan, Heather Voisey, and their colleagues trace how sustainable development was integrated into various European Union treaties, laws, policies, and initiatives.[102] Given this, it can be argued that sustainable development was a top down initiative.

Nevertheless, the theory and practice of sustainability quickly permeated most sectors, from sustainable agriculture, sustainable cities, sustainable rural communities, sustainable ecosystems, sustainable health, sustainable design, sustainability science, sustainable governance, sustainable business, sustainable communities, sustainable livelihoods, to sustainability education. This often involved reoriented professional training as well as inhouse professional workshops. When the United Nations Commission on Sustainable Development was established in 1992 to oversee programming related to Agenda 21, other organizations were established, such

Both this top down and bottom up dynamics enabled sustainability to comprise a paradigm shift or "revolution," enabling many to see the global linkages across all their niche work as well as seeing themselves as working in concert.[114] Expressing the thoughts of many globally, Richard Heinberg of the Post Carbon Institute says,

> [S]ustainability is more than a buzzword; it is a concept that defines what is needed for our civilization to survive . . . not only the ideas, but the courageous efforts of people around the world who are creating a way of life that is not only more survivable, but more just and satisfying as well.[115]

As this book has in part traced, Hawken explains that there were three roots to this vast movement: environmental activism, social justice activism, and the resistance of Indigenous groups to colonization, globalization, and ecological devastation.[116] The hope and promise for sustainability were evident not only in the halls of institutions, assembly rooms of governments, and boardrooms of organizations but also in the growing transnational counter-globalization movement at the turn of the millennium, with wide connections and substantial presence at key international summits.

The Integrative Promise of Sustainability Education

Arguably, the environmental movement and environmental education *were* successful in creating a shift in global thinking, which laid the groundwork for sustainability thinking to be taken up so extensively. It was now a more common belief that all societies and all sectors of society needed to concern themselves with the environmental impact of human activities. This was possible, as we experienced in North America, because Generation X[117] had already been schooled in awareness of and concern for environmental issues. Successful uptake of sustainability was also due to no pre-existing connections with an identifiable social movement that politicized discourse, leaving it open for many sectors to own the concept.

As a consequence, education scholar Peter Blaze Corcoran described in 2010, "sustainability has become the metanarrative of our time," even though it seems to be an ever "diminishing prospect" in terms of concrete attainment.[118] Corcoran claims that formal and nonformal educators *have* had success in *re*forming the field of education, though it has not been *trans*formed.[119] To further the work, he asserts "The nature of sustainability, and the prospect of unsustainability, require a fundamental change of epistemology, and therefore of education."[120] This firmly connects environmental and sustainability education with transformative education.

Similarly, environmental scientists Timothy O'Riordan and Heather Voisey describe that the concept of sustainability requires a societal transformation through a new way of knowing with "ecological, social, ethical and economic dimensions."[121] For Stephen Sterling, the concept of sustainability suggests a "shift in consciousness" through a "transformative level of learning."[122] For David Selby, sustainability has the potential to incorporate a "vernacular environmentalism" that dives deeply into the "lived, intimate reality of local place . . . fold[ing] together science, spirituality, and social justice."[123]

Such were the hopes for and requirements of sustainability education. However, British environmentalist Jonathan Porritt lays out the challenge in the foreword of one of the first textbooks on sustainability education by John Huckle and Stephen Sterling (1996).

> The educational system has been coopted by the language, values, and practice of the so called "free market" meekly renouncing progressive educational values in order to mould young people for induction into today's harsh, unforgiving, competitiveness-at-all-costs economy.[124]

Yet, he insists, "education for sustainability is perhaps the only way of getting back to those intrinsically liberal, progressive values that once underpinned our education system."[125] What has comprised the journey of sustainable development and sustainability education? We will begin with the story of "Education for Sustainable Development." As Joy Palmer describes it, this has been the global agenda, "emphasis[ing] the urgent need for world-wide environmental education, and its fundamental role in the transition towards sustainability."[126]

Sustainable Development and Education

In 1995, the International Environmental Education Programme (IEEP) of both UNESCO and UNEP, in place since 1975, was ending. For the UN, education was considered one of the key mechanisms and processes for manifesting sustainable development. While education became the forgotten priority among Rio policymakers, not discussed again until 1996,[127] educators had quickly adopted the concept of sustainability, given its integrative promise and broad currency. In the early years, multiple monikers were used: Education for Sustainable Development (ESD), Education for Sustainability (EfS), Learning for Sustainability (LfS), Sustainability Education (SE), Education for Sustainable Futures, Education for Sustainable Living, and Environmental Education for Sustainability, to name just a few.[128]

In 1992, Rio participants agreed that education was a key enabler for achieving the goal of sustainable development across the spectrum of lifelong learning.[129] UNESCO also maintained, "Education is UNESCO's top priority because it is a basic human right and the foundation on which to build peace and drive sustainable development."[130] Therefore, Chapter 36 of *Agenda 21* entitled *Promoting Education, Public Awareness and Training* was key in building early ideas about education and sustainability. Chapter 36 specifically focussed on the critical role of education for capacity-building, enabling "human beings and societies [to] reach their fullest potential."[131] Given this instrumental approach, the term became "Education for Sustainable Development" (ESD) rather than Sustainable Development Education. In other words, there were multiple goals within sustainable development, which education would enable.

Environmental education was considered one foundation from which to meet these global challenges. The recommendations in the 1978 *Tbilisi Declaration*[132] were still considered highly relevant in suggesting environmental education address a "broad spectrum of environmental, social, ethical, and cultural dimensions."[133] However, the Tbilisi goals had never been fully implemented or achieved. Then, with the release of the 1988 Brundtland Report, UNESCO-UNEP proposed that it was the concept of "sustainable development" (for a time called ecodevelopment) which could ensure considerations of ecological balance and resource conservation as well as social justice.

In 1992, UNESCO then formally proposed the concept "Education for Sustainable Development"[134] or ESD. It became the adopted term at the international level, understood as a "single, unifying concept."[135] In the spirit of Brundtland, aiming to strengthen global cooperation, it was expected that ESD could contribute to the "common end of improving the material and spiritual living conditions of the present generation without denying a decent life for generations to come."[136] In this view, education was integral to sustainable development and sustainable development to education.

In 1998, UNESCO asserted "[E]ducation, in its broadest sense, must be a vital part of all efforts to imagine and create new relations among people and to foster greater respect for the needs of the environment."[137] The concepts related to Education for Sustainable Development were driven by UN organizations in consultation with a broad base of experts. Yet, educators and education scholars also stepped forward to craft the documents and work through the concepts and approaches. In this way, educators were not merely functionaries to international bodies and government ministries but also active co-creators.[138]

In a parallel Rio gathering, the NGO *Alternative Treaty on Environmental Education* was signed. They chose to replace the problematic word "development" in "sustainable development" with "equity," to eliminate contradictory messaging, given that development is often equated with growth. It was out of this discussion among educators that the World Environmental Education Congress (WEEC) was established, quickly becoming the international network responding to the environmental education principles identified in Rio.

Since 2003, WEEC has been connecting international actors across multiple sectors every 2 years. With a permanent secretariat in Italy, WEEC attracts up to several thousand participants from hundreds of nations for its conferences. A plurality of perspectives and approaches have developed, including a spectrum of views on EE and ongoing debates regarding ESD.[139]

Another result from Rio and Agenda 21 was the desire for an ethical vision to guide the transition to sustainable societies. The result was the *Earth Charter*, a declaration signed by thousands of civil society organizations in 2000, making it the most significant global civil society document ever signed.[140] The primary purpose for the Earth Charter was to identify the "fundamental ethical principles for building a just, sustainable, and peaceful global society in the 21st century."[141] It is now a global movement, guided by Earth Charter International, which assists in educating, networking, and collaboration around the Earth Charter.[142]

The Earth Charter includes four interlocking pillars, as seen in the left column of Figure 5.2:

- respect and care for the community of life;
- ecological integrity;
- social and economic justice; and
- democracy, nonviolence, and peace.

FIGURE 5.2 Earth Charter—Turning Conscience into Action

Credit: Copyright held by Earth Charter International. www.earthcharter.org. Reproduced with permission.

The Earth Charter has 16 principles: interdependence of all life; love and responsibility; democracy and freedom; justice across generations; protect Earth's diversity; prevent ecological harm; sustainable lifestyles; share knowledge; eradicate poverty; equitable human development; gender equality and equity; dignity, inclusion, and well-being; transparency and participation; integrate values into education; respect all living beings; and nonviolence and peace. The Earth Charter is a document that integrates both the principles and especially values of sustainable development in a holistic way.

The Earth Charter is a valuable teaching document which led to the establishment of the Earth Charter Center for Education for Sustainable Development, affiliated with and on the campus of the UN University for Peace in San José, Costa Rica. The Earth Charter Education Center offers courses, masterclasses, and research opportunities that link to more recent goal statements, such as the Sustainable Development Goals (SDGs) for 2030.[143] As the Rector for the University of Peace Franciso Rojas Aravena states, the importance of the Earth Charter lies in the fact that it is a space for

> [R]eflection and dialogue, capable of listening and understanding the visions of different actors, coming from different fields of thought. The Earth Charter is a place where this plurality seeks to find shared values to design timely solutions to protect the planet.[144]

Toh Swee-Hin and Virginia Floresca Cawagas, in particular, illustrate the synergies, interconnections, and mutually enhancing ethics among peace education, the ESD movement, and the Earth Charter.[145]

The "Education for Sustainable Development" Debate

Five years after Rio in 1997, UNESCO hosted the Thessaloniki Conference to address the education goals related to the Earth Summit agreement. It endeavoured to build UNESCO's responsibility for "reshaping education so as to promote attitudes and behaviours conducive to a culture of sustainability."[146] It was agreed again that the Tbilisi agreement was still valid though underachieved. It was also agreed that there was an insufficient progress on education for implementing "sustainability, together with [the other mechanisms of] legislation, economy, and technology."[147]

The Thessaloniki conference triggered a heated debate about the connection between environmental education (EE) and education for sustainability development (ESD). Some argued that ESD is an important goal of EE, as it finally addressed the social and political dimensions of environmental issues.[148] Some advocated that environmental education be phased out and education for sustainability (EfS) replace it.[149] Others argued that EE should become ESD, given the failures of EE,[150] setting up a vociferous debate.

In rigorous defence of EE as a field, the intruder "sustainable development" was charged with being a "vague slogan"[151] and an illogical concept suffering from irreconcilable meanings.[152] For ESD opponents, it was seen to enable those with technocratic solutions as well as those with ecologically based solutions to use the same language yet work on nonaligned purposes. Still others considered ESD to be a tool for persuasion or mere "training," desiccating the very purpose and processes of what it means to be educated.[153]

The terms sustainable and development were individually and jointly critiqued. The term "sustain" is related etymologically to "sustenance" or "that which helps to endure." Yet, a general understanding of sustainability is to "maintain as is," and thus the intention was unclear. Questions then arose regarding what is to be maintained, by who, and how?[154] Some argued that sustainability was the goal and that perhaps sustainable development was the process; others argued the reverse. Some rejected the term sustainable development, preferring the term sustainability as it better conveyed a process, rather than an instrumental approach. For instance, Sterling claimed, "We believe that education for sustainability is a process, which is relevant to all people and that, like sustainable development itself, it is a process rather than a fixed goal."[155] It was argued that such a process view of sustainability allowed for differences in local community context, flexibility of community interpretation, and local level enactment rather than top down strategies.[156]

While a worthy debate that has persisted over years, the 1997 *Declaration of Thessaloniki* adopted the goals of "education for sustainable development," given growing currency, urgency, and acknowledgement of the concerns of both hemispheres. The conference moved ahead to determine the key components that formalized Education for Sustainable Development (ESD). Several other terms remained in usage, particularly education for sustainability (EfS) and sustainability education (SE). Eventually, plural perspectives to ESD would be encouraged.[157] In sum, the UN approach has been to consider sustainable development through the mechanism of education as a pathway to sustainability.[158]

Pushing back at sustainable development proponents, the World Environmental Education Congress (WEEC) held its first international conference in 2003, reemphasizing its presence, claiming conceptual territoriality, and distancing from ESD, for a time. By 2015, however, some EE proponents would engage environmental education *and* sustainability education in an integrated way, calling it environmental and sustainability education (ESE).[159]

Concept and Implementation of ESD

The goal of sustainable development was to balance "poverty alleviation and inequality reduction while ensuring intra- and intergenerational equity."[160] As well, sustainable development via ESD was seen as assisting in creating "environmentally aware decision-makers with indispensable support of a knowledgeable public."[161] Education was the long-term "change agent," and its "reorientation" could enhance "getting the message across" to the public as well as building the societal capacity to address environmental issues.[162]

The levels of educational attainment in a populace were not necessarily the issue. It was clear that highly educated populations were major contributors to unsustainable practices and lifestyles. Therefore, the goal was about reorienting education. In Northern hemisphere nations, for instance, this meant addressing the need for "more sustainable production and consumption patterns."[163] In Southern hemisphere nations, increasing education levels would enable the creation of jobs, businesses, agriculture, and industries with a lower environmental impact while enhancing quality of life. In this context, the concept of ESD "aimed to empower and equip current and future generations to meet their needs using a balanced and integrated approach to the economic, social and environmental dimensions of sustainable development."[164]

The transition to sustainability could be advanced through shifts in curriculum, teacher training, institutional operations, and community lifelong learning. Yet, the primary UNESCO goal was to prepare formal educators in schooling systems. So, for instance, the International Network of Teacher Education Institutions was established in 1998 for the purpose of reorienting teacher education to address sustainable development.[165] Erin Redman and colleagues offered recommendations of how to engage practicing teachers, including an initial intense intervention with long-term follow-up.[166] The role of UNESCO was as the coordinating agency who raised awareness and convened countries for discussion, capacity building, and decision-making, as they had the access to the ministries of education and other government decision-makers.[167]

As Rosalyn McKeown explains, the eventual Decade of Education for Sustainable Development (DESD) starting from 2015 was not just a top-down effort through staff members. Progress would not have been possible unless "many individuals and organizations tried something and then shared their success." She goes on to say, "The DESD ignited the imaginations of people of all different disciplines in formal, nonformal, and informal education. The groundswell of activity was *important* for moving the DESD forward."[168] Yet, this groundswell of activity had started well before the start of the DESD. It helped provide the evidence that "ESD worked" and was not "just a theoretical concept." As discussed later, "both top down and bottom up were necessary for the success of the UNDESD."[169]

However, ESD was also aimed increasingly at nonformal educators who were helping communities, organizations, and governments develop sustainability goals and implement local solutions. ESD helped establish higher education and research with a sustainability focus, assisting in the training of sustainability professionals and aiding local innovation, creativity, and new ways of thinking about practices. Finally, it was increasingly noted that ESD needed to be developed in a way that was locally relevant and culturally appropriate.[170]

Decade of Education for Sustainable Development (DESD) 2005–2014: Reorienting Education

UNESCO consistently placed education for sustainable development at the heart of the project for planetary education and human development. Thus, at Rio+10 in Johannesburg (2002), UNESCO announced the *Decade of Education for Sustainable Development* (DESD), to run from 2005 to 2014.

The DESD was designed to go well beyond ecological concerns, towards "a different vision of the world."[171] The basic vision of the DESD is "a world where everyone has the opportunity to benefit from education and learn the values, behaviour and lifestyles required for a sustainable future and for positive societal transformation."[172] As Moacir Gadotti explains, the primary aim was the reorientation of the current formal education curriculum.[173]

BOX 5.1 PRIMARY GOAL OF DESD

The *overall goal* of the DESD is to integrate the principles, values, and practices of sustainable development into all aspects of education and learning in order to encourage changes in behaviour that will create a more sustainable future in terms of environmental integrity, economic viability, and a just society for present and future generations.[174]

This mandated a global shift towards Education for Sustainable Development (ESD) to be undertaken by all Member Nations during the decade.[175] UNESCO's function became a "laboratory of ideas, standard-setter, clearing house, capacity-builder, and promoter of international cooperation."[176] Their intention was to build a broad partnership among international, regional, national, and subnational spheres to manifest this agenda. The DESD subgoals and objectives are listed next.

BOX 5.2 DESD SUBGOALS AND OBJECTIVES

Within the broad goals established by the General Assembly, the *subgoals* for the DESD at the national level are to:

- Provide an opportunity for refining and promoting the vision of and transition to sustainable development—through all forms of education, public awareness, and training.
- Give an enhanced profile to the important role of education and learning in sustainable development.

The *objectives* for the DESD are to:

- facilitate networking, linkages, exchange, and interaction among stakeholders in ESD;
- foster an increased quality of teaching and learning in education for sustainable development;
- help countries make progress towards and attain the millennium development goals through ESD efforts; and
- provide countries with new opportunities to incorporate ESD into education reform efforts.

The strategic document focussed on what nations committed to achieve through the DESD, under UNESCO's leadership. The central milestone was "clearly identifiable plans and/or activities in place in Member States."[177] The UNESCO commitment to education for sustainable development has not wavered, although the concept of ESD has evolved over time, and the thinking about educational pathways has certainly changed over the years.

Education has been considered in two ways, not only "an end in itself," but also the "motor for change" in achieving other global goals.[178] Thus, the DESD was formally linked to the eight millennium development goals (MDGs) for 2000 to 2015 (see Chapter 3) as well as to the Education for All movement and the United Nations Literacy Decade. It was to work synergistically with the MDGs[179] to reduce poverty and provide universal access to education. It was understood that ESD would subsume[180] but go beyond environmental education,[181] given these broader goals, which was the UNESCO response to the original EE/ESD debates.

Therefore, in this integrated, international approach to the global MDGs, first laid out in Agenda 21, additional priorities for ESD were:

- first, improving basic education (definitions based on national context with a focus on female literacy and participation);
- second, reorienting existing education (at all educational levels including higher education);
- third, enhancing public awareness and understanding (including critical media literacy); and
- fourth, a priority on training (environmental management skills to enable citizens to implement sustainability plans wherever they were located).

This approach thereby linked formal, nonformal, and informal education in the push towards sustainability.[182] The following were the *essential characteristics* of ESD, combining many aspects of earlier EE documents.

BOX 5.3 ESSENTIAL CHARACTERISTICS OF ESD

According to UNESCO, Education for Sustainable Development would:

- be based on the principles and values that underlie sustainable development;
- deal with the well-being of all three realms of sustainability—environment, society, and economy;
- promote life-long learning;
- be locally relevant and culturally appropriate;
- be based on local needs, perceptions, and conditions, while acknowledging that fulfilling local needs often has international effects and consequences;
- engage formal, non-formal, and informal education;
- accommodate the evolving nature of the concept of sustainability;
- address content, taking into account context, global issues, and local priorities;
- build civil capacity for community-based decision-making, social tolerance; environmental stewardship, an adaptable workforce, and quality of life;
- be interdisciplinary where no one discipline can claim ESD for its own, but all disciplines can contribute to ESD; and
- use a variety of pedagogical techniques that promote participatory learning and higher-order thinking skills.

The Implementation Toolkit focussed on the reorientation of education towards sustainable development, involving knowledge, issues, skills, perspectives, and values. It specifically ensured that values and ethics of environmental awareness were consistent with the knowledge, skills, and perspectives related to sustainable development.

However, similar to EE conundrums, it was assumed that ESD could not just be another educational "add-on." As Rosalyn McKeown *et al.* have said, "deciding what to leave out—what does not contribute to sustainability or is obsolete—is an integral part of the reorienting process."[183] It was similarly noted that the "holistic and interdisciplinary nature of ESD" would make implementation "more difficult in traditional school settings."[184] Thus, reorienting education encouraged integrated use of all the disciplines—science, social science, and the humanities. McKeown offers one example of what this integration of knowledge across the three sustainability pillars looks like in Figure 5.3.

The Implementation Toolkit advocated for a strengths model. Rather than focussing on retraining nearly 60 million teachers globally, the idea was to offer

	Environment	*Economy*	*Society*
Knowledge*	Hydrologic cycle	Supply and demand	Conflict
Issues	Protecting and managing freshwater; managing hazardous wastes	Combating poverty	Changing consumption patterns
Skills	The ability to acquire, manage, and analyse data	The ability to identify components of full-cost accounting	The ability to think critically about value issues
Perspectives	Linkage/interrelationship between/among contemporary global environmental issues	Look beyond local and national boundaries	Universal attributes of being human
Values	Ecological value of undisturbed land	Value of a sustainable livelihood	Economic value, religious value, and societal value compete

FIGURE 5.3 Integration of Knowledge with Three Sustainability Pillars

* The integration of knowledge in the three sectors is important to show human–environmental interactions and impacts.

Credit: Rosalyn McKeown, *Education for Sustainable Development Toolkit,* version 2, 2002. Available online at http://esdtoolkit.org. Knoxville: University of Tennessee, Waste Management Research and Education Institute. Reprinted with permission.

innovative and strengths-based approaches for preservice and in-service teachers. One aspect of the strengths model was finding the natural convergence points across subject curriculum as well as building on existing inter/transdisciplinary approaches. Another aspect was building a holistic understanding of sustainability from pre-existing awareness. A third strength was to build on the existence of adjectival educations in many nations. Drawing lessons from EE, ESD implementation was conceptualized in this way:

> The contributions of environmental education and science education communities to the environmental strand of ESD have been well-documented in the literature; however, equal attention has not been focused on the social and economic strands. Yet, the efforts of schools to create more just, peaceable, and equitable societies suggest that the social strand appears to be well-developed in many countries. In fact, schools that have programs in multicultural education, anti-racist education, gender equity, anti-bullying, and peace education can contribute substantially to the social strand of ESD.[185]

Drawing another lesson from the EE experience, implementation plans asserted that "true education is not indoctrination or inculcation" but requires dialogue and adaptation in each specific contexts.[186]

From this, seven implementation strategies were identified for individual nations to use in crafting an ESD plan.

BOX 5.4 DESD IMPLEMENTATION STRATEGIES

The implementation strategies for Member Nations included:

- vision-building and advocacy,
- consultation and ownership,
- partnership and networks,
- capacity-building and training,
- research and innovation,
- use of Information and Communication Technologies (ICTs), and
- monitoring and evaluation.[187]

A significant aspect of implementation for Member Nations was creating locally relevant and culturally appropriate values for ESD, informed by the principles inherent in sustainable development. Regional Centres of Expertise on ESD were created as an innovative global network of networks that could identify regional challenges, engage regional research, and provide learning spaces for sustainable development. Forty Regional Centres of Expertise were established in Africa, Asia, the Americas, Europe, the Middle East, and the Pacific, with regular reports offered through the new *Journal of Education for Sustainable Development*. This journal was established in 2007 under the leadership of Kartikeya Sarabhai, who has been the ongoing Editor-in-Chief as well as Director for the Centre for Environment Education in India. Fadeeva and Mochizuki illustrate the collaborative linkages and dynamic partnerships that emerged from regional centres of expertise in Figure 5.4.

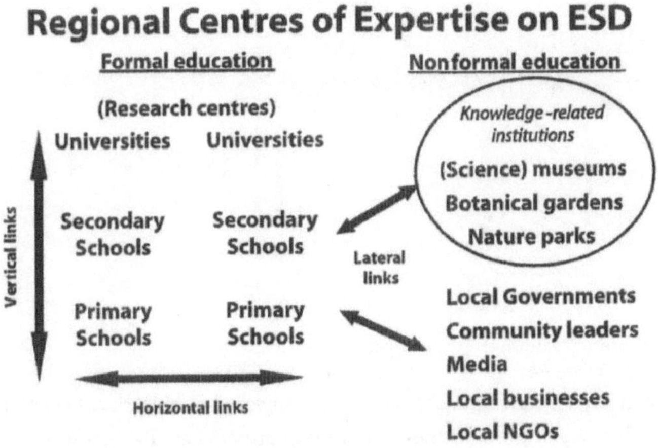

FIGURE 5.4 Regional Centres of Expertise on ESD

Credit: Used with permission of Sage Publications India, from "Regional Centres of Expertise: Innovative Networking for Education for Sustainable Development," Zinaida Fadeeva and Yoko Mochizuki, *Journal of Education for Sustainable Development*, Vol. 1, No. 2, 2007, pp. 229–237; permission conveyed through Copyright Clearance Centre, Inc.

Each Regional Centre helped to build a common regional and subregional agenda for ESD within its unique context of challenges and its regional and national goals for sustainable development. These centres assisted in building regional partnerships and international cooperation as well as assisting nations in developing their own goals, priorities, and actions.

The Neoliberal Turn in Sustainability Discourse

Neoliberalism became more entrenched in the new millennium, shifting UNESCO dialogue. In the Implementation Toolkit, McKeown et al. acknowledged that the neoliberal turn was now evident in the shift of language and tone throughout UNESCO documents. Language no longer necessarily referred to education as a social investment but as an economic investment that could develop "creativity, productivity, and competitiveness."[188] UNESCO discourse cohered with the policies of the World Bank and the International Monetary Fund (IMF).

The Overton Window[189] as seen in Figure 5.5 is useful to understand these ideological tensions, helping explain this shift in the acceptability of some discourses. From the 1980s, multiple right-wing economists and think tanks released economic and social ideas once considered "very extreme," proposing them as logical solutions to the global economic crisis.[190] The notion that governments should *not* be involved in the provision of education was built on reconstructed notions of "public" and "private," attempting to change the relation between government and citizens. Regarding schooling, previously unacceptable ideas such as offering choice between full public-funded and independent private-funded schools slowly gained acceptance.

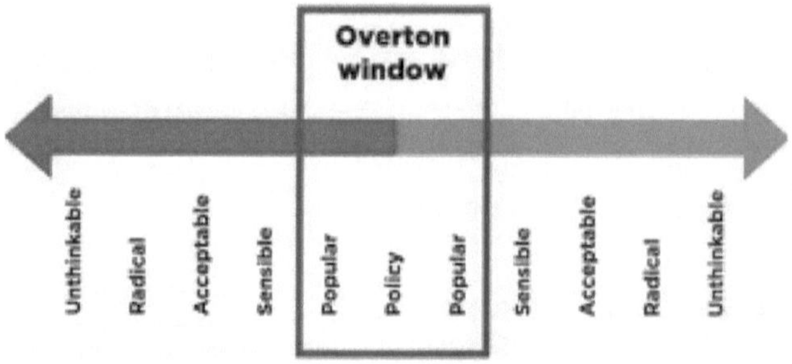

FIGURE 5.5 Overton Window

Image Credit: Toronto Guardian. Reprinted with permission.

This effectively undermined the concept of universal public education by promoting class-based educational programmes, where you are educated according to the tuition you are able to pay. As Jickling and Wals observe, "while the public sector becomes more privatized, the private sector is being reframed as essential for public wellbeing."[191] Consequently, environmental organizations and environmental education programmes were becoming increasingly dependent on private funding, often with negative impacts.

The idea that educational programmes die because they are unsustainable or fail to adapt, such as liberal, democratic citizenship education, was neoliberal logic that deliberately ignored the changes in funding and discourse designed to perpetrate a crisis in public schooling.[192] Further, niche marketing of specialization schools, whether for the arts or sports, was pursued as the foothold into education operating on the market principle. Policies that previously protected public schooling, such as fully public funding arrangements, mandatory school attendance, and educator-generated curricular foci, were losing ground and slowly becoming less acceptable in this era of consumer choice and individual rights over values of the public good.

Professor of Environmental Education Pablo Ángel Meira Cartea explains,

> Just like Penelope's cloth, what is sewn by day by the protagonists of EE is unsewn by the market at all hours through the powerful apparatus of symbolic inculcation that serves to feed the production-consumption cycle, stimulating oversatisfaction in basic needs and creating other needs—above all, in the most subjective terrain, that of desires—accompanied by new products and services to satisfy them, in a never-ending consumerist cycle.[193]

Similarly, David Orr asserts that environmental educators are doing good work, but it is like walking north on a southbound train accelerating in the opposite direction.[194] John Holford has argued that it is time "to reappropriate forms of education created in more democratic times, and for better purposes, but also provide a deeper understanding of how such approaches have been suppressed or driven to the margins."[195] McKeown and her UNESCO team agreed that while education for economic viability is important, "ESD will need to catch the wave of educational reform"[196] to ensure that the vital environmental issues and prospects for sustainability will be addressed on equal footing in education systems and programmes across the globe.

DESD Midterm and Final Reports

The global financial crisis emerged in 2008, early in the Decade of Education for Sustainable Development. While the *Bonn Declaration* of the 2009 World Conference on Education for Sustainable Development declared the need for "values, knowledge, skills, and competencies for sustainable living and participation in society and decent work,"[197] economic instability would overshadow it. The Bonn gathering reiterated that ESD was an integral component of *all* quality education for the 21st century, again connecting to the urgency of the United Nation's Millennium Development Goals (MDGs).[198]

This 2009 World Conference more clearly described ESD as "an approach to teaching and learning based on the ideals and principles that underlie sustainability."[199] It was expected that ESD would address human rights, poverty reduction, sustainable livelihoods, climate change, gender equality, peace and human security, protection of Indigenous cultures, and quality education.[200] Yet, questions were raised about whether ESD was even possible within existing societal and educational structures. Nevertheless, they advocated that ESD be more integrative, interdisciplinary, and holistic.[201]

Once the DESD midterm report was compiled in 2009,[202] it was apparent that trying to find a consensus on the meaning of ESD became less important and more a context for hosting national dialogues. For some, the dialogue appeared to be moving beyond the disciplinary debate between EE and ESD[203] and towards engaging the debate more productively by exploring meanings together[204] while acknowledging how regional contexts shape uptake.

Dutch professor of transformative learning for socio-ecological sustainability, Arjen Wals, found in his 2009 review that there were three kinds of relationships developing between environmental education (EE) and education for sustainable development (ESD),[205] dependent on national context. The three variations were:

- where the environmental education field was *narrowly conceived*, education for sustainable development updated or reformed it;
- where the environmental education field was *broadly conceived*, education for sustainable development was understood as synonymous, and terms were used strategically for funding purposes and initiatives utilizing overlapping infrastructures; and
- where the environmental education field was *relatively undeveloped*, education for sustainable development provided the concept and processes for initiating sustainability education nationally.[206]

There was also more recognition of a spectrum of approaches within ESD—from the more instrumental approach aiming for behaviour change; to a "pedagogical" approach emphasizing social learning; to a more critical approach using social analysis, imaginal thinking, and participatory learning. However, gains were clearly diminished by the devastating consequences of the global financial crisis in 2008–2009, primarily given funding lapses.

Overall, the findings of the midterm report for the Decade found that ESD *was* proliferating, albeit unevenly. Martha Monroe asserted that one priority for ESD research must be strategies for communication and education, which reach and include adult citizens, not just school children, supporting them in problem solving, behaviour change, as well as action taking, especially related to support of legislative mandates.[207] Thus, there was renewed emphasis on lifelong learning, or rather learning across the age spectrum outside of schooling. DESD promoted four nonschool learning sites for ESD activities: vocational training and continuing education; nongovernment organizations (NGOs); social movements and social media; and workplaces, professional associations, and unions. It particularly considered ESD to have an important role within civil society organizations such as NGOs and social movements.

With the growing climate crisis, ESD proponents began to emphasize climate change education, biodiversity issues, and disaster risk reduction, again responsive to regional contexts.[208] Once again, it included a more fulsome call to respect cultural diversity and Indigenous knowledge systems. Finally, this assessment of the first 5 years of the Decade led to a global monitoring and evaluation framework.

In 2012, at Rio+20, it was resolved that sustainable development should continue to be actively integrated into education *beyond* the years of the Decade. Education was considered to be "absolutely central" in achieving sustainable development, alongside technology, political, and financial solutions.[209] In the document *Shaping the Future We Want*, which built upon the Rio+20 document *The Future We Want*, Arjen Wals found ESD becoming a "unifying theme for many types of education that focus on different aspects of sustainability."[210] By 2012, ESD also appeared to be a catalyst for educational innovation, blurring traditional boundaries between educational institutions and nonformal educational agents.[211]

In the 2014 Final Report for the *Decade of Education for Sustainable Development*,[212] ESD was found to be a significant "enabler" for sustainable development. UNESCO had helped activate many international and national strategies which mandated ESD across formal education institutions. It helped establish extensive partnerships and networks globally. Further, UNESCO had found that strong political leadership and multi-stakeholder partnerships were instrumental in institutionalizing Education for Sustainable Development. In sum, there was now a strengthening global agenda for ESD and a foundation for scalability. Community engagement with ESD was growing and pedagogical innovation was advancing across all levels and areas of education, towards the central goal of learning to live and work sustainably. Yet, short-term electoral mandates continued to make progress difficult. With constantly shifting national commitments, the practice and funding of ESD lagged far behind policy commitments.

Post-Decade ESD and the SDGs 2015–2030: Transforming Education

Transformative Approaches

In 2013, a *Global Action Programme* for ESD was announced by the UNESCO General Conference, to be undertaken post-decade from 2015 to 2019. This was called the *Aichi-Nagoya Declaration*,[213] which desired to build upon the successes of the DESD, including improved conceptual understandings of ESD, the body of good practice that was developing, and the high-profile agenda for ESD in support of the new sustainable development goals (SDGs).

The General Assembly reaffirmed ESD as a critical and integral element of sustainable development, with ESD was a key enabler of all the other Sustainable Development Goals.[214] Thus, the new 2030 Agenda entitled *Transforming our World* represented the merger of the Millennium Development Goals (MDGs) with Education for Sustainable Development (ESD). The Sustainable Development Goals or SDGs for 2015–2030 were a blueprint to achieve a better and more sustainable future for all.[215] Figure 5.6 details the SDGs.

Goal 4.7 of the SDGs was dedicated to quality education, stipulating that inclusive and equal quality education as well as lifelong learning opportunities for all must:

> ensure that all learners acquire the knowledge and skills needed to promote sustainable development, including, among others, through education for sustainable development and sustainable lifestyles, human rights, gender equality, promotion of a culture of peace and non-violence, global citizenship, and appreciation of cultural diversity and of culture's contribution to sustainable development.[216]

The 17 indivisible goals are a plan of action for people, planet, and prosperity within a transformational vision meant to be far-reaching and historic.

FIGURE 5.6 UN Sustainable Development Goals (SDGs)

Credit: United Nations Sustainable Development Goals https://www.un.org/sustainabledevelopment/ Reproduced with permission. The content of this publication has not been approved by the United Nations and does not reflect the views of the United Nations or its officials or Member States.

For Stefania Giannini, Assistant Director-General of Education for UNESCO, this was a significant conceptual turn in ESD. Increasingly the questions became: is what people learn truly relevant to their lives? Is what they learn helping to ensure the survival of our planet? Finnish education scholar Maria Hofman advocated shifting the focus from "what teachers are supposed to teach within ESD [to] . . . what is an ESD supposed to achieve?"[217] In other words, "knowing how to act is the base for all ESD."[218] The SDGs are a "concerted supreme effort to change our historical course."[219] Sterling agrees that integrated perspectives and radical changes *are* needed in the education system, especially rethinking the "purposes and values that underpin education,"[220] if sustainable development is to be achieved.[221]

The UNESCO goal for ESD became that of providing the knowledge, awareness, and action that empower people to transform themselves and transform societies.[222] This was an explicit turn reorienting education for sustainable development towards an action-oriented and transformative pedagogy, a major shift in thinking about ESD. Kartikeya Sarabhai asserts that not only the paradigm of development must change, but so also must the paradigm of education.[223] One of the UNESCO Chairs, Arjen Wals, called this transformation "gestaltswitching," derived from the German concept of *gestalt*.[224] Through "transformative disruptions" in everyday experiences, school students and especially higher education students can become aware of their assumptions and learn to "switch back and forth between different mindsets."[225] In addition, he describes that the greater the plurality of backgrounds students bring to classroom dialogue, the greater the creativity of thinking. He advocates that "education for sustainability above all means the creation of space for transformative social learning."[226]

Competences for Sustainable Development

A second part of the shift in ESD thinking was the focus on cross-cutting competencies. As Mochizuki and Fadeeva argue, the competence-based approach emerged, given the perceived lack of relevance and "need to produce change agents"[227] for positive societal change. They distinguish "competences" from "competencies," where the latter is rooted in narrow vocational and workplace performance approaches within a neoliberal ideology. Rather, they say, SD competences "are conceptualized as alternative to competencies for 'modernization.'"[228] These alternatives include sustainable livelihoods, Traditional knowledges, interdependence worldviews, spiritual and perceptual awakening, and a deconstruction of consumerism within whole person learning. They define competences as intended learning outcomes "written as verbs, as the ability to do something."[229] Leaning on the 1996 Delors Report, they identify personal competences, social competences, knowledge domain competences, and methodological or action competences.

German ESD scholar, Marco Rieckmann, identified numerous "competency" frameworks, compiling them into one approach. From this integration, he determined eight core competencies for ESD:

- critical thinking,
- systems thinking,
- anticipatory competency (ability to evaluate multiple futures),
- normative competency (reflection on norms and values)
- strategic competency (implementing innovative actions),
- collaborative decision-making,
- self-awareness competency, and
- integrated problem solving.[230]

As Rieckmann clarifies, "while competencies describe the capacity or disposition to act to address complex challenges, they do not necessarily imply that an individual will act in a certain way in a specific situation."[231] Thus, he argues that "sustainability performance depends on the interplay of knowledge and skills, values and motivational drivers, and opportunities" which together influence personal behaviour, as demonstrated in Figure 5.7 by Rieckmann.

In 2022, the European Commission completed what they call the GreenComp or the European Sustainability Competence Framework,[232] which responds to the

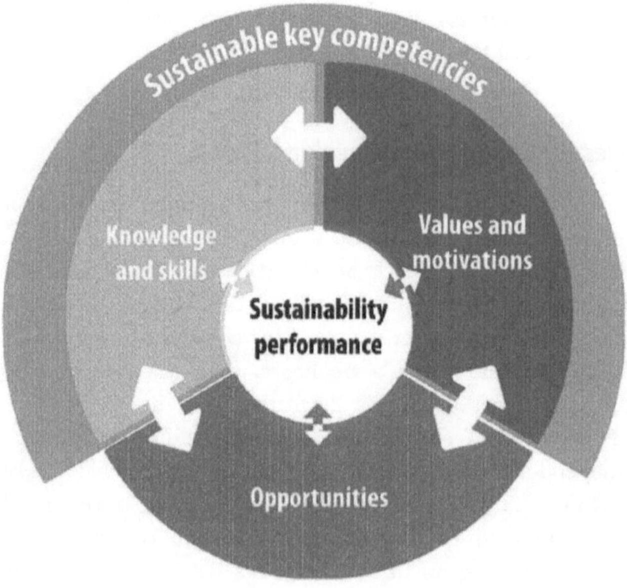

FIGURE 5.7 Key Competencies and Performance of Sustainability Citizens

Credit: Learning to Transform the World: Key Competencies in Education for Sustainable Development by Marco Rieckmann, 2018, p. 46. Reprinted by permission of UNESCO.

European Green Deal. The goal was to determine "what sustainability as a competence entails."[233] They identified 12 competences.

BOX 5.5 GREENCOMP—12 SUSTAINABILITY COMPETENCES

Embodying sustainability values includes the competences of:

- valuing sustainability
- supporting fairness
- promoting nature

Embracing complexity in sustainability includes the competences of:

- systems thinking
- critical thinking
- problem framing

Envisioning sustainable futures includes the competences of:

- futures literacy
- adaptability
- exploratory thinking

Acting for sustainability includes the competences of:

- political agency
- collective action
- individual initiative

They consider these competences as being relevant to the whole spectrum of lifelong learning, with the ability to foster systemic and critical thinking as well as agency for action. One example is The International Handprint Network begun in 2016 which, inspired by the ecological footprint concept, developed the Carbon Handprint and the Environmental Handprint. They encourage children and youth to examine the impact of their positive sustainability actions undertaken over 21 Global Action Days.[234]

Whole School/Whole Institution Approaches

A third part of the conceptual shift in ESD thinking was based on the finding that "whole school" and "whole institution" approaches were most effective, where "the institution itself functions as a role model for the learners."[235] Four aspects of an institution can be reoriented with the goal of creating an experiential place for learning sustainability, including: governance and policy, facilities and school operations, curriculum/teaching/learning, and community partnerships and relationships.

New pedagogical approaches for ESD include learner-centred approaches that build from existing individual and especially institutional strengths. It was advocated

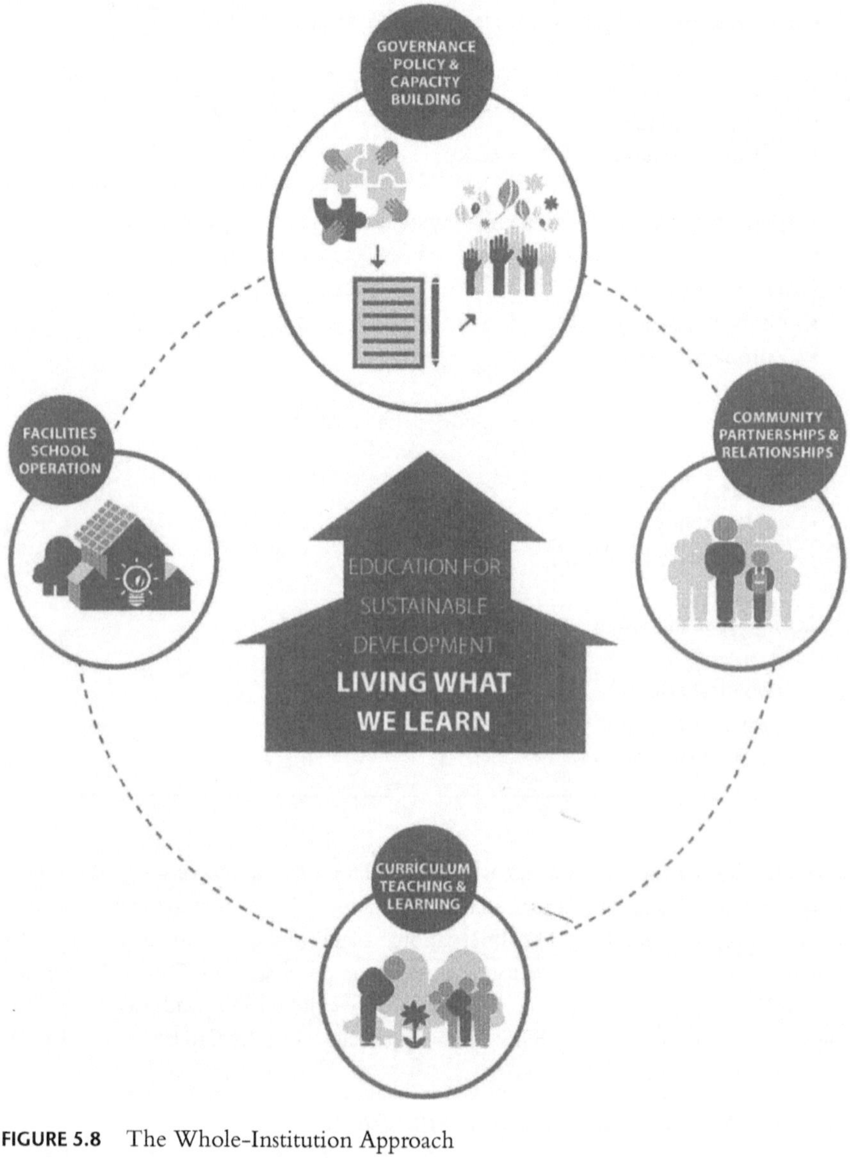

FIGURE 5.8 The Whole-Institution Approach

Credit: Learning to Transform the World: Key Competencies in Education for Sustainable Development by Marco Rieckmann, 2018, p. 47. Reprinted by permission of UNESCO.

that a more fulsome sustainability worldview would result from a constructivist approach to knowledge, an action-oriented learning approach linking knowledge and learning with innovation and action, and, finally, a transformative approach that questions current assumptions.[236] Green Schools and EcoSchools are just two global examples which structure an educational institution holistically around sustainability.[237] Many national initiatives exist as well, such as the Canadian Sustainable Schools programme and Australian Sustainability Schools.[238]

Critique of ESD

ESD as Faux/Weak/Technocentric Sustainability

Similar to the first two decades of intense effort in environmental education, the UN Decade for Education for Sustainable Development too fell short of expectations. Critiques arose from various quarters—some conceptual, some theoretical and philosophical, some implementation-related, and some pedagogical.

Conceptual critiques of sustainable development, detailed in Chapter 3, were reiterated by educators. The primary question was "How could current economic principles be used for rapid economic growth to address poverty in 'developing' countries, while at the same time greening the growth of 'developed' countries ostensibly protecting affluent lifestyles?" Sustainable development was considered an "oxymoron" in that development itself is inherently unsustainable.[239] Continued "development" was seen to give tacit approval to growth, affluence, and consumption, however moderated. As well, sustainable development largely appeared to focus on the power of market and technological solutions, such as providing substitutions for disappearing natural "resources." In this way, it was argued that humanity never faces the excesses of the current economic model and its ecological impacts.

As global educator David Selby explains, "the coming together of the environmental and development lobbies around the flag of [sustainable development]" was questionable as it seemed to imply that the West's middle class wanted to sustain their comfort level and that they could do so without destroying the " 'resource base' that is needed to do so."[240] Many concluded that too much compromise had evidently occurred between the environmental and development lobbies, leading to unworkable premises.[241] In response to this critique, UNESCO documents began to integrate sustainable consumption and lifestyles as part of their programming.

From a theoretical angle, the Brundtland Report was seen as using conservationist and utilitarian logic, thereby privileging human needs. In the same vein, ESD was seen as perpetuating anthropocentrism which continues to threaten the carrying capacity of the Earth (defined as the level at which the total population multiplied by the resource-use level will exceed the ecosystem's capacity to renew itself). In ESD, the natural world seemed to only be considered in terms of human welfare, leaving the existing economic model untouched.[242]

Further, ESD often neglected to build relationships between learners and the natural world. Despite worthy intentions, ESD was seen to replicate the underlying patterns creating the crises. Even plural approaches to ESD, Helen Kopnina suggests, still run the risk of maintaining an "anthropocentric paradigm."[243] Rolf Jucker believes that ESD cannot be mainstreamed as it is a modernist ideology with principles, values, and understandings of education diametrically opposed to sustainability, particularly the moral mandate.[244] Jucker proposed that learning for sustainability (LfS) needs to meet real needs in communities, utilize what we know about effective learning particularly face-to-face interaction, and move to communal learning in real time in a real place.

British education philosopher Michael Bonnett purported the motive of sustainable development policy as an attempt to escape the scarcity associated with a subsistence economy; yet, that may be the most sustainable way of living.[245] He further suggests that the

> [A]ttitude of mind that sanctions the injustice of exploitation and oppression, whether it be towards other humans or nature, is essentially the same and is thus the common enemy of developmentalism and environmentalism, . . . properly understood, [they] can be mutually supporting.[246]

However, this will require a "radical transformation" in which "sustainability [is] conceived not as a policy but as a *frame of mind*"[247] that shapes a certain way of being in the world. In his estimation, this revolves around the question, "what constitutes a right relationship with nature . . . [meaning] the self arising in the material/spiritual world of which we are a part?"[248] The current Western worldview precludes such a frame of mind. He describes that "transformation needs to be understood as a retrieval and regeneration . . . and reconnection . . . rather than some external import or imposition,"[249] which is an instrumental conception he sees evident in current ESD formulations.

From the preservationist view, environmental protection is not compatible with development or resource substitutability. There are, in fact, nonreplaceable aspects and functions of "critical natural capital."[250] Only by proscribing limits and restrictions on human behaviour can ecosystem health be prioritized. The precautionary principle has been advocated by many so that risk-laden activities are not undertaken until consequences can be fully identified.

Another theoretical critique has centred on development theory, which itself had reached an impasse.[251] Southern theorists argued that development *creates* inequality, cultural loss, environmental degradation, and economic dependency.[252] Development has long been suspected as a form of neocolonialism.[253] From the impasse, development theory began diverging into alternative discourses such as postdevelopment theory, Indigenous and postcolonial discourses, Southern theory, degrowth and end of growth ideas, as well as civilization collapse theories.[254]

From the implementation angle, sustainable development was seen as a top down endeavour. Such prescriptive constructions run the risk of homogenizing rather than diversifying approaches.[255] As Orr suggests, in such a technocratic approach, global-scale management occurs through a centralized international elite group of experts and promoters,[256] which can treat educators as technicians rather than participatory creators. Thus, William Scott argued for the liberty of educators not to do the bidding of global policy but to enact their freedom to deliberate and make choices about sustainability education in their own contexts. He described this as walking a tightrope between not neglecting the pressing realities and not engaging indoctrination, seeing it rather as a stimulating of endeavour rather than adherence to prescription.[257]

Some charge that ESD stays within an "instrumental rationality" in promoting "knowledge acquisition" approaches. Even when taking a participatory approach, Sauvé says, standard problem-solving models tend to be primarily reactive.[258] In determining how to evaluate ESD programmes, Kopnina argued that ESD does not have "clear aims and measurable objectives." Accountability tends to be towards the " 'big players,' including The World Bank, the IMF, and governments of the neoliberal consumerist societies."[259] Pablo Ángel Meira Cartea suggested that environmental educators do not need to accept the UN discourse as quasi natural or the institutional strength and resources behind ESD.[260] For him, ESD is not the evolutionary continuation of EE, and it does not necessarily offer new responses to the environmental crisis. Regula Kyburz-Graber charges that the critical discourses around ESD have "increasingly vanished" as the competence and measurement orientations have prevailed, particularly in research discourse.[261] She advocates for more emphasis on critically based real-life learning.

Violeta Orlović Lovren and Katarina Popović, in Walter Leal Filho's 2018 *Handbook of Lifelong Learning for Sustainable Development*, identify the loss of adult education within the discourse of lifelong learning for sustainable development. They assert that the discourse of lifelong learning has eclipsed the field of adult education despite "overwhelming evidence of its transformative power."[262] While the typical narrative is that lifelong learning replaced the term "adult education," they argue that adults are neglected in UNESCO documents except for concerns about the literacy of youth and adults or vocational learning and training for employability, illustrating the prevailing economic paradigm. Yet, it was clear from the MDGs and DESD that adult education, especially community-based adult education, has significant power to achieve the SDGs. They argue for re-establishing the importance of adult education within lifelong learning to recognize it as a specific space for sustainability education and needed programme funding.

Adult Learning and Education or ALE, as the new terminology, has been accepted within the lifelong learning discourse. In association with PASCAL (Place and Social Capital and Learning),[263] a new organization PIMA (Promoting, Interrogating, Mobilizing Adult Learning and Education), located in Australia, has been established as a network of adult and lifelong learning educators, activists, and scholars who are dedicated to resolving pressing issues, including climate change education (discussed later), while encouraging social, economic and ecological justice.[264]

Summing up some of these critiques, Canadian environmental educator scholar Lucie Sauvé et al. concluded that ESD was generally part of an instrumental view of education, a resourcist concept of the environment, and an economist view of development.[265] Combined, these perspectives are now considered *faux, weak, or technocentric sustainability*.[266]

Charges that weak sustainability was short on a critical and ethical pedagogy have resulted in calls for deeper change that could lead more firmly to societal transformation.[267] How education is approached has been considered vital. As Lucie Sauvé asserted, "Education is and must remain a space of liberty, a space where we can and need to critically explore the many dimensions of 'being humans on Earth' or, stated differently, "what it means *to be*, environmentally."[268]

Other Approaches to Sustainability Education

As previous Director of the Paulo Freire Institute, Brazilian Moacir Gadotti contends, "the concept of sustainability is complex and it goes far beyond sustainable development."[269] He cites Brandão who explains,

> [S]ustainability means a new egalitarian way, a free, fair, inclusive, and solidarity way to get people together in order to build their social living world at the same time that they handle, manage or transform the natural sustainable environments where they live and on which they depend to live and be together.[270]

This is not just an "improvement" in the current educational model but implies a qualitative different structure and process as well.[271] Exemplifying this is Gadotti's Earth Pedagogy or ecopedagogy, which was reviewed in Chapter 4. Glimpses into other approaches to sustainability education are provided in later chapters. While far from comprehensive, these intriguing glimpses can reveal specific strands of thinking, approaches to sustainability education theory and pedagogy, and sites of practice.

Sustainability in Higher Education

Since the 1970s and especially the 1980s, the green or sustainable campus movement gained a significant momentum. According to Tony Alabaster and Derek Blair, the first wave was comprised of new environmental courses in higher education in response to public demands, growing into full programmes and degrees and later workforce requirements. While the green campus movement started as the integration of environmental studies into higher education, it then expanded to assessing the ecological implications of institutional operations and the need for cultural change within institutions.[272]

Alabaster and Blair consider that "the ultimate aim of greening [is] to inculcate a natural 'culture' of environmental responsibility in FHE [further and higher education] institutions." In the sustainable campus movement, campuses are considered a micro society of sustainability and can serve as exemplary leaders as well as act as "a critical leverage point for change."[273] Additional facets of campus sustainability now include healthy and safety, physical and mental well-being, student and university governance, physical and financial accessibility, recreation, as well as cultural and social diversity.

Universities have any number of entry points into sustainability practice, including: change of existing curricula, new courses, new programmes and degrees, student clubs, new system-wide initiatives and policies, building construction or renovation, purchasing policies, waste and recycling, energy and water consumption, transportation, food, landscaping and stewardship of the natural environment on campus, education and outreach, health and safety, sustainability learning communities, financial priorities and accounting practices, and research including inter- and transdisciplinary research. As many case studies have identified, significant change occurs when there is a convergence of initiatives.

One of the founders of Second Nature, Anthony Cortese notes the responsibility that universities have for leading society into a sustainable future as well as for modelling sustainability.[274] One of the early achievements in higher education, was the signing of numerous declarations and agreements, including:

- the *Magna Carta of European Universities* (1988),[275]
- the *Tallories Declaration* (1990) from the Association of University Leaders for a Sustainable Future who also established the Campus Ecology Program (1989) with the National Wildlife Federation,[276]
- the *Halifax Declaration* (1991) from the International Universities Bureau which formalized Higher Education and Research for Sustainable Development (HESD),[277]
- the *University Charter for Sustainable Development* (1994) from COPERNICUS (CO-operation Programme in Europe for Research on Nature and Industry through Coordinated University Studies),[278] and the more recent
- *American College & University Presidents' Climate Commitment* (2006) which addresses the move to zero carbon and resilience practices.[279]

Although such broad declarations do not necessarily catalyse change, they do assist in networking and agenda setting. Student organizations also signed related declarations or joined networks such as the Sierra Youth Coalition's Sustainable Campuses Network.[280] In essence, a movement within a movement developed, the sustainable campus movement within the environmental movement, where student activists often took the lead in pressuring and educating for change.[281]

By the 1990s, there was a push for "cross-curricular greening" so that all higher education students acquired an environmental education to play a role as an "environmentally literate workforce and citizenry."[282] New associations quickly developed including the Association for the Advancement of Sustainability in Higher Education (AASHE, 2005),[283] Second Nature (1993) which is a nonprofit committed to sustainability in higher education,[284] and the Global Universities Partnership on Environment and Sustainability (GUPES, 2010) fostered by UNEP.[285] Other supports came from new journals, such as the *International Journal of Sustainability in Higher Education* (2000) edited by Walter Leal Filho and published in Germany, and the *International Journal of Sustainability in Teacher Education* (2007), which responds to the UNESCO Guidelines for reorienting teacher education to address sustainability. In Filho's *Encyclopedia of Sustainability in Higher Education*, Elizabeth Potter-Nelson and Joy Kcenia Polanco O'Neil describe the reorientation of teachers towards ESD and more broadly sustainability literacy, including the role of nonprofit agencies in this task, such as the Cloud Institute and Center for Ecoliteracy.[286]

Another early endeavour was developing sustainability mission statements and policies in higher education then undertaking a baseline audit with the aim of working towards institutionalizing a campus sustainability assessment and tracking system. For instance, the Sierra Club developed a sustainability audit system called the Campus Sustainability Assessment Framework, which provided indicators and benchmarks for multiple facets of university operations and life while promoting integrated environmental, social, and economic benefits. The Sustainability Tracking and Assessment and Tracking System (STARS) offered by AASHE is another which offers an independent evaluation of campus activities and produces a comparative College Sustainability Report Card.[287]

It has been demonstrated that the strongest elements in successful programmes have been capturing broad engagement across a campus, wide participation in initiatives, and building a sense of community and commitment around sustainability initiatives. According to Peggy Bartlett and Geoffrey Chase, the most successful changes have occurred when they involve both top down (senior administration) and bottom up (student, faculty, and staff) initiatives working in concert. Typically, the momentum is built by change agents or "champions" who work across levels promoting sustainability through their relationships, particularly to tap latent support. Thus, a diversity of approaches, systemic changes, and broad participation builds support and action and diffuses change across a campus.

The common barrier has been that universities are "highly resistant to change,"[288] because they are:

- organized along disciplinary boundaries with narrow specializations,
- utilize siloed knowledges creating difficulty for dialogue around curricular integration and dialogue,[289]
- lack systemic integration,[290]
- hold common misconceptions around the concept of sustainability,[291]
- lack interdisciplinary collaboration and capacity for effective advocacy,[292]
- feel financial pressures from "liaisons with corporations and government agencies,"
- experience the pressures of multiple stakeholders,
- experience cultural barriers to uptake,[293] and
- lack connectedness with the surrounding ecosystem.[294]

One practice to address barriers has been developing a centre that wholistically exemplifies sustainability, such as the EcoCenter at Griffith University (Australia) with John Fien as Director,[295] the Adam Joseph Lewis Center at Oberlin College (USA) with David Orr as Director emerging as a learning organization,[296] or The Genesis Project at Somerset College of the Arts and Technology (UK) which is an example of reschooling society, says John Blewitt.[297]

Overall, the assessment of Stephen Sterling, Larch Maxey, and Heather Luna in 2012 was that "the global higher education system as a whole remains maladapted to the conditions that we face."[298] Stephen Gough and William Scott examine the purposes of universities in particular, asserting that universities have an essential role in manifesting sustainability, but the challenge lies in finding a place for that role, an appropriate conception of sustainability, while maintaining a critical distance from this mission, all problematic.[299]

Newer directions in sustainability in higher education now include the Engaged University and Socially Responsible University,[300] the turn towards transdisciplinarity, and the overall transformation of higher education.[301]

Sustainability Literacy

In 2009, Arran Stibbe edited *The Handbook of Sustainability Literacy*.[302] For him, knowledge *about* sustainability is not enough, as it requires skills, attitudes, competencies, dispositions, and values that are/will be necessary for "surviving and thriving in the declining conditions of the world."[303] In this broad context, "literacy" refers to "the collection of skills that allow for effective participation and influence in diverse areas of social life."[304] It is also a resistance to 21st-century skills as denoted by corporate leaders and economic think tanks.

Preparing sustainability-literate teachers, particularly preservice teachers, is vital. Victor Nolet offers a framework that "represents a new paradigm for the preparation of teachers."[305] He identifies nine themes which offer a multidimensional construct of sustainability literacy, and he examines various field settings for teaching sustainability literacy.

In *Education for Sustainability*, Paul Clark claims that our "education is not fit for purpose if the purpose is sustainable living."[306] Our schools remain solidly of the industrial model rather than schools of sustainability. He explores "how to live, how to be, to think, to do" as a "reculturing for the time we live in."[307] To do this, particularly in urban places, he suggests that our guide for this process must be the natural world, which he calls becoming "Naturally Smart." He proposes a new social design for schools within their communities, including learning far more intensely from local ecosystems. Community food-growing projects and forest schools are examples of establishing multiple linkages within the local community. In this way, schools can be

> transformed into playing new roles within their communities, as educational facilities, libraries, resource centers, centers of excellence, food hubs,

> horticulture centers, arts events venues for conferences and public debates, and business and cooperative developments of all kinds.[308]

A school of sustainability can be at the heart of a sustainable learning community, as "a hub for connection of the urban and the rural, a connection of the mind-scape with the landscape."[309] Engaging sustainability in this way can foster a spontaneous leap, or emergence, into a new level of social organization that manifests sustainability.

Personal Sustainability/Sustainable Livelihoods

Through a workshop dialogue with David Orr, Paul Murray began asking the question, "why are we not changing as people?"[310] as part of embracing sustainability. He began to see sustainability less as a professional issue and more of a personal one for adults. His book *The Sustainable Self* tackles six facets of the sustainable self, as depicted in Figure 5.9.

He identifies the attributes which comprise sustainable living as: *awareness* in understanding the need for change, the complexity of sustainability, and that we as individuals matter; *motivation* as a deep intention to align our core values with behaviour that lends itself to sustainability; *empowerment* where we override our internal barriers to change including self-limiting beliefs; *knowledge* which includes

FIGURE 5.9 The Sustainable Self

Credit: Used with permission of Routledge. From *The Sustainable Self: A Personal Approach to Sustainability Education*, Paul Murray, 2011; permission conveyed through Copyright Clearance Center, Inc.

baseline understandings of sustainability principles and core subject matter; *skilful means* which includes various thinking and interpersonal competencies; and *practice* which integrates values, intentions, beliefs, knowledge, and skills towards positive ends. His sustainability training programme is aimed at postsecondary students, industry professionals, and the general public who are interested in sustainability but lack confidence and fluency.

Murray's work connects to my own work in offering a university extension course on sustainability for adults experiencing dislocation in the economy. My approach worked in the reverse from Murray's by starting with the issues that were most pressing for individuals which they brought into the learning space, such as job loss or high stress lives. Through multiple arts-based approaches, adults represented the daily pressures they were under. The course slowly introduced how sustainability as a concept and life practice might help transform their lives to be more life-giving for themselves and their families as well as enable them to find places of work or create new work that more closely aligned with their deepening goals and values.[311] Visits to exemplary sustainability leaders in the community, an examination of their values through their monetary and life energy expenditures as well as carrying out community observation and communicative research with significant others, led to profound transformative moments and community action projects with sustainability at the heart.[312]

John Blewitt, a lifelong and sustainability education scholar as well as Distinguished Fellow of the Schumacher Institute, described his approach to sustainability for everyday life in *The Ecology of Learning*.[313] Taking up the challenge of sustainability education in all possible learning spaces—formal, nonformal, informal—and drawing from transformative learning scholarship, he claims "All learning really becomes meaningful when there is some resonance with the everyday lifeworld of the learner."[314] He examines potential learning from leisure activities, neighbourhood activity, community action projects, and the media.

One last example is Heather Burns, at the University of Portland, who designed her graduate courses using ecological principles. She calls this the Burns model of sustainability pedagogy that comprises five dimensions: content (thematic approach to sustainability issues), perspectives (perspective taking, critical perspectives), process (experiential, active participation), context (natural and human environment), and design (evoking a transformative learning experience).[315] Pedagogically, it importantly includes fostering personal qualities that manifest sustainability, such as mindfulness for self-compassion, kindness, and presencing.[316]

Strong/Systems-Based/Ecocentric Sustainability Education

Growing numbers of educators have rejected the term sustainable development and ESD. It has been argued that "sustainability" without the word "development" is much more attuned to ecological sustainability.[317] The term sustainability has also been regarded as better suited to addressing bioequity among species[318] as well as the needed ethical improvement of humanity. It challenges anthropomorphism by bringing human existence into a lived balance with the rest of the natural world.[319]

The term sustainability also appeared to imply that Southern nations could develop a green trajectory rather than follow the old style industrialization. This aligned with critical development theorists such as Martin Carnoy and dependency theorists Manuel Castells and Fernando Cardoso, who recommended that developing nations do not need to adopt the hierarchical, industrial model. They argued that the industrial model was now inappropriate to the "informatization" of the world economy, as the new economic revolution.[320] Developing nations would do best, they said, to shift to flexible, nimble information-processing activities rather than production of material goods. In such "a [new] multipolar system of economic power," the concept of a "Third World" disappears.[321]

In Sterling's typology of education *about*, *for*, and *as* sustainability, it is education *about* sustainability (or sustainable development) that "sustains unsustainability."[322] Victor Nolet has agreed that "education is both the cause of unsustainability and the strategy for achieving sustainability."[323] Education *about* sustainability remains content focussed. It relies on first-order change in Bateson's typology, which is simply adaptive or accommodatory learning, maintaining the foundations of the current rapacious system. Basic values and assumptions remain unchanged. Specialized knowledge silos in higher education are able to persist and thus do not respond to the complexity of social and environmental issues. Therefore, education *about* sustainability is weak sustainability.

As Sterling suggests, education *for* sustainability "raises questions of philosophy and value about the nature of education . . . and beyond this, about the nature of being human."[324] So while there is an instrumental aspect in education towards some predetermined changes, it also focusses on purpose—in terms of what education is for. Further, it profiles the quality of learning—in terms of what education is.[325]

Sterling asserted that "sustainability education" shifts the foci and purpose of education towards "nurturing and realizing inherent [individual] potential" as well as "building a sustainable society."[326] In these ways, education *for* sustainability can be considered strong sustainability, implying a paradigm shift in several ways. First, it is second-order change which is reflective learning about assumptions and perspectives. Second, it is reformist in learning for social change, moving beyond the assumptions of an outdated worldview.

The strong sustainability perspective critiques root assumptions and prevailing myths, such as unlimited growth.[327] For instance, "sustainable development" still assumes "history as an elevator"[328] with no end, inspired by the Judeo-Christian linear cosmology. As Selby asserts, ESD is built upon the normalized view of human "separation from and dominance over nature."[329] ESD perspectives often assume an objective, rational empiricism or rather an attempt to stand outside reality from an Archimedean viewpoint.[330] Yet, humans realistically do not have the capacity to coordinate activities at a global scale while foreseeing multitudes of interacting factors.

As Victor Nolet[331] suggests, a "sustainability worldview" can be seen as a constellation of commitments: ecocentric ethics, life-affirming values, technical skills to create solutions or think across disciplines, and a sense of agency to make change both individually and collectively.[332] Some theorists, however, go beyond strong sustainability to problematize the entire framework of Western epistemology.[333] Strong sustainability takes steps in a good direction, but it can go deeper.

Deep/Strongest/Transformative Sustainability Education

Stephen Sterling and many others claim that humans require "relearning on a grand scale . . . necessitating a metamorphosis of our many current education and learning constructs."[334] He presents a third approach, that of education *as* sustainability or third-order change. In his view, education *as* sustainability is transformative in that it encourages the exploration of "epistemic change as a collaborative inquiry."[335] Transformative learning addresses epistemic change by questioning the dominant system of knowing. Tobias Schnitzler also identifies this recent interest in bridging transformative learning and sustainability education, drawing attention to building collaborative atmospheres and learning spaces, as a "core element of social transformation."[336]

In the case of deep or transformative sustainability education, "knowing is seen as approximate, relational and provisional."[337] Learning is continual and bound up in practice. In other words, education *as* sustainability is learning *as change*, engaging the "whole person and the whole learning institution."[338] Wendy Griswold sees a close alignment between radical or strong/deep sustainability education and ecojustice adult education, with many exemplars in the Spring 2017 issue of *New Directions for Adult and Continuing Education*.[339]

Christine Wamsler, at the Lund University for Sustainability Studies in Sweden, also calls for transformative learning towards sustainability.[340] She asserts that inner transformation has been vastly neglected in sustainability education and research, which should be engaged alongside ecosystem study and holistic pedagogies.[341] She examines the socioemotional dimensions of whole person learning, particularly mindfulness in building compassion, increasing personal resilience in times of climate change, and developing more worldview-sensitive approaches to sustainability.[342] She advocates contemplative approaches to assist in "walking the talk" of sustainability.[343]

In a special issue on Transformative Sustainability Education in the *Journal of Transformative Education* edited by Elizabeth A. Lange and Joy Kcenia Polanco O'Neil, scholars Jonathan Dawson, Stephen Sterling, and Paul Warwick identify deep sustainability as best pursued outside of regular institutions. The philosophy of knowing in deep, transformative sustainability conflicts with most existing structures of education. They discuss the exemplar of Schumacher College, a private "ecological college" which straddles the public/private divide by its association with a fully accredited public institution, Plymouth College.[344] It exists on the "creative edge," providing "radically democratic, enquiry-based, experiential education"[345] through a wholistic pedagogy. They conclude that the college is implementing a change in paradigm that incorporates an intentional change in purpose throughout the institution. Such a change in purpose "can be a harbinger of a deeper cultural shift, especially when aligned with and connected to growing progressive and reconstructive movements in civil society."[346]

Yona Sipos, Bryce Battisti, and Kurt Grimm at the University of British Columbia also call for the linking of sustainability education with transformative learning.[347] They describe a holistic approach as engaging the head (cognitive), hands (psychomotor), and the heart (affective domains), as an organizational principle.[348] While recognizing the call for a "complete overhaul of education," they offer learning objectives and a framework that can be integrated across existing curricula, which "exhibit an emergent property they call transformative sustainability learning (TSL)."[349] As demonstrated in Figure 5.10, the head represents engagement, hands the enactment, and the heart as enablement. Pedagogically, they blend action learning, community service learning, critical emancipatory pedagogy, environmental education, participatory action research, ecojustice pedagogy, problem-based learning, and Traditional Ecological Knowledge (TEK).

In much more detail, Satish Kumar with Pavel Cenkl describe the head, heart, and hands transformative learning of Schumacher College in their 2021 book. They describe their intentions to create a place outside of the "mainstream educational

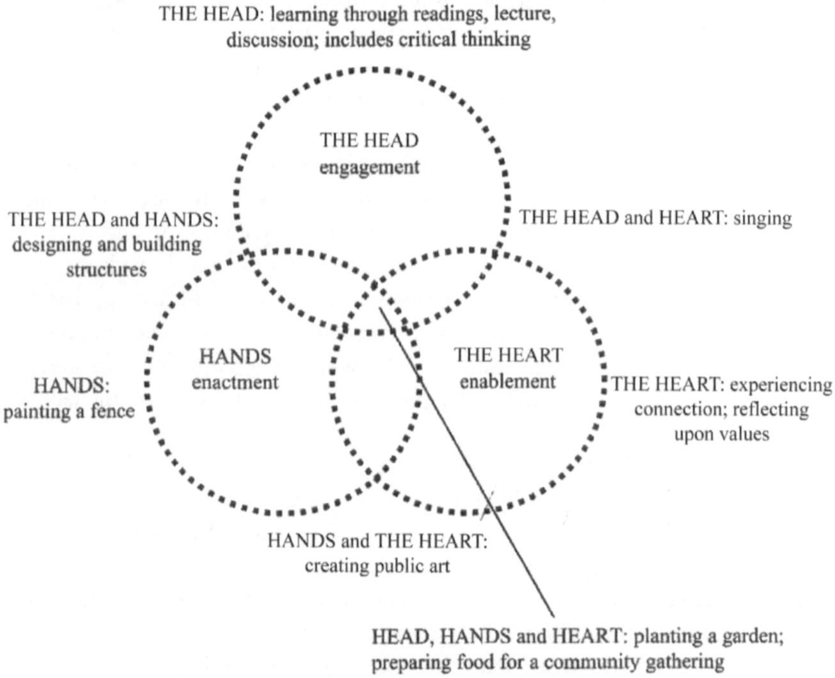

FIGURE 5.10 The Head, Heart, and Hands as TSL Pedagogy

Credit: Used with permission of Emerald Group Publishing Limited. From "Achieving Transformative Sustainability Learning: Engaging Head, Hands, and Heart," Yona Sipos, Bryce Battisti and Kurt Grimm, *International Journal of Sustainability in Higher Education*, Volume 9, No. 1, 2000, p. 75; permission conveyed through Copyright Clearance Center, Inc.

system . . . dominated by economic and materialist values" that can build upon "Nature [as] the source of life itself . . . [in which] humans are an integral part of Nature."[350]

This echoes my own findings when integrating sustainability education into a continuing education programme.[351] A programme that is accredited but not part of a degree programme offers far more freedom for a holistic pedagogy, while the linkage with an existing institution provides legitimacy and accreditation. The conditions necessary for transformative learning are easier to establish, from extensive field trips and retreats as well as experiential and participatory learning, not only towards an epistemological transformation but also an ontological transformation.[352] Furthermore, links to actors in social movements can more easily build motivational drivers, agency, and activism well beyond the timeframe of educational engagement.[353]

Transformative Sustainability Education scholar-practitioner Joy Kcenia Polanco O'Neil[354] agrees that an ontological shift in "being" is part of (be)coming sustainability. In O'Neil's view, ecological sustainability is epistemology-centric. In other words, ecological thinking only involves instrumental knowledge or a transformation in the way of knowing. Also needed is an ontological transformation (in the way of being) where one is *"(be)coming* sustainability."[355] She calls it "transformative learning *as* sustainability [given it] focuses on the process and dynamics of learning as change, towards living and learning ecologically sustainable lives."[356] Thus, there is a transformation in how we know, in what we know, as well as our way of being or rather *(be)coming* in the world. In this way, sustainability education is process-based.

O'Neil further suggests that strong or ecological sustainability still assumes reality as comprising separate entities, connected by interaction. Bringing the work of physicist Karen Barad and posthumanist agential realism into Transformative Sustainability Education, she explains that perceptions must be shifted from entities to relations, from a material and static realist position to an interactive material discursive intra-action. In this philosophy, all entities (human and the natural world) entangle to create agency and meaning. It is "performative" in that we are constantly acting our relations, called "relationing" herein, which is constantly co-emerging into new meaning patterns, what she calls "performative transformation."[357] The proposal of this ontological transformation within sustainability education is the hope that we feel intra-dependence, making a sustained transformative shift towards deep human and nonhuman relations, manifesting in a sustainable future.[358]

Since the Enlightenment, the dominant paradigm has been one of separation and mechanism, where reality comprises discrete entities that relate mechanistically. This has informed everything from the theory of planetary movement to scientific management theories. However, as the New Physics determined, human and nonhuman beings are not discrete entities but intrinsically entangled. This is a radical shift in the perceptual paradigm.[359] Entities cannot be teased apart and studied through static reductionist methods.[360] This has exciting but sometimes confounding implications for research and knowledge generation, explored in the next chapter, Chapter 6.[361]

Katie Ross[362] adds to O'Neil's analysis by suggesting that there are additional aspects in this movement beyond the mechanism and separation paradigm. It is not only about ways of knowing and ways of being, but also about cosmology (beliefs about the universe) and axiology (beliefs about what is ethical and valuable). She asserts that there needs to be a transformation towards relational cosmo-onto-axi-epistem-ologies, providing "a new constellation of performing meaning."[363]

In questioning the whole of Western intellectual inheritance, there is a parallel questioning of ways of educating. Up to now, the Eurocentric intellectual heritage has considered itself to be the *only* philosophy and the *only* science, blind to or denigrating marginalized with nonwestern expressions.[364] There is now a reclaiming and inclusion of many other longstanding philosophical and scientific knowing traditions.[365] Theorists are critiquing the anthropocentric, patriarchal, and colonizing assumptions about nature, women, and other marginalized, racialized, colonized groups, assumptions that have existed for several thousand years.[366]

One educational vision offered by Edmund O'Sullivan merges deep ecology with critical ecopsychology, ecospirituality, and ecofeminism resulting in a comprehensive cosmology, which he argues contemporary education lacks.[367] Such a cosmological approach fosters a planetary consciousness where the "universe is the primary university."[368] Drawing from cultural historian Riane Eisler, O'Sullivan suggests moving beyond dominator systems—embedded in science, modern warfare, environmental degradation, patriarchy, male violence, and nuclear armament—which all "maintain a deeply adversarial view of life permeated with struggles for power."[369] The three stages of survival, critique, and creation, based on Thomas Berry, are part of O'Sullivan's educational vision, where ecology is not just a subject but is regarded as our vital living context.[370]

Canadian transformative ecological educator Mark Hathaway, together with Brazilian liberation theologian Leonardo Boff, offer a similar cosmological approach in *The Tao of Liberation: Exploring the Ecology of Transformation*. They illustrate that the emerging scientific cosmology is compatible with the spiritual dimension of liberation.[371]

> Our cosmology encompasses our understanding of the origin, evolution, and purpose of the universe and the place of human beings within it. The way we experience and understand the cosmos—our 'cosmovision'—lies at the very core of our beliefs about the nature of transformation.[372]

They assert that our ability to imagine otherwise has been substantially restricted. The New Science illustrates a view of the cosmos that is constantly unfolding and evolving a new understanding of change dynamics. This brings the "importance of intuition, spirituality, and ancient wisdom traditions"[373] to the fore.

Edgar Morin,[374] Erich Jantsch,[375] Basarab Nicolescu,[376] and Margaret Somerville[377] are all addressing the need to take knowledge out of disciplinary silos to "create the integrated knowledge required to address complex social issues."[378] Transdisciplinarity is most prevalent in the process of question-asking, where disciplines "are deeply embedded in each other"[379] with no boundaries. This is reflected in more holistic questions, or metaquestions, which have increased potential to address complex sociopolitical realities. It does so by violating disciplinary boundaries but not fragmenting knowledge. Furthermore, it does not take a "fix-it" attitude but moves within collaborative open-ended explorations. For some, it is part of "postnormal science" in breaking free of reductionism, mechanism, certainty, and expertism.[380]

Not least, Transformative Sustainability Education promotes the "transformation of current social or economic systems"[381] in moving beyond industrialism, fossil fuels, and the capitalist market. Moacir Gadotti draws out the educational implications. "Simply improving the current model of education is to continue to follow the educational model that has been destroying the planet since the nineteenth century."[382] Continuing to educate young people and adults for acclimatizing to existing social and economic systems is to fail the task of human survival.

As John Ikerd *et al.* summarize,

> Deep sustainability is rooted in a worldview that is fundamentally different from the conventional worldview that supports contemporary thought and even the instrumental worldview that underpins most public and private sustainability initiatives and educational programs today. . . . Ultimately, we must radically redesign our economic and social system based on a new worldview and new paradigms to achieve authentic sustainability.[383]

Sustainability cannot be reached through resource efficiencies and technological substitutions or, rather, weak sustainability. Recognizing that "natural capital" must not be degraded and that ecological limits are necessary is strong sustainability, but it does not go far enough. Sustainability cannot be reached by just changing our way of thinking and worldview, which only goes part of the way to addressing the roots of unsustainability. Rather, sustainability education must address the economic, epistemological, "ethical, philosophical, and spiritual roots of human well-being"[384] in ways that give space for the regenerative powers of the natural world and assist in a renewal of societies. As Haydn Washington concludes, *strongest* sustainability "accepts the intrinsic value of Nature and is ecocentric."[385] The implication is that humans return to elements of Old Sustainability (see Chapter 2), where humans value the principle of harmony and work to maintain balance with the natural world.

Deep and strongest sustainability therefore must concern our ways of "knowing, learning, and being in the world."[386] It must draw from nonwestern sources, breaking open the current box of acceptability and legitimacy. Historian Paul Kennedy called for "nothing less than the reeducation of humanity."[387] This entails building an Earth-honouring culture. It is "educating *as life*,"[388] understanding the intelligibility of all life around us from which we learn and participate in co-creativity, discussed further in the next chapter.

Just Sustainabilities/Decolonizing Sustainability

Working in the American context, professor of urban and environmental policy and planning Julian Agyeman has consistently noted the "equity deficit" in mainstream "green" or "environmental" discourse. As Bob Bullard, Bob Evans, and Agyeman proposed,

> A truly sustainable society is one where wider questions of social needs and welfare, and economic opportunity are integrally related to environmental limits imposed by supporting ecosystems.[389]

The notion of just sustainabilities arose in the early 2000s. Agyeman highlights the plural form of "just sustainabilities," which goes beyond prescriptiveness to plural and culturally relevant conceptions. There is not only one just sustainability.

Drawing from Richard Wilkinson and Kate Pickett's book *The Spirit Level: Why Greater Equality Makes Societies Stronger*,[390] Agyeman asserts that every measure of social and environmental well-being is stronger when there is less gap between the wealthiest and the lowest income earners. This generates more trust in society, more pro-environmental behaviours, and more adaptability and innovation.

Agyeman defines just and sustainable communities as attaining four conditions: quality of life; meeting inter- and intragenerational needs; recognition of racial, cultural, diasporic, and ethnic minority groups; and living within ecosystem limits.[391] Moreover, he advocates for a "guided transition to a different economic system" that allows for collaborative forms of consumption, innovative organizational models such as cooperatives, blurred lines between producer and consumer, resource sharing, fair distribution of work and income, reinventing work, green taxes, and growth in autonomy.[392]

Plamen Makariev[393] explores "recognition" in Zvi Bekerman and Ezra Kopelwitz's book *Cultural Education-Cultural Sustainability: Minority, Diaspora, Indigenous, and Ethno-Religious Groups in Multicultural Societies*,[394] explaining that the mixing of cultures promotes innovation, creative energy, and originality, derived from respectful cultural syntheses.[395] Drawing from Canadian philosopher Charles Taylor, individual and group identity is defined as dialogical, where we define ourselves through dialogue from within our relationalities. "Due recognition is not just a courtesy we owe people. It is a vital human need."[396] In other words, we approach all people with an "presumption of equal worth."[397] Despite how we regard their cultural products and aesthetic, it is the recognition of the meaningfulness of identity, autonomy, and moral meaning for that group which is of central importance.

Maori Lewis Williams[398] asserts that Indigenous worldviews still remain at the margins of scientific, environmental, and educational discourses. The need for recognition and cultural continuance is of vital importance to Indigenous cultures. Those "factionalized both by colonization and in many ways, in terms of the community itself" are working to restore cultural health in relation to the land. "The land is us," says Jeannette Armstrong of the *Syilx Okanagan* peoples.[399] Armstrong says that living in Canada yet unable to obtain basics, such as clean drinking water, profiles a key indicator of equal access, including, to food, water, and shelter as well as pleasure and enjoyment of life. Such circumstances highlight how often it is the minority groups who reflect and express what is going wrong in a society, which often eventually impacts the whole society. Thus, the principle of balance between minority groups and majority groups must be honoured.

Armstrong describes an Indigenous practice where people are tasked with speaking for specific groups during decision-making, such as land speakers, water speakers, elder speakers, and medicine people speakers. It is their collective responsibility to share how a particular action will impact each of these human, nonhuman, and more-than-human groups, within all their relations. Anthropologist and linguist Mark Turin adds that language revitalization and decoupling colonizer languages from landscapes are vital, not only for reclaiming Traditional territories but also for reclaiming cultural knowledge.[400]

Lewis describes how the *Maori* worldview overcomes the reductionism of modern worldviews. The *Maori* worldview is multidimensional in accepting spiritual understandings, shamanic ways of knowing, intuitive and dream knowing, and the reality of life force energy that courses through all living beings. She recommends that these ways of knowing and worldviews become part of Transformative Sustainability Education. When we see ourselves as co-participants in reality, through these understandings, continual transformations of our imagination and consciousness will occur.

Naomi Mumbi Maina-Okori, Jada Renee Koushik, and Alexandria Wilson address intersectionality[401] from feminist and decolonizing frameworks in an analysis of environmental and sustainability education (ESE). As they explain, "intersectional analysis originates from Black feminist perspectives on how social identities/subjectivities collide and collude to reproduce systemic and unique forms of oppression." This extends now to Indigenous knowledge systems,[402] with intersectionality including spirituality, cultural alignment, heritage, lineage and connection to the land, in addition to gender, race, sexual orientation, socioeconomic status/class, and ability.[403] In examining the level of the ESE field's inclusion of minority women voices and nonwestern voices as well as the extent to which ESE addresses social justice issues, they acknowledge that there has been an uptake through posthuman and biocentric frameworks. However, they also call for "a broadening of ESE to include Black feminist and Indigenous approaches and a refocus on social justice and Indigenous knowledge systems that address sovereignty and entail a land-based approach to education."[404] One place to start, they say, is for those in the field to "reflect on the intersections inherent in their own experiences."[405]

South African Distinguished Professor of global change and environmental learning research, Heila Lotz-Sisitka also engages with the dynamics of decolonization, "not just of land, resources, and minds," but also of market forces, global capital, consumption, privatization of the commons, as well as all forms of "neoliberal looting."[406] She advises a clearer framing of "absence *and* emergence"[407] particularly in the African context. Drawing from legal and social theorist De Sousa Santos, this is identifying not only what is not there but also what is as well as what can be.

This takes a stance of radical co-presencing with the historically silenced, who can make the invisible visible.

In Northern Columbia, Martha Chaves with her colleagues Thomas Macintyre, Sofia Villa-Barajas, and Andres Makú-Pardo examined decolonial pedagogies that seek *buen vivir*, meaning a responsible "good life." For them, a *buen vivir* pedagogy is "community-centric, ecologically-balanced, and [incorporates] culturally-sensitive worldviews whose plural nature exposes the possibilities for bridging cultures and knowledge systems."[408] They too point to the importance of cosmologies, such as *Quechua* cosmology, an acceptance of Traditional knowledge, transmodern ways of knowing, including intuition, spirituality, ceremony, dreams, as well as multiple forms of self-development and horizontal forms of power which negotiates between different worldviews. These are just a few glimpses into invigorating new strands of theory and practice.

Climate Change Education

As Chapter 3 recounted in detail, the neoliberal "structural adjustment" of national economies and societies incorporated a backlash against climate science. Yet, the science from the Intergovernmental Panel on Climate Change (IPCC), established in 1988 and releasing its first report in 1990, was irrefutable, given the involvement of 2,000–2,500 scientists assessing multiple facets of climate change. That said, the 1990s saw little headway in global dialogues, given what has now been termed "junk, counter, or pseudo science" using disinformation in the attempt to discredit the evidence of environmental risk and human causation.

The Earth Summit in 1992 included a Convention on Climate Change with a pathway towards reducing emissions of greenhouse gases, particularly carbon, and it offered monies to the South to help them reach climate goals. In 1997, while the Kyoto Protocol included legally binding targets, various nations in the so-called Carbon Club including the United States walked away from the Protocol, devised unilateral responses, or found loopholes to satisfy corporate interests. Emissions continued to rise quickly. The COP (Conference of the Parties) meetings each year inched ahead though not making significant progress which cohered with scientific recommendations. It was the IPCC reports in 2001 and 2007, which profiled climate change as part of the sustainable development and sustainability education agenda. It clearly put "the development agenda into the climate change arena,"[409] given far worse impacts for vulnerable groups and nations.

By 2015 and the Paris Agreement, CO_2 had increased by 40% from pre-industrial rates, and the Earth's surface had warmed by .85°, leading to some of the impacts which have now become familiar—more intense storms, flooding, wild fires, sea-level rise, and drought. The Climate Justice movement became part of the transnational counter globalization movement calling for radical and alternative measures for the well-being of humanity and unborn generations.

With the 2018 IPCC Special Report,[410] it was clear that global warming could reach 1.5°C as early as 2030. Further modelling indicated that a rise in Earth's temperature to 2°C would be catastrophic and should be avoided at all costs. They specified that global coordination was required to reach the goals of reducing carbon 50% by 2030 and of reaching zero carbon emissions from human activities by 2050. The 2021 IPCC Report emphasized that we are now at "Code Red," said the UN Secretary-General António Guterres. He added that "the alarm bells are deafening, and the evidence is irrefutable," in which the 1.5°C threshold is perilously close.[411]

By 2010, more educators were turning their attention to climate change education or "climate education" for short, as part of sustainability/ESD education. Climate change denial hindered previous efforts, given the lack of mass public support. According to a 2017 study of climate education plans and policies across Canada,[412] it was apparent that while there had been significant focus on energy efficiency conversions for schools, there were few education policies that reflected national climate policies, despite rhetoric. There was little teacher professional development, integration of climate science and disaster risk reduction into curricular areas, and very little attention to the multiple dimensions of climate justice, with several exceptions.

As David Selby and Fumiyo Kagawa from Sustainability Frontiers[413] asserted, to which many agree, climate change is the defining ecological impact that reemphasizes that we require a fundamental "Copernican-type revolution"[414] for human survival. In 2010, they proposed a "transformative educational agenda" that addressed runaway climate change. For them, this included: confronting denial, despair, and grief; offering alternative conceptions of the "good life;" embracing human embeddedness in the natural world; drawing useful insights from peace and social justice education; and localizing focus and action.

Selby and Kagawa argued that Climate Change Education (CCE) ought to remain integrated within sustainability education, including education in disaster risk reduction, "adapting" to a changing climate, and finding ways of reducing or "mitigating" carbon emissions institutionally and personally.[415] In 2017, building from Matthew Harmin, MJ Barrett, and Carolyn Hoessler,[416] Selby and Kagawa called for a " 'transformative sustainability learning' [which] needed to be framed as an 'interdisciplinary and intercultural academic field' marked by 'epistemological stretching' that opens the way for processes of paradigm reconstruction and shift."[417] Vital for our "suicidal planet," they advocated subversive learning that interrogates the economic growth ideology and consumption lifestyles head on as the "elephant in the room."[418] They also described the need for restorative learning that restores vernacular nature learning toward the cultivation of deep affiliation.

In 2012, calling it "Climate Change Education for Sustainable Development," Allison Anderson called for education systems and schools that are "climate-proofed and resilient as well as sustainable and green."[419] As a fellow of the Brookings Institution, she also advocated for teaching not only the knowledge about causes of climate change but also relevant skills for problem solving and risk reduction. She suggested Climate Change Education be part of formal education as well as technical and vocational education and training.

Adult and Continuing Education professor, Shirley Walters described her work as an "embedded activist researcher" addressing climate justice during the 2017/2018 drought in South Africa and then during the COVID pandemic 2020/2022.[420] Using ecofeminism as a transformation framework for climate justice helped to illustrate the multidimensionality of such crises. It also foregrounded the needs of those most impacted by climate-related catastrophes. Rather than initiatives designed according to middle class biases, they prioritized providing food and water to urban, working-class Black women.

Sustainababble, Failure, and Postsustainability?

To close, Engelman suggests we are in the era of "sustainababble" where sustainability discourse has been emptied of meaning in some forums.[421] While sustainability as a concept has reached strong global currency, it is also "an adjective that can condone business as usual."[422] This has only increased the contestation around the term and political rifts among proponents. There has been a search for a replacement concept, such as regenerative, "vibrant, resilient, thriveable, or even flourishing" societies.[423]

Like Paul Hawken, some consider sustainability to be a social movement[424] that has helped raise public awareness as well as give bottom-up momentum to global initiatives. Moreover, the term has become a signifier that points to a broad consensus about what has value, despite the debates.[425] In many ways, it has been a fluid concept that continues to evolve in relation to history and context.[426]

Adult educator John Holford suggests that while sustainability has become common sense, overall it has failed, given the "dead hand of neoliberalism."[427] Unfortunately, ESD has not engaged in critically "unpacking the neo-liberal growth agenda"[428] or challenging corporatist objectives in education. It also has not encouraged a paradigm analysis of the field.[429] ESD continues to situate sustainability as an issue of values and culture while ignoring the deeper roots of unsustainability, the capitalist political economy which actively works against sustainability.

As Heather Farley and Zachary Smith ask, what is it that "keeps this concept [of sustainability] from bringing about transformational change"?[430] Over this time frame of global sustainability discourse, neoliberalism systematized global markets, weakened national borders, created large regional trading blocs, accelerated financial exchange and massive wealth accumulation, and encouraged the reach of corporate sector principles throughout most parts of society.[431] Despite the global reach of ESD, education overall has been refocussed onto skills training and employability for a global market, as part of top-down economic globalization. In this way, the ESD agenda has been constantly undermined by neoliberal influences on education.

Some thinkers, like Jickling and Sterling, suggest we are moving into post-sustainability,[432] on the assertion that mainstream approaches to sustainability have failed. Failure for them is evidenced in the deteriorating global condition. González-Gaudiano claims that ESD is an "empty signifier" that operates like a myth or salvation narrative.[433] Thus, he calls for new constructs to overcome deficiencies in the old, while leaving what is worthy intact. He asserts that the "manic quest to be new and innovative—to be the first"[434] is all part of the problem. The real question is "what should we do next?"[435] to assist human survival and even thriveability.

Jickling and Sterling suggest that instead of debates over signifiers, we rethink education itself. For them, the term post-sustainability encapsulates the need to "frame a whole vision for education aligned to our extraordinary times."[436] They propose that deconstructing ESD can lead to a reconstruction of education which "respond[s] to the educational imperatives of our time, particularly as they relate to ecological crises and human/nature relationships."[437] I take up this challenge in the next chapter.

Notes and References

Chapter line drawing by Canadian artist Erin Marie Lange Christensen. www.erinchristensen.com. Reproduced with permission.

1 David Orr, *Earth in Mind: On Education, Environment, and the Human Prospect* (Washington, DC: Island Press, 1994), p. 12
2 Joy Palmer, *Environmental Education in the 21st Century: Theory, Practice, Progress, and Promise* (London: Routledge, 1998), p. 17.
3 Palmer, *Environmental Education in the 21st Century*, p. 18.
4 John Huckle, "Realizing Sustainability in Changing Times," in John Huckle and Stephen Sterling (Eds.), *Education for Sustainability* (London: Earthscan, 1996), p. 9.
5 Huckle, "Realizing Sustainability in Changing Times," p. 7.
6 Paraphrased from Stephen Sterling, *Sustainable Education: Re-visioning Learning and Change*, Schumacher Briefings (Cambridge, UK: Green Books, 2001), p. 21.
7 Karen Mundy and Lynn Murphy, "Transnational Advocacy, Global Civil Society? Emerging Evidence from the Field of Education," in Hugh Lauder, Phillip Brown, Jo-Anne Dillabough, and A. H. Halsey (Eds.), *Education, Globalization & Social Change* (Oxford: Oxford University Press, 2006), p. 1004.
8 Sterling, *Sustainable Education*, p. 23.
9 Peter Mayo, *Gramsci, Freire & Adult Education: Possibilities for Transformative Action* (London: Zed Books, 1999), p. 177.
10 Refers to the Trilateral Commission. See M. Patricia Marchak, *The Integrated Circus: The New Right and the Restructuring of Global Markets* (Montreal: McGill-Queen's University Press, 1991), p. xi.
11 David Harvey, *A Brief History of Neoliberalism* (Oxford: Oxford University Press, 2005).
12 In Canada, this was first experienced in the province of Alberta. Citizens stood on the legislative steps protesting these education and social service cuts for years, to no avail. For the Alberta experience, see Trevor Harrison and Gordon Laxer (Eds.), *The Trojan Horse: Alberta and the Future of Canada* (Montreal: Black Rose Books, 1995 and Trevor Harrison (Ed.), *The Return of the Trojan Horse: Alberta and the New World (Dis)Order* (Montreal: Black Rose Books, 2005).
13 Ortega y Gasset in "The Revolt of the Masses," as cited in Benjamin Barber, "Jihad vs McWorld," *The Atlantic Monthly*, March 1992, p. 62.
14 Stephen Ball, *Education plc: Understanding Private Sector Participation in Public Sector Education* (London: Routledge, 2007), p. 6.
15 Mundy and Murphy, "Transnational Advocacy, Global Civil Society?" p. 1004.
16 Sterling, *Sustainable Education*, p. 40.
17 Maude Barlow and Heather-Jane Robertson, *Class Warfare: The Assault on Canada's Schools* (Toronto: Key Porter Books, 1994).
18 Barlow and Robertson, *Class Warfare*, pp. vi–vii.
19 Ball, *Education plc*, p. 6.
20 Peter Sacks, *Generation X Goes to College: An Eye-Opening Account of Teaching in Postmodern America* (Chicago: Open Court, 1996), p. 157.

21 Gil Forer, Prianka Srinivasan, and Marcie Merriman, *Beyond COVID-19: The Gen Z Perspective* (London: Ernest & Young Global Limited, 2020).
22 For example, see www.c21canada.org/wp-content/uploads/2015/04/Shifting-Minds-LEARNING-INDEX-Sept-29.pdf.
23 For example see www.conferenceboard.ca/docs/default-source/educ-public/esp2000.pdf.
24 See Mayo, *Gramsci, Freire & Adult Education.*
25 Joy Kcenia Polanco O'Neil, "Institutional Fragility: Structures of Dominance in American Higher Educational Institutions Inhibiting Sustainable Education," *Frontiers in Sustainability*, June 28, 2021. www.frontiersin.org/articles/10.3389/frsus.2021.662527/full. See also Iain Stewart, Victoria Hurth, and Stephen Sterling (Eds.), *Re-Purposing Universities for Sustainable Human Progress* (Lausanne: Frontiers Media SA, 2022). https://doi.org/10.3389/978-2-88974-858-7. www.frontiersin.org/research-topics/14778/re-purposing-universities-for-sustainable-human-progress.
26 Orr, *Earth in Mind*, p. 8.
27 Alfonso Montuori, "Forward," in Edgar Morin (Ed.), *On Complexity* (Cresskill, NJ: Hampton Press, Inc, 2008), p. xvi.
28 Montuori, "Forward," p. ix.
29 Edward Weber and Anne Khademian, "Wicked Problems, Knowledge Challenges, and Collaborative Capacity Builders in Network Settings," *Public Administration Review*, March/April 2008, pp. 334–349.
30 David Orr, "Foreword," in Michael Stone and Zenobia Barlow (Eds.), *Ecological Literacy: Educating our Children for a Sustainable World* (San Francisco: Sierra Club Books, 2005), p. 8.
31 Orr in Stone and Barlow, *Ecological Literacy*, x.
32 Orr, *Earth in Mind*, p. 12.
33 Orr, *Earth in Mind*, p. 14.
34 Sterling, *Sustainable Education*, p. 21.
35 Sterling, *Sustainable Education*, p. 29.
36 David Orr, "What Is Education For?" in *Earth in Mind*, pp. 7–15.
37 Orr, "What Is Education For?"
38 Richard Kahn, *Critical Pedagogy, Ecoliteracy & Planetary Crisis: The Ecopedagogy Movement* (New York: Peter Lang, 2010).
39 Stephen Sterling, "Education in Change," in J. Huckle and Stephen Sterling (Eds.), *Education for Sustainability* (London: Earthscan Publications Ltd, 1996).
40 Sterling, *Sustainable Education*, p. 30.
41 Sterling, *Sustainable Education*, p. 23.
42 Lucie Sauvé, "Environmental Education between Modernity and Postmodernity: Searching for an Integrating Educational Framework," *Canadian Journal of Environmental Education*, Vol. 4 (Summer 1999), p. 14.
43 Sterling, *Sustainable Education*, p. 14.
44 Tara Goldstein and David Selby (Eds.), *Weaving Connections: Educating for Peace, Social and Environmental Justice* (Toronto: Sumach Press, 2000).
45 Betty Reardon, *Comprehensive Peace Education: Educating for Global Responsibility* (New York: Teachers College Press, 1988), pp. 2–3.
46 See https://amshq.org/About-Montessori/History-of-Montessori.
47 Reardon, *Comprehensive Peace Education,* p. 1.
48 Reardon, *Comprehensive Peace Education*, p. 4.
49 Johan Galtung, "Violence, Peace, and Peace Research," *Journal of Peace Research*, Vol. 6, No. 3 (1969), pp. 167–191 as found in Catherine Baille Abidi, *Pedagogies for Building Cultures of Peace: Challenging Constructions of an Enemy* (Leiden: Brill/Sense, 2018), p. 7.
50 In Reardon, *Comprehensive Peace Education*, p. 6.
51 Toh Swee-Hin and Virginia Floresca-Cawagas, *Peace Education: A Framework for the Philippines* (Quezon City: Phoenix Publishing House, Inc., 1987), p. 1.
52 Toh and Floresca-Cawagas, *Peace Education.*

53 See www.upeace.org/.
54 Baillie Abidi, *Pedagogies for Building Cultures of Peace: Challenging Constructions of an Enemy* (Leiden: Brill/Sense, 2018), pp. 4–8.
55 Abidi, *Pedagogies for Building Cultures of Peace*, pp. 8–10.
56 Goldstein and Selby, *Weaving Connections*.
57 See Richard Peet and Elaine Hartwick, *Theories of Development: Contentions, Arguments, Alternatives* (New York: The Guilford Press, 2015).
58 Wolfgang Sachs (Ed.), *The Development Dictionary: A Guide to Knowledge as Power* (London: Zed Books, 1993).
59 Louella Cronkhite, "Development Education: Making Connections North and South," in Goldstein and Selby, *Weaving Connections*, pp. 150.
60 Cronkhite, "Development Education," pp. 146–167.
61 Paulo Freire, *Pedagogy of the Oppressed* (New York: Seabury Press, 1970). See Elizabeth A. Lange, "Fragmented Ethics of Justice: Freire, Liberation Theology and Pedagogies for the Non-Poor," *Convergence*, Vol. XXXI, No. 1–2 (1998), pp. 81–94.
62 Rick Arnold and Bev Burke, *A Popular Education Handbook: An Education Experience Taken from Central America and Adapted to the Canadian Context* (Toronto: CUSO and OISE-Adult Education Department, 1983).
63 Diane Gerin-Lajoie, *Educator's Discourses on Student Diversity in Canada: Context, Policy, and Practice* (Toronto: Canadian Scholars' Press, 2008), p. 9.
64 Gerin-Lajoie, *Educator's Discourses on Student Diversity in Canada*, p. 16.
65 Christine Sleeter, *Empowerment through Multicultural Education* (Albany: SUNY Press, 1991), p. 3.
66 Sleeter, *Empowerment through Multicultural Education*, p. 5.
67 Sleeter, *Empowerment through Multicultural Education*.
68 Gerin-Lajoie, *Educator's Discourses on Student Diversity in Canada*, p. 18.
69 Sleeter, *Empowerment through Multicultural Education*, p. 2.
70 Reardon, *Comprehensive Peace Education*.
71 Sleeter, *Empowerment through Multicultural Education*, p. 7.
72 Gerin-Lajoie, *Educator's Discourses on Student Diversity in Canada*, p. 10.
73 Gerin-Lajoie, *Educator's Discourses on Student Diversity in Canada*, p. 10.
74 Gerin-Lajoie, *Educator's Discourses on Student Diversity in Canada*, pp. 14–15.
75 Lyndsay Moffat, "Education for Gender Equity: Origins and Development," in Goldstein and Selby, *Weaving Connections*, p. 168.
76 Mary Field Belenky, Blythe McVicker Clinchy, Nancy Rule Goldberger and Jill Mattuck Tarule, *Women's Ways of Knowing: The Development of Self, Voice, and Mind* (New York: Basic Books, 1986).
77 Carol Gilligan, *In a Different Voice: Psychological Theory and Women's Development* (Cambridge: Harvard University Press, 1982). Nel Noddings, *Caring: A Feminine Approach to Ethics and Moral Education* (Berkeley: University of California Press, 1984/2003).
78 Cited in Moffat, "Education for Gender Equity," pp. 178–179.
79 Kathleen Weiler, *Women Teaching for Change: Gender, Class & Power* (New York: Bergin & Garvey Publishers, 1988).
80 Cited in Moffat, "Education for Gender Equity," pp. 180–181.
81 Meaning lesbian, gay, transexual, queer and/or questioning, intersex, asexual, two spirited, and others.
82 André Grace, "Critical Adult Education," in Tara Fenwick, Tom Nesbit, and Bruce Spencer (Eds.), *Contexts of Adult Education: Canadian Perspectives* (Toronto: Thompson Educational Publishing, 2006), pp. 128–139.
83 Grace, "Critical Adult Education," p. 138.
84 Tim McCaskell and Vanessa Russell, "Anti-Homophobia Initiatives at the Former Toronto Board of Education," in Goldstein and Selby, *Weaving Connections*, pp. 27–56.
85 John Egan and Anthony Flavell, "Towards Celebration through Education: Queer Canadian Adult Education," in Tara Fenwick, Tom Nesbit, and Bruce Spencer (Eds.),

Contexts of Adult Education: Canadian Perspectives (Toronto: Thompson Educational Publishing, 2006), pp. 260–269.

86 Lee Anderson, "Why Should American Education Be Globalized? It's a Nonsensical Question," *Theory into Practice*, Vol. 21, No. 3 (1982), pp. 154–159.

87 Barbara Tye and Kenneth Tye, *Global Education: A Study of School Change* (New York: State University of New York Press, 1992), pp. xvi–xvii.

88 See for instance R. Hanvey, *An Attainable Global Perspective* (New York: The American Forum for Global Education, 1976/1982).

89 American President's "Commission on Foreign Lange and International Studies," *Geographical Education*, Vol. 5, No. 3 (1979), pp. 12–18.

90 Graham Pike and David Selby, *Global Teacher, Global Learner* (London: Hodder & Stoughton, 1988).

91 Pike and Selby, *Global Teacher, Global Learner*, p. 2.

92 Julian Agyeman, *Sustainable Communities and the Challenge of Environmental Justice* (New York: New York University Press, 2005), p. 6.

93 Goldstein and Selby, *Weaving Connections*.

94 Sandra Garren and Robert Brinkmann, "Sustainability Definitions, Historical Context, and Frameworks," in Robert Brinkmann and Sandra Garren (Eds.), *The Palgrave Handbook of Sustainability* (London: Palgrave Macmillan, 2018), pp. 1–18. Also see Deborah Turner, Figure 1: Pillars of Sustainability, "Sustainability and Library Management Education," *Journal of Sustainability Education*, Vol. 7 (December 2014).

95 Helen Kopnina and Eleanor Shoreman-Ouimet, *Sustainability: Key Issues* (London: Earthscan, 2015), p. 11.

96 Gro Brundtland, *Our Common Future* (New York: UN World Commission on Environment and Development and Oxford University Press, 1987), p. 42. Downloaded from www.un-documents.net/ocf-02.htm#I.

97 Brundtland, *Our Common Future*, p. 41.

98 Kopnina and Shoreman-Ouimet, *Sustainability*, p. 11.

99 David Selby and Fumiyo Kagawa, *Sustainability Frontiers* (Opladen, Germany: Barbara Budrich Publishers, 2015). See also Peter Rogers, Kazi Jalal, and John Boyd, *An Introduction to Sustainable Development* (London: Earthscan, 2008), p. 42.

100 Simon Dresner, *The Principles of Sustainability* (London, UK: Earthscan (2002).

101 Lucie Sauvé, "Environmental Education between Modernity and Postmodernity: Searching for an Integrating Educational Framework," *Canadian Journal of Environmental Education*, Vol. 4 (Summer 1999), pp. 9–35.

102 Timothy O'Riordan and Heather Voisey (Eds.), *The Transition to Sustainability: The Politics of Agenda 21 in Europe* (London: Earthscan, 1998).

103 Asian Development Bank, *World Sustainable Development Timeline* (Manila: ADB, March 2012). http://hdl.handle.net/11540/708.

104 https://www.unsdsn.org/.

105 https://www.iisd.org/mission-and-goals.

106 See for instance Simon Dresnner, *The Principles of Sustainability* (London: Earthscan, 2002), Andres Edwards, *The Sustainability Revolution: Portrait of a Paradigm Shift* (Gabriola Island: New Society Publishers, 2005) as well as William Blackburn, *The Sustainability Handbook: The Complete Management Guide to Achieving Social, Economic, and Environmental Responsibility* (London: Earthscan, 2007).

107 See https://iclei.org/.

108 Garren and Brinkmann, "Sustainability Definitions, Historical Context, and Frameworks," p. 12. See https://resilientcitiesnetwork.org/.

109 Andre Edwards, *The Sustainability Revolution: Portrait of a Paradigm Shift* (Gabriola Island, BC: New Society Publishers, 2005).

110 Tony Hodge, "A Systemic Approach to Assessing Progress toward Sustainability," in Anne Dale and John Robinson (Eds.), *Achieving Sustainable Development* (Vancouver: UBC Press, 1996), pp. 267–292.

111 Paul Hawken, *Blessed Unrest: How the Largest Movement in the World Came into Being and Why No One Saw It Coming* (New York: Viking, 2007).
112 Hawken, *Blessed Unrest*, p. 3.
113 Hawken, *Blessed Unrest*, p. 3.
114 Edwards, *The Sustainability Revolution*.
115 Richard Heinberg, "Advance Praise," in Edwards, *The Sustainability Revolution*.
116 Hawken, *Blessed Unrest*, p. 12.
117 Baby Boomers are considered those born between 1946 and 1964. Gen X was born between 1965 and 1980. Gen Y or the Millennials were born between 1981 and 1996. Gen Z was born between 1997 and 2012. Gen A or Generation Alpha starts with children born in 2012 and will continue at least through 2025, and maybe later. PewResearch, *WhereMillenialsEndandGenerationZBegins*.January17,2019.www.pewresearch.org/fact-tank/2019/01/17/where-millennials-end-and-generation-z-begins/.
118 Peter Blaze Corcoran, "Foreword," in Paula Jones, David Selby, and Stephen Sterling (Eds.), *Sustainability Education in Higher Education: Perspectives and Practice Across Higher Education* (London: Earthscan, 2010), pp. xiii–xv.
119 Corcoran, "Foreword," p. xiii.
120 Corcoran, "Foreword," p. xiii.
121 T. O'Riordan and H. Voisey, *The Politics of Agenda 21 in Europe* (London: Earthscan, 1998).
122 Sterling, *Sustainable Education*, p. 15.
123 Selby, "Thoughts from a Darkened Corner," p. 33.
124 Jonathan Porritt, "Foreword," in John Huckle and Stephen Sterling (Eds.), *Education for Sustainability* (London: Earthscan, 1996), p. xi.
125 Porritt, "Foreword," p. xii.
126 Palmer, *Environmental Education in the 21st Century*, p. 77.
127 D. Tilbury, R. Stevenson, J. Fine, and D. Schreuder, *Education and Sustainability: Responding to the Global Challenge* (Gland, Switzerland: International Union for Conservation of Nature and Natural Resources, 2002), p. 7.
128 Tilbury et al., *Education and Sustainability*, p. 8.
129 UNESCO, *Agenda 21: The United Nations Programme of Action from Rio, 1992* (New York, NY: United Nations). https://sustainabledevelopment.un.org/outcomedocuments/agenda21.
130 Alexander Leicht, Julia Heiss, and Won Jung Byun (Eds.), *Issues and Trends in Education for Sustainable Development* (Paris: UNESCO, 2018).
131 UNESCO, *Agenda 21 Chapter 36 Promoting Education, Public Awareness and Training* (Paris: UNESCO, 1992). www.un-documents.net/a21-36.htm.
132 UNESCO, *Intergovernmental Conference on Environmental Education: Final Report* (Paris: UNESCO, 1978). www.gdrc.org/uem/ee/tbilisi.html.
133 Leicht, Heiss, and Byun, *Issues and Trends in Education for Sustainable Development*, p. 26.
134 UNESCO, *Reshaping Education: Towards Sustainable Development* (London: UNESCO Environment and Development Briefs, 1992). https://unesdoc.unesco.org/ark:/48223/pf0000108542?posInSet=5&queryId=65c0ca0e-22a9-44af-941d-1e9d77addc5c.
135 Alexander Leicht, Bernard Combes, Wong Jung Byun, and Adesuwa Vanessa Agbedahin, "From Agenda 21 to Target 4.7: The Development of Education for Sustainable Development," in Leicht, Heiss and Byun, *Issues and Trends in Education for Sustainable Development*, pp. 25–38.
136 UNESCO, *Reshaping Education*, p. 1.
137 UNESCO, *Environment and Society: Education and Public Awareness for Sustainability*. Thessaloniki Conference Discussion Paper (Paris: UNESCO, 1998), p. 15. https://digitallibrary.un.org/record/252859?ln=en.
138 For a description of the process, see *International Implementation Scheme* (Paris: UNESCO, 2005). https://unesdoc.unesco.org/ark:/48223/pf0000148654.
139 Helen Kopnina and Frans Meijers, "Education for Sustainable Development (ESD): Exploring Theoretical and Practical Challenges," *International Journal of Sustainability in Higher Education*, Vol. 15, No. 2 (2014), pp. 188–207.

140 Earth Charter Initiative (2012). Downloaded from www.earthcharterinaction.org/content/.
141 https://earthcharter.org/about-the-earth-charter/faqs/.
142 https://earthcharter.org/about-the-earth-charter/earth-charter-international/.
143 https://earthcharter.org/library/earth-charter-center-education-sustainable-development-brochure-2017/.
144 Mirian Vilela and Alicia Jimenez, *Earth Charter, Education, and the Sustainable Development Goal 4.7: Research, Experiences, and Reflections* (San Jose: UPEACE Press with Earth Charter International, 2020), p. ii.
145 Toh Swee-Hin and Virginia Floresca Cawagas, "Peace Education, ESD and the Earth Charter: Interconnections and Synergies," *Journal of Education for Sustainable Development*, Vol. 4, No. 2 (2010), pp. 167–180.
146 UNESCO, *Declaration of Thessaloniki* (Thessaloniki: UNESCO, 1997). https://digitallibrary.un.org/record/252859?ln=en.
147 UNESCO, *Declaration of Thessaloniki.*
148 In Helen Kopnina, "Education for Sustainable Development (ESD): The Turn away from 'Environment' in Environmental Education?" *Environmental Education Research*, Vol. 18, No. 5 (2012), pp. 699–717. https://doi.org/10.1080/13504622.2012.658028.
149 P. Martin, "A WWF View of Education and the Role of NGOs," in J. Huckle and S. Sterling (Eds.), *Education for Sustainability* (London, UK: Earthscan, 1996).
150 See Sterling, *Sustainable Education* or Martha Monroe and John Fien, "The Role Environmental Education Might Play in the Decade for Education for Sustainable Development," *Applied Environmental Education & Communication*, Vol. 4, No. 3 (2005), pp. 203–206 as well as E. Ärlemalm-Hagsér and A. Sandberg, "Sustainable Development in Early Childhood Education," *Environmental Education Research*, Vol. 17, No. 2 (2010), pp. 187–200.
151 Bob Jickling, "Why I Don't Want My Children Educated for Sustainable Development," *Journal of Environmental Education*, Vol. 23, No. 4 (1992), pp. 5–8.
152 J. Huckle, "Education for Sustainability: Pathways to the Future," *Australian Journal of Environmental Education*, Vol. 7 (1991), pp. 43–62.
153 See Jickling, "Why I Don't Want My Children Educated for Sustainable Development."
154 Heather Farley and Zachary Smith, *Sustainability: If it's Everything, Is it Nothing?* (London: Routledge, 2020), p. 12.
155 Stephen Sterling/EDET Group, *Good Earth-Keeping: Education, Training and Awareness for a Sustainable Future*. Environmental Development Education and Training Group (London: UNEP-UK, 1992).
156 Chris Maser, *Resolving Environmental Conflict* (Delray Beach, FL: St. Lucie Press, 1996).
157 Steven Gough and William Scott, *Higher Education and Sustainable Development: Paradox and Possibility* (London: Routledge, 2007).
158 McKeown, Hopkins, Rizzi, and Chrystalbridge, *Education for Sustainable Development Toolkit*, p. 9.
159 See for instance, Peter Blaze Corcoran, Joseph Weakland, and Arjen Wals, *Envisioning Futures for Environmental and Sustainability Education* (Wageningen: Wageningen Academic Publishers, 2017).
160 Renata Buriti, " 'Deep' or 'Strong' Sustainability," in W. Leal Filho (Ed.), *Encyclopedia of Sustainability in Higher Education* (London: Springer Nature Switzerland AG, 2019).
161 UNESCO-UNEP, "Sustainable Development via Environmental Education," *Connect: UNESCO-UNEP Environmental Education Newsletter*, Vol. 13, No. 2 (1988).
162 Sterling, *Sustainable Education*, pp. 35, 37.
163 McKeown, Hopkins, Rizzi, and Chrystalbridge, *Education for Sustainable Development Toolkit*, p. 13.
164 Leicht, Heiss, and Byun, *Issues and Trends in Education for Sustainable Development*, p. 7.
165 UNESCO, *Guidelines and Recommendations for Reorienting Teacher Education to Address Sustainability* (Paris: UNESCO, October 2005).

166 Erin Redman, Arnim Wiek, and Aaron Redman, "Continuing Professional Development in Sustainability Education for K-12 Teachers: Principles, Programme, Applications, Outlook," *Journal of Education for Sustainable Development*, Vol. 12, No. 1 (2018), pp. 59–80.
167 Information provided by Rosalyn McKeown, personal communication, July 29, 2002.
168 Rosalyn McKeown, personal communication, July 29, 2002.
169 Rosalyn McKeown, personal communication, July 29, 2002.
170 McKeown, Hopkins, Rizzi, and Chrystalbridge, *Education for Sustainable Development Toolkit*, p. 14.
171 UNESCO, *UN Decade of Education for Sustainable Development 2005–2014: The DESD at a Glance* (Paris: UNESCO, 2005). ED/2005/PEQ/EDS/3. https://unesdoc.unesco.org/ark:/48223/pf0000141629.
172 UNESCO, *International Implementation Scheme* (Paris: UNESCO Education Sector, October 2005), p. 6.
173 Moacir Gadotti, "Reorienting Education Practices towards Sustainability," *Journal of Education for Sustainable Development*, Vol. 4, No. 2 (2010), pp. 203–211.
174 UNESCO, *UN Decade of Education for Sustainable Development* (Paris: UNESCO, 2005). Emphasis added.
175 Lucie Sauvé, Tom Berryman, and Renee Brunelle, "Three Decades of International Guidelines for Environmental-Related Education: A Critical Hermeneutic of the United Nations Discourse," *Canadian Journal of Environmental Education,* Vol. 12 (2007), p. 33.
176 UNESCO, *Draft International Implementation Scheme for the UN Decade of Education for Sustainable Development 2005–2014* (PARIS: UNESCO, August 11, 2005).
177 UNESCO, *International Implementation Scheme*, p. 16.
178 UNESCO, *UN Decade of Education for Sustainable Development*, p. 3.
179 UNESCO, *UN Decade of Education for Sustainable Development*, p. 5.
180 Sauvé et al., *Three Decades of International Guidelines,* p. 34.
181 UNESCO, *UN Decade of Education for Sustainable Development*, p. 8.
182 Rosalyn McKeown, Charles Hopkins, Regina Rizzi, and Marianne Chrystalbridge, *Education for Sustainable Development Toolkit* (Paris: UNESCO, October 2006).
183 McKeown, Hopkins, Rizzi, and Chrystalbridge, *Education for Sustainable Development Toolkit*, p. 18.
184 McKeown, Hopkins, Rizzi, and Chrystalbridge, *Education for Sustainable Development Toolkit*, p. 32.
185 McKeown, Hopkins, Rizzi, and Chrystalbridge, *Education for Sustainable Development Toolkit*, p. 26.
186 McKeown, Hopkins, Rizzi, and Chrystalbridge, *Education for Sustainable Development Toolkit*.
187 For more information, see UNESCO, *UN Decade of Education for Sustainable Development* (Paris: UNESCO, 2005).
188 UNESCO, *Reshaping Education*, p. 14.
189 Joseph Overton, *The Overton Window*. Mackinac Center for Public Policy. www.mackinac.org/OvertonWindow. Accessed 7 November, 2021.
190 Marchak, *The Integrated Circus*, p. 93.
191 Bob Jickling and Arjen Wals, "Globalization and Environmental Education: Looking beyond Sustainable Development," *Journal of Curriculum Studies*, Vol. 40, No. 1 (2008), p. 2.
192 John Holford, "The Misuses of Sustainability: Adult Education, Citizenship and the Dead Hand of Neoliberalism," *International Review of Education*, Vol. 62 (2016), pp. 541–561.
193 Pablo Ángel Meira Cartea, "In Praise of Environmental Education," *Policy Futures in Education*, Vol. 3, No. 3 (2005), p. 290.
194 David Orr, "Walking North on a Southbound Train," *Conservation Biology*, Vol. 17, No. 2 (2003), pp. 348–351 cited in Jickling and Wals, "Globalization and Environmental Education," p. 3. (The use of north and south here does not refer to global hemispheres.)

195 Holford, "The Misuses of Sustainability," p. 543.
196 McKeown, Hopkins, Rizzi, and Chrystalbridge, *Education for Sustainable Development Toolkit*, p. 30.
197 UNESCO, *Bonn Declaration* (Bonn: UNESCO World Conference on Education for Sustainable Development, 2009). www.iau-hesd.net/sites/default/files/documents/2009_-_bonn_declaration_en.pdf.
198 Chris Duke and Heribert Hinzen, "Adult Education and Lifelong Learning within UNESCO," *Adult Learning*, Vol. 22. No. 4 (2011), pp. 18–23.
199 Kopnina and Meijers, "Education for Sustainable Development (ESD): Exploring Theoretical and Practical Challenges," p. 189.
200 Kopnina and Meijers, "Education for Sustainable Development (ESD): Exploring Theoretical and Practical Challenges," p. 189.
201 Wynn Calder and Alexander Leicht, "International Workshop on the UNDESD, Bonn, Germany," *Journal of Education for Sustainable Development*, Vol. 1, No. 1 (2007), pp. 143–145.
202 UNESCO, *Review of Contexts and Structures for Education for Sustainable Development* (Paris: UNESCO, Division of the Coordination of United Nations Priorities in Education, 2009).
203 Rosalyn McKeown and Charles Hopkins, "Moving Beyond the EE and ESD Disciplinary Debate in Formal Education," *Journal of Education for Sustainable Development*, Vol. 1, No. 1 (2007), pp. 17–26.
204 Heila Lotz-Sisitka, "Reflecting on the 2007 World Environmental Education Congress," *Journal of Education for Sustainable Development*, Vol. 1, No. 2 (2007), pp. 181–188.
205 Arjen Wals, *Review of Contexts and Structures for Education for Sustainable Development Learning for a Sustainable World 2009* (Paris: UNESCO, 2009). https://unesdoc.unesco.org/ark:/48223/pf0000184944.
206 Wals, *Review of Contexts and Structures for Education for Sustainable Development Learning for a Sustainable World*.
207 Martha Monroe, "A Priority for ESD Research: Influencing Adult Citizens," *Journal of Education for Sustainable Development*, Vol. 1, No. 1 (2007), pp. 107–113.
208 Leicht, Combes, Byun, and Agbedahin, "From Agenda 21 to Target 4.7," p. 36.
209 Leicht, Combes, Byun, and Agbedahin, "From Agenda 21 to Target 4.7," pp. 28–29.
210 Arjen Wals and Cathy Nolan (abridged version). *Shaping the Education of Tomorrow: 2012 Report on the UN Decade of Education for Sustainable Development* (Paris: UNESCO, 2012). https://unesdoc.unesco.org/ark:/48223/pf0000216606.
211 Wals and Nolan, *Shaping the Education of Tomorrow*. https://unesdoc.unesco.org/ark:/48223/pf0000216606.
212 UNESCO, *Shaping the Future We Want: UN Decade on Education for Sustainable Development (2005–2014)*. Final Report (Paris: UNESCO, 2014). https://unesdoc.unesco.org/ark:/48223/pf0000230302.
213 UNESCO, *Aichi-Nagoya Declaration on Education for Sustainable Development* (Paris: UNESCO, 2014). www.unesco.org › fileadmin › ERI › pdf › Aichi-Nagoya_Declaration_EN.
214 *UN General Assembly Resolution* 72/222 (2017). https://documents-dds-ny.un.org/doc/UNDOC/GEN/N17/466/61/PDF/N1746661.pdf?OpenElement.
215 See https://sdgs.un.org/2030agenda and https://sdgs.un.org/goals.
216 UNESCO, *Transforming Our World: The 2030 Agenda for Sustainable Development* (Paris: UNESCO). https://sdgs.un.org/2030agenda.
217 Maria Hofman, "What is an Education for Sustainable Development Supposed to Achieve – A Question of What, How and Why?" *Journal of Education for Sustainable Development*, Vol. 9, No. 2 (2015), p. 214.
218 Hofman, "What is an Education for Sustainable Development Supposed to Achieve," p. 219.
219 Stephen Sterling, "A Commentary on Education and Sustainable Development Goals," *Journal of Education for Sustainable Development*, Vol. 10, No. 2 (2016), pp. 208–213.

220 Sterling, "A Commentary on Education and Sustainable Development Goals," p. 211.
221 Stephen Sterling, "Sustainable Education," in Donald Gray, Laura Colucci-Gray, and Elena Camino (Eds.), *Science, Society, and Sustainability: Education and Empowerment for an Uncertain World* (New York: Routledge, 2009), pp. 105–118 cited in Hofman, "What Is an Education for Sustainable Development Supposed to Achieve," p. 223.
222 Stefania Giannini, "Foreword," *Education for Sustainable Development: A Roadmap* (Paris: UNESCO, 2020), p. iii.
223 Kartikeya Sarabhai, "Foreword," in Rajeswari Namagiri Gorana and Preeti Rawat Kanaujia (Eds.), *Schooling for Sustainable Development* (Netherlands: Springer Netherlands, 2016), p. viii. www.springer.com/series/8635.
224 Gestalt means that the whole has unique qualities and aspects that result from within unified whole than just the sum of the parts.
Arjen Wals, "Mirroring, Gestaltswitching and Transformative Social Learning: Stepping Stones for Developing Sustainability Competence," *International Journal of Sustainability in Higher Education*, Vol. 11, No. 4 (2010), pp. 380–390.
225 Wals, "Mirroring, Gestaltswitching and Transformative Social Learning," p. 386.
226 Wals, "Mirroring, Gestaltswitching and Transformative Social Learning," p. 388.
227 Yoko Mochizuki and Zinaida Fadeeva, "Competences for Sustainable Development and Sustainability: Significance and Challenges for ESD," *International Journal of Sustainability in Higher Education*, Vol. 11, No. 4 (2010), pp. 391–403.
228 Mochizuki and Fadeeva, "Competences for Sustainable Development and Sustainability," p. 394.
229 Mochizuki and Fadeeva, "Competences for Sustainable Development and Sustainability," p. 392.
230 Marco Rieckmann, "Learning to Transform the World: Key Competencies for Education for Sustainable Development," in Leicht, Heiss, and Byun (Eds.), *Issues and Trends in Education for Sustainable Development* (2018), pp. 44–45. https://unesdoc.unesco.org/ark:/48223/pf0000261445.
231 Rieckmann, "Learning to Transform the World," p. 45.
232 Guia Bianchi, Ulrike Pisiotis, and Marcelino Cabrera Giraldez, *GreenComp– The European Sustainability Competence Framework*, Margarita Bacigalupo and Yves Punie (Eds.) (Luxembourg: Publications Office of the European Union, 2022). https://publications.jrc.ec.europa.eu/repository/handle/JRC128040.
233 Bianchi, Pisiotis, and Cabrera Giraldez, *GreenComp*, p. 2.
234 Kartikeya Sarabhai, "Editorial," *Journal of Education for Sustainable Development*, Vol. 12, No. 2 (2018), pp. v–vi. www.handprint.in/global_handprint_network.
235 Rieckmann, "Learning to Transform the World," p. 46.
236 Rieckmann, "Learning to Transform the World," p. 49.
237 See https://centerforgreenschools.org/ and www.ecoschools.global/. See also Ailin Iwan, Nirmala Rao, and Kenneth Poon, "Characteristics of Green Schools: Observations of Award-Winning Green Preschools in Bali, Berkeley, and Hong Kong," *Journal of Education for Sustainable Development*, Vol. 12, No. 2 (2018), pp. 140–159.
238 See https://sustainableschools.ca/ and Sonja Kuzich, Elisabeth Taylor, and Peter Charles Taylor, "When Policy and Infrastructure Provisions are Exemplary but still Insufficient: Paradoxes Affecting Education for Sustainability (EfS) in a Custom-Designed Sustainability School," *Journal of Education for Sustainable Development*, Vol. 9, No. 2 (2015), pp. 179–195.
239 See D. Orr (1992) and J. Disinger (1990), "Environmental Education for Sustainable Development?" *Journal of Environmental Education*, Vol. 21, No. 4, pp. 3–6.
240 David Selby, "Thoughts from a Darkened Corner: Transformative Learning for the Gathering Storm," in David Selby and Fumiyo Kagawa (Ed.), *Sustainability Frontiers: Critical and Transformative Voices from the Borderlands of Sustainability Education* (Opladen, Germany: Barbara Budrich Publishers, 2015), pp. 21–42.
241 Selby and Kagawa, *Sustainability Frontiers*.

242 Michael Bonnet, "Environmental Education and the Issue of Nature," *Journal of Curriculum Studies*, Vol. 39, No. 6 (2007), pp. 707–721.
243 Helen Kopnina, "Education for Sustainable Development (ESD): The Turn away from 'Environmental' in Environmental Education," *Environmental Education Research*, Vol. 18, No. 5 (2012), p. 700.
244 Rolf Jucker, *Do We Know What We Are Doing? Reflections on Learning, Knowledge, Economics, Community, and Sustainability* (Cambridge: Cambridge Scholars Publishing, 2014). Review by Michael Carroll, *Journal of Education for Sustainable Development*, Vol. 9, No. 2 (2015), pp. 229–232.
245 Michael Bonnett, "Education for Sustainable Development: Sustainability as a Frame of Mind," *Journal of Philosophy of Education*, Vol. 37, No. 4 (2003), pp. 675–690.
246 Bonnett, "Education for Sustainable Development," p. 681.
247 Bonnett, "Education for Sustainable Development," p. 683. Emphasis in original.
248 Bonnett, "Education for Sustainable Development," pp. 683, 684.
249 Bonnett, "Education for Sustainable Development," pp. 685, 686.
250 Buriti, " 'Deep' or 'Strong' Sustainability," p. 3.
251 See Frans Schuurman (Ed.), *Beyond the Impasse: New Directions in Development Theory* (London: ZED Books, 1993).
252 On dependency theory, see Richard Peet, *Theories of Development*, and Jorge Larrain, *Theories of Development*, to start.
253 See Gustavo Esteva, "Development," in Wolfgang Sachs (Ed.), *The Development Dictionary: A Guide to Knowledge and Power* (London: ZED Books, 1992).
254 See Majid Rahnema, *The Post-Development Reader*; Raewyn Connell, *Southern Theory*; Herman Daly, *Beyond Growth*; Richard Heinberg, *The End of Growth*; Tim Jackson, *Prosperity without Growth*; Jared Diamond, *Collapse*; Joseph Tainter, *The Collapse of Complex Societies*; to name a few.
255 Jickling and Wals, "Globalization and Environmental Education," p. 4.
256 David Orr, *Ecological Literacy: Education and the Transition to a Postmodern World* (Albany, NY: SUNY, 1992).
257 William Scott, "Education and Sustainable Development: Challenges, Responsibilities, and Frames of Mind," *Trumpeter*, Vol. 18, No. 1 (2002), np. https://trumpeter.athabascau.ca/index.php/trumpet/article/view/123/134.
258 Sauvé et al., *Three Decades of International Guidelines*, p. 18.
259 Kopnina and Meijers, "Education for Sustainable Development (ESD)," pp. 191, 192.
260 Pablo Angel Meira Cartea, "In Praise of Environmental Education," *Policy Futures in Education*, Vol. 3, No. 3 (2005), pp. 284–295.
261 Regula Kyburz-Graber, "50 Years of Environmental Research from a European Perspective," *The Journal of Environmental Education*, Vol. 50, Nos. 4–6 (2019), pp. 378–385. https://doi.org/10.1080/00958964.2019.1704559.
262 Violeta Orlović Lovren and Katarina Popović, "Lifelong Learning for Sustainable Development—Is Adult Education Left Behind?" in W. Leal Filho, Mark Mifsud, and Paul Pace (Eds.), *Handbook of Lifelong Learning for Sustainable Development*, World Sustainability Series (Berlin: Springer, 2018). https://doi.org/10.1007/978-3-319-63534-7.
263 http://pascalobservatory.org/about/introduction.
264 www.pimanetwork.com/ (Shirley Walters is the current President).
265 Sauvé et al., *Three Decades of International Guidelines*.
266 Simon Dresner, *The Principles of Sustainability* (London: Earthscan, 2002).
267 Richard Kahn, *Critical Pedagogy, Ecoliteracy & Planetary Crisis* (New York: Peter Lang, 2010).
268 Sauvé et al., *Three Decades of International Guidelines,* p. 51. Emphasis is mine.
269 Gadotti, "Reorienting Education Practices towards Sustainability," p. 204.
270 Carlos Rodrigues Brandão, *My Home, the World* (Aparecida SP: Ideias e Letras, 2008), cited in Gadotti, "Reorienting Education Practices towards Sustainability," p. 210.
271 Gadotti, "Reorienting Education Practices towards Sustainability," p. 210.

272 See Peggy Barlett and Geoffrey Chase, *Sustainability on Campus: Stories and Strategies for Change* (Cambridge: The MIT Press, 2005) and
Tony Alabaster and Derek Blair, "Greening the University," in John Huckle and Stephen Sterling (Eds.), *Education for Sustainability* (London: Earthscan Publications, 2001).

273 Barlett and Chase, *Sustainability on Campus*, p. 5.

274 Anthony Cortese, "The Critical Role of Higher Education in Creating a Sustainable Future," *Planning for Higher Education*, Vol. 31, No. 3 (2003), pp. 15–22.

275 See www.magna-charta.org/ See also Walter Leal Filho, "Dealing With Misconceptions on the Concept of Sustainability," *International Journal of Sustainability in Higher Education*, Vol. 1, No. 1 (2000), p. 9.

276 See http://ulsf.org/talloires-declaration/.

277 See https://talloiresnetwork.tufts.edu/ and www.iau-hesd.net/documentation/30-halifax-declaration.html. See also www.nwf.org/EcoLeaders/Campus-Ecology-Resource-Center.

278 See www.iau-hesd.net/sites/default/files/documents/copernicus.pdf.

279 See https://secondnature.org/our-history/.

280 See https://ulsf.org/the-sierra-youth-coalition-and-the-sustainable-campuses-project/.

281 See Michael Shriberg, "Is the 'Maize-and-Blue' Turning Green? Sustainability at the University of Michigan," *International Journal of Sustainability in Higher Education*, Vol. 4, No. 3 (2003), pp. 263–276. www.emerald.com/insight/content/doi/10.1108/14676370310485465/full/html. Also see Willaim Mansfield, III, "Taking the University to Task," *World Watch*, Vol. 11, No. 3 (1998), pp. 24–30. https://eric.ed.gov/?id=EJ572488. Also see, Elizabeth A. Lange and Aaron Chubb, "Critical Environmental Adult Education in Canada: Student Environmental Activism," in Patricia Cranton and Leona English (Eds.), *Reaching Out Across the Border: Canadian Perspectives in Adult Education* (pp. 61–72). New Directions for Adult and Continuing Education, No. 124 (2006). San Francisco: Jossey-Bass.

282 Alabaster and Derek Blair, "Greening the University," p. 97.

283 See www.aashe.org/.

284 See https://secondnature.org/signatory-handbook/the-commitments/ and https://secondnature.org/climate-action-guidance/network/.

285 Mahesh Pradhan and Ayombi Mariam, "The Global Universities Partnership on Environmental and Sustainability (GUPES): Networking of Higher Educational Institutions in Facilitating Implementation of the UN Decade of Education for Sustainable Development 2005–2014," *Journal of Education for Sustainable Development*, Vol. 8, No. 2 (2014), pp. 171–175.

286 Elizabeth Potter-Nelson and Joy Kcenia O'Neil, "Role of Teachers on Education for Sustainable Development," in Walter Leal Filho (Ed.), *Encyclopedia of Sustainability in Higher Education* (Springer Nature Switzerland, 2019). https://link.springer.com/referencework/10.1007/978-3-319-63951-2.

287 See https://stars.aashe.org/.

288 Barlett and Chase, *Sustainability on Campus*, p. 9.

289 Sarah Holdsworth and Ian Thomas, "Framework for Introducing Education for Sustainable Development into University Curriculum," *Journal of Education for Sustainable Development*, Vol. 9, No. 2 (2015), pp. 137–159.

290 Adam Cheeseman, Tarah Sharon Alexandra Wright, Jaylene Murray, and Marcia McKenzie, "Taking Stock of Sustainability in Higher Education: A Review of the Policy Literature," *Environmental Education Research*, Vol. 25, No. 12 (2019), pp. 1697–1712. https://doi.org/10.1080/13504622.2019.1616164.

291 Walter Leal Filho, "Dealing With Misconceptions on the Concept of Sustainability," *International Journal of Sustainability in Higher Education*, Vol. 1, No. 1 (2000), p. 9.

292 Anne Sibbel, "Pathways towards Sustainability through Higher Education," *International Journal of Sustainability in Higher Education*, Vol. 10, No. 1 (2009), pp. 68–82. https://doi.org/10.1108/14676370910925262.

293 Paul Sylvestre, Tarah Wright, and Kate Sherren, "Exploring Faculty Conceptualizations of Sustainability in Higher Education: Cultural Barriers to Organizational Change and Potential Resolutions," *Journal of Education for Sustainable Development*, Vol. 7, No. 2 (2013), pp. 223–244.
294 Barlett and Chase, *Sustainability on Campus*, p. 13.
295 See www.griffith.edu.au/griffith-sciences/ecocentre and John Fien, Innovative Practice Example: The Griffith University EcoCentre, Brisbane, Australia, ESD Toolkit. www.esdtoolkit.org/discussion/change.htm among other examples.
296 David Orr, "Can Educational Institutions Learn? The Creation of the Adam Joseph Lewis Center at Oberlin College," in Barlett and Chase, *Sustainability on Campus*, pp. 159–175. See also www.buildingsandcities.org/insights/research-pathways/building-pedagogy.html and David Orr, "Architecture as Pedagogy," in *Earth in Mind: On Education, Environmental, and the Human Prospect* (Washington, DC: Island Press, 1994), pp. 112–116. A tour is available at www.buildingsandcities.org/insights/research-pathways/building-pedagogy.html.
297 John Blewitt, *The Ecology of Learning: Sustainability, Lifelong Learning, and Everyday Life* (London: Earthscan from Routledge, 2006), pp. 209–213.
298 Stephen Sterling, Larch Maxey, and Heather Luna (Eds.), *The Sustainable University: Progress and Prospects* (Abingdon: Routledge, 2012), p. xxii.
299 Stephen Gough and William Scott, *Higher Education and Sustainable Development: Paradox and Possibility* (London: Routledge, 2007).
300 See *International Journal of Sustainability in Higher Education*, 2021 special issue on European Perspectives on the Socially Responsible and Engaged University. www.emerald.com/insight/publication/issn/1467-6370/vol/23/iss/4. For instance, see Ronald Barnett and Carolina Guzman-Valenzuela, "The Socially Responsible University: A Challenging Project," *International Journal of Sustainability in Higher Education*, Vol. 23, No. 4 (2021), pp. 752–766. https://doi.org/10.1108/IJSHE-03-2021-0100.
301 Christian Rammel and Oliver Vettori, "Dealing with the Intangible: Using the Analytical Lens of Hidden Curricula for a Transformative Paradigm of Sustainable Higher Education," *Journal of Education for Sustainable Development*, Vol. 15, No. 2 (2021), pp. 234–249.

Also see Janet Moore, "Seven Recommendations for Creating Sustainability Education at the University Level: A Guide for Change Agents," *International Journal of Sustainability in Higher Education*, Vol. 6, No. 4 (2005), pp. 326–339.
302 Arran Stibbe (Ed.), *The Handbook of Sustainability Literacy: Skills for a Changing World* (Totnes: Green Books, 2009).
303 Arran Stibbe and Heather Luna, "Introduction," in Arran Stibbe (Ed.), *The Handbook of Sustainability Literacy: Skills for a Changing World* (Totnes: Green Books, 2009), p. 10.
304 Stibbe and Heather Luna, "Introduction," p. 11.
305 Victor Nolet, "Preparing Sustainability-Literate Teachers," *Teachers College Record* (2009), pp. 409–442. https://doi.org/10.1177/016146810911100207.
306 Paul Clarke, *Education for Sustainability: Becoming Naturally Smart* (London: Routledge, 2012), p. 43.
307 Clarke, *Education for Sustainability*, pp. xvii, 3.
308 Clarke, *Education for Sustainability*, pp. 48.
309 Clarke, *Education for Sustainability*, pp. 121.
310 Paul Murray, *The Sustainable Self: A Personal Approach to Sustainability Education* (London: Earthscan, 2011), p. ix.
311 See a brief description in Elizabeth A. Lange, "Transformative and Restorative Learning: A Vital Dialectic for Sustainable Societies," *Adult Education Quarterly*, Vol. 54, No. 2 (2004), pp. 121–139. https://doi.org/10.1177/071713603260276.
312 Also see Elizabeth A. Lange, "Fostering a Learning Sanctuary for Transformation in Adult Sustainability Education," in Jack Mezirow and Edward Taylor (Eds.), *Transformative Learning in Practice* (San Francisco: Jossey-Bass Publishers, 2009), pp. 193–204.

313 John Blewitt, *The Ecology of Learning: Sustainability, Lifelong Learning, and Everyday Life* (London: Earthscan from Routledge, 2006).
314 Blewitt, *The Ecology of Learning*, p. 10.
315 Heather Burns, "Teaching for Transformation: (Re)Designing Sustainability Courses Based on Ecological Principles," *Journal of Sustainability Education*, Vol. 2 (2011). Also see https://drheatherburns.wordpress.com/the-burns-model/.

See two other descriptions: Heather Burns, "Meaningful Sustainability Learning: A Study of Sustainability Pedagogy in Two University Courses," *International Journal of Teaching and Learning in Higher Education*, Vol. 25, No. 2 (2013), pp. 166–175.

Heather Burns, Sybil Kelley, and Heather Spalding, "Teaching Sustainability: Recommendations for Best Pedagogical Practices," *Journal of Sustainability Education*, Vol. 19 (2019).
316 Heather Burns, "Fostering Resiliency and Care Integrating Self Compassion into a Graduate Course," *Journal of Contemplative Inquiry*, Vol. 8, No. 1 (2021), pp. 200–216.
317 Victor Nolet, "Preparing Sustainability-Literate Teachers," *Teachers College Record*, Vol. 111, No. 2 (2009). www.tcrecord.org. ID Number: 15177.
318 Robert Taylor, *Taking Sides: Clashing Views in Sustainability* (McGraw-Hill Education, 2013).
319 Joy Kcenia O'Neil, *"Cooking to Learn" while "Learning to Cook": "(Be)coming and (Re) membering Sustainability* (Prescott, AZ: Prescott College, 2015), p. 55.
320 Martin Carnoy, Manuel Castells, Stephen Cohen and Fernando Henrique Cardoso, *The New Global Economy in the Information Age: Reflections on our Changing World* (University Park, PE: The Pennsylvania State University Press, 1993).
321 Carnoy et al., *The New Global Economy in the Information Age*, pp. 6, 7.
322 Sterling, *Sustainable Education*, p. 14.
323 Victor Nolet, *Educating for Sustainability: Principles and Practices for Teachers* (New York: Routledge, 2016), p. 7.
324 Sterling, *Sustainable Education*, p. 24.
325 Paraphrased from Sterling, *Sustainable Education*, p. 26.
326 Sterling, *Sustainable Education*, p. 26
327 Orr, *Ecological Literacy*.
328 Selby, *Thoughts from a Darkened Corner*, p. 29.
329 Selby, *Thoughts from a Darkened Corner*, p. 29.
330 "The Archimedean point is a point 'outside' from which a different, perhaps objective or 'true' picture of something is obtainable." Oxford Dictionary of Philosophy (Oxford: Oxford University Press, 2008). https://doi.org/10.1093/acref/9780199541430.001.0001.
331 Nolet, *Education for Sustainability*, p. 68.
332 Nolet, *Education for Sustainability*, pp. 62–71.
333 Edgar Morin, *On Complexity* (Cresskill, NJ: Hampton Press, Inc, 2008).
334 Morgan Williams, Foreword in See *Change: Learning and Education for Sustainability* (Wellington, NZ: Parliamentary Commissions for the Environment, 2004), p. 4 www.pce.parliament.nz/media/pdfs/See_change_report.pdf in Stephen Sterling, "Transformative Learning and Sustainability: Sketching the Conceptual Ground," *Learning and Teaching in Higher Education*, No. 5 (2010–2011), p. 19.
335 Sterling, "Transformative Learning and Sustainability," p. 27.
336 Tobias Schnitzler, "The Bridge between Education for Sustainable Development and Transformative Learning: Towards New Collaborative Learning Spaces," *Journal of Education for Sustainable Development*, No. 2 (2019), pp. 13, 242–253.
337 Sterling, *Sustainable Education*, p. 61.
338 Sterling, *Sustainable Education*, p. 61.
339 Wendy Griswold, "Sustainability, Ecojustice, and Adult Education," *New Directions for Adult and Continuing Education*, No. 153 (Spring 2017). https://doi.org/10.1002/ace.20217.

340 Christine Wamsler, "Education for Sustainability: Fostering a More Conscious Society and Transformation towards Sustainability," *International Journal of Sustainability in Higher Education*, Vo. 21, No. 1 (2020), pp. 112–130.
341 Wamsler, "Education for Sustainability," p. 113.
342 Wamsler, "Education for Sustainability," pp. 115–116.
343 Wamsler, "Education for Sustainability," p. 123.
344 Jonathon Dawson, Stephen Sterling and Paul Warwick, "Transforming Sustainability Education at the Creative Edge of the Mainstream: A Case Study of Schumacher College," *Journal of Transformative Education*, Vol. 16, No. 4 (2018), pp. 323–343.
345 Dawson et al., "Transforming Sustainability Education at the Creative Edge of the Mainstream."
346 Dawson et al., "Transforming Sustainability Education at the Creative Edge of the Mainstream."
347 Yona Sipos, Bryce Battisti, and Kurt Grimm, "Achieving Transformative Sustainability Learning: Engaging Head, Hands, and Heart," *International Journal of Sustainability in Higher Education*, Vol. 9, No. 1 (2008), pp. 68–86, 75.
348 Sipos, Battisti, and Grimm, "Achieving Transformative Sustainability Learning."
349 Sipos, Battisti, and Grimm, "Achieving Transformative Sustainability Learning," p. 68.
350 Satish Kumar and Pavel Cenkl, *Transformative Learning: Reflections on 30 Years of Head, Heart, and Hands at Schumacher College* (Gabriola Island: New Society Publishers, 2021), p. xii.
351 Elizabeth A. Lange, "Transformative and Restorative Learning: A Vital Dialectic for Sustainable Societies," *Adult Education Quarterly*, Vol. 54, No. 2 (2004), pp. 121–139. https://doi.org/10.1177/071713603260276.
352 Elizabeth A. Lange, "Transforming Transformative Learning through Ontologies of Relationality," *Journal of Transformative Education*, Vol. 16 (2018), pp. 280–301. https://doi.org/10.1177/1541344618786452.
353 Lange, "Transformative and Restorative Learning."
354 Joy Kcenia Polanco O'Neil, "Cooking to Learning" while "Learning to Cook."
355 Joy Kcenia Polanco O'Neil, "*(Be)Coming* and *(Re)Membering* as Sustainability: An Innovative Living Systems Model for Higher Education," in Walter Leal Filo (Ed.), *Handbook of Theory and Practice of Sustainable Development in Higher Education* (New York: Springer, 2017), pp. 317–333.
356 O'Neil, "'Cooking to Learning' while 'Learning to Cook'," p. 19.
357 Elizabeth A. Lange, Joy Kcenia Polanco O'Neil, and Katie E. Ross, "Educating during the Great Transformation: Relationality and Transformative Sustainability Education," *Studies in Adult Education and Learning*, Vol. 27, No. 1 (2021), p. 27. https://doi.org/10.4312/as/9692.
358 Joy Kcenia Polanco O'Neil, "Transformative Sustainability Learning Within a Material-Discursive Ontology," *Journal of Transformative Education*, Vol. 16, No. 4 (2018), pp. 365–387. https://doi.org/10.1177/1541344618792823.
359 Capra and Luisi, *The Systems View of Life*.
360 O'Neil, "'Cooking to Learning' while 'Learning to Cook'," p. 21.
361 See Elizabeth St. Pierre, "Post Qualitative Research: The Critique and the Coming After," in N. K. Denzin and Y. S. Lincoln (Eds.), *The SAGE Handbook of Qualitative Research* (Los Angeles, CA: SAGE, 2011), pp. 611–625.

Also, Elizabeth St. Pierre, "A Brief and Personal History of Post Qualitative Research Toward 'Post Inquiry'," *Journal of Curriculum Theorizing*, Vol. 30, No. 2 (2014), pp. 2–19.

Basarab Nicolescu, *Manifesto of Transdisciplinary* (K.-C. Voss, Trans., D. Appelbaum, Ed.) (New York: State University of New York Press, 2002). Patti Lather and Elizabeth St. Pierre, "Post-qualitative Research," *Journal of Qualitative Studies in Education*, Vol. 26, No. 6 (2014), pp. 629–633. https://doi.org/10.1080/09518398.2013.788752. Rosi Braidotti, *The Posthuman* (Cambridge, UK: Polity Books, 2013).

362 Katie E. Ross, *Transforming the Ways We Create Change: Experiencing and Cultivating Transformative Sustainability Learning* (Sydney: Institute of Sustainable Futures, University of Technology Sydney, 2020).
363 Ross, *Transforming the Ways We Create Change.*
364 Ben-Ami Scharfstein, *The Western Blindness to Non-Western Philosophy. Paper presented at the 20th World Congress of Philosophy – Comparative Philosophy*. Boston: Paper retrieved from *The Paideia Archive.* www.bu.edu/wcp/Papers/Comp/CompScha.htm.
365 See the *Diversity and Inclusiveness Syllabus Collection* of the American Philosophical Associationathttps://www/apaonline.org/members/group_content_view.asp?group+110430&id+380970.
366 See the classics/founding mothers: Vandana Shiva, *Staying Alive: Women, Ecology, Development* (Berkeley, CA: North Atlantic Books, 1988/2015); Carolyn Merchant, *The Death of Nature: Women, Ecology and the Scientific Revolution* (New York: HarperOne, 1980/2020); Marilyn Waring, *Counting for Nothing: What Men Value and What Women are Worth* (Wellington, NZ: Allen & Unwin, 1988/1999); Greta Gaard, *Ecofeminism: Women, Animals, Nature* (Philadelphia: Temple University Press, 1993); Rosemary Radford Ruether, *Gaia & God: An Ecofeminist Theology of Earth Healing* (San Francisco: HarperSanFrancisco, 1992). Also see Robert Doyle Bullard, *Unequal Protection: Environmental Justice and Communities of Color* (San Francisco: Sierra Club Books, 1994).
367 Edmund Sullivan, *Transformative Learning: Educational Vision for the 21st Century* (London: Zed Books, 1999), p. 3.
368 Thomas Berry, "Foreword," in Sullivan, *Transformative Learning*, pp. xiii–xiv.
369 Sullivan, *Transformative Learning*, p. 142.
370 Sullivan, *Transformative Learning.*
371 Hathaway and Boff, *The Tao of Liberation*, p. 10.
372 Mark Hathaway and Leonardo Boff, *The Tao of Liberation: Exploring the Ecology of Transformation* (Maryknoll: Orbis Books, 2009), p. 10.
373 Hathaway and Leonardo Boff, *The Tao of Liberation*, p. 10.
374 Edgar Morin, *On Complexity* (Cresskill, NJ: Hampton Press, Inc, 2008).
375 Erich Jantsch, *The Evolutionary Vision: Toward A Unifying Paradigm of Physical, Biological and Sociocultural Evolution* (Boulder, CO: Westview Press, 1981).
376 Basarab Nicolescu, *Manifesto of Transdisciplinarity* (New York: SUNY Press, 2002).
377 Margaret Somerville and David Rapport, *Transdisciplinarity: Recreating Integrated Knowledge* (Oxford: EOLSS Publishers, 2000).
378 Somerville and Rapport, *Transdisciplinarity*, p. xiii.
379 Somerville and Rapport, *Transdisciplinarity,* p. xiiv.
380 Anthony McMichael, "Transdisciplinarity in Science," in Margaret Somerville and David Rapport (Eds.), *Transdisciplinarity: Recreating Integrated Knowledge* (Oxford: EOLSS Publishers, 2000), pp. 203–209.
381 Fein and Tilbury, *The Global Challenge of Sustainability*, p. 3.
382 Moacir Gadotti, "Reorienting Education Practices toward Sustainability," *Journal of Education for Sustainable Development*, Vol. 4, No. 2 (2010), pp. 203–211.
383 John Ikerd, Lonnie Gamble, and Travis Cox, *Deep Sustainability: The Essentials*. Creative Commons. https://sites.google.com/site/sustainabilitydeep/.
384 Ikerd, *Deep Sustainability*, n.p.
385 Haydn Washington, *Demystifying Sustainability: Towards Real Solutions* (Abingdon: Earthscan from Routledge, 2015), p. 42.
386 Ikerd, *Deep Sustainability,* n.p.
387 Paul Kennedy in Orr, *Earth in Mind.*
388 Elizabeth A. Lange, Joy Kcenia Polanco O'Neil, and Katie E. Ross, "Educating during the Great Transformation: Relationality and Transformative Sustainability Education," *Studies in Adult Education and Learning*, Vol. 27, No. 1 (2021), pp. 23–46. https://doi.org/10.4312/12/9692.

389 Julian Agyeman, *Introducing Just Sustainabilities: Policy, Planning, and Practice* (London: Zed Books, 2013), p. 4.
390 Richard Wilkinson and Kate Pickett, *The Spirit Level: Why Greater Equality Makes Societies Stronger* (New York: Bloomsbury Press, 2009).
391 Agyeman, *Introducing Just Sustainabilities*, p. 7.
392 Agyeman, *Introducing Just Sustainabilities*, p. 10.
393 Plamen Makariev, "Cultural Education and the Paradigm of Recognition," in Zvi Bekerman and Ezra Kopelwitz (Eds.), *Cultural Education-Cultural Sustainability: Minority, Diaspora, Indigenous, and Ethno-Religious Groups in Multicultural Societies* (New York: Routledge, 2008), pp. 403–416.
394 Zvi Bekerman and Ezra Kopelwitz, *Cultural Education-Cultural Sustainability: Minority, Diaspora, Indigenous, and Ethno-Religious Groups in Multicultural Societies* (New York: Routledge, 2008).
395 See also Lynn Payne and Joy Kcenia Polanco O'Neil, "Cultural Sustainability in Higher Education," in Walter Leal Filho (Ed.), *Encyclopedia of Sustainability in Higher Education* (London: Springer Nature Switzerland, 2019). https://doi.org/10.1007/978-3-319-63951-2_109-1.
396 Charles Taylor, "The Politics of Recognition," in Amy Gutmann (Ed.), *Multiculturalism: Examining the Politics of Recognition* (Princeton: Princeton University Press, 1994), pp. 26 cited in Makariev, "Cultural Education and the Paradigm of Recognition," p. 407.
397 Makariev, "Cultural Education and the Paradigm of Recognition," p. 408.
398 Lewis Williams, "Deepening Ecological Relationality Through Critical OntoEpistemological Inquiry: Where Transformative Learning Meets Sustainable Science," *Journal of Transformative Education*, Vol. 11, No. 2 (2013), pp. 95–113.
399 Jeannette Armstrong, "An Okanagan Worldview of Society," in Sam Mickey, Mary Evelyn Tucker, and John Grim (Eds.), *Living Earth Community: Multiple Ways of Being and Knowing* (Open Book Publishers, 2020), pp. 163–170.
400 Mark Turin, "Indigenous Language Resurgence and the Living Earth Community," in Mickey, Tucker, and Grim, *Living Earth Community*, pp. 171–184.
401 Intersectionality as a term originated from the works of Kimberlé Crenshaw in 1989 as a political analysis that attends to the many forms of power relations, particularly how different aspects of a person's social and political identities together create differential forms of discrimination and privilege.
402 Naomi Mumbi Maina-Okori, Jada Renee Koushik, and Alexandria Wilson, "Reimagining Intersectionality in Environmental and Sustainability Education: A Critical Literature Review," *The Journal of Environmental Education*, Vo. 49, No. 4 (2018), pp. 286–296.
403 Alex Wilson, "How We Find Ourselves: Identity Development and Two Spirit People," *Harvard Educational Review*, Vol. 66, No. 2 (1996), pp. 303–318.
404 Maina-Okori, Koushik, and Wilson, "Reimagining Intersectionality in Environmental and Sustainability Education," p. 292.
405 Maina-Okori, Koushik, and Wilson, "Reimagining Intersectionality in Environmental and Sustainability Education," p. 293.
406 Heila Lotz-Sisitka, "Decolonization as Future Frame for Environmental and Sustainability Education: Embracing the Commons with Absence and Emergence," in Peter Blaze Corcoran, Joseph Weakland, and Arjen Wals (Eds.), *Envisioning Futures for Environmental and Sustainability Education* (Wageningen: Wageningen Academic Publishers, 2017), pp. 45–62.
407 Lotz-Sisitka, "Decolonization as Future Frame for Environmental and Sustainability Education," p. 53.
408 Thomas Macintyre, Martha Chaves, Sofia Villa-Barajas, and Andres Makú-Pardo, "Educating for Development of Educating for the Good Life? *Buen Vivir* Imaginaries and the Creation of One's Own Myth," in Peter Blaze Corcoran, Joseph Weakland,

and Arjen Wals (Eds.), *Envisioning Futures for Environmental and Sustainability Education* (Wageningen: Wageningen Academic Publishers, 2017), pp. 193–204.
409 Jennifer Elliott, *An Introduction to Sustainable Development* (Abingdon: Routledge, 2013), p. 107.
410 www.ipcc.ch/sr15/.
411 https://news.un.org/en/story/2021/08/1097362.
412 Andrew Bieler, Randoph Haluza-Delay, Ann Dale, and Marcia McKenzie, "A National Overview of Climate Change Education Policy: Policy Coherence between Subnational Climate and Education Policies in Canada (K-12)," *Journal of Education for Sustainable Development*, Vol. 11, No. 2 (2017), pp. 63–85.
413 www.sustainabilityfrontiers.org/.
414 David Selby and Fumiyo Kagawa, "Runaway Climate Change as Challenge to the 'Closing Circle' of Education for Sustainable Development," *Journal of Education for Sustainable Development*, Vol. 4. No. 1 (2010), pp. 37–50.
415 Fumiyo Kagawa and David Selby, "Ready for the Storm: Education for Disaster Risk Reduction and Climate Change Adaptation and Mitigation," *Journal of Education for Sustainable Development*, Vol. 6, No. 2 (2012), pp. 207–217.
416 Matthew Harmin, M. J. Barrett, and Carolyn Hoessler, "Stretching the Boundaries of Transformative Sustainability Learning: On the Importance of Decolonizing Ways of Knowing and Relations with the More-than-human," *Environmental Education Research*, Vol. 23, No. 10 (2016), pp. 1489–1500. https://doi.org/10.1080/13504622.2016.1263279.
417 David Selby and Fumiyo Kagawa, "Teetering on the Brink: Subversive and Restorative Learning in Times of Climate Turmoil and Disaster," *Journal of Transformative Education*, Vol. 16, No. 4 (2018), pp. 302–322. https://doi.org/10.1177/1541344618782441.
418 Selby and Kagawa are citing Mayer Hillman, Tina Fawcett, and Sudhir Chella Rajan, *Suicidal Planet: How to Prevent the Global Climate Catastrophe* (New York: St. Martin's Press, 2017).
419 Allison Anderson, "Climate Education for Mitigation and Adaptation," *Journal of Education for Sustainable Development*, Vol. 6, No. 2 (2012), pp. 191–206.
420 Shirley Walters, "Learning about Climate Justice in Times of Drought and Covid-19," *International Journal of Lifelong Education*, Vol. 41, No. 1 (2022), pp. 45–59. https://doi.org/10.1080/02601370.2021.1982038.
421 Robert Engelman, "Beyond Sustainbabble," in The Worldwatch Institute (Ed.), *Is Sustainability Still Possible?* (Washington, DC: Island Press, 2013).
422 Engelman, "Beyond Sustainababble," p. 5.
423 Elizabeth A. Lange, "Transformative Sustainability Education: From Sustainababble to a Civilization Leap," in M. Milana et al. (Eds.), *The Palgrave International Handbook on Adult and Lifelong Education and Learning* (London: Palgrave Macmillan, 2018), p. 403. https://doi.org/10.1057/978-1-137-55783-4_21
424 Paul Hawken, *Blessed Unrest: How the Largest Movement in the World Came into Being and Why No One Saw It Coming* (New York: Viking, 2007).
425 Simon Dresner, *The Principles of Sustainability* (London: Earthscan, 2002).
426 Daniela Tillbury, Robert Stevenson, John Fien, and Danie Schreuder, *Education and Sustainability: Responding to the Global Challenge* (Gland, Switzerland: International Union for Conservation of Nature and Natural Resources (IUCN) and The World Conservation Union (IUCN), 2002).
427 John Holford, "The Misuses of Sustainability: Adult Education, Citizenship and the Dead Hand of Neoliberalism," *International Review of Education*, Vol. 62 (2016), pp. 541–561.
428 Selby, *Thoughts from a Darkened Corner*, p. 25.
429 Kopnina, *Education for Sustainable Development (ESD)*.
430 Heather Farley and Zachary Smith, *Sustainability: If It's Everything, Is It Nothing?* (London: Routledge, 2020), p. 3.

431 Carlos Torres, *Education and Neoliberal Globalization* (New York: Routledge, 2009) in Selby, *Thoughts from a Darkened Corner*, p. 26.

432 Bob Jickling and Stephen Sterling, *Post-Sustainability and Environmental Education: Remaking Education for the Future* (Cham, Switzerland: Palgrave Macmillan, 2017).

433 Edgar González-Gaudiano, "Education for Sustainable Development," *Policy Futures in Education*, Vol. 3, No. 3 (2005), pp. 243–250.

434 Jickling and Sterling, *Post-Sustainability and Environmental Education*, p. 4.

435 Jickling and Sterling, *Post-Sustainability and Environmental Education*, p. 5.

436 Jickling, "Preface," in Jickling and Sterling, *Post-Sustainability and Environmental Education*, p. xiii.

437 Jickling and Sterling, *Post-Sustainability and Environmental Education*, p. 1.

6

TRANSFORMATIVE SUSTAINABILITY EDUCATION

The Not New and Improved

It's all a question of story. We are in trouble just now because we do not have a good story. We are in between stories. The Old Story—the account of how the world came to be and how we fit into it—is not functioning properly, and we have not learned the New Story. The Old Story sustained us for a long time. It shaped our emotional attitudes, provided us with life purpose, energized action.

DOI: 10.4324/9781003159643-6

> It consecrated suffering, integrated knowledge, guided education. . . . Today, however, our traditional story is nonfunctional in its larger social dimensions, though some persons believe it firmly and act according to its dictates. . . .
>
> Where can we begin? My suggestion is that we begin where everything begins in human affairs, with the basic story, the account of how things came to be at all, how they came to be as they are, and how the future of life of humanity can be given some satisfying direction. We need a story that will educate human beings, heal them, guide them.[1]

We are at a pivotal historical moment where many agree with Thomas Berry that our current societal story no longer serves us. Ecological linguist Arran Stibbe asserts that there is now a historic search for new stories that we can live by,[2] not only to survive but eventually thrive in healthy relation to our home, the Earth. It is a time for deep questions, including, who are we as a species, what is our purpose, where have we come from, and where are we going?[3] The Modern story had one set of answers to these questions, some of which I have traced in earlier chapters.

American author Charles Eisenstein suggests that the task immediately before us is guiding ourselves through the "empty space between stories and into a new story."[4] He continues,

> Like the crisis, the transition we face goes all the way to the bottom. Internally, it is nothing less than a transformation in the experience of being alive. Externally, it is nothing less than a transformation of humanity's role on planet Earth.[5]

This is neither a technocratic task nor one based on highflying rhetoric. Rather, it is a reimagining that is grounded in ancient wisdom, newer discoveries in science, and inspiring social practices being carried out in quiet places that we know can carry us into this new story. This task of transformation calls on educators to integrate the best of what they know and actuate it in this moment. Many of you are doing this, already.

However, as education professor Paul Clarke says, we also know that the field of education, as currently structured, is "not fit for purpose" as it is one of the "main carriers of the existing dysfunction."[6] Therefore, we need to rethink and reimagine education from the ground up in terms of purpose and structure. Transformation concerns what story we are seeing/perceiving, what story we are telling/teaching, what story we are enacting each day, what story we are imagining/leaning into, what story we (individually and collectively) are being/becoming, all within the story that is emerging as part of our commoning, or the way a community is designing itself daily.[7] This requires mutually constituting cultural and structural transformation. These hopeful stories guide the structure of the next three chapters.

As educators, our task is to carefully enable ourselves and then our learners to loosen their grip on the old stories while fixing their eyes for glimpses of the new story on the horizon. Every act of learning through a transformative process takes one more transition step into a regenerative and sustainable society. Knowing and understanding the transformative process and the role educators can play as transformative leaders are crucial. Finding the open spaces for such a transformative practice is also crucial.

As sketched in previous chapters, environmental education and sustainability education (ESE) including ESD constitute an ecology of approaches and known effective practices. We need to build from what we know and "stretch" into what we don't know, as sustainability educator Katie Ross says.[8] Environmental educators Matthew Harmin, MJ Barrett, and Carolyn Hoessler discuss "epistemological stretching" as pedagogy that expands on ways of knowing. One example is expanding perspectives on "relations," through both nonwestern ways of knowing and knowing with the more-than-human.[9] Katie Ross goes further to discuss the need to stretch from separatist logics into relational meaning-making along multiple facets—cosmological, ontological, anthropological, self, spiritual, and epistemological including axiology, aesthetics, rhetorology, causality, time, beauty, and societal vision.[10] She calls this "cosmo-onto-axi-epistem-ological stretching."[11]

Mexican researcher on "EE for sustainability," Edgar González-Gaudiano, asserts that environmental education remains an "open, unstable, multiple, relational discourse"[12] where "divergence and antagonism"[13] keep the field evolving, including ongoing contestation over ESD. Environmental and sustainability educators in Latin America and the Caribbean, have all advocated for a comprehensive approach and action-oriented integration of social justice, equity, diversity, and ecology within sustainability education.[14] Yet there is still ambivalence around priorities, or the heart of practice, polarizing the priorities either of environmental degradation or degrading human circumstances such as poverty. Framing sustainability within this dualistic dilemma limits thinking by staying within a reductionist perspective and ongoing debates about trade-offs.[15]

ESD is rife with contestations. González-Gaudiano says, ESD was "offered to us as the new integral solution that will overcome the deficiencies and imperfections of previous solutions."[16] Mary Paden reports, in all her global conversations on every continent, environmental educators struggled with the concept of sustainability. Sustainability was yet another new concept which did not emanate from a "field of study," yet was concerned about similar issues. Paden differentiated EE from ESD by insisting that a *systemic perspective* must be taken, enabling us to "get beyond the new and exciting management/technical approaches . . . to grapple with the more complex issue of how to promote all three E's together [environment, economy, equity]."[17] Others argue that we also need a *process way of thinking* to stretch beyond existing conceptualizations and debates.[18]

The larger historical context within which sustainability education has resided is the neoliberal political economic regime which deliberately created backlash to environmental education and undermined sustainability education in myriad ways. While often a technocratic approach, ESD *did* make strides in crafting a global education agenda and progressing towards sustainability goals. However, the deepest roots of unsustainability remain unaddressed, threatening to "business as usual." That said, the ecology of approaches in what is now called environmental and sustainability education (ESE) has always been important in continuously moving public commitment and action towards sustainability. No effort is wasted, with the result, as stated by Peter Blaze Corcoran in the previous Chapter 5, that environmental and social "sustainability has become the metanarrative of our time."[19]

This book does *not* offer a new and improved solution, to be "part of the manic quest to be new and innovative,"[20] as this too is one of the root causes of unsustainability. Most of the resources we require for transformative sustainability education are already available. So, this book integrates evidence-based transformative learning practices with deep sustainability education from a Relationality worldview. This chapter examines this integration, while the next chapter, Chapter 7, examines the old imaginary story of education across modern history) and the imaginal work needed for epochal shift. Finally, Chapter 8 describes old wisdom practices as well as hopeful new social and educational exemplars around the globe, which challenge some of the deepest roots of unsustainability. The goal is to paint a polyarchy of integrated possibilities for transformation, sustainable societies, and education/learning that educators can adapt into their unique contexts.

Introduction to Transformative Learning

Starting in 1975, Jack Mezirow and his wife Edee offered foundational theorizing on transformative learning for the adult education field, shifting the field away from the predominance of behaviourism and towards humanist constructivism where meaning-making is central. His theory-in-progress was an amalgam of influences, from Thomas Kuhn on paradigms, social scientists like Jurgen Habermas on critical theory, educators like Paulo Freire on *conscientization*, or rather, a deepening critical awareness of sociocultural reality, as well as psychologists like Stephen Gould on psychological humanism and difficult life transitions. Mezirow worked across personal and social dimensions to account for the cultural messaging system which shapes individual frames of reference. For him, "the formative learning of childhood becomes transformative learning in adulthood," or ought to.

The term *trans* means to "go across" and formation, and *formus* or *morpheus* means "morphing" or "to take on a new shape."[21] More than just a phase of adult development, transformation involves a fundamental structural change.[22] Perspective transformation, as per Mezirow's theory,[23] involves questioning and changing one's assumptions when they no longer serve. Mezirow considered a change in one's personal frame of reference as a structural change in how they view themselves and the world. He said there is a range of transformative learning: transforming assumptions on technical knowledge (through instrumental reflection); transforming assumptions on communication and meaning-making processes (through practical reflection on norms, values, and other meanings); and transforming assumptions about social systems (critical or emancipatory reflection on power relations). Alongside Mezirow's psycho-critical approach to transformative learning, there have been two other primary theoretical approaches, the social-emancipatory approach based on Paulo Freire's work[24] and the psychoanalytic approach drawing from Carl Jung's work.[25] For more detailed typologies, see Edward Taylor.[26]

Mezirow considered transformative learning a quintessential process for adults in the Modern Age. Transformative learning appears at that time in adult life when an individual realizes they have "uncritically assimilated ideas"[27] or, rather, approved ways of seeing and understanding they inherited from childhood and the culture around them. They realize, often through a disorienting dilemma or personal crisis, that "applying old ways of knowing" is no longer working.[28] A disorienting dilemma can be triggered by an unexpected event, person, or idea which one cannot make sense of within their existing framework of understanding. This can be triggered by the death of someone close, divorce, serious illness, or accident, travel, conflict, burnout, addiction, or trauma. It can also occur through engagement with diverse perspectives as well as new ideas provided through an educational course, book, or film.

Often, adults will attend an educational programme after a disorienting dilemma has already occurred, further loosening their old ways of knowing. According to Mezirow, the overall purpose of employing transformative learning in an educational space is to assist adults in developing their moral decision-making and deliberative practices for democratic citizenship. The goal of a transformative learning process is to help adults develop "more inclusive, discriminating, permeable, better validated, and autonomous perspectives."[29]

Although individuals may be presented with a transformative opportunity, it does not mean that they will engage. Habits of expectation and frames of reference "determine the scope of our attention" and thus can be "self fulfilling prophecies."[30] Learners may also respond by maintaining "rigid and highly defended thought patterns."[31] Mezirow suggests this is a form of "hardening of the categories."

Mezirow identified nine phases of change in addition to the *disorienting dilemma*, which is the first phase,[32] although the universality of these nine phases has been widely contested. Nevertheless, there are some core elements that educators can recognize and facilitate.[33] The first is *individual experience*, where the life experience an adult brings to a learning situation becomes the "primary medium of transformative learning" as the "starting point" for "critical examination of normative assumptions" underpinning value judgements.[34]

A second core element is *critical reflection* which promotes a process of questioning deeply held assumptions and beliefs from prior experience. It is digging to understand the assumptions upon which existing knowledge and beliefs are based. To guide educators, Mezirow suggests variations of epistemic questions such as: What is your frame of reference [on this topic]? Which assumptions support your frame of reference? Where does the knowledge it represents come from? How have your assumptions employed taken-for-granted definitions of the issue? Are some dimensions of the issue left unconsidered? What and how have your assumptions been influenced by significant others? What and how have assumptions been influenced by fear, anxiety [and any other emotions], intuition, habit, or self-concept?[35]

The third core element is *dialogue*. The role of a facilitator is to provide a dialogical learning space where learners can examine their assumptions and belief frameworks in a supportive environment. This requires high levels of trust and compassion as learners hear and test new meanings, explore new ideas and knowledge, experiment with new behaviours, and develop new capacities. In this way, an educator can provide a "learning sanctuary" or a protected space outside of the rough-and-tumble of daily life, which enables safe questioning and exploration.[36] Patricia Cranton adds that encouraging ways of being that are authentic[37] and genuine with one another also assists in creating mutuality in the learning space. Offering practices to learners for engaging conflict, managing emotion, and showing empathy can be part of adult developmental learning. With time, learners eventually integrate all these learnings into a new way of knowing and autonomous thought, sometimes beyond the formal learning.

As is evident, Mezirow's theory stays at the cognitive level, meaning that it refers to a change in how one thinks. It is also largely rational, in assuming that people use reason to rethink their belief system. However, transformative theory has expanded significantly, especially in considering more *holistic modalities for learning*, the fourth core element. Research has demonstrated that emotion and relationships are vital aspects of the transformative process. Thinking frameworks and beliefs are often associated with emotions, sometimes carried somatically in the body, or carried psychically, emerging through dreams and meditation.

Similarly, the process of examining beliefs can be highly emotional, dependent on the topic. Research has demonstrated that pedagogical engagement can be less about reasoned discussion and more about using art, movement, music, drama, myth, poetry, contemplative practices, psychic material in dreams, and other expressive modes of exploration, as part of whole person learning and sense-making that digs under rationality. Types of relationships found to be impactful in a transformative process include love relationships, memory relationships, imaginal relationships (envisioning) as well as peer relationships. Good peer learning in an educational space includes, when possible, nonhierarchical relations, feedback without evaluation, voluntary participation, freedom in partner selection, authentic expression, and mutual goal setting.[38] These features establish the conditions for social learning or rather learning from observation and modelling.[39]

The fifth core element is *awareness of context* in terms of understanding the larger context of society at any one moment as well as all the contexts brought into the learning space with each participant. Prior individual experiences and their current events and issues can shape a learner's predisposition for change as well as offer "pedagogical entry points"[40] through which to engage learners in high interest learning.[41]

The most holistic definition of transformative learning, which integrates the three dominant theories and best relates to sustainability, comes from Edmund O'Sullivan and the Ontario Institute for Studies in Education at the University of Toronto:

> Transformative learning involves experiencing a deep, structural shift in basic premises of thought, feelings, and actions. It is a shift of consciousness that dramatically and permanently alters our way of being in the world. This shift includes our understanding of ourselves, our self-locations, and our relationships with other humans and with the natural world. It also involves our understanding of power relations in interlocking structures of class, race, and gender, our body awareness, our visions of alternative approaches to living, and our sense of possibilities for social justice, peace, and personal joy.[42]

O'Sullivan takes an "integral" approach to transformative education for what he calls "integral planetary education," considered necessary for this "transitional period of history."[43]

His effort is to place learning within the "totality of life's context" while building from Thomas Berry's framework of survive, critique, and create. In other words, we need to first *survive* the coming epochal transition while using educational processes that deal with denial, despair, and grief. We need to engage in a transformative *critique* of our failing dominant cultural form by attending to the "deep ontological basis of Western European thinking" and the modern scientific-industrial worldview. Following Canadian philosopher John Ralston Saul's assertion, our society is largely unconsciousness and only superficially democratic, given that the current knowledge systems and ideological systems, particularly corporatist ideology, camouflage reality and overwhelm us with useless information.[44] Part of critique is to deconstruct hierarchical power which includes patriarchy and imperialism. Then, just as "living systems adapt by transforming themselves, and learning occurs," so can human cultures *create* through "education for integral creativity."[45]

Such an education considers the "dynamic wholeness that encompasses the entire universe and [the] vital consciousness residing both within us and, at the same time, all around us in the world," towards an "ecological selfhood" and "sense of place."[46] He concludes:

> [We] must understand that transformative education fundamentally questions the wisdom of our current educational ventures. Our present educational institutions, which are in line with, and feed into, industrialism, nationalism, competitive transnationalism, individualism, and patriarchy, must be put into question.[47]

For O'Sullivan, the pathway is a "major revolution in our view of the world that came with the paradigm of modernism," including a much wider cosmology, a conclusion shared in this book.

In deep sustainability education, then, transformative learning refers to questioning and transforming both personal belief structures as well as societal ideational and material structures. As O'Sullivan contends, a transformative sustainability education examines power relations embedded in political and economic constructs as well as in cultural belief systems that reproduce these relations. It is digging below the surface of daily thinking and acting habits to uncover the "story" that currently drives understandings and habits.

Despite Mezirow's incorporation of critical theorists, what was missing in his work was a deeper accounting of these power structures within the modern political economy. Modernity has been a new form of violence, illustrated in the military industrial complex and new global forms of exploitative capitalism, oppressive politics, more efficient and deadly means of war, environmental destruction, and unjust social structures. As Transformative Learning professor Chad Hoggan has asserted, transformative learning theory can shift the patterns of *how* we think, not just *what* we think, as a metatheory.[48] Shifting how we think helps address underlying causes of systemic violence.

Within the context of Transformative Sustainability Education, transformative learning undertakes a critique of Newtonian ontology and Cartesian epistemology, disrupting some of our deepest embedded assumptions. As quantum scientists carried out their research which came to constitute the New Science, many of them experienced an existential disorientation and deep transformation that took them into a profoundly new worldview. So must we. The following describes this shift towards Relationality, a core feature of Transformative Sustainability Education.

Relationality: Pathways into Transformation

> We are profoundly relational beings who have been living—with some difficulty—in anti-relational systems of thought and ways of doing things. Finally, all that has begun to change.[49]

As cultural historian Charlene Spretnak[50] describes, we are now undergoing a Reality Revolution[51] from the mechanist way of knowing and being towards a relational way of knowing and being. She calls this the Relational Shift. This is less a shift to something new, than a shift back to wisdom that is already known by humanity, but was eclipsed, now rediscovered through the New Science. This relational shift has much potential to address the crisis *in* education and constitutes a transformative approach to sustainability education. Relaxing our current parameters of thinking and inching towards a relational understanding, described later in the chapter, educators can begin creating deeply transformative learning spaces.

Conceptualizing Relationality

There tends to be a misunderstanding about the terms relational and relationality. They are *not* referring to social relationships or even ecological relations, although they are a part of them. A relational ontology is a relational understanding of reality and existence.[52] A relationing way of seeing perceives reality as an organic web of relations, where the cosmos is connected as a vast sea of energy, and all things are connected within a living system. Scientists in various disciplines, such as Fritjof Capra, Ervin Laszlo, Elisabet Sahtouris, Francisco Varela, and others, turned to Indigenous and Eastern philosophy, poetry, and mystical spirituality for the language to explain this "new map" of reality, this new seeing and perceiving.[53]

While science gave us reductionist thinking with its insights and benefits, Ervin Laszlo says that "science [is now becoming] an instrument for the recovery of the wholeness of the universe, and with it the wholeness of all that exists in it, including ourselves, our thoughts, our feelings, our dreams, our fears and our hopes—and, above all, our visions and our creativity."[54] Laszlo expands,

> This vision of a self-creating and self-recording cosmos, evolving in a continuous if highly differentiated sweep from primeval unity to its present deeply linked diversity, corresponds to an intuition that was present in human consciousness from time immemorial. The creation myths of the widest variety of cultures agree that the things and beings of the observable universe came about as a concretization or distillation of the basic energy of the cosmos, descending from its original source. The physical world is a reflection of energy vibrations from more subtle worlds which, in turn, are reflections of still more subtle energy fields. This leitmotif runs through most mystical teachings. [. . . For instance, in] the Indian *Upanishads*, the original Source is an energy-dense space that came into being with the cosmos.[55]

This understanding that is emerging through physics provides a different cosmology, a cosmology of wholeness and purposiveness. "The universe's hospitality to life is due neither to a special act of creation nor to blind chance: it is due to progressive cosmic evolution, unfolding across a long series of interconnected evolutional cycles."[56] This story is best captured by Brian Swimme and Thomas Berry in *The Universe Story: From the Primordial Flaring Forth to the Ecozoic Era, A Celebration of the Unfolding of the Cosmos*.[57]

This relational cosmology is a different understanding of matter and space. Matter is no longer primary, the energy field is primary, expressing matter. Matter condenses out of the primordial energy field and will sink back into it. Laszlo continues,

> Space-time has not only a geometry, á la Einstein, but a basic physical reality. It is a plenum, a filled medium that can be perturbed—one that can create patterns and waves. Light and sound are travelling waves in this continuous energy field, and tables and trees, rocks and swallows, and other seemingly solid objects, are standing waves in it.[58]

To capture the sense of flowing dynamism in the cosmos and the *living* web of Earth life, Sahtouris calls this *EarthDance*. As a Greek, she went back into the Greek creation myth of Gaia, merging it with the New Science of James Lovelock called Gaia theory evidenced by NASA's images, as in Figure 6.1. This is how she describes it:

> The story of Gaia's dance begins with an image of swirling mist in the black nothingness called Chaos by the ancient Greeks—an image reminding us of modern photos of galaxies swirling in space. In the myth it is the dancing goddess Gaia, swathed in white veils as she whirls through the darkness. As she becomes visible and her dance grows ever more lively, her body forms itself into mountains and valleys; then sweat pours from her to pool into seas, and finally her flying arms stir up a windy sky she calls Ouranos—still the Greek word for sky—which she wraps around herself as protector and mate. Though she later banishes Ouranos—Uranus, in Latin—to her depths for claiming credit for creation, their fertile union as Earth and Heaven brings forth forests and creatures including the giant Titans in human form, who in turn give rise to the gods and goddesses and finally to mortal humans. . . .

FIGURE 6.1 Grand Swirls from NASA's Hubble Telescope

Credit: ESA/Hubble & NASA. Acknowledgement: Flickr user Det58.

> To satisfy [human] curiosity, Gaia let her knowledge and wisdom leak from cracks in the Earth at places such as Delphi where her priestesses interpreted it for people. . . .
>
> The new scientific story of Gaian creation has other parallels to the ancient myth. We now recognize the Earth as a single self-creating being that came alive in its whirling dance through space, its crust transforming itself into mountains and valleys, the hot moisture pouring from its body to form seas. As its crust became ever more lively with bacteria, it created its own atmosphere, and the advent of sexual partnership finally did produce the larger life forms—the trees and animals and people. . . . And in a sense, Gaia's knowledge and wisdom are still leaking from her body—not just at Delphi, but everywhere we care to look in a scientific study of our living planet.[59]

So how did this new scientific story emerge, paralleling various cosmological creation stories?

Quantum Worldview

From the early 20th century, the findings of relativity and quantum physics initiated a deep revision in the understanding of the universe. The emerging findings challenged most modern assumptions. For Einstein, the laws of classical mechanics could no longer be reconciled with the findings in electromagnetics. In developing relativity theory, he proposed that space at the quantum level comprised energy fields of differing strengths, a force field. Matter was understood more as a bound form of energy, like a standing wave in the ocean.

The fundamental reality is that the "substructure of the universe is a sea of quantum fields."[60] In other words, there is a medium that interconnects all of reality, hence the notion of relationality as fundamental reality. Space is not a vacuum or emptiness but a complex network of energy fields. The universe is connected by waves of energy, continuously moving out over time and space. As Laszlo described it, the substrate of the universe is a cosmic plenum,[61] a term Aristotle originally used.[62] Others call it the Zero Point Field,[63] the place where all fields merge, a matrix or "field of fields."[64] The implication is that there is a huge store of energy in this field, more than all the energy contained within matter.[65] This is a crucial *first step* in relationality thinking.

At the sub-atomic level, "particles are not objects or things in the materialist sense."[66] Quanta, the smallest building blocks, are bundles of energy. However, they can also appear as particles.[67] Particles, such as electrons, can behave like waves as in light waves or as matter. Matter and energy were now considered interchangeable, a wave/particle duality.[68] Einstein's famous equation $E = mc^2$ means that there is a mass–energy equivalence. In other words, at a certain point of concentration, energy expresses itself as mass. The equation describes how much energy is needed to create the appearance of mass.[69]

In experiments, scientists found that matter could break down into waves, then re-emerge as particles. When observed, the wave congealed into a material entity, thereby becoming visible. It was also deduced that observation changes the phenomenon,[70] making objectivity or a neutral standpoint impossible. Heisenberg called this the *observer effect*. He stated, "The world thus appears as a complicated tissue of events, in which connections of different kinds alternate or overlap or combine and thereby determine the texture of the whole."[71] The context of relations now becomes a fundamental characteristic, a *second step* in relationality thinking.[72]

At the subatomic level, quantum physics also found that the properties of the universe were uncertainty, nonfixedness, and nonlinearity. While matter *looks* solid, at the quantum level it is actually fluid, malleable, and in constant creative flux. No particle is ever at rest; it is always vibrating or trembling to some degree. "All elementary particles interact by exchanging energy mediated by other virtual quantum particles"[73] which exist in a fleeting way, causing fluctuations of energy. Heisenberg concluded that the precise properties of a particle could never be fully known because of the constant fluctuation, called the *uncertainty principle*.

Further, in developing quantum mechanics beyond Einstein, scientists discovered that motion could occur in leaps from one energy state to another, known as a *quantum leap*. Predicting particle behaviour through precise measurement was no longer possible but could only be expressed in terms of probabilities, called *probability theory* in mathematics. Particles not only appear unpredictably as matter or energy, but they could also appear randomly in various times and places, even simultaneously. There were no longer causal, deterministic links between particles, as traditionally thought in Newtonian terms, only properties that emerge dependent on context.

It also became clear that nonlocal entities, or rather entities at a distance, can instantaneously impact each other and that this can happen faster than the speed of light.[74] With the discovery of nonlocal coherence, or nonlocality, any illusion of separation disappeared. Space as traditionally conceived by science was no longer accurate. Subatomic particles could only be understood through an indivisible and complex web of relations.[75]

BOX 6.1 KEY ELEMENTS OF A QUANTUM WORLDVIEW

A summary of key elements of this new quantum worldview includes:

- a quantum object can be at more than one place at the same time, which is a wave property;
- a quantum object is only in ordinary space–time when it collapses from a wave into a particle, where we can view it;

- a quantum object can make a quantum jump, ceasing to exist in one place and simultaneously appearing over in another place, without travelling through space as ordinarily conceived;
- there is a quasi-instant connection among all the parts of a thing, or the whole, called nonlocal coherence, without adhering to time as ordinarily conceived;[76]
- at the subatomic level, particles can behave as matter or as energy waves;
- there is a basic unity underlying the universe;
- matter and energy cannot be separated;
- analysing one elementary particle to understand the whole is no longer possible;
- there is a mutual interrelatedness of all things, impacting each other synergistically;
- time is relative and in motion, flowing at different rates in different dimensions;
- space is not fixed and three-dimensional but curved with a possible 13 dimensions; and
- the view of separation and autonomy, whether of atoms or humans, is only a delusion of our consciousness.

These scientists discovered that form, matter, time, and space were not static and predictable as understood in the Newtonian paradigm but dynamic and constantly changing. In sum, the universe is highly relational not mechanistic like a clockwork universe as Newton described.[77]

Einstein maintained that our task now is to free ourselves from the prison of mind that only perceives separation, as a *third step* in relationality thinking.

> A human being is part of the whole, called by us the universe. A part limited in time and space. He (sic) experiences himself, his thoughts, and feelings, as something separate from the rest, a kind of optical delusion of his consciousness. This delusion is a kind of prison for us, restricting us to our personal desires and to affection for a few persons nearest to us. Our task must be to free ourselves from this prison by widening our circle of compassion to embrace all living creatures.[78]

Gaia Theory

In the 1960s, working with NASA,[79] James Lovelock was part of a team trying to derive ways to detect the possibility of life on Mars. He realized there were few theories he could use to define life or that offered criteria by which life could be identified.[80] He blamed this lapse on the division between the natural and life sciences and the growing multiplicity of disciplines and specializations.

Drawing from physicists, he went on to define life as a process where there is an abundant flow of energy. Entropy is the naturally occurring loss of energy leading to disorder over time. Living forms continuously draw in energy from their environment to reduce entropy and thereby keep living.

The question of life then revolved around the core features of an appropriate atmosphere that could support life. How did the Earth not absorb too much solar energy, or reflect too much energy back into space, thereby creating a sweet spot for the conditions of life on Earth? With the help of microbiologist Lynn Margulis, he developed a hypothesis that all the living matter on Earth regulated favourable conditions for itself. In other words, it is the biosphere itself, containing all the life forms on Earth, which keeps Earth's atmosphere within a comfortable range to suit general needs.[81]

It was Lovelock's writer friend, William Golding, who furnished the concept of Gaia from Greek mythology. Gaia was the goddess of the Earth and one of the most revered goddesses, with multiple names across ancient cultures. In 1972, the Gaia hypothesis became a shorthand reference for the biosphere as a self-regulating entity, whose composite identity is far more than the sum of the parts.[82] Later in the 1970s, through multidisciplinary research, Lovelock and associates were able to illustrate the interactions between the organic and inorganic parts of the Earth, particularly oceans, land masses, ice, and air which help keep the blanket of atmosphere within a stable range. It is now understood that the totality of organisms, surface rocks, oceans, and atmosphere are a tightly coupled evolving system, maintaining Earth as a self-regulating living system.[83]

In the 1980s, his research team developed computer modelling that illustrated these dynamics of self-regulation, called Daisyworld, where the whole system of organic and inorganic forms regulated temperature at a level close to optimal for plant growth. Moreover, he stipulated that this was a purposeful system. The ongoing purpose is to regulate surface conditions so that it remains favourable to Life.

The theory included 10 predictions, which have been born out, as further evidence of self-regulation. Of course, Lovelock fell short of describing the Earth as alive or sentient. He said that we currently do not have the science for testing this idea.[84] Nevertheless, he identified the reactive gases of oxygen and methane in our atmosphere, which are needed for life to exist as the signature of a life-sustaining planet. Their "disequilibrium radiates an infrared signal," which Lovelock later described as an "unceasing song of life . . . audible to anyone with a receiver, even from outside the solar system."[85]

The Gaia hypothesis was originally dismissed, eventually acknowledged, and then raised to the status of a theory. Lovelock would go on to receive many awards and entry into the Royal Society. While Gaia theory has been critiqued largely from reductionist and discipline-specific perspectives, Gaia theory has not been falsified. In fact, it is clear that "Life evolves in response to environmental change, but the environment also evolves in response to biological change."[86]

Lovelock still holds to the name Gaia, even though it led to early dismissiveness. As he says, "Gaia theory is a holistic, whole-system theory, and as such cannot be modelled using the concepts of the Earth or life sciences separately."[87] In other words, the Earth is a single living entity which cannot be examined in isolated ways through disciplinary silos. His work sparked a whole new field, called Earth system science, and a new generation of research.

Lovelock passed away at 103 years old in 2022, but he remained active until his death, describing that Gaia has been resisting human interventionism through multiple negative feedback loops. One could argue that the emergence of COVID-19 as well as progressively more extreme weather events are negative feedback to arrest human behaviour and rebalance Gaia life processes. Lovelock says that if Gaia cannot maintain processes within the existing balance, and moves into a hot state, Gaia will survive but as a diminished biosphere.[88] All life forms and processes will need to adjust or die. "This is how Gaia keeps a habitable planet: species that improve habitability flourish and those that foul the environment are set back or go extinct."[89]

Public perceptions of the Earth changed significantly in the 1960s and 1970s as the astronauts brought back photos of our gleaming azure marble planet against the deep blackness of space. It prompted an instinctive recognition that there is a unity of life forms and processes on this planet. Lovelock says this vision can return us back to our Gaian sensibilities. As "a partner in Gaia,"[90] he explains, we need to balance our human needs with the needs of other life forms. He concludes,

> From a Gaian viewpoint, all attempts to rationalize a subjugated biosphere with [humans] in charge are as doomed to failure as the similar concept of benevolent colonialism. They all assume that [humans are] the possessor of this planet. . . . The Gaia hypothesis implies that the stable state of our planet includes [humans] as a part of, or partner in, a very democratic entity. . . . It may be that the destiny of [humankind] is to become tamed, so that the fierce, destructive, and greedy forces of tribalism and nationalism are fused into a compulsive urge to belong to the commonwealth of all creatures which constitutes Gaia.[91]

James Lovelock, Elisabet Sahtouris, and Ervin Laszlo all describe the dynamics of the universe as a cosmic dance, which Laszlo calls "the sacred dance of matter, life, and field."[92] Lovelock adopted this metaphor, because, he says,

> [A]ny model we make of nature is at heart metaphorical in that it begins with some image or formula familiar to us humans and used to represent the complexities of nature in simply, understandable, and useful ways. No metaphor should be mistaken for reality . . . I am increasingly impressed by scientists and philosophers who find non-mechanical metaphors for natural systems useful in interpreting Gaia theory.[93]

The implication of Gaia theory, says Sahtouris, is that "we see ourselves in the context of our planet's biological evolution"[94] so that we no longer miss the clues we are given for understanding the world and cosmos around us. She argues that we have missed these clues previously because "we have not understood ourselves as living beings within a larger being. . . . Earth [as] a live planet rather than a planet with life upon it."[95] Charles Eisenstein says that this shifting understanding "shows how close we all are, all of us ordinary humans, to a profound transformation of consciousness and being."[96]

Living Systems Worldview

Life sciences have also been putting the pieces back together, rather than continuing to probe pieces in isolation as reductionist science. In the 1920s, organismic biologists proposed that living organisms were integrated wholes.[97] This reversed the Cartesian notion that every complex system could be understood by the behaviour of the individual parts as well as the assumption that living systems were dead and lifeless, operating mechanically. Rather, the parts could only be understood by the organizational properties of the whole. Systems were no longer just mechanical parts with a function but *living* systems that were self-maintaining.

In exploring thermodynamics, Ilya Prigogine found that closed, isolated systems like machines maintain a near-equilibrium state. However, open systems, like human bodies and Earth's climate, exist in a state far-from-equilibrium, but which are stable.[98] It appeared that this state characterizes living systems. In this far-from-equilibrium state, negative feedback loops act as a self-stabilizing dynamic, maintaining a balance of forces, within liveable parameters. When fluctuations occur at a size and rate which cannot be contained within existing parameters, usually caused by positive feedback loops which are self-amplifying (like climate change), the system is brought to the threshold of critical instability. In this threshold state, the system bifurcates: either the system completely breaks down, or it spontaneously generates a higher level of organization. This latter process is called emergence.

Most biological and social systems are governed by nonlinear, chaotic relationships. In bifurcation, a living system could follow any number of possible trajectories that are not fully predictable, as seen in Figure 6.2. Linear mathematical formulas could no longer fully capture these patterns of chaotic living systems. Eventually, with enhanced computing ability using nonlinear equations, possible trajectories could be traced, yet no trajectory could be known definitely.

Scientists discovered that patterns of behaviour are each determined by a particular kind of attractor. In chaotic systems, these are called strange attractors meaning that repeated patterns are nearly similar, but not exactly, generating fractal patterns. So, what appears as chaotic has an underlying pattern, predictable within a range. Visually, these fractal patterns are repeated throughout the living world, as seen in Figure 6.3.

While conventional science and consequently modern social systems have attended to categories, quantities, and measurement, nonlinear dynamics attend to qualities and patterns.[99] Further, living systems feature patterns of complexity where small changes over time can escalate to large-scale impact, called butterfly effects.

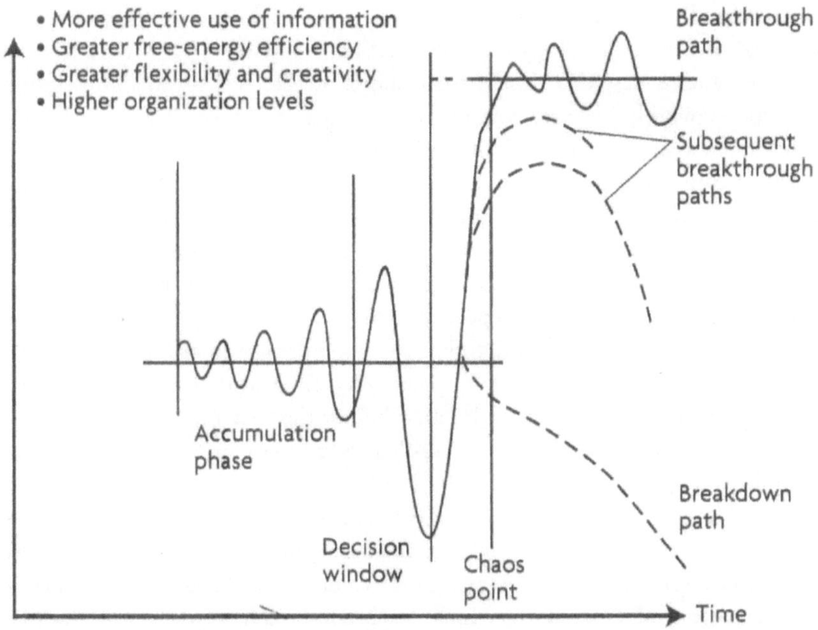

FIGURE 6.2 Bifurcation Diagram

Credit: Quantum Shift in the Global Brain by Ervin Laszlo published by Inner Traditions International and Bear & Company, © 2008 (p. 19). All rights reserved. http://www.Innertraditions.com. Reprinted with permission of publisher.

FIGURE 6.3 Fractal, Illustration

Credit: SAKKMESTERKE/SCIENCE PHOTO LIBRARY F030/7171. Reprinted with permission.

In the mid-20th century, these developments in biology were followed by the new field of ecology. Ecology proposed that Earth's living systems were all interlinked as systems nested within systems, either at the level of organisms relating to the outer world, or larger communities of organisms relating to their physical environment as an ecological whole.[100] Different levels of systems were recognized: parts of organisms, organisms, and communities of organisms. Each living system has two characteristics: not only integrating within a larger system, but also self-organizing within itself,[101] whether the human body or an ecosystem. The properties of each were emergent, in that they only emerged at that particular level of organization, not within the component parts.

Living systems therefore came to be understood as a web of life that is a living network interacting with other living networks, all in a nonhierarchical manner. This network pattern holds at all levels, whether the metabolic functions among cells or food webs in an ecosystem. This led to "thinking in terms of connectedness, relationships, patterns, and context."[102]

In studying colour perception, Chilean neuroscientist Humberto Maturana[103] discovered that the nervous system has a circular organization, which he considered a feature of all living systems, allowing for evolutionary change. He saw this as a network pattern of processes in which "each component is to help produce and transform other components while maintaining the overall circularity of the network . . . the basic 'organization of the living.'"[104] Together with Francisco Varela, they called this a process of self-organization or *autopoesis*—*auto* meaning "self" and *poiesis* meaning "making" or "self-making."[105] They explain that "In a living system, the product of its operation is its own organization."[106]

Central to the maintenance of a living system, Maturana said, was cognition. For instance, a cell's cognition enables it to interact with its environment, maintaining dynamic balance. It uses cognition internally to ensure a state of health. It uses cognition externally to maintain its individuality through functional boundary making. This self-making is another key characteristic of a living system.[107] Thus, the process of cognition *is* the process of life.[108] "Life . . . may be seen as a system of interlocked autopoietic systems."[109]

Maturana also asserted that "the nervous system is not only self-organizing but also continually self-referring," so that perception and cognition in general, "do not represent an external reality" that is independent of the organism, but "rather *specify* one through [this] circular organization."[110] From there, he reasoned that "Living systems are cognitive systems, and living as a process is a process of cognition. This statement is valid for all organisms, with and without a nervous system."[111] As Capra summarizes, Maturana identified "cognition with the process of life."[112]

Varela also resolved the traditional enmity between objectivism and subjectivism by suggesting that cognition is embodied action or "enaction." By embodied he means:

> [F]irst, that cognition depends upon the kinds of experience that comes from having a body with various sensorimotor capacities and second, that these

> individual sensorimotor capacities are embedded in a more encompassing biological, psychological, and cultural context. By using the term action we mean to emphasize once again that sensory and motor processes, perception, and action, are fundamentally inseparable in lived cognition . . . they have also evolved together.[113]

Varela and Maturana agree with philosopher Merleau-Ponty "that the organism both initiates and is shaped by the environment."[114]

In sum, quantum physicists, Gaia theorists, and organismic biologists experienced a related conceptual revolution that shifted perception from the parts and functions to the whole of patterns and relationships. This has led to what is now known as living systems thinking or systems thinking, in short. Teaching for a quantum and living systems worldview is a key task of Transformative Sustainability Education. Various facets and implications are discussed later in the chapter as interweaving strands of a whole.

Facets of Relationality for Transformative Sustainability Education

Radical Wholeness

As the New Science has revealed, the substrate of the universe is an unbroken wholeness. Relationality is "the complicated web of relations between the various parts of a unified whole."[115] The perceptual shift is moving from the *parts to the whole*. With any system, the whole is different from the sum of the individual parts. By shifting focus from the parts to the whole, we can better grasp relations.

From this New Science, physicist David Bohm[116] developed a new model of reality, which he called a radical theory of wholeness. He suggested that instead of starting with the parts, we ought to start from the multidimensional whole. Bohm suggested if we wish to move away from the partial understandings of science towards a 'nonfragmentary worldview', we need to begin with the assumption that reality is undivided, unbroken, and without borders; that reality flows, enfolds, and unfolds, containing both "internal" consciousness and "external" reality.[117] This is possible through our imaginal capacities. Interestingly, in the early 1800s, as part of the Romantic movement, poet and artist William Blake had intuited the need to reunify reason with imagination, saying "the enemy of whole vision is reasoning power's divorce from the imagination."[118]

Bohm proposed that the explicate order or material world that we live in unfolds from the implicate order which is the underlying flow of energy. The material world, the explicate order, then at some point disintegrates back into the implicate order.[119] In the implicate order, any one element contains the totality of the universe, called a holograph. In Greek, *holos* means "whole," and so holograph is an image of the whole.

An image of this can be found in the Buddhist metaphor of Indra's net, where every pearl holds the reflection of all the other pearls. In other words, the universe is an undivided wholeness enfolded into an infinite, timeless background source. It continually unfolds into visible and temporal material and then enfolds and recedes back into the background source.

Conscious thought can grasp the unfolded and material, but only a consciousness *beyond rational thought* can experience the enfoldedness of the implicate order. The challenge, Bohm says, is to transform our selves to live in time as well as in the timeless, simultaneously. This points us to the importance of spirituality and other expressive mechanisms, such as art and poetry. Fritjof Capra concludes,

> If physics leads us today to a world view which is essentially mystical, it is not only based on intuition, but also on experiments of great precision and sophistication, and on a rigorous and consistent mathematical formalism.[120]

So, a flowing between scientific description and spiritual sensibility helps express the relational nature of reality, the beginnings of a reenchantment of the universe.[121] Rather than see spiritualities as superstition, as religion used for purposes of oppression, or as myth which has been thought of as fantastical tales, all of which "muddied" science, this reenchantment involves a metaspirituality, blending truths common across spiritual traditions, scientific insights, and a broader range of human knowing and living patterns.

Realistically, however, we can only discuss reality in fragmentary ways, but we can do so understanding that is it part of a whole and that this whole is its own reality. One way to understand this is through the image of nested systems. Our circulatory system is nested in our body system which is nested in social systems which is nested in the living systems of Earth which is, in turn, nested in a cosmos system. This approach is not describing reality "as it is" as much as it is offering insights and images into a reality that is implicit, not fully describeable or specifiable.[122]

Paradox helps us to understand knowledge as whole *and* partial, reality as real *and* constructed, the cosmos as unity *and* diversity, thinking as reason *and* imagination, and human growth as autonomy *and* relatedness. Thus, one part of the challenge is to take up *both/and* thinking rather than *either/or* thinking. Bohm suggests that we *choose* to see wholeness: "if you say wholeness is ultimate, you always leave room for the partiality where it is called for. Then your attitude is going to be one of bringing about wholeness ultimately."[123]

The uptake of the emerging quantum and living systems paradigm has been slow, given the complexity of the concepts and the radical departure from common perceptions. However, from the 1970s to 1990s, this relational shift expanded into the fields of neuroscience, consciousness studies, psychology, archeology, philosophy, cosmology, politics, ecology, feminist theory, and theology.[124] Out of this development was the move towards *transdisciplinarity*, which is founded on a relational logic and perception.[125]

As professor of transformative inquiry Alfonso Montuori explains, transdisciplinarity is "inquiry-driven not discipline driven."[126] It is not multidisciplinary in examining a problem from the lens of different disciplines, and it is not interdisciplinary in using methods from one discipline to inform another.[127] Transdisciplinarity is a new approach to inquiry that is metaparadigmatic in examining the paradigmatic assumptions used in the construction of knowledge and generating metaquestions. It uses complexity thinking which considers context and relationship. It strives to overcome paradigmatic blind spots and projections that drive disciplinary territoriality.[128] It also integrates "the knower in the process of inquiry" including body, emotions, and mind, as a process of self-inquiry. One other feature is that it avoids the "dualism of opposing binary pairs"[129] which is typical in scholarly inquiry. As part of systems thinking, it examines the "excluded middle" between pairs, examining the relation between.[130]

Romanian philosopher Basarab Nicolescu suggests that transdisciplinarity is far more capable of addressing the pressing complex issues facing us by reconciling the sciences, social sciences, and humanities. For instance globalization is a process of homogenization generated through both cultural and economic contact, while generating profound conflict and reactionary defense postures. Another gulf is between science and spirituality, he says. By studying the space between, a new attitude towards inquiry as well as a new vision emerges, revealing a fruitful relationship. In this case what is found is a different definition of the sacred:

> The sacred is first of all an experience; it is transmitted by a feeling—that "religious" feeling—of that which links beings and things and, in consequence, induces in the very depths of the human being an absolute respect for the others, to whom he (sic) is linked by their all sharing a common life on one and the same Earth. . . . In fact, the presence of the sacred is our own human transpresence in the world.[131]

Nicolescu calls a view of the totality of levels of perception "transperception" which can yield intense "moment[s] of scientific and artistic creation."[132] These are the transdisciplinary dialogues he suggests we now need to have to help us traverse the decline of civilization.

Systems Thinking

Seeing whole requires systems thinking.[133] Just as the quantum physicists had to struggle to achieve this conceptual shift,[134] so must we find footholds to anchor these new understandings. This conceptual revolution requires several shifts in perception, which lead to different ways of teaching and learning, different ways of organizing institutions and society, and different social change dynamics. As articulated by *Center for Ecoliteracy* cofounder Fritjof Capra,[135] these shifts are not either/or alternatives but rather movements along a multifaceted continuum, a stretching[136] towards understanding.

As Donella Meadows explains in *Thinking in Systems*, there are "many fractious schools of systems thought."[137] Living systems theory draws more from the disciplines of biology, chemistry, ecology, physics, and cognition, discussed at length by Fritjof Capra and Pier Luigi Luisi in *The Systems View of Life*.[138] Modern systems theory often draws from computers, cybernetics, and complexity mathematics. Meadows and other living systems theorists use the language and symbols of system dynamics. This approach is based on the MIT System Dynamics group with Jay Forrester as one of the leads. The purpose of systems thinking is to create "a basic ability to understand and deal with complex systems."[139] Meadows describes a system as

> [A] set of things—people, cells, molecules, or whatever—interconnected in such a way that they produce their own pattern of behaviour over time. The system may be buffeted, constricted, triggered, or driven by outside forces. But the system's response to these forces is characteristic of itself, and that response is seldom simple in the real world. . . . The system, to a large extent, causes its own behaviour![140]

All living beings are complex systems whether they are a forest, organization, or humans. Thus, "a system is an interconnected set of elements that is coherently organized in a way that achieves something."[141] To examine a system is to look for "elements, interconnections [relationships], and a function [nonhuman system] or purpose [human system]."[142] These mechanisms maintain the integrity of the system.

Systems thinking requires distancing from our rational abilities towards use of our other abilities. It requires distancing from our tendency to think in linear ways, like cause to effect. It requires distancing from looking "out there" or "in here," binaries which simplify that which is complex.

"Seeing" the universe, Earth, and human communities as an interconnected web of relations and nested systems with unique properties and *patterns* of organization is part of this shift. Moving from "What is it made of?" to "What is its pattern?" revives the sense of a living universe in which patterns of relationships exist within an organized harmonious whole.[143]

Within systems, certain configurations of relationship appear again and again in patterns. Understanding how a pattern works in one natural or social system helps us to understand other systems that manifest a similar pattern. For instance understanding how flows of energy affect a natural ecosystem may illuminate how flows of information affect a social system.

As one example, historian Jeremy Lent[144] has examined his discipline from this pattern-finding perspective. To understand the trajectory of human history, Lent blends a materialist analysis (historical changes which come from geographical, technological, and environmental causes) with an idealist analysis (historical changes which come from the human mind constructing reality yielding ideas and behaviours within specific physical and environmental contexts) to understand the variable ways in which social change occurs and how we might facilitate it. In taking a "cognitive approach to history" and adopting the concept of feedback loops, he says "the cognitive frames through which different cultures perceive reality . . . [which] create a pattern of meaning in people's lives . . . have had a profound effect on their historical direction."[145] In Western modernism, "the cognitive framework went something like this; the Truth has been discovered by Science, which leads to continual Progress as a result of Man using his unique faculty of Reason for the benefit of all applied to all cultures."[146]

However, Lent says, this is not a one-way street of causality. Humans also change their environments. It has been "reciprocal loops [that] transform societies and ultimately shape the course of history."[147] Humans not only adapt to their environment to develop certain kinds of livelihoods, but they also actively construct a niche for their own needs, developing specialized attributes and a set of shared symbols and practices.

> Each society shapes the cognitive structures of individuals growing up in its culture through imprinting its own patterns of meaning on each infant's developing mind . . . [this] establishes structures of cognition that influence us to perceive, understand and think about the world according to certain patterns . . . ultimately affect[ing] the actions and choices they make in the world. When aggregated to an entire civilization, these patterns of meaning shape history and fundamentally alter the world around us.[148]

Similarly, linguist George Lakoff suggests, "new metaphors have the power to create a new reality," just as new realities generate new metaphors.

Capra and Jakobsen indicate, "The basic pattern of organization of a living system is the *network*."[149] All ecosystems are networks of relationship, which comprise the web of life. The primary relationships are cooperation, partnership, and networking. Thus, mapping relations is a vital capacity for understanding a system as an integrated whole, done with humility as relations can never be fully disentangled and accounted for.[150]

When system theorists suggest "seeing whole," they look for the interconnections or relationships between elements, not the elements or dissected pieces of elements. In the systems view, the "objects" of study are the networks of *relationships*. An ecosystem is not just a collection of species but living things interacting with each other and their nonliving environment. Systems thinking then considers the "relationship between structure and behaviour."[151] To do this is to watch for the relationships that comprise a system and how a system behaves, not its rhetoric or goals.[152]

In the living systems view, *nested systems* are integrated wholes whose properties cannot be reduced to smaller units. Thus, the reductionist approach of understanding a system by the individual components cannot understand the patterns of complex systems where certain properties only emerge at certain levels of organization. It is a shift of figure to ground, from fragment to whole, and from static to dynamic.

Living systems exist at all ecological scales, whether forest or ocean ecosystems, or social systems such as a family or work organizations. Systems have multileveled structures nested within each other, such as cells and organs nested in the human body, which is nested within the economy, nested within the society, nested within ecosystems, nested within the Earth system (Gaia), which is in turn nested within the cosmic system. Each of these levels forms an integrated whole, while at the same time is part of an even larger whole.

Western science has focussed on "things" that can be measured and quantified. This implies that the phenomena which can be measured and quantified are important —with the implication that what cannot be measured and quantified does not count or does not exist. Rather than measurement which focusses on quantity, systems thinking is looking for *qualities* of relationships in networks.

As systems are nested within other systems, there are usually "purposes within purposes."[153] "Keeping subpurposes and overall system purposes in harmony is an essential function of successful systems." To change a system, changing the relationships is far more powerful than changing the elements. Changing the purpose for the relations can be the most profound level of change. Yet, elements, interconnections, and purposes are equally important. Thinking in this way can avoid some of the current conundrums in sustainability.

Shifting focus from the parts to the whole implies shifting from analytical thinking to *contextual thinking*. As Capra and Jakobsen explain, "every organism is constituted by its connections to other organisms."[154] Further, these relationships between organisms constitute a system, which itself is embedded in a larger system or context. Thus, "systems thinking is always contextual thinking."[155] Rather than a "reduced view," this is a "unifying view across multiple scales, levels, and perspectives."[156]

Living systems develop and evolve. Understanding these systems requires a shift in focus from structure to *process*, such as evolution, renewal, and change. Organisms all develop as part of various processes, all interwoven. With the continual processes of energy *flow* and matter *cycles*, normal development, decay, and regeneration occur. Further, all earth systems and living beings maintain *dynamic balance* through feedback loops. Positive feedback loops accelerate the direction of change, and negative feedback loops counter the direction of change.

Whether organisms or ecosystems, all life changes over time through cognition and learning. Learning cultures lead to creativity and flexibility.[157] Joy Kcenia Polanco O'Neil draws from Fritjof Capra's ecological living principles, bringing them into Transformative Sustainability Education. In learning spaces, she calls this a "*living learning system*."[158] As we live and learn key ecological principles within "education *as* sustainability," the focus is on the processes of learning, *as* change. This is a nondual view of education that does not divide educator and learner into subject and object. In an applied approach, using kitchen-based learning to teach about Sustainable Food Systems, she views "food, cooking tools, students, instructor entangled co-evolving, co-creating material, and the cycle of learning activities (transforming food, eating, sensing, discussion) as iterative stabilizing and destabilizing process of learning."[159] Learning is performed, emerging out of the entangled relations, dynamics, and processes.

The tension between environmental education and sustainability education occurs given that they are currently structured as a binary pair. This can be overcome by looking to "the middle" in terms of the relationship between them. Understanding them both as part of a larger system and context can reorganize our way of thinking away from competitive territorialities. As Montuori says, "disciplinary fragmentation . . . actually *frames* knowledge [and] creates blind spots by framing the world in a discipline-driven way that actually prevents certain subjects from being 'seen.'"[160]

In sum, relationality entails *perceptual shifts* towards: the whole, relationships, context, quality, process, and patterns.[161] Similarly, learning the "*language of nature*"[162] or rather the principles of ecology as seen in Figure 6.4, involves six dynamics: networks, nested systems, cycles, flows, development, and dynamic balance.[163] In the view of Capra and Luisi, "a sustainable society must be designed in such a way that its ways of life . . . do not interfere with nature's inherent ability to sustain life."[164] "From the systemic point of view, then, the only viable solutions are those that *are* sustainable."[165]

To build a society that respects ecological processes, an understanding of the basic facts of nature's patterns and processes is vital, called ecoliteracy. Systems thinking, as part of ecoliteracy, mimics natural processes, but so must we mimic natural design in our all activities, so that we are not interfering negatively with the processes of Life. This is called "biomimicry" by Janine Benyus,[166] where *bios* means life and *mimesis* means imitation. The principles of biomimicry are:

- *nature as model*, where science studies the ways life solves problems and executes tasks, then designs accordingly;
- *nature as measure*, where nature is the standard by which to judge what works and what will last, emulating "life's genius;" and
- *nature as mentor*, where we learn from the living world rather than extract from it.[167]

Ecopioneers, bioneers,[168] and other nature-based innovators are part of the biomimcry revolution remaking how "we grow food, make materials, harness energy, heal ourselves, store information, and conduct business."[169] Subprinciples for biomimicry include: nature runs on sunlight, use only the energy needed, fit form to function, recycle everything, waste is food for another, reward cooperation, bank on diversity, demand local expertise, curb excess, and tap the power of limits.[170] Innovative ecotechnologies include glues mimicking kelp holdfasts, fibres mimicking spider webs, solar cells mimicking leaf photosynthesis, and sewage treatment through living machines mimicking natural percolation and bacterial action.

This is a corrective response to Gregory Bateson's question, "Why do schools teach almost nothing of the pattern that connects?"[171] In response, Joanna Macy calls her book *Coming Back to Life*, the "work that reconnects," in assisting workshop

Principles of Ecology

Core concepts derived from patterns and processes by which nature sustains life:

Networks

All members of an ecological community are interconnected in a vast and intricate network of relationships, the web of life. They derive their esential properties and, in fact, their very existence from these relationships.

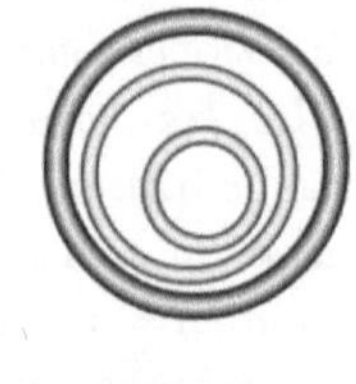

Nested Systems

Througout nature we find multi-leveled structures of systems nesting within systems. Each of these forms is an integrated whole within a boundary while at the same time being part of a larger whole.

Cycles

The interactions among the members of an ecological community involve the exchange of energy and resources in continual cycles. The cycles in an ecosystem intersect with larger cycles in the bioregion and in the planetary biosphere.

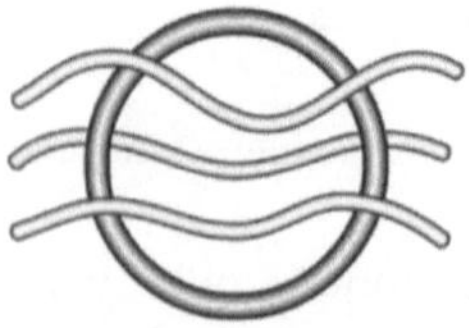

Flows

All organisms are open systems, which means they need to feed on a continual flow of energy and resources to stay alive. The constant flow of solar energy sustains life and drives all ecological cycles.

Development

The unfolding of life, manifesting as development and learning at the individual level and as evolution at the species level, involves an interplay of creativity and mutual adaptation in which organisms and environment coevolve.

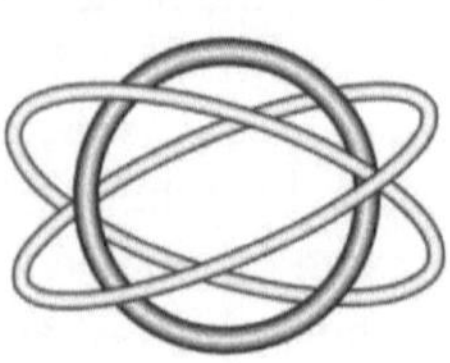

Dynamic Balance

All ecological cycles act as feedback loops, so that the ecological community regulates and organizes itself, maintaining a state of dynamic balance characterized by continual fluctuations.

—Fritjof Capra

FIGURE 6.4 The Language of Nature

Credit: Principles of Ecology diagram by Fritjof Capra was originally published by the Center for Ecoliteracy. For more information, visit www.ecoliteracy.org.

learners in systems thinking and pattern recognition as well as unblocking the grief and pain that emerges when we understand the aliveness of the Earth and what we have wrought upon such complexity and beauty.

One example of systems thinking is the emerging research on trees and forests. As the Irish poet, medicine woman, and botanist/medical biochemist, Diana Beresford-Kroeger, so beautifully says,

> Each leaf of every tree makes up the global forest. This forest is the environment that drives and fulfills the dream of each leaf in a vast rhythmic cycle called life. Nothing is outside. We are all of it in a unity that transcends the whole.[172]

Conventional science has seen trees to be isolated entities in competition with other trees and plants. It has understood the forest as a battleground of species competing for water, nutrients, and sunlight. This whole understanding has profoundly shifted in the last 25 years through ecological systems thinking.

Beresford-Kroeger says each species of tree is responsible for sustaining about 40 species of insect. Trees enable biodiversity as a habitat for all the other species that contribute to the life support system on which we depend. Trees act as reservoirs of water, helping create rain, clean the air, and anchor soil. They generate oxygen benefitting all of us who are oxygen breathers.

Trees are connected in a vast underground network of roots and fungal threads, called mycorrhizal networks. Underneath the soil, the mycorrhizae fungi provide the connective tissue between trees and many other plants, not only sharing nutrition but acting as conduits of communication as well. When trees put down young roots, they search for moisture and all the nutrients they need for growth. As they feel around with their roots searching for food, their hair-like roots meet up with microscopic fungal filaments, which is the white mass around the roots, forming a network. The tree finds the nutrients it needs, and the fungi receive the excess sugar the leaves are producing for the tree.

Through forest ecologist Suzanne Simard's research,[173] we now know that there are "mother trees" who nurture their young in multiple ways, as kinship networks. Often because of their size and the size of neighbouring trees, saplings can be too shaded. Mother trees recognize the saplings from their own seed and will send them carbon and sugar to assist their growth. They will also send nutrients to assist in fighting diseases. They may shrink their roots to allow room for their young to grow. When a big legacy or old growth tree is dying, it will send its knowledge chemically to all the trees around it. If it was a "well-loved mother tree," the trees around it will continue to send sugars to the stump and roots, keeping the stump alive for decades.

German forester Peter Wohlleben once found a beech stump 500 years old that still had green chlorophyll in it. Wohlleben says, "when beeches do this, they remind me of elephants . . . they are reluctant to abandon their dead, especially when it's a big, old, revered matriarch."[174] While trees don't have nervous systems, they can feel, and when a tree is cut, it sends electrical signals like wounded human tissue. Trees also use airborne chemicals to warn neighbouring trees of an insect infestation or heavy mammal browsing. In response, neighbours have time to generate chemicals to discourage harm. Trees are now considered more alert and sophisticated than were originally thought.

Wohlleben sees a forest as a superorganism of diverse species that includes kinship networks and alliances. There are alliances with other plants, sometimes called plant guilds which are often mimicked in permaculture gardening, where some plants attract pollinators, repel predators, attract seed dispersers, bring up minerals from deep in the earth, or fix nitrogen for the trees. Each species has a function and place in the ecosystem.

Biologist David Haskell suggests that you can smell the different aromas of trees, an aerial language. Haskell also suggests trees have songs.

> I turned my ear to trees . . . I found . . . living memories of trees, manifest in their songs, tell of life's community, a net of relations. We humans belong within this conversation, as bloodkin and incarnate members. . . . We're all—trees, humans, insects, birds, bacteria—pluralities. Life is embodied network. . . . In all these places, tree songs emerge from relationship. Although tree trunks seemingly stand as detached individuals, their lives subvert this atomistic view.[175]

Trees are now understood as the "great connectors."[176] In sum, forests are complex systems—of elements, interrelationships, and purposes—irreplaceable by humans.

Using forests as an exemplar, an immigrant serving organization is exploring how they can flow into transformative leadership following forest processes. This pertains not only to their ancestral cultural roots mingled within the human ancestral story, but also the canopy that is formed through their work across ethnocultural groups, which can be sheltering and healing. Understanding the flow of energy and ideas through their "trunk" helps them stay nourished and courageous in work that exists at the intersection of armed conflict experiences, racism, structural and institutional violence, migration, existential crisis, and family crisis.

The intransigence of global issues that we face currently—poverty, war, economic injustice, climate change—are known as "wicked problems." Wicked problems are complex problems that are "unstructured, crosscutting and relentless"[177] and thereby resistant to traditional problem solving. Simple formulas, disciplinary silos, and cause-effect thinking cannot address the dynamic complexity of causes and effects in systems which have a logic and life force of their own. The competitive model is a distinct disadvantage when dealing with wicked problems. In all such complex systems, the self-referring system itself is a source of the problem.[178]

Wicked problems are better addressed by networking across sectors and policy domains, through a constant intake of knowledge and information, as well as through engaging new ways of relating and collaborating across professional and disciplinary boundaries. For example Fritjof Capra and Ove Daniel Jakobsen carried out a systems analysis of the capitalist economy. Drawing from Marjorie Kelly, they advocate "private ownership for the common good" which includes worker-owned businesses, cooperatives, nonprofits, and customer-owned banks, all part of a "solidarity economy."[179] In other words, the "economy becomes the servant of nature, not the master of nature."[180] The economy should be a living system nested in other living systems, including societal systems, ecosystems, and the Gaia system, as in Figure 6.5.

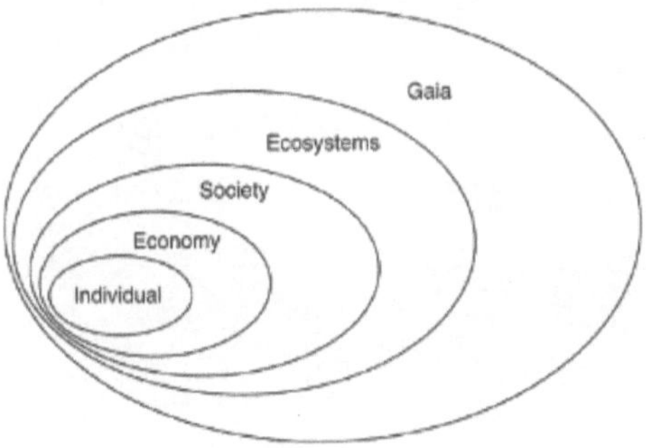

FIGURE 6.5 The Economy as a Nested System

Credit: Ove Jakobsen and Fritjof Capra. Reprinted with permission.

Becoming part of the "organic network of reality—the web of life,"[181] the economy would follow basic natural patterns. The economy would emerge out of quality, nonexploitative relationships among participants, rather than a competitive game of besting others and "selling" ourselves on the market. An economy would go beyond the "traditional one-dimensional monetary scale"[182] to include local alternative currencies and barter systems. Money systems should be connected to the "real economy" rather than operating as an independent self-referencing system. Outrageous accumulation of wealth and financial schemes that prey on people would be seen as a pathology. Decentralized economic structures at local levels could stimulate creativity and local resilience in ways that are restorative and regenerative of both human and natural systems. A circular economy ensures that resource consumption is minimized, and waste is returned to the natural world or recycled for another use. Rather than a system "fuelled by materialism and greed," limits would be recognized so that human activities do not cross certain thresholds, beyond which irreversible impacts are set in motion.

In other words, the economy should operate within planetary boundaries, called a "safe operating space." Global activities would have limits established by governance bodies relating to: climate change, loss of biosphere integrity, land system change, freshwater use, biochemical flows, ocean acidification, atmospheric aerosol loading, stratospheric ozone depletion, and novel entities.[183] This is best captured in Kate Raworth's *Doughnut Economics*.[184] Building from "internationally agreed social standards and in Earth-system science," the doughnut is a visual illustration of the social and ecological boundaries as seen in Figure 6.6. The inner section are 12 dimensions of the "shortfalls in wellbeing, such as hunger, ill health, illiteracy, and energy poverty." The outer section are nine dimensions which are "an ecological ceiling,

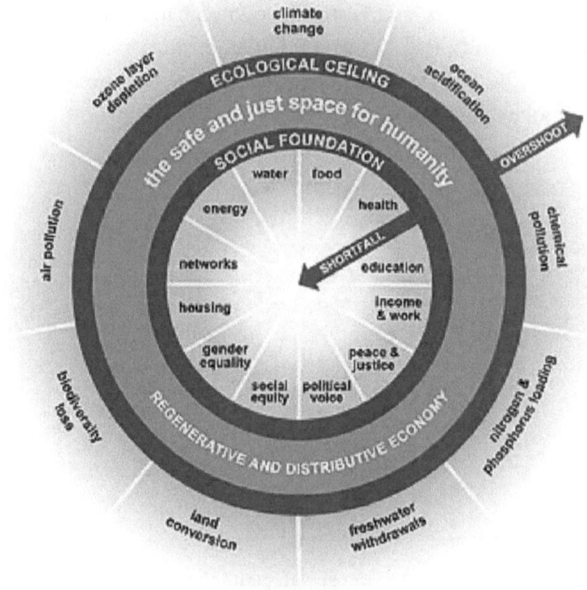

FIGURE 6.6 The Doughnut of Social and Planetary Boundaries

Credit: The Doughnut of Social and Planetary Boundaries. Kate Raworth and Christian Guthier. Reprinted with permission. CC-BY-SA 4.0.

beyond which lies an overshoot of pressure on Earth's life-supporting systems, such as climate change, ocean acidification, and biodiversity loss." When humanity stays between these two sections, they are living within an "ecologically safe and socially just space in which all of humanity has the chance to thrive."[185] The doughnut is an excellent visual that emerges out of the creativity of "the included middle," between traditional ideas of environmental education and sustainability education.[186]

Another aspect of setting limits is the *precautionary principle*, which stipulates "When an activity raises threats of harm to human health or the environment, precautionary measures should be taken even if some cause-effect relationships are not fully established scientifically."[187] Rather than forging ahead on an action, more time would be taken to assess the negative ramifications to people and planet. In these ways, basic ecological principles are applied to the purpose of reconstructing the economy and society for the common good using cooperative rather than competitive relations.

Process Thinking and Energy Fields

Another facet of relationality is the focus on movement rather than static entities by examining flow patterns. Mihaly Csikszentmihalyi, a Hungarian-American psychologist, talks about "finding flow' as those exceptional moments where we are so engrossed in an experience, that it becomes "effortless action," "being in the zone," "peak experience," or "ecstasy."[188] One moves beyond thinking to a sense of flow where a different awareness emerges. In this zone, a new type of seeing reveals the path with all of one's skills on tap.

This flow state occurs in peak sports performance or other complex tasks like surgery or artistic creation, as well as in mystical states.[189] Time slows down because of the intense concentration. When one is in the moment, there is little sense of time passing. In a state of heightened relaxation, a responsiveness emerges that balances flowing with acting. A flow experience can be exhilarating, providing a moment of intense living, which is why some people are addicted to mountain climbing and other extreme sports. Happiness cannot be approached directly as assumed in modernity, but rather the byproduct of flow experiences is deep happiness.

As another facet of flow, David Bohm discusses a process approach to knowing, where knowledge and knowing are not static to be consumed, but always moving and revealing.[190] He says, "what we need to do is go beyond thinking in order that things should flow through us more naturally."[191] Many innovators and researchers know that when one leaves off rational thought, a flash of insight or intuitive leap can occur, providing a pathway of potentiality. As Louis Pasteur suggested, "chance favours only the prepared mind."[192]

In the present moment, Bohm says, we are thinking. As the moment passes, the thinking then becomes thought or rather, information contained in memory. Too often, we approach our daily lives mechanically, where we use the presuppositions of past thought to frame our current responses. Much of psychological therapy is based on this premise. Bohm agrees we need to discover these presuppositions that have become habit and free ourselves of them, as they block fresh thinking in the moment.[193] When we do this, "a new idea appears as a sort of seed deep from the implicate order,"[194] which you can unfold into your thinking and living.

Thus, we need to develop "a deep intention that goes beyond thought."[195] The purpose of thinking is to not use your old thoughts to explain new situations. Rather, ideas are always circulating, emerging out of the implicate order. When you are open, questioning, or curious, these new ideas may emerge for you. Suspending the old thoughts leaves room for new ones. Thoughts from the thinking mind, emotions from the psyche, and physical impulses from the body are constantly flowing and dancing with each other, teaching us.

Psychologist Stanislav Grof says that "each of us contains the information about the entire universe or all of existence, [each of us] has potential experiential access to all its parts, and in a sense *is* the whole cosmic network."[196] Intriguingly, Laszlo affirms

> Everything that goes on in our mind leaves its wave-traces in the zero-point field of the quantum vacuum, and everything that goes on in our mind can be received by those who know how to "tune in" to the subtle traces that propagate there.[197]

Letting go of emotional identification with old ideas, we can explore new ones. Erik Erikson says, as we age, we have a choice to either move into the realm of stagnation or into generativity.[198] Bohm calls this process the dance between our thinking and our thought, where we can examine the categories our thought patterns have established over time. It is relaxing these thought patterns which can make the boundaries of our categories more permeable, allowing us to see anew.

Further, there must be a force that emerges, an energy, to push an idea out of the implicate into explicate consciousness and then to break through the barriers of conditioning which are the typical grooves of thought.[199] As adults, we need to explore our metaphysical ideas about the nature of reality that we developed in childhood. Even if we turned away from these childhood ideas, they still form one foundation of our thought.[200] With this knowledge, it is easier to manifest an openness to the energy pushing an idea our way. While we cannot ultimately know reality, we can orient ourselves in these ways towards the creativity of this "unbroken wholeness."[201]

In this way, knowing and being are not separate, but a whole. Bohm has described this as experiencing "the totality of existence as an undivided flowing movement without borders . . . Thus, the wholeness permeates all that is being discussed, from the very outset."[202] Bohm sums up this orientation:

> It takes the mind in its totality to perceive the totality of anything. The fragmented mind inevitably sees in parts, the person who begins with the part thereby fragments his [sic] mind. In a sense, to perceive the totality of understanding, one must be a totality.[203]

For Bohm, there is no gap between mind and matter, psyche, and body, as the underlying energy field shapes all. As discussed earlier, cognition is present even at quantum levels. Thus, *knowing* is a verb, always in process,[204] not an achievement or acquisition. When, through our intelligence, we have an experience of perception that reveals underlying patterns, this does not only come from our way of knowing but from our way of being or the ontological dimension. This ontological experience is part of the processes of unfoldment and enfoldment, which operate below the level of abstract thought at the level of intuition and other states of consciousness. We cannot reason ourselves into deeper ontological insight, but we can be open to receiving these kinds of ontological experiences.

Coming to understanding ourselves as flows and processes requires a transformation of our structural and mechanical understanding of reality. Transformative educators can teach about the New Science not only as content but also through mindfulness activities and energy work that assists learners in perceiving themselves and the cosmos differently. Through the development of the *witness self*, we can become attuned to the flowing of energy as part of our individual processes as well as in social processes. Many ecological identity activities offer pathways to perceive ourselves as a literal part of the Earth, with perforated boundaries through which energy flows. Our identities expand to include all that we are in relation with. Thus, we move beyond transformative learning that is solely concerned with what is seen with past thought, and with static entities.

This leads us to a relational understanding of belonging, mapping our place in the world. The modern question of "who am I?" no longer need carry such angst. We can recognize our multitude of relations that accompany us into any educational space—personal, familial, cultural, economic, political, social, spiritual, and natural relations. Explicitly acknowledging these relations as educators and learners leads us into a more complex understanding of the energies and perceptions at work in a learning space.

David Selby suggests that one transformative task for educators and learners revolves around this radical interconnectedness, as a much-needed transformation of cognition in this historical moment.[205] Like Selby, we can be constantly looking for the signature of the whole in the midst of increasing disconnection and environmental losses. He suggests that we imagine ourselves as "momentary configurations of energy,"[206] as these "structures out of which biological entities are made are transient, unstable entities with constantly changing molecules dependent on a flow of energy to maintain form and structure."[207] He sums up by saying,

> [W]e would be wise to countenance the implicate order in our understanding of our macroworld—to see that in a profound and very real way, everything is embedded in everything else and that things or objects are ontologically subordinate to flows and patterns. Everything is thus the signature of the whole.[208]

The implication here is that consciousness and matter are integral to each other.

Communal Individuality

As Laszlo explains, this

> [T]ransforms everything we know also of life and the world of the living. The emerging vision gives us the image of an interconnected web of nature that produces all the things that we observe and feel in a continuous, organic process of self-creation.[209]

In a conventional Western understanding, when a person acts, they are like the proverbial billiard ball that impacts other self-contained beings, generating a linear impact and response. In a relationality view, no person is self-contained or truly autonomous. In other words, the central phenomena of Life *is* relationship, which forms the ground of our being. Our knowing flows from this relationality.

To provide an image of relationality in the human world, we are born into certain relations—someone who mothers us and/or who fathers us. We become a child who acts, reacts, and relates from within this primary set of relations. This is just one relation that shapes us from the beginning. We are a multitude of social and natural relations—the water that quenches our thirst, the soil that begets the food which nourishes us, the bodies of those caretaking us, the language they speak and the way their bodies move, their energy field, the smell and feeling of clothes on skin, and the natural and social world that we hear and see, *in utero* and beyond. This is our matrix of life. This matrix makes us. All these elements *constitute* our identity. Our being is immersed from the beginning of our embodiment.

Before we develop language or a separate sense of self, at around 2 years of age, our relationships constitute who we are and our being in the world.[210] These relations also shape our coming to know. A common assertion is: relations precede knowing.[211] Our individuality emerges out of this relational context, neither fully constructing and containing us, nor rendering us fully separate and autonomous. Our life purpose is to understand who we are in this matrix of relations in which we exist, learning how we wish to "be" in these relations.

What relationality refers to then, says Spretnak, is that we are thoroughly relational beings of great complexity, who are both composed of and nested within contextual networks of dynamic and reciprocal relationships.[212] Relationality does not refer then to an *inter*relational nature of relationships, as that focus remains on the subjects having the relation.

As physicist Karen Barad explains, relationality refers to *intra-action* where the focus is on the relation, the entangling which performs a world into being.

> The neologism "intra-action" *signifies the mutual constitution of entangled agencies*. That is, in contrast to the usual "interaction," which assumes that there are separate individual agencies that precede their interaction, the notion of intra-action recognizes that distinct agencies do not precede, but rather emerge through, their intra-action.[213]

This is a

> radical reworking of the traditional notion of causality . . . [which leads to] a lively new ontology: the world's radical aliveness comes to light in an entirely nontraditional way that reworks the nature of both relationality and aliveness (vitality, dynamism, agency).[214]

In physics, this means that both the objects being observed and the agencies of observation are constituted within the phenomena that occur.[215] The subjects of the relation do not pre-exist the relation, but the relation brings the subjects into being.

Barad calls this a *performative* understanding of being, as our inextricable relations perform us into being. "A performative account insists on understanding thinking, observing, and theorizing as practices of engagement with, and as part of, the world in which we have our being."[216] In the case of a young child, the nature of those relations—whether trusting or not, affirming or not, compassionate or not—shapes not only our basic beliefs about the world, but we continue to act in response to these early relations that established a certain set of neural pathways as well.

In a break from modernist thinking, we cannot be "liberated" from these relations, especially the ones we consider a constraint. Our human and nonhuman kin have already shaped us, there is no escape. Rather, the task is to become embedded more meaningfully in our relations. Within communal individuality, we *do* have self-determination. We emerge as an individual from within this web of relations by determining how to respond. As Gould[217] explains, individuals are fundamentally individuals-in-social-relations rather than liberal, autonomous, interior individuals who relate to an external world. Rather, we are constituted by other humans, history, society, and the Earth. This is the meaning of *communal individuality*.

Current individualist notions of society separate us from community and our environs and, importantly, separate ethics from politics, science, and the economy. Yet, all these phenomena are integrally related.[218] The liberal notion of individualism can be seen as a phase before communal individuality, where we understand ourselves as beings in and for ourselves, rather than selling ourselves to others for our job, or being in relations where we are used instrumentally for what we can do or give. In communal individuality, we self-consciously reroot ourselves in our social and natural relations. Moving in this direction can address the profound alienation that has been part of modern liberal, capitalist societies.[219]

Distributed Cognition and Relational Becoming

In a posthumanist interpretation of Bohr, Karen Barad builds from the view of reality as being in flux, constantly revealing itself. This goes one step further from John Dewey and his transactional understanding and the constructivist concepts of Barbara Thayer-Bacon. To Barad, the body is much more a performance than a thing, a performative self.[220] Life is always in the process of becoming, emerging out of the matrix of performing relations. As she says, "Humans are part of the world-body space . . . In contrast to the spectator theory of knowledge . . . knowing is a part of being."[221]

Rather than coming to fully formed knowledge as an independent knower, we engage the world and thus come to know. Our knowledge emerges from intra-acting with other humans, life forms, and natural phenomena embedded with us in the same reality. Knowing then is becoming together, where we perform each other into being as we observe, relate, and interact. In this way, knowing is distributed across relations, human and nonhuman.

For Barad, the human becomes decentred, or, rather, humans are no longer the centre of reality and a dominant concern. Humans are not the only knowers or the only ones with agency or, rather, purposeful responsiveness. Matter is "not a thing, but a doing, a congealing of agency."[222] Thus, other living beings and material systems also act and respond. Hence the title she gave her book, "meeting the universe halfway." All life forms and phenomena become teachers and learners.

Entangled Ethics

The moral muteness of science is no longer possible if we are radically kin within the web of life. By being in relation, we are already entangled and therefore have an obligation to the other which is part of ourselves. Thus, for Barad, the ethical, ontological, and epistemological aspects cannot be separate as traditionally conceptualized. She calls this ethico-onto-epistemology as "this way of thinking ontology, epistemology, and ethics together makes for a world that is always already an ethical manner."[223]

Our part in mutual obligation is reciprocity, where we account for our actions in this tangled web of relations. We also account for what we help to bring into being through our attentions, intentions, and actions. Ethics also concern whole biotic communities in their nestedness. Thus, we must examine how we appropriate parts of the living world for human desires. We must do so with honour and gratitude or choose not to appropriate, out of respect.

We must also account for how our entanglements "reconfigure our beings, our psyches, our imaginations, our institutions, and our societies."[224] Not only do we act, matter also acts on us. We do not stand above, or apart from, all our relations. When we see the living—the vitality and the responsive aspects of the world around us—we are perceiving relationally. Responsibility then becomes the core of our subjectivity, interactive within all our connections.[225]

Charles Eisenstein suggests that the transformation in our stories and ways of perceiving reality "is nothing less than a transformation of humanity's role on planet Earth."[226] When one perceives all things are related, then the responsibility—ability to respond—is in ethical acts of reciprocal caring, part of a moral order that we find already exists in traditional Indigenous culture. As Barad says, science can no longer be outside ethics, as being and meaning are caught up in practices, doings, and actions, so nothing lies outside of ethical consideration.[227] Transformative learning helps learners place themselves within a much larger cosmic and Earth-based reality than an individualist and anthropocentric reality. It is bringing humans back into their natural relations as well as social relations, allowing for communication, participation, and responsibility.

One example of radical relationality ethics is Barbara McClintock's genetic research on corn. Contrary to the detachment of the scientific enterprise, she advocated her students to "take the time and look," to be patient to hear what corn has to say, and to see the "hidden complexity."[228] Through what she called having "a feeling for the organism," she grew a sympathetic understanding of corn. She sharpened her discernment of what was going on for each plant, their story over time. Thus, corn plants became subjects in their own right.[229] She was clear that science needs a deep emotional investment of this kind, an awe for the ingenuity of nature, and a respect for the mysteries of living forms. While she was dismissed within her field of biology for many years, she was eventually awarded a Nobel prize.

Similarly, eminent biologist E.O. Wilson profiles the need for *biophilia* as a fostering of the human bond with other species.[230] Biophilia is the natural affinity for life in which we cease our "predatory actions toward each other" and eliminate ideas that we have "conquered the world" or understood it.[231] He suggests these "prevailing myths [are] obsolete, unreliable, and destructive." He provocatively asserts that "splendor awaits . . . within walking distance of where you sit." When we affiliate with other organisms, "they are the matrix in which the human mind originated and is permanently rooted."[232] Bringing learners to these small and grand wonders just out the door is in part the response-ability of educators.

Mind–Body Problem and the Issue of Consciousness

One key challenge in modern mechanistic science has been the great difficulty of explaining mind and consciousness, called the mind–body problem. The quantum paradigm generates important explanations about mind and consciousness. These new theories have tremendous potential for understanding the relations between mind, body, and consciousness as well as the relation between knowing and being. They hold vital significance for learning and education as well.[233]

Three approaches have traditionally explained the existence of the mind in relation to the body. *Dualism* proposed that mind and matter are two irreconcilable realms, with little explanation for how mind emerges or, what Descartes called, the thinking substance. On the other hand, for *materialists*, the mind emerges from the operation of the brain and its neural processing, physical in nature.[234] In this view, mind depends on matter for existence. Yet, there is still no real understanding of how mind, which is not physical, could exist. For *Idealists*, the mind is primary and either creates or mediates what we perceive as matter. All matter then originates in the mind as it is the perceptual organizing tool. In this view, the natural world is not a material entity, but illusion generated by the mind, still leaving the origin of mind unexplained. Philosopher Christian de Quincey suggests that all three explanations require some sort of "miracle" to account for mind.[235]

There is a fourth approach, called the *panpsychism*[236] approach, which simply means that all physical things have consciousness. This idea existed not only as far back as Ancient Greece in Heraclitus but also in the works of Spinoza (the early Enlightenment) and in those of Whitehead and Bohm (20th century). This book relies on this vein of thought, suggesting that both mind and matter are equally real and natural. The wave/particle duality at the quantum level offers a powerful parallel for understanding the mind/body relationship.[237] The mind and body are two sides of our human existence, which are internally related, called relational holism.[238]

In this view, consciousness exists at the most primary level of existence and is like a wave. Our body as the physical aspect is more like a particle. The mental and physical aspects have a common origin in the quantum vacuum. Mind is not in the brain, but many of its physical elements hold memories and respond to stimuli. Consciousness is "the between" of mind and body. Thus, consciousness is relational, not held by any one aspect. Consciousness exists in "the between" of relations.

In the quantum view, consciousness is the foundation of all that exists. In his book *Radical Nature*, de Quincey explains that "consciousness goes all the way down" to all levels of reality.[239] Similarly, Bohm, Zohar, Barad, Maturana, Varela, Capra, and Laszlo all suggest that consciousness is a property of *all* living systems, not just humans. Meaning is not just human-based but part of the world's intelligibility.[240] Further, consciousness is not only the basis of awareness in all beings. It is also that which fosters relationships of mutual cooperation between all beings and systems.[241]

Thus, there is a profound participatory aspect to consciousness. As quantum theologian Diarmuid O'Murchu describes it, through interacting, giving, and receiving, a resonance emerges where the individual parts lose an independent identity. They become a "quantum self." They form interdependent relationships within the new whole. This could be anything from "the marriage of two people to a newly felt bond with the universe itself."[242] Humans no longer perceive themselves to be masters of the clocklike universe and the only thinking being. We see ourselves as co-participants in a living system that is a coherent conscious whole.

Relational Knowing and the BodyMind

Education philosopher Barbara Thayer-Bacon suggests that we need to raise the status of the body equal to the mind, as a wholistic *bodymind*.[243] We need to not only overcome the body/mind split, but we also need to overcome the duality of rationality/irrationality which has informed the gender binary. Further, she says, our claims to knowledge need to be "provisional and approximate" as all knowledge is socially situated and contextual.[244] In moving from spectator ideas of learning towards learner agency, we acknowledge our own embeddedness in a context as well as the active nature of learning.[245] In other words, we come to know through our relations with others. "When we understand we are part of the oneness of the universe, then we can begin to understand how connected we are as knowers—not only to each other, but to our products as well as our knowledge."[246]

In other words, our knowing emanates from our position as beings-in-relation-to-others. "All knowing develops out of relational practices we participate in from the moment of conception."[247] The first 3 years of life are the psychological birth of the child, establishing a core identity through their social and interactive relations, which are largely preverbal and prerational.[248] Even though there is little language, there is an abundance of communication whereby the parent is caring for the child's needs and the child responds.

This reciprocity within caring relations of parent and child and then that of the child with all other care providers and peers ensures the child's continued life, growth, and sense of social belonging. Gestures, vocalization, and pre-cognitive language all mimic care providers, developing meaning. Later with language, a mind and self-consciousness develop from within the context of these relations. As sociologist Margaret Mead suggested, we then develop a sense of a generalized other which prescribes acceptable behaviours within a culture, as part of the socialization process.

Later, in the transformative processes of adulthood, we depart from this socialization as we discover "faults, cracks, and ruptures in the current symbolic universe and [transform] the explanations."[249] We develop a differentiated self, often through a transformative process, but it is still in reference to all these personal and social relations. Drawing from Buddhism, Mead says,

> [T]he biggest illusion our mind invents is the illusion of a separate self, the belief in the existence of a persistent, continuously existing, single, independent ego . . . the paradox about the self is that the individual self leads away from an awareness of the w/holeness of the universe, and yet we gain consciousness of the universal through the individual self.[250]

In this nondualist ontology, we overcome the sense of separateness of self and appreciate we are an integral part of the flow of reality. Through transformative learning, our ego boundaries start to soften, helping us to embrace wholeness and radical relatedness.

In ecological relations, Thayer-Bacon explains that we understand ourselves as "one with nature,"[251] a radical relationality. We are not separate entities but part of the natural world. While we establish boundaries between human and nature, this separation does not really exist. From within such a deep relationality, we would be unable to exploit and dominate, as this compromises our sense of larger self.

We and others are whole beings—body, mind, emotion, spirit, soul/psyche, instinct, and will. As educator Paul Hart suggests, relational knowing is moving towards an integrative mind.[252] Just as the aforementioned scientists describe, it allows for a variety of ways of knowing, including contemplative, empathic, aesthetic, embodied/somatic, and imaginative. It understands and uses our five levels of consciousness with distinct brain waves, such as delta (in restorative sleep), theta (the meditation gateway between waking and sleeping), alpha (calm alertness in the flowing now), beta (normal waking state), and gamma (expanded consciousness evoking higher virtues). Seeing whole allows the creative, artistic right brain to become an equal partner in consciousness with the reasoning left brain. We can work deliberately with many of these states of consciousness as well as with the many ways of knowing.

Four Gifts of Knowing

As part of radical relationality, de Quincey asserts that consciousness is the "very fabric of all our relationships."[253] Consciousness is not an individual possession but an "inbetween,"[254] emerging in relationship between subjects. It is the true meaning of intersubjectivity as "mind-to-mind" communication.[255] Such communication has a feeling base to it and is mostly transverbal.[256] This type of communication includes all life forms and not just humans.

Religious studies professor Christopher Bache suggests working with two kinds of consciousness.[257] Christopher Bache calls reason our daytime consciousness founded on physical senses and our nighttime consciousness founded on dreams, intuition, mystical experiences as well as other non-ordinary states of consciousness.[258] In the Western world, reason has prevailed, but it relied on categorization which divided genders, cultures, and the natural world according to these two ways of knowing. It divided truth from wisdom and intellectual knowing from intuitive knowing. It is time now to acknowledge all the different gifts of knowing in an integrated way.

Christian de Quincey, in *Radical Knowing*, suggests that there are four ways of knowing, different ways of learning about ourselves and making meaning in the world, each with its own transformative path.[259] The four ways of knowing are the philosopher's gift of reason, the scientist's gift of the senses (empiricism), the shaman's gift of participatory feeling, and the mystic's gift of direct experience with the sacred. Each of these ways of knowing is a tool needed for different purposes. Healing the rift between them and understanding the context in which to use them are part of the task of this age.

Philosopher's Reason: The philosopher's gift of knowing is rational cognition, analytic logic, and precise language to derive abstract ideas as well as concepts from symbols or experience, arriving at meaning. Yet, reason takes us only so far. Preverbal feelings, intuition, and imagination all lie outside its ken.

Scientist's Senses: The scientist's gift of knowing uses the senses via empirical methodology to understand the world and derive organized knowledge. Through observation, we understand the small and large patterns around us, which have enabled us to not only survive but also thrive as a species in multiple environments. Now, using empirical procedures, scientists meticulously observe, record, analyse, report, and retest findings to arrive at theories. While this has been a heroic knowledge quest, this conventional way of knowing has been limited to the physical world. Only now it is expanding to examine the invisible matrix of energy that envelops us, including gravitational, electromagnetic, and electrochemical energy.

Shaman's Embodied Feeling: The shaman's gift of knowing is through embodied feeling and communication with the subtle realm through altered states of consciousness. Through participatory engagement, shamans enter the cosmic energy flow and engage in communication between the visible and invisible worlds, serving as a 'messenger of the gods,' the explicit role of shamans. To explain this, De Quincey relies on anthropologist Jeremy Narby's work in the Peruvian Amazon.[260] Photons are the light that is emitted through the universe, constituting all things, carrying messages. "Shamans take their consciousness down to the molecular level and gain access to information related to DNA, which they call "animate essences" or "spirits."[261] It is Narby's belief that shamans are receiving the "biophotons emitted by all the cells of the world."[262] They can "pick up images and information from any living being anywhere in the global network."[263] "Light, or photons—exchanged between DNA and cells—form the ultimate basis of the entire global economy of life. Life, quite literally, *is* light."[264] Cosmic intelligence, then, is carried by photons.

Temporarily freed from the body through shamanic ritual, the spirit can explore the database of the natural world and spirit world—through dreams, visions, shamanic journeying, and other means. Thus, energy work and work with light are ways of knowing.

Mystic's Direct Experience: The mystic's gift of knowing is direct transcendental experience through sacred silence and intuition, in directly apprehending the spirit world. They commune or cross the threshold into the realms of silence, returning to Source. By slowing down the incessant internal dialogue, an awareness of life in the here and now is possible by entering the stream of life and consciousness. You can only follow the stream, you cannot grasp it, define it, or give it language, as intuition can only point to wholeness. We can, however, experience this "communion of knowing and being," which is the experience of "Source."[265] Generally, we perceive dualities as being part of our material and embodied life, yet both the material and abstract emerge out of a nondual reality, the Oneness.[266] Many spiritual texts have described this process. For instance Tao or "The Way" is following "the flow of time, the course of life [to] the source of everything and nothing . . . the communion of knowing and being . . . the ineffable awareness of the ineffable."[267]

Radical Dialogue

David Bohm, as one of the great physicists and thinkers, has identified numerous concepts that have aided in a deeper understanding of the quantum reality. One of his contributions is Bohmian dialogue. Along with Indian philosopher and educator Jiddu Krishnamurti,[268] Bohm believed that the problems in our thinking are at the collective level. Drawing from physics, he wondered what implications the paradox of the observer and observed had for commonly accepted processes such as reflection.

Bohm says that Western thought is literal thought, whereas Bohmian dialogue is participatory thought, working at the much deeper relational level. Bohm offers a brief description:

> Dialogue is really aimed at going into the whole thought process and changing the way the thought process occurs collectively. We haven't really paid much attention to thought as a process.
>
> We have engaged in thoughts, but we have only paid attention to the content, not to the process. . . . Dialogue can be considered as a free flow of meaning between people in communication, in the sense of a stream that flows between banks.
>
> The spirit of dialogue is, in short, the ability to hold many points of view in suspension, along with a primary interest in the creation of a common meaning. . . . A key difference between a dialogue and an ordinary discussion

> is that, within the latter, people usually hold relatively fixed positions and argue in favour of their views as they try to convince others to change. At best this may produce agreement or compromise, but it does not give rise to anything creative. Moreover, whenever anything of fundamental significance is involved, then positions tend to be rigidly nonnegotiable and talk degenerates either into a confrontation in which there is no solution, or into a polite avoidance of the issues. Both these outcomes are extremely harmful, for they prevent the free play of thought in communication and therefore impede creativity.[269]

Bohm suggests that this process of dialogue offers a more creative way "out of collective difficulties."[270] Such dialogue and thinking skills tap the implicate order, leading to wisdom.

Moving Towards Wisdom: A Spiritual Pathway

De Quincey suggests that the rational mind, when deeply connected to feeling and intuition, yields wisdom. It is not just logic or sensory experience but understanding meaning systems which can illuminate the human condition. While "science and philosophy seek *truth*, spirituality seeks *wisdom*."[271]

According to de Quincey, the goal of wisdom is integrating the following: contextual knowing about a situation, feeling knowing as empathy and compassion, perceptual knowing as instinct, embodied knowing, spiritual knowing for insight, and pragmatic knowing for action.[272] Rather than seeing them as opposed, truth and wisdom are complementary. As de Quincey emphasizes, expressing the highest forms of wisdom lays beyond reason, and even beyond language. We can only use story, music, poetry, and art to approximate the knowing that comes from wisdom. Wise ones, such as Indigenous Elders, generally teach through story, letting learners struggle to process their experiences through the story, developing their own meanings, especially by revisiting the story cyclically over a lifetime.

Capra explains that "Before the seventeenth century, the goals of science were wisdom, understanding the natural order, and living in harmony with it."[273] After a detour of 300+ years, these are the three elements to which we can return. For too long, wisdom has resided outside of science, social science, and education, considered nondefinable and indescribable, given its concern with the ineffable.

Exciting new brain research provides evidence that wisdom has, partially, a biological basis in the brain. Wisdom, say neuroscientists Dilip Jeste and Scott Lafee, is a process associated with aging, where life experiences are processed into lessons and habits that we identify as wisdom. After studying the regions of the brain associated with various capacities, as well as mapping the many spiritual, psychological, and descriptions of wisdom across multiple cultures, they identified six characteristics most closely associated with wisdom.[274] Testing these characteristics across various populations ensured resonance.

Prosocial attitudes and behaviours: High levels of empathy, compassion, and altruism are prosocial attitudes and behaviours. Wise individuals understand the feelings of others and translate empathy into helpful behaviour, where egoism gives way to altruism. They feel a strong connectedness to others, manifesting relationality.

Emotional stability with happiness: Wise individuals demonstrate a high degree of self-control and emotional regulation.

Balancing decisiveness with acceptance of uncertainty: Individuals exhibiting wisdom consider perspectives and desires, different from their own, as being equally valid. They do not easily succumb to fear or ascribe negative traits or motivations to others. They navigate with calmness in the face of uncertainty and ambiguity.

Self-reflection and understanding: Wise individuals use intuition and insight to guide themselves through difficult situations. They regularly reflect on their own perspectives, habits, and motivations to achieve an ever-increasing level of self-awareness and openness.

Social decision-making and pragmatic knowledge for life: Wise individuals offer sound reasoning and advice, sharing their knowledge and skills in ways that benefit others.

Spirituality: Wise individuals have a basic belief in something larger than the individual and society—the ineffable—that there is a deep purposefulness in the universe.

Throughout human history, transformation has necessarily involved spirituality as part of human wholeness. The ancient practices of transformation understand wisdom as being part of spiritual awakening, providing a more expansive context for learning and living. Transformative learning can be learning about spiritual, psychological, emotional, and physical flows, part of exploring the wholeness of the individual within the social collective. With changes in consciousness, there is "an innate knowing of energy" where our unified, embodied "self" communicates with the "sentient energy" in the universe in "two-way communication."[275] We learn to sense the world of energy that is pulsing around us constantly. Attuning to the energy matrix through which information constantly travels, we can be open to and sense the collective consciousness out of which wisdom emerges, including in learning spaces.[276]

Reimagining Transformative Learning

Transformative learning is not just a transformation of *form*. Transformation is a process as well, which directs the shape of the form. The New Science is a process ontology, where entities, including objects and humans, are *not* fundamental. Process is an ontology of experiencing events and "organisms in process" as entities emerge and recede.[277] Events and relations are the focus then and relations become the "very condition for education."[278]

Education is a learning engagement that sinks into the intelligibility of the universe in which we are entangled.[279] The energy and information that flow through the permeable boundaries of our mind and body to all other beings and elements can be directed by intention. Education is located within these moving happenings.

We ourselves are bundles of experiencing and feeling which form "energetic patterns-within-patterns, self-organizing cascades of creativity."[280] In process thinking, relationality is constantly making us across past, present, and future. We are this "invisible dance of communication" where a sense of respectful relationality ripples outward.[281]

One of the "hallmarks of life" is the "spontaneous emergence of order at critical points of instability—referred to simply as 'emergence.'"[282] New properties "emerge" at a higher level of self-organization, which were not present previously, through nonlinear causality. As we live our lives within the current civilization, constant choice points occur, which feed either habitual or novel patterns. At times, an accumulation of perturbations leads to increasing chaos. A tipping point is reached—an ultrasensitive point where even the smallest of changes force the system into a bifurcation, where one of several trajectories are chosen. When the system bifurcates, it either breaks through and moves towards a higher level of organization in which new properties are emergent, or the system moves towards breakdown.[283] The choice points humanity faces regarding climate change are such a moment. Through transformative learning as process, we have a possibility of emergence, both individually and collectively.

In this historical moment, transformative learning can be an expansion of perceptions of reality and, thus, a continual questioning and broadening of cosmological, ontological, axiological as well as epistemological assumptions. This conceptual revolution is itself the transformative aspect and generator of deep sustainability. When glimpsed, one moves towards living in an integrated way, honouring both spirit and body/material life. This is a richer understanding of, and purpose for, transformative learning as the process of sustainability education and sustainability education as the process of transformation. It is transformative in intent for education systems and for sustainable societies to which we turn in Chapters 7 and 8.

Transformative Sustainability Education: Flowing into Relationality

The primary focus of this chapter has been exploring the facets of relationality in Transformative Sustainability Education, which involves what Katie Ross calls a "cosmo-onto-epi-axiological stretching."[284] Joy Kcenia Polanco O'Neil, Katie Ross, and I collaboratively developed 15 principles of transformative sustainability education, capturing some key features of Transformative Sustainability Education from a relationality perspective, which require such stretching. As explained by Ross in the original article, each facet of Transformative Sustainability Education is "inextricably linked," yet out of necessity, the principles are "presented . . . in categories as we are bound to a linear method of writing."[285] As Joy Kcenia Polanco O'Neil explains, the word "as" in the descriptions connotes that they are about an "emergent way of knowing" that involves not only an epistemological but also an ontological shift.[286] The word "educating" rather than education uses process-based languaging.

Together, these enacted principles are not only transformative content but also create a much more profoundly relational *ontos*, or way of being in the process of learning together. The 15 principles are:

- educating as Life;
- educating as energy flow;
- educating as consciousness going all the way down;
- educating as spirituality;
- educating as relationing, dynamic process, nonlinear movement, emergence;
- educating as material agency;
- educating as self-in-community;
- educating as radical relatedness and kinship;
- educating as diversity, inclusion, and collaboration;
- educating as transdisciplinarity;
- educating as emergent curriculum;
- educating as patternseeking;
- educating as multiple modalities of meaning;
- educating the whole, embodied person; and
- educating as sustainability.

For further discussion and exemplars, please see our original article.[287] Chapter 8 expands on my interpretation of these principles and possibilities for practice.

Conclusion

To come full circle, De Quincey (2010) says, stories matter—especially stories about matter, mind, and body. The modern story of inert/passive/dead matter, a mechanistic universe, and humans as the only sentient beings is giving way to a post-anthropocentric, post-Newtonian, and post-Cartesian story, once again recognizing a cosmos that is alive, sentient, and feeling. This old story no longer serves us. The new story is one of Relationality. Mythological language is apt for conveying this aliveness of Earth, and the dynamic kinship of all beings, evoking an ethics of care.

Big Stories have powerful transformative capacity to shift consciousness and elevate humanity. These stories are a "matrix of narratives, agreements, and symbolic systems that comprise the answers our culture offers to life's most basic questions."[288] It is the Western Big Story, shaped by science, that is now being transformed by science itself. However, the "new" story has been intuited by mystics over human history. Loosening the grip on the old story while stretching into various veins of new thinking, reimagining, and reembodying, we can play a role as transformative leaders for deep sustainability, which can enable us to flow into a new civilization.

Notes and References

Chapter line drawing by Canadian artist Erin Marie Lange Christensen. www.erinchristensen.com. Reproduced with permission.

1 Thomas Berry, "The New Story: Comments on the Origin, Identification, and Transmission of Values," in Arthur Fabel and Donald St. John (Eds.), *Teilhard in the 21st Century: The Emerging Spirit of Earth* (Maryknoll, NY: Orbis Books, 2003), pp. 77–88.

2 Arran Stibbe, "Education for Sustainability and the Search for New Stories to Live By," in Joan Armon, Stephen Scoffham, and Chara Armon (Eds.), *Prioritizing Sustainability Education: A Comprehensive Approach* (London: Earthscan from Routledge, 2020), pp. 232–245.

3 Charles Eisenstein, *The More Beautiful World our Hearts Know is Possible* (Berkeley: North Atlantic Books, 2013), pp. 5–6.

4 Eisenstein, *The More Beautiful World our Hearts Know Is Possible*, p. 6.

5 Eisenstein, *The More Beautiful World our Hearts Know Is Possible*, p. 6.

6 Paul Clarke, *Education for Sustainability: Becoming Naturally Smart* (London: Routledge, 2012), pp. 43, 30.

7 Arturo Escobar, *Designs for the Pluriverse: Radical Interdependence, Autonomy, and the Making of Worlds* (Durham: Duke University Press, 2018), p. 5.

8 Katie Ross, *Transforming the Ways We Create Change: Experiencing and Cultivating Transformative Sustainability Learning* (Sydney: University of Technology Sydney, 2020). http://hdl.handle.net/10453/149105.

9 Matthew Harmin, M. J. Barrett, and Carolyn Hoessler, "Stretching the Boundaries of Transformative Sustainability Learning: On the Importance of Decolonizing Ways of Knowing and Relations with the More-than-human," *Environmental Education Research*, Vol. 23, No. 10 (2016), pp. 1489–1500. https://doi.org/10.1080/13504622.2016.1263279.

10 Ross, *Transforming the Ways We Create Change.*

11 Ross, *Transforming the Ways We Create Change.*

12 Edgar González-Gaudiano, "Education for Sustainable Development: Configuration and Meaning," *Policy Futures in Education*, Vol. 3, No. 3 (2005), pp. 243–250. doi.org/10.2304/pfie.2005.3.3.2.

13 Edgar González-Gaudiano, "Environmental Education: A Field in Tension or in Transition?" *Environmental Education Research*, Vol. 12, Nos. 3–4 (2006), pp. 291–300. https://doi.org/10.1080/13504620600799042.

14 González-Gaudiano, "Education for Sustainable Development," p. 243.

15 Mary Paden gives some examples of these trade-offs, such as pursuing environmental preservation but at the cost of economic prosperity or pursuing social justice at the cost of environmental preservation. Mary Paden, "Education for Sustainability and Environmental Education," in Keith Wheeler and Anne Perraca Bijur (Eds.), *Education for a Sustainable Future: A Paradigm of Hope for the 21st Century* (New York: Kluwer Academic/Plenum Publishers, 2000), p. 9.

16 González-Gaudiano, "Education for Sustainable Development," p. 247.

17 Mary Paden, "Education for Sustainability and Environmental Education," in Keith Wheeler and Anne Perraca Bijur (Eds.), *Education for a Sustainable Future: A Paradigm of Hope for the 21st Century* (New York: Kluwer Academic/Plenum Publishers, 2000), p. 9.

18 For instance, see Ross, *Transforming the Ways We Create Change*; Joy Kcenia O'Neil, "(Be)Coming and (Re)Membering Through Kitchen Based Learning as Sustainability: An Innovative Living Learning Systems Model for Higher Education," in W. Leal Filho et al. (Eds.), *Handbook of Theory and Practice of Sustainable Development in Higher Education*, World Sustainability Series (New York: Springer International Publishing, 2017), pp. 317–333. https://doi.org/10.1007/978-3-319-47889-0_23 as well as Elizabeth A. Lange, Joy Kcenia Polanco O'Neil, and Katie E. Ross, "Educating During

the Great Transformation: Relationality and Transformative Sustainability Education," *Studies in Adult Education and Learning*, Vol. 27, No. 1 (2021), p. 34. Also see process theorists Alfred North Whitehead via Robert Mesle, *Process-Relational Philosophy: An Introduction to Alfred North Whitehead* (West Conshohocken, PA: Templeton Press, 2008).

19 Peter Blaze Corcoran, "Foreword," in Paula Jones, David Selby, and Stephen Sterling (Eds.), *Sustainability Education in Higher Education: Perspectives and Practice Across Higher Education* (London: Earthscan, 2010), pp. xiii–xv.

20 González-Gaudiano, *Environmental Education.*

21 Elizabeth A. Lange, "Interrogating Transformative Learning: Canadian Contributions," in Tom Nesbit, Susan Brigham, Nancy Taber, and Tara Gibb (Eds.), *Building on Critical Traditions: Adult Education and Learning in Canada* (Toronto: Thompson Educational Publishing, 2013), p. 107.

22 Lange, "Interrogating Transformative Learning," p. 108.

23 Chad Hoggan, "Transformative Learning as a Metatheory: Definition, Criteria, and Typology," *Adult Education Quarterly*, Vol. 66, No. 1, pp: 57–75. https://doi.org/10.1177/0741713615611216.

24 Paulo Freire, *Pedagogy of the Oppressed* (New York: The Seabury Press, 1970).

25 For further description, see Lange, "Interrogating Transformative Learning."

26 Edward Taylor, "Transformative Learning Theory," in Anna Laros, Thomas Fuhr, and Edward Taylor (Eds.), *Transformative Learning Meets* Bildung (Rotterdam: Sense Publishers, 2017), pp. 17–29.

27 Jack Mezirow, "Learning to Think Like an Adult: Core Concepts of Transformation Transformative Theory," in Jack Mezirow & Associates (Eds.), *Learning as Transformation* (San Francisco: Jossey-Bass Publishers, 1991), p. 8.

28 Jack Mezirow, *Transformative Dimensions of Adult Learning* (San Francisco: Jossey-Bass Publishers, 1991), p. 3.

29 Mezirow, *Transformative Dimensions of Adult Learning*, p. 155, and Learning as Transformation.

30 Mezirow, *Transformative Dimensions of Adult Learning*, p. 50.

31 Mezirow, *Transformative Dimensions of Adult Learning*, p. 156.

32 Jack Mezirow, "Transformative Learning Theory," in Jack Mezirow and Edward Taylor (Eds.), *Transformative Learning in Practice: Insights from Community, Workplace and Higher Education* (San Francisco; John Wiley & Sons, 2009), pp. 18–31.

33 These are paraphrased from Edward Taylor, "Fostering Transformative Learning," in Jack Mezirow and Edward Taylor (Eds.), *Transformative Learning in Practice: Insights from Community, Workplace and Higher Education* (San Francisco; John Wiley & Sons, 2009), pp. 4–13.

34 Edward Taylor, "Fostering Transformative Learning," in Jack Mezirow and Edward Taylor (Eds.), *Transformative Learning in Practice: Insights from Community, Workplace and Higher Education* (San Francisco; John Wiley & Sons, 2009), pp. 3–17.

35 Mezirow, "Transformative Learning Theory," p. 21.

36 Elizabeth A. Lange, "Fostering a Learning Sanctuary for Transformation in Adult Sustainability Education," in Jack Mezirow and Edward Taylor (Eds.), *Transformative Learning in Practice* (San Francisco: Jossey-Bass Publishers, 2009), pp. 193–204.

37 Patricia Cranton, *Understanding and Promoting Transformative Learning* (San Francisco: Jossey-Bass, 2006).

38 Taylor, "Transformative Learning Theory," p. 24.

39 For more on social learning, see Alfred Bandura's *Social Learning Theory* (London: Pearson Publishers, 1976). For more on social learning in relation to sustainability, see Arjen Wals, *Social Learning towards a Sustainable World* (Wageningen: Wageningen Academic Publishers, 2007).

40 Elizabeth A. Lange, "Transformative and Restorative Learning: A Vital Dialectic for Sustainable Societies," *Adult Education Quarterly*, Vol. 52 (2004), pp. 121–139.

41 Lange, "Transformative and Restorative Learning."
42 Edmund O'Sullivan, Amish Morrell, and Mary Ann O'Connor (Eds.), *Expanding the Boundaries of Transformative Learning: Essays on Theory and Praxis* (New York: Palgrave, 2002), p. xvii.
43 Edmund O'Sullivan, "The Project and Vision of Transformative Education: Integral Transformative Learning," in Edmund O'Sullivan, Amish Morrell, and Mary Ann O'Connor (Eds.), *Expanding the Boundaries of Transformative Learning: Essays on Theory and Praxis* (New York: Palgrave, 2002), pp. 1–22.
44 John Ralston Saul, *The Unconscious Civilization* (Toronto: House of Anansi Press, 1995).
45 O'Sullivan, "The Project and Vision of Transformative Education," p. 3.
46 O'Sullivan, "The Project and Vision of Transformative Education," pp. 8–9.
47 O'Sullivan, "The Project and Vision of Transformative Education," p. 10.
48 Chad Hoggan, "Transformative Learning as a Metatheory: Definition, Criteria, and Typology," *Adult Education Quarterly*, Vol. 66, No. 1, pp: 57–75. https://doi.org/10.1177/0741713615611216.
49 Charlene Spretnak, *Relational Reality: New Discoveries of Interrelatedness That Are Transforming the Modern World* (Topsham, ME: Green Horizon Books, 2011), p. 6.
50 Spretnak, *Relational Reality*.
51 Ervin Laszlo, *Quantum Shift in the Global Brain* (Rochester, VR: Inner Traditions, 2008), p. 1.
52 Ervin Laszlo, *The Whispering Pond: A Personal Guide to the Emerging Vision of Science* (Boston: Element, 1996).
53 Fritjof Capra, *The Tao of Physics: An Exploration of the Parallels between Modern Physics and Eastern Mysticism* (Boulder, CO: Shambhala, 2010). Ervin Laszlo, *The Whispering Pond: A Personal Guide to the Emerging Vision of Science* (Boston: Element, 1996). Elisabeth Sahtouris, *EarthDance: Living Systems in Evolution* (Lincoln, NE: iUniversity Press, 2000). Francisco Varela, Evan Thompson, and Eleanor Rosch, *The Embodied Mind: Cognitive Science and the Human Experience* (Cambridge: MIT Press, 1991).
54 David Loye, "Foreword I," in Ervin Laszlo (Ed.), *The Whispering Pond: A Personal Guide to the Emerging Vision of Science* (Boston: Element, 1996), p. xi.
55 Laszlo, *The Whispering Pond*, p. 202.
56 Laszlo, *The Whispering Pond*, p. 203.
57 Brian Swimme and Thomas Berry in The Universe Story: From the Primordial Flaring Forth to the Ecozoic Era, *A Celebration of the Unfolding of the Cosmos* (New York: HarperCollins, 1992).
58 Laszlo, *The Whispering Pond*, pp. 205–206.
59 Elisabet Sahtouris, *EarthDance: Living Systems in Evolution* (San Jose: iUniversity Press, 2000), pp. 6–7.
60 Lynn McTaggart, *The Field* (New York: Harper, 2008), p. 11.
61 Laszlo, *Quantum Shift in the Global Brain*.
62 McTaggart, *The Field*, p. 22.
63 McTaggart, *The Field*, p. 23.
64 McTaggart, *The Field*, p. 23
65 McTaggart, *The Field*, p. 19.
66 Alexander Wendt, *Quantum Mind and Social Science: Unifying Physical and Social Ontology* (Cambridge: Cambridge University Press, 2015), p. 60.
67 Diarmuid Murchu, *Quantum Theology: Spiritual Implications of the New Physics* (New York: Crossroad Publishing Company, 2004).
68 Danah Zohar, *The Quantum Self: Human Nature and Consciousness Defined by the New Physics* (New York: Quill/William Morrow, 1990), p. 9.
69 McTaggart, *The Field*, p. 33.
70 Wendt, *Quantum Mind and Social Science*, p. 29.
71 Heisenberg in Capra, *The Web of Life*.
72 Carolyn Merchant, *The Death of Nature* (New York: Harper & Row, 1989), p. xvii.

73 MacTaggart, *The Field*, p. 19.
74 MacTaggart, *The Field*, p. 11.
75 MacTaggart, *The Field*, p. 11.
76 Murchu, *Quantum Theology*; Laszlo, *Quantum Shift in the Global Brain*; Zohar, *The Quantum Self*.
77 Spretnak, *Relational Reality*, p. 5.
78 Einstein in David Suzuki, *The Sacred Balance: Rediscovering our Place in Nature* (Vancouver: Greystone Books, 1997), p. 26. Exclusive language in original.
79 North American Space Agency.
80 James Lovelock, *Gaia* (Oxford: Oxford University Press, 1987).
81 Paraphrased from Lovelock, *Gaia*, p. 9.
82 Lovelock, *Gaia*, p. xii.
83 James Lovelock, *The Vanishing Face of Gaia* (New York: Perseus Books, 2009), p. 255.
84 Lovelock, *The Vanishing Face of Gaia*.
85 Tim Radford, "Lovelock at 100: The Gaia Saga," *Nature*, Vol. 570 (2019), pp. 441–442.
86 Radford, "Lovelock at 100."
87 Lovelock, *The Vanishing Face of Gaia*, p. 198.
88 Lovelock, *The Vanishing Face of Gaia*, p. 180.
89 Lovelock, *The Vanishing Face of Gaia*, p. 181.
90 Lovelock, *Gaia*, p. 143.
91 Lovelock, *Gaia*, pp. 145, 148. Exclusive language in original.
92 Laszlo, *The Whispering Pond*, p. 199.
93 James Lovelock, "Foreword," in Sahtouris, *EarthDance*, p. xvi.
94 Sahtouris, *EarthDance*, p. 1.
95 Sahtouris, *EarthDance*, pp. 2, 4.
96 Eisenstein, *The More Beautiful World our Hearts Know is Possible*, p. 7.
97 Fritjof Capra and Pier Luigi Luisi, *The Systems View of Life: A Unifying Vision* (Cambridge: Cambridge University Press, 2014), pp. 63–65.
98 Prigogine in Capra, *The Web of Life*, p. 86.
99 Capra and Luisi, *The Systems View of Life*, p. 114.
100 Capra and Luisi, *The Systems View of Life*, p. 67.
101 Capra and Luisi, *The Systems View of Life*, p. 67.
102 Capra and Luisi, *The Systems View of Life*, p. 65.
103 Maturana in Capra, *The Web of Life*, p. 96.
104 Maturana in Capra, *The Web of Life*, p. 96.
105 Capra, *The Web of Life*, p. 97.
106 Maturana and Varela in Capra, *The Web of Life*, p. 98.
107 Capra and Luisi, *The Systems View of Life Capra*, p. 129.
108 Capra, *The Web of Life*.
109 Capra and Luisi, *The Systems View of Life*, p. 135.
110 Capra, *The Web of Life*, pp. 96–97. Emphasis in original.
111 Maturana in Capra, *The Web of Life*, p. 97.
112 Capra, *The Web of Life*, p. 97.
113 Francisco Varela, Evan Thompson, and Eleanor Rosch, *The Embodied Mind* (Cambridge, MA: The MIT Press, 1991), p. 173.
114 Varela, Thomspon, and Rosch, *The Embodied Mind*, p. 174.
115 Fritjof Capra, *The Tao of Physics: An Exploration of the Parallels between Modern Physics and Eastern Mysticism* (Boulder, CA: Shambala, 1975/2010), p. 138.
116 David Bohm, *Wholeness and the Implicate Order* (London: Routledge, 1980).
117 Bohm, *Wholeness and the Implicate Order*, p. x.
118 Blake in Lauren Artress, *Walking a Sacred Path: Rediscovering the Labyrinth as a Spiritual Tool* (New York: Riverhead Books, 1995/2006), p. 112.
119 Merchant, *The Death of Nature*, p. xvii.
120 Capra, *The Tao of Physics*.

121 It was Max Weber who first coined the term disenchantment in association with modernism. See Richard Jenkins, "Disenchantment, Enchantment, and Re-Enchantment," *Max Weber Studies*, Vol. 1 (2000), pp. 11–32.
122 Bohm, *Wholeness and the Implicate Order*, p. 17.
123 David Bohm, *Unfolding Meaning* (London: Routledge, 1985), p. 52.
124 For more description, see Elizabeth A. Lange, "Transforming Transformative Learning through Ontologies of Relationality," *Journal of Transformative Education*, Vol. 16, No. 4 (2018), pp. 280–301. https://doi.org/10.1177/1541344618786452.
125 Elizabeth A. Lange, Joy Kcenia Polanco O'Neil, and Katie E. Ross, "Educating during the Great Transformation: Relationality and Transformative Sustainability Education," *Studies in Adult Education and Learning*, Vol. 27, No. 1 (2021), p. 34.
126 Alfonso Montuori, "Foreword: Transdisciplinarity," in Basarab Nicolescu (Ed.), *Transdisciplinarity: Theory and Practice* (Cresskill, NJ: Hampton Press, 2008), p. xi.
127 Paraphrased from Montuori, "Foreword: Transdisciplinarity," p. ix.
128 Paraphrased from Montuori, "Foreword: Transdisciplinarity," p. xv.
129 Paraphrased from Montuori, "Foreword: Transdisciplinarity," p. xii.
130 Diane Laflamme, "Ethics and the Interplay Between the Logic of the Excluded Middle and the Logic of the Included Middle," in Basarab Nicolescu (Ed.), *Transdisciplinarity: Theory and Practice* (Cresskill, NJ: Hampton Press, Inc. (2008), pp. 147–154. Thanks to Katie E. Ross for drawing my attention to this concept.
131 Basarab Nicolescu, "In Vitro and In Vivo Knowledge – Methodology of Transdisciplinarity," in Basarab Nicolescu (Ed.), *Transdisciplinarity: Theory and Practice* (Cresskill, NJ: Hampton Press, 2008), p. 16.
132 Nicolescu, "In Vitro and In Vivo Knowledge – Methodology of Transdisciplinarity," p. 15.
133 This paragraph summarizes the key characteristics of systems thinking from Capra and Luisi, *The Systems View of Life*, pp. 80–83.
134 Capra and Luisi, *The Systems View of Life*, p. 79.
135 Capra and Luisi, *The Systems View of Life*, pp. 80–83.
136 As used by Katie Ross in: *Transforming the Ways We Create Change: Experiencing and Cultivating Transformative Sustainability Learning* (Sydney: University of Technology Sydney, 2020). http://hdl.handle.net/10453/149105.
137 Donella Meadows, *Thinking in Systems: A Primer* (White River Junction, VT: Chelsea Green Publishing, 2008), p. ix.
138 Capra and Luisi, *The Systems View of Life*.
139 Meadows, *Thinking in Systems*, p. x.
140 Meadows, *Thinking in Systems*, p. 2.
141 Meadows, *Thinking in Systems*, p. 11.
142 Meadows, *Thinking in Systems*, p. 11.
143 Charlene Spretnak, *The Resurgence of the Real: Body, Nature, and Place in a Hypermodern World* (New York: Routledge, 1999).
144 Jeremy Lent, *The Patterning Instinct* (Amherst, NY: Prometheus Books. 2017).
145 Lent, *The Patterning Instinct*, p. 19.
146 Lent, *The Patterning Instinct*, p. 16.
147 Lent, *The Patterning Instinct*, p. 20.
148 Lent, *The Patterning Instinct*, p. 22.
149 Fritjof Capra and Ove Jakobsen, "A Conceptual Framework for Ecological Economics Based on the Systemic Principles of Life," *International Journal of Social Economics*, Vol. 44, No. 6 (2017), pp. 835. https://doi.org/10.1108/IJSE-05-2016-0136.
150 Michel Alhadeff-Jones, "Transformative Learning and the Challenges of Complexity," in Edward Taylor, Patricia Cranton, & Associates (Eds.), *The Handbook of Transformative Learning* (San Francisco, CA: John Wiley, 2012), pp. 178–194.
151 Meadows, *Thinking in Systems*, p. 1.
152 Paraphrased from Meadows, *Thinking in Systems*, p. 14.
153 Paraphrased from Meadows, *Thinking in Systems*, p. 15.

154 Capra and Jakobsen, "A Conceptual Framework for Ecological Economics Based on the Systemic Principles of Life," p. 834.
155 Capra and Jakobsen, "A Conceptual Framework for Ecological Economics Based on the Systemic Principles of Life," p. 834.
156 Lange, O'Neil, and Ross, *Educating for the Great Transformation*, p. 34.
157 Stephen Sterling, *Sustainable Education: Re-visioning Learning and Change* (Cambridge: Green Books, 2001), p. 46.
158 Joy Kcenia O'Neil, "(Be)Coming and (Re)Membering Through Kitchen Based Learning as Sustainability: An Innovative Living Learning Systems Model for Higher Education," in W. Leal Filho et al. (Eds.), *Handbook of Theory and Practice of Sustainable Development in Higher Education*, World Sustainability Series (New York: Springer International Publishing, 2017), pp. 317–333. https://doi.org/10.1007/978-3-319-47889-0_23.
159 Joy Kcenia O'Neil, "(Be)Coming and (Re)Membering Through Kitchen Based Learning as Sustainability," p. 326.
160 Montuori, "Foreword: Transdisciplinarity," p. xv.
161 Fritjof Capra, "Speaking Nature's Language: Principles for Sustainability," in Michael Stone and Zenobia Barlow (Eds.), *Ecological Literacy: Educating our Children for a Sustainable World* (San Francisco: Sierra Club Books, 2005), pp. 20–22.
162 Fritjof Capra, "Speaking Nature's Language," pp. 22–29. See also Barbara Widhalm, "Educators as Architects of Living Systems: Designing Vibrant Learning Experiences Beyond Sustainability and Systems Thinking," *Journal of Sustainability Education*, Vol. 2 (March 2011). www.jsedimensions.org/wordpress/wp-content/uploads/2011/03/Widhalm2011.pdf. This chart was provided by the Centre for Ecoliteracy and draws from Stone and Barlow, *Ecological Literacy*. Another rendition comes from Linda Booth Sweeney, "12 Living Systems Principles," see footnote 164. www.lindaboothsweeney.net/index.php/thinking/principles/principles.html.
163 Michael Stone, *Applying Ecological Principles* (Berkeley, CA: Center for Ecoliteracy), 25 July 2012. Downloaded from www.ecoliteracy.org/article/applying-ecological-principles. Combined with Fritjof Capra, "Speaking Nature's Language," pp. 18–29.

In contrast, Linda Booth Sweeney lists 12 principles of living systems: interdependence, system integrity, biodiversity, cooperation and partnership, rightness of size, living cycles, waste = food, feedback, nonlinearity, emergent properties, flux, and the commons. www.lindaboothsweeney.net/index.php/thinking/principles/principles.html.
164 Capra and Luisi, *The Systems View of Life*, p. xi.
165 Capra and Luisi, *The Systems View of Life*, p. xi, my emphasis added.
166 Janine Benyus, *Biomimicry: Innovation Inspired by Nature* (New York: Quill/William Morrow, 1997).
167 Benyus, *Biomimicry*, pp. i, 2.
168 See Bioneers at https://bioneers.org/ and Steve Lerner, *Eco-pioneers: Practical Visionaries Solving Today's Environmental Problems* (Cambridge: The MIT Press, 1997), as an early example.
169 Benyus, *Biomimicry*, p. 2.
170 Benyus, *Biomimicry*, p. 7.
171 Gregory Bateson, *Mind and Nature: A Necessary Unity* (New York: Bantam, 1979), p. 8.
172 Diana Beresford-Kroeger, *The Global Forest* (New York: Viking Group, 2010), pp. 4–5.
173 Suzanne Simard, *Finding the Mother Tree: Discovering the Wisdom of the Forest* (New York: Random House, 2021).
174 Peter Wohlleben, *The Hidden Life of Trees: What They Feel, How They Communicate—Discoveries from A Secret World* (Vancouver: Greystone Books, 2016).
175 David George Haskell, *The Songs of Trees: Stories from Nature's Great Connectors* (New York: Penguin Books, 2017), pp. vii–viii.
176 Haskell, *The Songs of Trees*.
177 Edgar Weber and Anne Khademian, "Wicked Problems, Knowledge Challenges, and Collaborative Capacity Builders in Network Settings," *Public Administration Review* (March/April 2008), pp. 334–349. https://doi.org/10.1111/j.1540-6210.2007.00866.x.

178 Paraphrased from Meadows, *Thinking in Systems*, p. 2.
179 Capra and Jakobsen, *A Conceptual Framework for Ecological Economics Based on the Systemic Principles of Life*, pp. 831–844.
Marjorie Kelly's book is *Owning our Future: The Emerging Ownership Revolution, Journeys to a Generative Economy* (Oakland: Berrett-Koehler, 2012).
180 Capra and Jakobsen, "A Conceptual Framework for Ecological Economics based on the Systemic Principles of Life," p. 836.
181 Capra and Jakobsen, "A Conceptual Framework for Ecological Economics based on the Systemic Principles of Life," p. 837.
182 Capra and Jakobsen, "A Conceptual Framework for Ecological Economics based on the Systemic Principles of Life," p. 837.
183 Johan Rockstrom, Will Steffen, Kevin Noone, et al., "A Safe Operating Space for Humanity," *Nature*, Vol. 461 (2009), pp. 472–475. https://doi.org/10.1038/461472a.
184 Kate Raworth, *Doughnut Economics*. www.kateraworth.com/doughnut/.
185 Kate Raworth, "A Doughnut for the Anthropocene: Humanity's Compass in the 21st Century," *The Lancet*, Vol. 1 (May 2017). www.thelancet.com/planetary-health.
186 Kate Raworth, *Doughnut Economics: Seven Ways to Think Like a 21st Century Economist* (London: Penguin Random House, 2017).
187 In Victor Nolet, *Educating for Sustainability: Principles and Practices for* Teachers (New York: Routledge, 2016), p. 72.
188 Mihaly Csikszentmihalyi, *Finding Flow: The Psychology of Engagement with Everyday Life* (New York: Basic Books, 1997).
189 Csikszentmihalyi, *Finding Flow*, pp. 29–30.
190 David Bohm, *Thought as a System* (London: Routledge, 1994), p. 167.
191 Bohm, *Thought as a System*, p. 36.
192 www.lindau-nobel.org/chance-favors-the-prepared-mind/.
193 Bohm, *Unfolding Meaning*, p. 37.
194 Bohm, *Unfolding Meaning*, p. 38.
195 Bohm, *Unfolding Meaning*, p. 39.
196 Stanislav Grof in Laszlo, *The Whispering Pond*, p. 219.
197 Laszlo, *The Whispering Pond*, p. 219.
198 Erik Erikson, *Childhood and Society* (New York: W. W. Norton & Company, 1950/1963), pp. 266–268.
199 Bohm, *Unfolding Meaning*, p. 42.
200 Bohm, *Unfolding Meaning*, p. 55.
201 Bohm, *Unfolding Meaning*.
202 David Bohm, *The Essential David Bohm* (London: Routledge, 2003), p. 80.
203 Bohm, *The Essential David Bohm*, p. 210.
204 Barbara Thayer-Bacon, *Relational "(e)pistemologies"* (New York: Peter Lang Publishing, 2003).
205 David Selby, "The Signature of the Whole: Radical Interconnectedness and Its Implication for Global and Environmental Education," in O'Sullivan, Morell, and O'Connor, *Expanding the Boundaries of Transformative Learning* (New York: Palgrave, 2002), pp. 77–94.
206 Selby, "The Signature of the Whole," p. 82.
207 Harold Morowitz, 1972 in Selby, "The Signature of the Whole," p. 82
208 Selby, "The Signature of the Whole," p. 82.
209 Laszlo, *The Whispering Pond*, pp. 206–207.
210 Thayer-Bacon, *Relational "(e)pistemologies."*
211 Thayer-Bacon, *Relational "(e)pistemologies."*
212 Spretnak, *Relational Reality*, p. 4.
213 Karen Barad, *Meeting the Universe Halfway: Quantum Physics and the Entanglement of Matter and Meaning* (Durham: Duke University Press, 2007), p. 33. Emphasis in original.
214 Barad, *Meeting the Universe Halfway*, p. 33.

215 Barad, *Meeting the Universe Halfway*, p. 197.
216 Barad, *Meeting the Universe Halfway*, p. 133.
217 Carol Gould, *Marx's Social Ontology* (Cambridge: MIT Press, 1978).
218 Discussed more extensively in E. Lange, "Is Freirean Transformative Learning the Trojan Horse of Globalization and Enemy of Sustainability Education? A Response to C.A. Bowers," *Journal of Transformative Education*, Vol. 10, No. 1 (2012), pp. 3–21. https://doi.org/10.1177/1541344612453880.
219 See Bertel Ollman, *Alienation* (Cambridge, MA: University Press, 1971).
220 Barad, *Meeting the Universe Halfway*, p. 376.
221 Barad, *Meeting the Universe Halfway*, p. 341.
222 Barad, *Meeting the Universe Halfway*, p. 336.
223 Karen Barad, *Matter Feels, Converses, Suffers, Desires, Yearns, and Remembers: An Interview with Karen Barad. 7th European Feminist Research Conference* (Utrecht University, Germany, 2009).
224 Barad, *Meeting the Universe Halfway*, p. 383.
225 Barad, *Meeting the Universe Halfway*, p. 391.
226 Eisenstein, *The More Beautiful World our Hearts Know is Possible*, p. 6.
227 Barad, *Meeting the Universe Halfway*, p. 132.
228 Evelyn Fox Keller, *A Feeling for the Organism: The Life and Work of Barbara McClintock* (New York: Henry Holt and Company, 1983), p. 206.
229 Keller, *A Feeling for the Organism*, p. 200.
230 Edward O. Wilson, *Biophilia* (Cambridge: Harvard University Press, 1984).
231 Wilson, *Biophilia*, p. 140.
232 Wilson, *Biophilia*, p. 139.
233 Wendt, *Quantum Mind and Social Science*, p. 31
234 Zohar, *The Quantum Self*, p. 95.
235 Christian De Quincey, *Radical Knowing: Understanding Consciousness through Relationship* (Rochester, VT: Park Street Press, 2005), p. 24.
236 Philip Goff, William Seager, and Sean Allen-Hermanson, "Panpsychism," in *Stanford Encyclopedia of Philosophy* (Stanford, CA: The Metaphysics Research Lab, Stanford University, 2016).
237 Zohar, *The Quantum Self*, p. 98.
238 Zohar, *The Quantum Self*, p. 100.
239 De Quincey, *Radical Knowing*, p. 165.
240 Barad, *Meeting the Universe Halfway*, p. 335.
241 Diarmuid O'Murchu, *Quantum Theology: Spiritual Implications of the New Physics* (New York: The Crossroad Publishing Company, 2004), p. 36.
242 O'Murchu, *Quantum Theology*, p. 36.
243 Thayer-Bacon, *Relational "(e)pistemologies"*, p. 27.
244 Thayer-Bacon, *Relational "(e)pistemologies"*, p. 29.
245 Thayer-Bacon, *Relational "(e)pistemologies"*, p. 29.
246 Thayer-Bacon, *Relational "(e)pistemologies"*, p. 29.
247 Thayer-Bacon, *Relational "(e)pistemologies"*, p. 101.
248 Thayer-Bacon, *Relational "(e)pistemologies"*, p. 111.
249 Thayer-Bacon, *Relational "(e)pistemologies"*, p. 83.
250 Thayer-Bacon, *Relational "(e)pistemologies"*, p. 87.
251 Thayer-Bacon, *Relational "(e)pistemologies"*, p. 90.
252 Paul Hart, *The Integrative Mind* (Lanham, MA: Rowman and Littlefield, 2014).
253 de Quincey, *Radical Knowing*.
254 de Quincey, *Radical Knowing*.
255 de Quincey, *Radical Knowing*, p. 252.
256 de Quincey, *Radical Knowing*, p. 249.
257 de Quincey, *Radical Knowing*, pp. 11–13.
258 Christopher Bache in De Quincey, *Radical Knowing*, p. 244.

259 de Quincey, *Radical Knowing*, p. 122.
260 See Jeremy Narby, *The Cosmic Serpent: DNA and the Origins of Knowledge* (Los Angeles: TarcherPerigee, 1999).
261 Jeremy Narby in de Quincey, *Radical Knowing*, p. 234.
262 Jeremy Narby in de Quincey, *Radical Knowing*, p. 234. See also Narby, *The Cosmic Serpent*.
263 Jeremy Narby in de Quincey, *Radical Knowing*, p. 235.
264 Jeremy Narby in de Quincey, *Radical Knowing*, p. 232.
265 Jeremy Narby in de Quincey, *Radical Knowing*, p. 239.
266 Aaron Cheak, *Alchemical Traditions* (Melbourne: Numen Books, 2013), p. 34.
267 de Quincey, *Radical Knowing*, pp. 238, 239.
268 For more, see Mousumi Mukherjee and Sandal Agrawal, "Decolonising Lifelong Education: Learning from J. Krishnamurti," *International Journal of Lifelong Education*, Vol. 40, No. 4 (July/August 2021), pp. 328–338.
269 www.bohmdialogue.org/ See also David Bohm, *On Dialogue* (New York: Routledge, 1996).
270 www.bohmdialogue.org/.
271 de Quincey, *Radical Knowing*, p. 35.
272 de Quincey, *Radical Knowing*, p. 36.
273 Capra, *The Tao of Physics*, p. 335.
274 Dilip Jeste and Scott Lafee, *Wiser: The Scientific Roots of Wisdom, Compassion, and What Makes Us Good* (Boulder: Sounds True, 2020), pp. 13–15.
275 de Quincey, *Radical Knowing*, pp. 120–121.
276 Christopher Bache, *The Living Classroom: Teaching and Collective Consciousness* (New York: SUNY, 2008).
277 de Quincey, *Radical Nature*, pp. 152, 154, 155.
278 Simon Ceder, *Cutting through Water: Towards a Posthuman Theory of Educational Relationality* (Lund, Sweden: Lund University, 2016), p. 27.
279 Elizabeth A. Lange, "Living Transformation: The Alchemy of Change in an Epochal Shift," in Aliki Nicolaides, Saskia Eschenbacher, Petra Buergelt, Yabome Gilpin-Jackson, Marguerite Welch, and Mitsunori Misawa (Eds.), *The Palgrave Handbook of Learning for Transformation* (London: Palgrave Macmillan, 2022), pp. 733–749. https://link.springer.com/book/10.1007/978-3-030-84694-7.
280 de Quincey, *Radical Nature*, p. 155.
281 Freya Mathews in Dawson Church, *Mind to Matter*, Kindle Edition (Carlsbad, CA: Hay House, 2018), p. 12.
282 Capra and Luisi, *The Systems View of Life*, p. 116.
283 Laszlo, *Quantum Shift in the Global Brain*.
284 Ross, *Transforming the Ways We Create Change: Experiencing and Cultivating Transformative Sustainability Learning*.
285 Lange, O'Neil, and Ross, "Educating during the Great Transformation," p. 40.
286 O'Neil, "(Be)Coming and (Re)Membering Through Kitchen based Learning as Sustainability," p. 321.
287 Original Source: Lange, O'Neil, and Ross, "Educating during the Great Transformation," pp. 23–46.
288 Eisenstein, *The More Beautiful World Our Hearts Know Is Possible*, p. 4.

7

THE MODERN STORY OF EDUCATION

Crisis *of* Education

> Education and [spirituality], especially, should awaken in the young an awareness of the world in which they live, how it functions, how the human fits into the larger community of life, the role that the human fulfills in the great story of the universe, and the historical sequence of development that have shaped our physical and cultural landscape. Along with this awareness of the past and present, education and [spirituality] should communicate some guidance concerning the future. The pathos of these times, however, is precisely the impasse that we witness in our educational and [spiritual] programs . . . unable to support or guide the transformation that is needed in its proper order of magnitude.[1]

The old mechanistic story no longer holds; we are in the midst of a Reality Revolution, tending towards relationality. We, as humans, are dance partners

DOI: 10.4324/9781003159643-7

in the material unfolding of the universe, part of the long cosmic evolution, and part of the self-creating and cognitive capacities of this dynamic cosmos. As discussed in the last chapter, Transformative Sustainability Education involves a cosmo-onto-axi-epistemological transformation towards relationality, as deep sustainability. In this way, Transformative Sustainability Education addresses the crisis *in* education through a third order change or epistemic transformation, expanding notions of how we come to know and what is valid knowledge about reality. With its many facets, Transformative Sustainability Education can help establish conditions for transformative learning towards relationality, through radical wholeness, systems thinking, process thinking, energy fields, communal individuality, entangled ethics, and wisdom. As Thomas Berry suggests in the opening quote, it is learning for supporting the life-giving capacities of Earth so that the whole community of Life is nourished by our human presence and can continue the dancing vitality of Life indefinitely, which is the fullest meaning of sustainability.

We now need to step back from our investments in *formal* education systems, whether schooling or higher education, and consider the crisis *of* education. Repurposing education in relation to society and reimagining the field of education are one pathway towards sustainable societies. This is no longer an option; it pertains to our very survival. The family of Life will continue in some form, with or without us.

This is not a call for a unitary approach but rather a call to educators to engage learning that creates conditions for the planet to heal, appropriate to each specific context. It is solving the urgent crisis—an integrated climate, environmental, economic, social and spiritual crisis—by learning to live relationally. It is creating a polyarchy of ways in which humans can be in "right relationship" with Earth[2]. As Hayden Washington describes, "sustainability is thus healing the planet along with our society and our maladjusted economy." This requires ideas, commitment, and real action.[3] One place to begin is with a new social imaginary.

Social Imaginaries

Canadian philosopher Charles Taylor describes the existing world, the modern Western world, as a social imaginary. A social imaginary means "the way our contemporaries imagine the societies they inhabit and sustain."[4] In fact, he says, it is numerous co-existing social imaginaries,[5] growing to become "self-evident," and thus invisible to succeeding generations. It includes our normal expectations, common understandings, and our collective social practices, more "unstructured and inarticulate"[6] than doctrines and theories. These multiple imaginaries are brought to life daily through images, stories, art, and iconography[7] often through media, comprising a mythology or "implicit map of social space"[8] with associated meanings. Introducing ideas and theories that "penetrate and transform the social imaginary" assists in helping these meaning systems become more visible and trigger improvisation towards a new imaginary.[9]

Every society has its mythologies or, rather, maps by which cultures populate a way of life[10]—religious or secular—as the carrier of the imaginary. When the mythology is mistaken for literal fact and becomes the only possibility, unquestioned, violence can ensue. The Western imaginary has been one of constant violence on many fronts, recounted in previous chapters, especially the reverberating impacts of the colonial/neocolonial eras, founded on a One Truth imaginary.

Today, this modern Western imaginary animates the social habits, norms, worldviews, and institutions we are most familiar with, specifically the industrial capitalist market economy and private property. As Erich Fromm brilliantly asserts,

> [O]ur judgements are extremely biased because we live in a society that rests on private property, profit, and power as the pillars of its existence. To acquire, to own, and to make a profit are the sacred and unalienable rights of the individual in the industrial society. What the sources of property are does not matter; nor does possession impose any obligations on the property owners.[11]

Private, from the Latin *privare*, means "to deprive of."[12] In other words, depriving others of something, by owning it, defines what it is to be powerful and successful. Using property without concern for the collective good is also another source of contemporary power. Fromm goes on to identify that private property norms "mold the character of its members . . . [through] the wish to acquire property, to keep it, and to increase it, i.e., to make a profit, and those who own property are admired and envied as superior beings." A key element of sustainability education is examining such property structures and how they manifest in education systems.

Fromm identified numerous property structures at work generally in Western societies—not only cherishing possessions and homes, but also utilizing universities and colleges as knowledge markets and marriage markets, later selling ourselves on the labour "market." A traditional property form has been owning living beings such as "man in a patriarchal society" who has, over Western history, been master of "his" wife, children, animals, any slaves or workers, and physical property.[13] It is this property form which has led to so many human and environmental violations.

As Fromm explains, property structures are tightly tied to individualism and thus identity. This ownership form can include

> friends, lovers, health, travel, art objects, God, one's own ego. . . . Individualism, which in its positive sense means liberation from social chains, means, in

> the negative sense, 'self-ownership.' The right—and the duty—to invest one's energy in the success of one's own person.
>
> Our ego is the most important object of our property feeling, for it comprises many things: our body, our name, our social status, our possessions (including our knowledge), the image we have of ourselves and the image we want others to have of us . . . this 'thing' is the basis of our sense of identity.[14]

Acquisition and throwing away, "a transitory having and using," offer a sense of identity and power, especially the freedom to grow tired and dispose of items alongside the eagerness to buy the new, latest, and greatest.[15] Feeding our ego in these multiple ways maintains these property structures in society and our sense of self within it.

A contemporary aspect of our social imaginary, then, is "seeing" society to be comprising these property structures. For instance, we view the economy as the primary sector, with its purpose of producing and exchanging goods, and the primary route to the goal of individual ownership of money and goods and national wealth. Security and economic prosperity have been the two goals of an economy. We understand our "value" according to where we sit on this stratified scale of ownership and wealth. Seeking to maintain or improve our prosperity drives global consumerism and thus inequality, says Franklin Obeng-Odoom.[16] We have difficulty imagining life without labouring, money accumulation, and acquisition of consumer goods, the triad that boundaries our collective and individual life. Yet, overlooking this history and property relations, say Obeng-Odoom, is to perpetuate unjust sustainability.

Education prepares us fittingly for successful participation in an industrial economy and engagement in forms of private property ownership and in the acquisitive consumer system. As Ivan Illich explains,

> [P]eople are taught all this not by the teacher but by the curriculum hidden in the structure of school. It does not matter what the teacher teaches so long as the pupil has to attend hundreds of hours of age-specific assemblies to engage in a routine decreed by the curriculum and is graded according to his ability to submit to it. People learn that they acquire more value in the market if they spend more hours in class. They learn to value progressive consumption of curricula. They learn that whatever a major institution produces has value . . . They learn to value grade advancement, passive submission . . . disciplined competition . . . [and being] holders of knowledge stock in the speciality in which they have made investments of their time. They learn to accept their place in society precisely in the class and career corresponding to the level at which they leave school and to the field of their academic specialization.[17]

One aspect of the crisis *in* education is this property form, embedded throughout educational structures and processes. By extension, education approaches that remain "uncritical of the market-based framing" of environmental sustainability are not addressing the root cause of "monism" or, rather, reducing all aspects of life including education to the economy.[18]

The other two components of the contemporary social imaginary, says Charles Taylor, are the public sphere as a meeting place where people imagine themselves deliberating on collective issues as well as the idea of self-governing people based on the power of elected assemblies. For instance public education was designed as a necessary process for preparing individuals with knowledge about the political system, capacities for reasoning and engaging democratic processes, and attitudes amenable to collective decision-making through representative democracy.

While, theoretically, education builds the skills for self-governance and provides space for citizens and citizens-in-training to make informed individual decisions according to their convictions, there are fundamental ethical contradictions between our economic and political systems. The original notion of individualism was socially rooted, where one's existence was dependent on family and community. Individualism was

> part of a rigorous morality that understood self-truth as embedded in a larger social responsibility. Individualism was meant to rest on the moral ideal that the individual would take the reflective time necessary to be connected to their inner moral voice and from this basis determine what kind of society they wished to live in, and then to participate in creating that social vision.[19]

In other words, individualism was meant to consider mutual benefit, but, as Taylor says, this has been lost from view as communities were eroded through the assumptions and practices of a competitive economic system.[20]

In liberal democracy, individualism became atomistic, with autonomous individuals who relate to an external world and whose rights are protected from encroachment by law, especially property rights. However, this was only possible once people became mobile and disembedded from their communities and understood themselves to be independent actors. Taylor contends that individualism became trivialized, where it is now "a centering on the self and a concomitant shutting out, or even unawareness, of the greater issues or concerns that transcend the self, be they religious, political, historical," or economic.[21]

The implication is that a study of land and property rights in relation to different conceptions of the individual is useful, with attention to the notion of a socially embedded individualism, or more specifically, an individual-in-relations. The rejection of hierarchy was an important aspect of individualism, where all are equal, but individualism should not now be a "solvent of community" or ignore the right of others to the means of life.[22]

Taylor argues that with a deepening cultural malaise, sense of alienation, and "excessive interest in private wealth and fortune,"[23] people live only for the moment and for themselves in seeking self fulfillment, some to the point of narcissistic self-absorption, rather than for their place in history or for posterity. Lifted out of communities by economic dynamics, geographic mobility, and ethical fragmentation means people are less able to effectively form common purpose among themselves and act collectively, except in small single-issue groups. As Taylor and Ralston Saul both argue, political theorist Tocqueville warned in the mid-1800s that an empty public square leaves space for increasing political oppression.[24] Canadian philosopher John Ralston Saul suggests that the loss of reasoned public debate and active participation in self-government has been surrendered to a passive acceptance of corporatist agendas, or "soft despotism."[25]

Another contradiction became the definition of society as primarily an economic matter, which shapes politics, rather than the other way around. In a profound ethical reversal, Adam Smith's invisible hand of competitive self-interest was to translate into the utilitarian "most good for the highest number." The original morality of frugality within the Protestant work ethic was lost, and the pursuit of long hours of disciplined work towards economic acquisition was not a valued ethic, rationalized as the contribution to national wealth production. The undergirding imaginary is that of the society as an economy and the individual right of pursuing economic self-interest, no questions asked. Making rational choices in one's best economic interest has been understood to be naturally contributing to the collective interest. Yet, corruption is defined as "purely self regarding actors."[26] Taken to natural ends, we can witness the gross accumulation of wealth at the degrading expense of millions of people and the living world. When we legitimize wealthy class interests and their domination of political systems, "the overall effects on the individual are passivity and conformity in those areas that matter and nonconformism in those which don't."[27] In other words, the privileging of the economy, which allows for the corporate takeover of meaningful and full democracy, and accepts the distortions of individualism to mean getting what one can for themselves have eroded the balance and meaningfulness of the three social imaginaries that comprise society—the economy, state, and civil society.[28]

Carol Gould distinguishes between abstract and concrete democracy, where liberal democracy is an abstract equality in that equality is protected in law but is not necessarily a concrete lived reality. An understanding of individuals-in-relations, or a relationality perspective, supports concrete equality with self-interest balanced with the interests of others, for the collective good, not only in theory but also in practice. The society that is then generated is fundamentally just. In this view, "Internal relations are those in which the individuals are changed by the relations to each other, that is, where these relations between individuals are such that both are reciprocally affected by the relation."[29] One is never outside the harm they may be creating for others. It is this process of objectification and alienation that enables those in liberal democracies to separate themselves and their decisions from their societal impact, including appropriation and degradation of the natural world and other people.

What If?

Of all our cognitive capacities, educational philosopher Maxine Greene explains, it is imagination that allows us to "break from the taken for granted, to set aside the familiar."[30] Recovering the imagination can break through the paralysis and restore "the sense that something can be done in the name of what is decent and humane."[31] Imagination can make us "wide awake to the world"[32] and give "credence to alternative realities."[33] Imagination can generate the energy to break through the anaesthetizing images to envision and enact new possibilities, in this case for education, as it relates to larger society. Yet, as American ecofeminist writer Rebecca Solnit explains, "Changing the story isn't enough in itself, but it has often been foundational to real changes."[34]

Some insist imagination leads to "pie-in-the-sky" or utopian thinking, simply not realistic, or perhaps interesting—but on too long a timeframe. Some environmental and sustainability advocates underline the urgent necessity for confronting the crass self-interests of politicians and powerful corporate lobbies to precipitate large-scale change. The urgency is undeniable. One element of social change has always been the power of asking "what if?" and "why not?" to ensure we are going deep enough and in a good direction with our social change efforts.

An education imaginary shapes why we educate, who is educated, where we educate, when we educate, what is learning, what is learned, how we teach, and which systems are put in place to facilitate education and learning. When we engage in a robust social re-imagining regarding the nature and structure of education, it can begin to shift. Imagining beyond current horizon lines in education, we can find "thresholds"[35] or openings beyond existing bounds to the "what if/as if" worlds being created by visionaries, who are thinkers, writers, poets, designers, leaders, and artists of all kinds.[36] But let us delve further into the historical faces of our modern social imaginary as it pertains to education, prior to imagining alternatives.

Faces of the Modern Social Imaginary in Education

In this age of "societies,"[37] education became "the social institution through which society provides its members with important knowledge, including basic facts, job skills, and cultural norms and values."[38] However organized, the education system reproduces the dominant social, economic, and cultural forms of a society, in this case, in accord with the modern social imaginary.

Overall, this modern education imaginary is the culmination of a mechanist cosmology, an *ontos*[39] of autonomous individualism and *homo economicus* (economic man who makes rational decisions to maximize gain) within private property relations,[40] an ontology of separation, anthropocentric humanist ethics, and a science-based, rationalist epistemology augmented by industrial/technological utopianism (the belief that technology can solve most production problems), and facilitated by a bureaucratic rationality as the most efficient organizational form. This constellation is what it means to be "modern," including modern education.

The initial goal of modern education has been the transformation of European and, then, all other cultures into modern societies. Following this, the primary goal now is the reproduction of this modern Western form of education, part of its conservative nature. Yet, when a large change is needed in a society, there is a collective turn towards educational institutions to carry out the change, whether nationally or internationally. This has made education a site of cultural struggle between the forces of reproduction and the forces of transformation, with its multiple faces and subsidiary currents. The following section describes, in very broad strokes, the historical faces of the modern education imaginary.

Educating for Proselytizing and Civilizing: A Control/Judge/Reward/Punish Logic

Over early Western history, formal education has typically been for the elites, through private academies, as in Classical Greece and Rome.[41] Private tutors and monastic schools in the Medieval age were the prime centres of extended learning for the aristocracy, evolving into the earliest Western universities, the first in Bologna, Italy, in 1088; then Oxford in England (1096); and the Sorbonne in Paris (1257).[42] For other classes, one could be apprenticed into a trade, ensuring a higher income, either through their parents/relatives or by joining a guild and working under guildmasters.[43] Grammar schools for boys developed in association with churches, guilds, and, sometimes, hospitals and orphanages, largely under the authority of the Roman Catholic Church.[44] While women were generally not educated, they had access to education through religious orders or apprenticeships in trades occupied by women.

After the Protestant Reformation and development of the printing press, schools developed around local parishes/congregations to provide basic literacy for reading the Bible. Given the "preoccupation [with] cultivation of faith," schools were a mechanism of spiritual reform and training, so that adherents could develop an intellectual understanding of the faith, rather than just piety and obedience.[45] The growth of humanism during the Renaissance and rediscovery of classical texts resuscitated educational ideas regarding a balance between physical education and scholastics, informing development of the "liberal arts."[46] Education expanded beyond the elites, beginning to incorporate notions of child development, and connecting "educational, social, and political problems."[47] As Max Weber notes, the expansion of education and uptake among the growing commercial or bourgeois class, often Protestant, were several pivotal factors to the development of early capitalism.[48]

As empires, European-style education was exported through colonialism.[49] Religious workers accompanied explorers and state functionaries, transporting the "conversion and civilizing" purpose of education to the colonies and colonists, bringing them or keeping them in into the European and Christian sphere of influence. Education was seen to yield more governable subjects.

With the Enlightenment and the age of nation states from the French Revolution, the "aim of education [shifted to] be the mastery of thinking and judgment rather than the mere assimilation of facts."[50] Czech philosopher John Comenius[51] argued for universal education, where "all men should be educated to full humanity—to rationality, morality, and happiness."[52] He further stated that "the whole of the human race may become educated, men of all ages, all conditions, both sexes and all nations."[53] Very slowly, the nation state began to undertake responsibility for supporting schools as preparation for citizenship. By the 18th century, ideas of secular education and progress arose.[54] Public universities and public libraries became widespread, and opportunities for "elementary" schooling grew. Education using the mother tongue for instruction rather than Latin increased rates of literacy throughout Europe.

As educational philosopher Barbara Thayer-Bacon[55] explains, the ontology undergirding early Euro-American education is heavily derived from the Judeo-Christian tradition. The Roman Catholic notion of original sin meant that idleness was the "devil's workshop."[56] In this social imaginary, children and young adults needed to be strictly supervised in controlled environments by recognized authorities, directed constantly into productive pursuits, then evaluated, while punished for straying off task and for curious questioning rather than obedience.

In the Protestant belief system, the notion that individual struggle and achievement on earth are rewarded in heaven, and that wealth and progress are signs of merit, resulted in an educational system of judge-and-reward or punish structures of obedience through extrinsic praise and marks. This judge-and-reward system was paralleled in the economic system. Happiness was expected to flow from the merit incurred through personal effort to control impulses and engage in hard work and productive activity in predetermined roles. This informed the "civilizing missions" in which all those not in the elites—the "inferior, sinners, primitive, less intelligent, backward, and even savage"—would come into Divine favour and participate in the "dominion over the Earth."[57]

In the colonies, while missionaries often preferred to teach in the "native" language, the British Empire, determined in 1830 that the English language should be the language of education, not only considered a "superior" language and canon of literature, but also integral to British imperialism.[58] This attitude was repeated by each colonizing European country.

Over time, however, it has been the English language which has prevailed as the language of trade and global communication. Euro-American imperialism continues to this day, no longer through coercion but according to an economic imperative originally established during the colonial era. Thus, this economic imperative is tied to the language imperative and the education imaginary of control, judge, and reward/punish.

Furthermore, it was literacy, not oralcy, that was accepted. Oral cultures were swept away in the drive towards universal literacy through schooling as the dominant educational form. As more evidence emerges in Canada about residential schools, generally operated by churches under the auspices of the state, established to "educate" Indigenous people, it is clear now how violent these attempts at genocide were, erasing:

- ways of being human including languages, clothing, traditional diet, and social structures;
- ways of thinking including myths, histories, spiritualities, and learning processes; and
- ways of living that did not adhere to definitions of "civilized."

So far, ground-penetrating radar is helping to find between 4,000 and 8,000 children and young people who died and were buried anonymously at these church and state-supported schools. The Indigenous social imaginary was considered backward, uncivilized, and ignorant, to be extinguished. Many children were.

Western education contributed to conquest and accumulation, for exploiting and controlling all life, founded on a Judeo-Christian cosmology and scientific rational empiricism. The "natural world" was considered imperfect and thus raw matter for improvement. No longer considered sacred, the natural world was an inert, secular "landscape" upon which the sacred Christian redemption story was lived out. We are now facing the existential consequences of this drive to conquer, convert, control, judge, improve, and amass.

Educating for Modernizing: A Schooled Logic

At the end of the colonial era and the early industrial era, then into the 20th century, publicly tax-funded education or rather "free or common schools"[59] expanded systematically to support fragile new democracies, industrializing economies, and colonial competitiveness. Social reformers advocated public education to provide not only Christian character and ethical training but also literacy and numeracy during this time of mass dislocation. Education was understood by reformers as emancipatory for the "lower classes." For others, it was a mechanism of social control.

From the late 1800s to first half of the 20th century, attitudes and expectations—from punctuality, obedience, to cleanliness—were literally beaten into children by hand or stick. Content was drilled into them always with the threat of violence, whether carried out by public or religious institutions. Soon a secular, public system, Euro-American schooling, was built upon an industrial, assembly-line model with

> drab factory-like buildings set apart from their surrounding communities with which they have little or no interaction; bland individual classrooms with their rows of desks and scant resources; the regulated routines of daily

> life (the bells, timed lessons, and work schedules); the physical segregation by age, ability, subject area, and activity.[60]

Schooling was designed within a bureaucratic structure to socialize the population towards: class and authority hierarchies, the market, abstract rules, clock time, objectivity, rationality, wage employment, technical and scientific authority (over traditional/religious authority), and geographic mobility. The goal was to shape people for waged work by generating discontent with existing ways of life and fostering ambitions of "being modern" being educated, and "progress-minded."

In this instrumentalist paradigm of education, teachers are technicians who must adhere to a systematized curriculum, bureaucratized administration, professionalized training, designated budgets, assessment protocols, and industrial style school spaces.

> Universal schooling was meant to detach role assignment from personal life history; it was meant to give everybody an equal chance to any office. Even now many people wrongly believe that school ensures the dependence of public trust on relevant learning achievements. However, instead of equalizing chances, the school system has monopolized their distribution.[61]

While most teachers are compassionate and genuinely concerned for their students, Illich considers the implication to be the creation of "needs." Learners are considered lacking and in a state of deficit, *needing* education. As functionaries, Ivan Illich suggests, teachers are professionals with specialized knowledge who diagnose needs and prescribe curricular and behavioural "treatments"[62] to students, a one-size-fits-all industrial approach. Through professional educators, learners gain the "power to behave according to economic rationality,"[63] the discipline to make money in the workplace to provide for their needs.

In sum, we have come to see education only as schooling. Scarcity ensures that those who do not have access to education are "marginal, not just to the national economy but to modern humanity itself. . . . any choice between alternative satisfactions, which is implied in the concept of economic need, is ruled out for them."[64]

Further, the education system and education profession disable people from meeting their own needs, now only possible through the purchase of the services of multiple specialists. Rather than relying on "the commons,"[65] we are beholden to the market. Eventually, we come to meet our needs exclusively through the market. The implication is that we trust only specialists who handle knowledge as commodity. Liberty is reduced to "consumer-enshrined rights," such as certification.[66] This continually undermines our desire to engage in self-learning and knowledge outside the education market.

This logic was particularly evident in international development programming. One example of agricultural development education is described by Ernest Sirolli,[67] an Italian participant in an early 1970s development project. Their development group was teaching Zambians in Chirundu to grow tomatoes as a cash crop. The money they paid local Zambians for land clearing and planting was not enough to encourage the locals to return to work every day for a 9 to 5 workday. Through their *commons*, the locals had ample fish in the river and game in the forest, so waged work was unnecessary. To habituate them to waged work however, these young men decided to entice them with consumer items they had brought with them, such as "sunglasses, watches, radios, beer and yes, whiskey."[68] He described that getting locals hooked on consumer items effectively "conquered" Chirundu. Six months later, when the team arrived to harvest the abundant and giant tomato plants though, they saw a trampled field and well-fed hippos sunning in the Zambezi River.

He explains the multiple ways that this agricultural learning was misguided, repeated around the globe as a domestication process, orienting people and their learning to the economic market and consumer goods. Whether field learning for adults or schooling for young people, the process focussed on

> raising educational and occupational aspirations, loosening adherence to traditional customs and beliefs, [fostering] an openness to new experiences, and [reducing] familism or family ties . . . just a few such modern orientations which result from the school experience.[69]

Now that the numbers of fish in the river and game in the forest have perilously declined as part of the ongoing sixth extinction,[70] nonwaged work and reliance on the commons is barely an option. Year by year, the profound losses of cultural diversity, social cohesion, environmental integrity, and economic self-sufficiency press in on the global human family.

In sum, Western education normalized social interventionism and the commodification of education. The role of Western education systems was to provide the knowledge, skills, and attitudes (KSAs) for the following functions, the:

- economic function of developing skills for productive waged employment;
- socially integrative function of peopling modern social institutions;
- distributive function of transforming local social psychologies into modern values, attitudes, and abilities, the basis upon which income is distributed and consumptive lifestyles are acquired; and the
- political function of instilling skills and values for representative democratic nationhood.

The Western education system used massification and institutionalization as part of the state apparatus to access all citizens, incrementally expanding mandatory primary and secondary education with the option of colleges and universities for tertiary education.

Schooling has been held out as the tool by which the global community can eliminate poverty and reduce environmental degradation, accepted goals of sustainable societies. There are limits, however, to how effective this can be within such an institutionalized structure, as the roots of poverty and environmental degradation go far deeper. In their classic book, *The Limits of Educational Reform*, Martin Carnoy and Henry Levin assert,

> The overriding theme is that the principal cause of failure to reform the social and economic system through educational institutions stems not from the incompetence of the reformers, but from the inability to alter a schooling system that is largely functional to the existing polity: the system reproduces the inequalities created by its system of production.[71]

In other words, reform will be constrained by the existing logic of the economic system. Sceptically, Carnoy and Levin suggest that not much is reformed in education unless it can increase profits and/or contribute to power accumulation for the elites, particularly when business and private investors reinforce the notion that current economic leaders are the best and natural "director[s] for the economy."[72] This principle bears out in international negotiations around climate and environment, where ideologues and profit makers shape the adoption of commitments. Currently, environmental and sustainability education has not escaped the impact of this logic, one reason for limited impact to date.

While education certainly has created new pathways to economic well-being for many social groups, it has simultaneously disvalued and deliberately dismantled non-White, Indigenous, tribal, village, and other kinds of political, economic, educational, and cultural structures.[73] From the beginning, starting with the enclosure movement in Britain, it dismantled the "commons" where notions of wealth were based on community, not individual self-reliance, and on shared resources of land, water, and forests, not on what money could buy individually.[74] Spheres of self-organized life for survival and flourishing have been violated[75] by ascribing economic "functions" and fitting individuals solely for the industrial system. Additionally, the underlying meritocratic principle of "survival of the fittest" has created a system whereby some survive and thrive, and others do not.

Wolfgang Sachs identifies the trinity of Western modernization to be capital, bureaucracy, and science.[76] This has been fused into the modern imaginary in education and its social engineering. These elements are now so much a part of our lives that we do not even realize that there are other ways of thinking about learning and organizing education. We have become prisoners of a schooled logic and corresponding set of laws, policies, and practices.[77] As Illich predicted, institutions often end up working in ways reverse to their original purpose.[78] Yet, more convivial forms of education are possible alongside schooling.

Educating for Globalizing: The Profit Logic

The neoliberal turn in the 1980s was a backlash against the social welfare state, which had been defined as "the historic compromise between capital, labour, other social classes and cultural groups, and the state."[79] The redistribution of wealth by social welfare states ran aground socially and economically.

The purpose of the social welfare state was to restrict the spheres from which profits could be made, yielding an economy of moral restraint. The state then provided social programmes, including education, through public tax monies as one mechanism for addressing poverty and social inequality. Under the siege of neoliberalism globally, many public programmes were revolutionized in favour of freedom for corporate and business powers to pursue profits wherever and whenever. This has fantastically enriched the top 1%, the global elites,[80] while shrinking the middle class and deeming many surplus to the economy altogether.

One example of neoliberalism in education has been Chile, where the repression of citizens created the perfect "climate needed to experiment with economic restructuring."[81] In the 1980s, the Chilean military regime rigidly applied the guidelines for Structural Adjustment Programs (SAPs) (see Chapter 3). Based on the principles of free markets, private management, and profit-oriented programmes, they reduced the Chilean public budget by more than 27%. This continued through the transition to democracy starting in 1989.

Public education programmes were left to deteriorate. The population was encouraged to migrate towards schools privately managed by profit-making owners. Chilean education was to be an example of fully operating on the free market principle. Application into private schools was heavily competitive and expensive, leaving the academically weaker or economically challenged students in the public system. This exacerbated social and economic inequality, part of the intention. As services and funding were withdrawn from the public system, schools became overcrowded with larger class sizes and poorer quality education.[82]

At the breaking point in 2006, over a million students reacted against increases in school bus fares and university admission test fees as part of the privatization of the education system. Growing into a national movement, they demanded "free education" and equal access to "quality education for all." Called the Penguin Revolution, given the look of high school uniforms (see Figure 7.1), students engaged in demonstrations and school takeovers, also rallying university students and the general public. National strikes, met with police brutality, further outraged the public.[83]

Continued mass mobilizations over several years continued, given the slow pace of reform, broadening their protest to all facets of neoliberalism and its systematic financial starvation of Chilean public programmes. In the end, student demands were met and new legislation for financing the public school system was proposed. This is just one example of what British sociologist of education, Stephen Ball, explains as the redesign of the public sector, including education. Education is "re-imagin[ed] . . . as a commodity"[84] rather than as an "opportunity or social good" outside the market.[85]

In fully industrialized nations, neoliberal redesign included opportunity for profit-making not only establishing niche educational institutions under private ownership but also, in public institutions, this included the sale of consumer goods

FIGURE 7.1 Penguin Revolution in Chile

Credit: Photo provided by Dr. Donna Chovanec.

under dedicated service contracts, including vending machines, food outlets, and media provision. This effectively blurred the lines between education, entertainment, and advertising,[86] for the purposes of profit-making.

In higher education in industrialized nations, neoliberal globalization proponents advocated for "new knowledge-based econom[ies] requiring large investments in education, science, and technology"[87] as a way to stay competitive globally. Commercialization of research and technology was promoted for additional profit-making opportunities. Universities were marketized in competition with one another. Through an audit culture, they were monitored for efficiency and outcomes, particularly looking for more student completions using less resources. By allowing colleges and polytechnics to call themselves universities, an additional competitive group was established.

Business leaders were hired as university presidents across both public and private higher education institutions. As British researcher of higher education management, Rosemary Deem explains, in the past

> the notion that the activities and cultures of universities either required managing or were, in any meaningful sense, 'managed,' would have been regarded

> as heretical. Universities were perceived as communities of scholars researching and teaching together in collegial ways; those running universities were regarded as academic leaders rather than as managers or chief executives.[88]

Scholarly autonomy and trust among academics were replaced with regulation, hierarchy, and a competitive culture. It was the implementation of a "reward and punish" regime.[89] The underlying assumption is that employees, including scholars, are "fundamentally untrustworthy and thus incapable of self-reform or change." They require impositional management with clear lines of authority requiring obedience.[90]

In public schools, private companies were hired to run national and subnational programmes—from literacy programming, school inspections, supply teachers, to school meals.[91] As Ball summarizes, "The private sector is now embedded in the heart and sinews of state education services at all levels, intertwined in the day-to-day business of decision making, infrastructural development, capacity building, and services delivery."[92]

A second aspect of neoliberal redesign was the incorporation of New Managerialism[93] which brought business and commercial values into schools and other educational institutions. Akin to the corporate world, bonuses were offered to administrators, and sometimes educators, for performance outcomes that met or exceeded targets and benchmarks. Performance management hinged on measurable criteria of student progress. The era of testing against national and then international standards regulated teacher relationships with the students, staff, "management," working conditions, and curriculum. Disciplines, which did not lend themselves to testing, were phased out as extraneous, including fine arts and physical education, favouring a focus on science, mathematics, and technology.

Further, there was pressure on parents to support their children through extra services no longer provided in the public system, leading to a large increase in e-learning, tutors, and other private education services, as additional sources of profits. Finally, corporations were allowed to fund aspects of schooling, such as buildings, programmes, or labs. Corporations were then invited to sit on curriculum boards and hiring boards, as well as supply curriculum resources from their perspective, to take pressure off dwindling public funding for basic resources. In these ways, the ideologies of conspicuous consumerism, competitive individualism, entrepreneurship, technical skilling, and the global division of labour were built right into all sectors of education. Further, state funds freed up from social and educational programmes were now redirected into investment and subsidies for the corporate sector.[94]

This cataclysmic shift, a narrowing of purpose and tightening of the relation between schooling and the economy, has had intergenerational impacts. The so-called failures of environmental and sustainability education must be understood in the context of this deeper shift in economic ideology and societal organization. Environmental and sustainability education was a small voice in a much larger education momentum moving in an opposite direction.

Educating for Production, Education as Production

With the technological revolution replacing many sectors of the economy with computers and robotics, communications professor Robert McChesney and journalist John Nichols explain that the world emerging now is one of a jobless economy and a citizenless democracy. In an economy that values profit-making over providing work, and the profit usefulness of un/under employment rather than full employment, they suggest we are returning to the "Gilded Age, back to a future of plutocrats and peasants, of masters and servants."[95] The privileged elite choose and support the politicians who run. They use smart technologies to distort messaging, hack opponents, and sway voters by giving them what they want to hear, not what they will actually do in policy making. The process of voting is now a democracy of spectatorship, not authentic deliberation by fully informed citizen voters.

Economist James Galbraith predicted that whole groups of people will become obsolete, not just unemployed.[96] Jeremy Rifkin contended that this third industrial revolution of the Information Age will permanently eliminate millions of workers and whole job categories from the economy,[97] leading to further rising economic polarities, crime, and violence. Already, on the global margins since the 1990s, there is a new social class called "disposable people" who are "surplus population"[98] to the global capitalist economy, say educationists Dana Stuchul, Madhu Suri Prakash, and Gustavo Esteva. They see a new "society of control" developing that accelerates environmental destruction and massive dispossession for production, as a form of "perpetual war."[99]

Concepts of production have changed over history and across cultures. As economic historian Karl Polanyi described in *The Great Transformation*,[100] the modern economy became an autonomous sphere with its own self-regulating dynamics over the 19th and 20th centuries. Eventually, the economy became the dominant sphere, promising prosperity and equality for all. But as economist Joseph Stiglitz comments, "Today, there is no respectable intellectual support for the proposition that markets, by themselves, lead to efficient, let alone equitable outcomes."[101] With neoliberalism, equity via the economy is not even a goal.

Interestingly, the Latin root of the word "production" is *producere*, meaning "to stretch, to spend, to prolong, to draw into visibility."[102] Thus, *producere* is the possibility of something moving from the invisible or latent to the visible and manifest. Ancient and Indigenous understandings are that Earth herself is the primary producer, the Mother, with whom humans must cooperate to meet their needs. The Earth bestows Her gifts, "anointing" humans, if they engage Her with respect, strength, and virtue.[103]

With the Christianization of Europe under the Roman Empire, another meaning of production emerged—that of a male monarchical God who is creator king. In this view, Earth becomes the queen of production, with humans as the "husbanders."[104] Then, in early Modernity, the goal was to free "man" from the constraints of nature, expressed both by Francis Bacon and later Rene Descartes. In this view, "nature" becomes servant and slave, no longer considered alive or with agency. "Man" becomes the possessor and manipulator of Earth for "his" own ends.[105] Production then becomes purely a human creation and an economic concept, in contrast to a previously reverent relationship with land.[106] The notion of a sacred exchange between humans and Earth, with livelihoods undertaken within the parameters of right relationship,[107] was lost.

For centuries, education was for the elites, in a variety of forms. Then, the design of public schooling was to create a "compliant and proficient labour force,"[108] education for industrial production. With neoliberal globalization, this evolved one step further. Now, education is considered production itself. A new right-wing critique of schooling emerged in the 1980s that applied production ideology to schooling, which had been kept outside the production sphere. It was now reasoned that applying fiscal austerity measures to social programmes, such as education, would encourage "doing more with less" and thereby improve organizational efficiencies. Even further, taking education out of the public sphere and into private ownership, it could become a new source of profit-making. The New Right cited a host of rationales for this educational restructuring: lack of accountability of the education system, the overweening power of teacher unions, the lack of parental choice, the need to build a competitive position in a global economy, (ostensibly) poor educational performance according to international standards, and the "out-of-control costs" of education per capita. By restricting education, especially through private schools and rising higher education tuitions, education soon becomes available only to elites once again.

The goal of schooling is not just *for* production now, it is a form *of* production, within the market system of supply and demand, with knowledge as the commodity. More than ever, education has become more tightly linked to planning and production, aligned with profit-making.

Yet, just as production dominates the education imaginary, growing numbers of people may never be included in the workplace. What kind of education is required for a workerless economy? For a citizenless democracy?

Recapitulation: Modern Social Imaginary in Education

In sum, the modern social imaginary in part comprises education via schooling—allied to the various purposes of proselytizing, civilizing, modernizing, globalizing, and producing. Adult and continuing education has been supplementary or complimentary to schooling and higher education as needed. Yet, these forms of education have been recently recalibrated to focus on employment, training, and commercialization of research to the detriment of the other elements of the imaginary—the public sphere and self-governance.

The goal of enfranchisement had been to expand the population who participate in collectively determining the structures of their lives. This expansion of democracy is now what neoliberal advocates are protesting and rolling back. Given the overweening influence of corporate lobbies on government, physicist and ecofeminist activist Vandana Shiva emphasizes that democracy is increasingly at risk.[109] The protection of democracy is vital. However, other forms of democracy are also possible where the economic sphere can find its right relation with "we, the people" and the living world.

What is less self-evident in this social imaginary is all the ways that the property form has shaped the structures and processes of human life. Educational systems have not only perpetuated the reproduction of the property form throughout all spheres of life, but educational systems are now a form of property. Knowledge, processes of "knowing," and structures of educational systems are part of the property form. Thus, more and more elements of our lives are enclosed within the property form.

Resistance will grow, especially as opportunities for good work, or any work, are increasingly lost and as climate change makes habitation in some places more difficult or even impossible. Over centuries of lifting learning out of community bonds and out of "vernacular villages,"[110] the now globalized " 'one world' seeks to substitute the thousands of real and living worlds with a single non-world, says Iranian scholar Majid Rahnema. The "one world" is now a totally acultural and amoral economic corporation whose only purpose is to serve the interests of its shareholders."[111] As many Indigenous, Southern, and other theorists have described, there has been no room in the framing of our social imaginary to actually "see" and thus value other kinds of relationships with land, forms of governance, forms of wealth, forms of *being* human, forms of human *knowing*, and forms of balancing individual and societal needs.[112] Part of this is the inability to "see" other ways of educating. Our very imaginations have been desiccated. Yet, as Stuchul, Prakash, and Esteva maintain, "The 'individual' as the basic cell of social organization as defined by Western modernity is losing its grip on the social imagination." It is time to decolonize our minds and stir to "wide-awakeness, to imaginative action, and to renewed consciousness of possibility."[113]

Reimagining Education and Learning

Interestingly, the etymology of the word "education" is from the Latin *educere* which intimates that learning is a remembering of the knowledge embedded in oneself, evoking it into consciousness. This is similar to many Indigenous understandings. The other related Latin word is *educare*, meaning the taking in of new information and shaping of a person into a cultural form.

> Craft [in 1984] noted that there are two different Latin roots of the English word "education." They are "educare," which means to train or to mold, and "educere," meaning to lead out. While the two meanings are quite different, they are both represented in the word "education." Thus, there is an etymological basis for many of the vociferous debates about education today.
>
> The opposing sides often use the same word to denote two very different concepts. One side uses education to mean the preservation and passing down of knowledge and the shaping of youths in the image of their parents.

> The other side sees education as preparing a new generation for the changes that are to come—readying them to create solutions to problems yet unknown. One calls for rote memorization and becoming good workers. The other requires questioning, thinking, and creating. To further complicate matters, some groups expect schooling to fulfill both functions, but allow only those activities promoting "educare" to be used. Balance in educational aims is a valid focus for educators.[114]

From these Latin derivatives, we can see the origins of the contestation in relation to the purposes for, and pedagogies of, learning and education. Education has been called upon to fulfil seemingly contradictory roles—both reproducing as well as transforming the existing social order. The form which becomes dominant has depended on the constellation of moral, political, economic, social, and cultural goals as well as the controlling socio-economic and ideological interests that lay behind these goals. One pathway to reimagining education is to engage "*educere*" for a new education imaginary. Where might we look for indications of a new social imaginary?

Starting in 1999, the World Social Forum has been the counter summit to the neoliberal agenda of the World Economic Forum. The Zapatista's slogan, "walk forward questioning," captures this spirit.[115] It started engaging the question of "what if?" and declaring "Another World is Possible."[116] The existing imaginary is no longer self-evident, responding to what Vandana Shiva calls "the philosophical and ethical bankruptcy of globalization."[117]

In Porto Alegre, Brazil, the World Social Forum envisioned "The World *We* Want" including new conceptualizations of democracy, human rights, social justice, and internationalism. "Perhaps the most consistent theme in the World Social Forum is the argument that social hierarchies such as gender, race, class, culture, and political power are not a legitimate form of organizing social and economic production or reproduction."[118]

As interesting as the ideas, are the processes used in the World Social Forum. While there have been some conflictual years across major differences,[119] a new start for the World Social Forum is taking shape. Boaventura de Sousa Santos, a Portuguese social scientist, has proposed the following processes, important for educators as well:

- common ground must be found among divergence;
- democracy must be defended, strengthened, and expanded;
- we must work harder to decolonize our minds;
- real concrete equality must be pursued;
- using (old) terminology/categories to define the future is slippery and conflictual; and
- the human and natural worlds must reach a balance.[120]

As educators, searching out new ideas and watching the innovative directions of related social movements and NGOs are vital for seeing the bigger picture and reinvigorating our work for sustainability education. Such imaginal space must be afforded in educational spaces as much as in social movement spaces, or preferably for educators *within* social movements spaces.

As educators Bernard Charlot and Paul Bélanger summarize from the World Education Forum that was part of the World Social Forum in 2001, the fundamental principles affirmed were that "Public education for everyone is an inalienable right guaranteed and paid for by the state. It must not be treated as a commodity. It must be radically democratic, egalitarian, and fair."[121] They further summarize that the victims of reducing education to a commodity and schools to market logic are "those most in need: poor people, immigrants, Indigenous people, ethnic, religious, and cultural minorities, as well as teachers themselves, because of bad working conditions and low morale."[122]

It is important to redefine the values that have been "appropriated by neoliberal thought" and the assumptions of the property form in education. It is "transforming public schools and many of their practices" as well as "educating teachers about principles of democratic organization" in relation to curriculum, management, team work, transdisciplinarity, and cross continental dialogue. They argue for creating spaces "in which alternatives and counterprojects can be tried out."[123] In sum, formal education can be liberated from these old imaginaries. Lifelong learning using *educere* must continue to develop

> the capacity of adults to develop their potential and release hitherto undeveloped creative forces throughout an individual's lifetime. Developing the intelligence of a society as a whole is a universal source of richness and, more than ever, something to be universally encouraged.[124]

In sum, it is constructing "a new citizenship project" founded on social inclusion and recognition of diversity as well as enabling other systems of knowledge, knowing, and governance to co-exist. As Hayden Washington argues, "the missing ingredient is collective engagement for political and structure change."[125] There comes a time when public education is so threatened that we need to be part of this collective engagement to enable sustainable societies.

There is one more challenge facing us. In questioning the morality and naturalness of our current order,[126] some argue that we are in the dying times of modernism. How might we educate through the epochal shift into societies premised upon Relationality?

Educating for Epochal Shift

This historical moment has been alternatively called The Great Turning,[127] The Great Transition,[128] or The Great Transformation,[129] even though the destination is still in the making. Jeremy Lent refers to this time as an epochal shift.[130] Joanna Macy suggests that there are three co-evolving stories of our time: the Great Unravelling, the Great Turning, and Business as Usual.[131] Others call these three worldviews as: the catastrophic leading to barbarization, the transformational creating the Great Transition, and the evolutionary maintaining the conventional world.[132] Macy asks which story will we choose to put our energy behind? Which intentions will garner our attention?

Dark Times for Empire

A "Dark Age"[133] often refers to an early period in a major transition, a time of immense difficulties. For us now, these signs include pandemics such as COVID-19 and dramatically accelerated climate chaos with the intensification and destructiveness of extreme water, fire, and wind events. Another early sign of decline, says urban scholar Jane Jacobs, is social and political chaos, including cultural xenophobia when a society begins to impose a fortress around itself as protection.[134] The rise of the New Right, its hijacking of conservatism, lynching of liberalism, and the deliberate generation of a rage/hatred industry promoting culture and race wars are examples of this decline. On many levels, there is now an implosion of familiar White middle-class assumptions and expectations, especially as climate costs and inflation shifts realities. This is an opportunity for deep unlearning.

Related, Jacobs says, fundamentalism in both religious and political forms takes hold. Fundamentalism together with authoritarianism can offer comfort to many through binary (black/white) thinking and a return to familiar class, caste, and role stratification, one way of dealing with an increasingly chaotic and unpredictable world.

It can easily be said that we live in a time of "bread and circuses," like the decaying Roman Empire. Public approval is achieved not by excellent governance and an active citizenry adhering to their ideals but by providing the most basic requirements such as food alongside entertaining but empty distractions such as circuses, in this case transmitted via television and myriad electronic devices. There is another parallel that Ancient Rome historian Rostovtzeff observes,

> a psychological change in those classes of society which had been up till then the creators of culture. Their creative power and creative energy dry up . . . their life is no longer an effort toward a creative ideal for the benefit of humanity; their minds are occupied either with material interests, or with ideals unconnected with life on earth.[135]

Regarding the fall of Ancient Rome, he says that it was "a long period of political and social anarchy" surrounded by apathy. In the end, while a few thinkers kept the "ideals of Greek freedom alive," the majority "came to distrust reason" and were occupied with political infighting.[136]

In the systems paradigm, complex systems evolve through bifurcation. It does not appear that we are experiencing the balancing of normal pendulum swings, which offer a course correction and thereby maintain stability.[137] Such changes are self-stabilizing through either negative feedback systems that dampen or buffer changes or positive feedback systems that amplify and enhance changes, both working to maintain a dynamic balance within liveable parameters.[138] Now, however, it appears that the trends we are experiencing could signal that we are heading towards a bifurcation point, leading either towards cultural breakdown or cultural breakthrough.[139]

Geographer Jared Diamond suggests that historical declines are related to the nature of a society's responses to its challenges.[140] Historically, decline is related to four challenges: environmental damage, climate change, rise of hostile neighbours, and loss of supportive partners, especially trading partners. Existing structures, ideas, and values are ineffective in responding. Decline sets in, and a Dark Age ensues—a time of accelerated cultural collapse.

For Jacobs, a Dark Age is a culture's dead end. Western civilization, if traced from Ancient Classical Greece, has lasted 2,500 years, the average timeframe for civilizations. Over the last 500 years, living cultures have been lost through an intense process of political and economic colonization, including slavery, through the European colonial and American politico-economic empires. The grief and pain still continue today, as the cultural loss continues.

Jacobs says that another element of decay in a Dark Age is mass amnesia. Eventually, people no longer remember even what has been lost. This is true for many cultures who were colonized by the Roman Empire 2,000 years ago. In this time of globalization, Hardt and Negri suggest this is no longer the imperialism of one state extending its borders. Now it is a "decentered and deterritorializing apparatus of rule that progressively incorporates the entire global realm."[141] It has no limits as little exists outside of it, warranting the description of "Empire," as it seeks to enclose all that still lays outside of it, fostering amnesia.

As historian Ronald Wright suggests, our belief in the naturalness of "irreversible changes in one direction . . . towards improvement"[142] has for too long framed our thinking. Regularly, people say "you can't stop progress" or "well, that's progress." The resignation in this linear progress thinking will "lead beyond reason to catastrophe,"[143] a looming dead end for Western culture. As the repetition in history reveals, many great cultures have fallen prey to the progress myth. Wright says that they became so successful that they fell victim to their own success, outstripping their environmental home.

As Apffel-Marglin describes, "The emergence of the individual who calculated his [sic] advantage while responding to impersonal invisible market forces, and who unilaterally acted upon both the land and the people for his own advantage *required* the disenchantment of nature."[144] The expressive dimensions of social life have been truncated to money-making and acquisitive behaviour[145] leading to cynicism and despair. The West has fallen into some of the emptiest expressions of being human. Reconstructing the natural world as dead and passive enabled human domination.[146] The dominant mind frame is one of profound disrespect towards all other life forms, elements, and forces, expressed through a myriad of daily thoughts and actions.

This is another cause for a lack of progress among leaders and influencers gathering on the international stage, as the peoples of the world watch and wait for meaningful action. In the end, they are largely jostling for votes, profits, and/or power. And, in some cases, they are recklessly fomenting the dissolution of civilization, as chaos makes significant money for the powerful. For those who feel insulated by individual wealth and technological innovation, history shows that there will be "no shield from chaos," especially in a Dark Age that may dwarf all others.[147] As Diamond says, "[I]n the long run, rich people do not secure their own interests and those of their children if they rule over a collapsing society and merely buy themselves the privilege of being the last to starve or die."[148] There is only "hunger, eviction, abject misery, and conditions of strict survival for many millions, perhaps the majority of people on Earth."[149]

Out of growing social chaos, however, people are taking back their lives. Especially in the Fourth World, but elsewhere too, they are rebuilding their lives outside the bounds of capitalism,[150] on the principles of concrete equality.[151] Hardt and Negri strenuously argue that the current task is not just to resist Empire but also to creatively redirect ourselves towards compelling alternatives that can invent new forms of power, democracy, and being, reappropriating knowledge and production, moving beyond Empire.

Beyond the Dark Age

Wright suggests that we need long-term thinking to mitigate the worst aspects of this Dark Age, in preparation for what is to come after. Historian Richard Tarnas agrees that we must consider the long arc of history. We need to examine the convictions and assumptions that have formed the foundation of modern thinking.[152] By retracing our steps, we can recover roots that were lost in the march to modernism and integrate these meaningful pieces back into a social imaginary for a new era.[153]

For signs of the new, Rostovtzeff suggests looking to where new energy of creativity and genius is located. It is emerging now among those recreating the commons and relearning "commoning"[154] in many small, localized spaces. It is found in the regenerative movement. Some are on ancestral land and others in humble urban spaces. In part, they have been inspired by, and are learning from, the Zapatistas in Chiapas, Mexico. In resistance to the North American Free Trade Agreement in 1994 as one face of neoliberalism, they declared their independence from the Mexican government. They have been recreating their economy, politics, social practices, educational practices, and cultural norms in ways that honour their Traditional ways of knowing and being. Gustavo Esteva and Arturo Escobar both agree that they are creating a mode of existence that is place-based and relational at its foundation. Let us turn to these new seeds, glimmers of hope on the horizon, in Chapter 8.

Notes and References

Chapter line drawing by Canadian artist Erin Marie Lange Christensen. www.erinchristensen.com. Reproduced with permission.

1 Thomas Berry, *The Great Work: Our Way into the Future* (New York: Bell Tower, 1999), p. 71. Note: I have replaced the word "religion" with the word "spirituality" as it captures a wider body of ideas and practices than formal religion.
2 These thoughts are inspired by Hayden Washington's "what sustainability should be?" in *Demystifying Sustainability: Toward Real Solutions* (London: Earthscan from Routledge, 2015), p. 195.
3 Washington, *Demystifying Sustainability*, p. 195.

4 Charles Taylor, *Modern Social Imaginaries* (Durham: Duke University Press, 2004), p. 6. Cited in Elizabeth A. Lange, "(Re)Igniting a Sociological Imagination in Adult Education: The Relevance of Classical Theory," *International Journal of Lifelong Education*, Vol. 34, No. 5 (2015), pp. 491–513. https://doi.org/10.1080/02601370.2015.1028574.
5 Charles Taylor, "Modern Social Imaginaries," *Public Culture*, Vol. 12, No. 1 (2002), pp. 91–124.
6 Charles Taylor, *Modern Social Imaginaries* (Durham: Duke University Press, 2004), p. 25.
7 Taylor, *Modern Social Imaginaries*, p. 23.
8 Taylor, *Modern Social Imaginaries*, p. 25.
9 Taylor, *Modern Social Imaginaries*, p. 29.
10 Ronald Wright, *A Short History of Progress* (Toronto: House of Anansi Press, 2004).
11 Erich Fromm, *To Have or To Be?* (New York: Continuum Books, 1976/1999), p. 69.
12 Fromm, *To Have or To Be?*, p. 69.
13 Fromm, *To Have or To Be?*, p. 70.
14 Fromm, *To Have or To Be?*, p. 71.
15 Fromm, *To Have or To Be?*, p. 72.
16 Franklin Obeng-Odoom, "Teaching Sustainability: From Monism and Pluralism to Citizenship," *Journal of Education for Sustainable Development*, Vol. 14, No. 2 (2020), p. 237.
17 Ivan Illich, *Tools for Conviviality* (London: Marion Boyars, 1973/2009), pp. 61–62.
18 Obeng-Odoom, "Teaching Sustainability," p. 240.
19 Elizabeth A. Lange, "Is Freirean Transformative Learning the Trojan Horse of Globalization and Enemy of Sustainability Education? A Response to C.A. Bowers," *Journal of Transformative Education*, Vol. 10, No. 3 (2012), pp. 3–21. https://doi.org/10.1177/1541344612453880.
20 Taylor, *Modern Social Imaginaries*, pp. 18, 76.
21 Charles Taylor, *The Malaise of Modernity* (Concord, ON: House of Anansi Press, 1991), p. 14.
22 Taylor, *Modern Social Imaginaries*, pp. 18, 21.
23 Taylor, *Modern Social Imaginaries*, p. 78.
24 Taylor, *The Malaise of Modernity*, p. 9.
25 John Ralston Saul, *The Unconscious Civilization* (Toronto: House of Anasi Press, 1995).
26 Taylor, *Modern Social Imaginaries*, p. 78.
27 Ralson Saul, *The Unconscious Civilization*, p. 2.
28 Taylor, *Modern Social Imaginaries*, p. 79.
29 Carol Gould, *Marx's Social Ontology: Individuality and Community in Marx's Theory of Social Reality* (Cambridge: The MIT Press, 1978), p. 37.
30 Maxine Greene, *Releasing the Imagination: Essays on Education, the Arts, and Social Change* (San Francisco: Jossey-Bass, 1995), p. 3.
31 Greene, *Releasing the Imagination*, p. 35.
32 Greene, *Releasing the Imagination*, p. 4.
33 Greene, *Releasing the Imagination*, p. 3.
34 Solnit, *Hope in the Dark*, p. xvi.
35 MJ Barrett, Matthew Harmin, Bryan Maracle, Molly Patterson, Christina Thomson, Michelle Flowers, and Kirk Bors, "Shifting Relations with the More-than-human: Six Threshold Concepts for Transformative Sustainability Learning," *Environmental Education Research*, Vol. 23, No. 1 (2017), pp. 131–143. https://doi.org/10.1080/13504622.2015.1121378.
36 Greene, *Releasing the Imagination*, p. 4.
37 The term society arose in the era of the nation-state. It refers to "people who interact in a defined territory and share a culture." Society as a topic of research and study did not begin until the early 1800s with Comte, known as the first sociologist, although the term originated in the English language in the 15th century. From John Macionis and Linda Gerber, *Sociology*, Eighth Canadian Edition (Toronto: Pearson Education, Inc.), p. 86. 58.
38 Macionis and Gerber, *Sociology*, p. 506.

39 I am adopting the differentiation between *ontos* and ontology where *ontos* is about a way of being and ontology is beliefs about reality. From Chris Beeman and Sean Blenkinsop, "Environmental End Game: *ontos*," *Environmental Education Research*, Vol. 26, Nos. 9–10 (2020), pp. 1479–1490. https://doi.org/10.1080/13504622.2019.1663792.
40 This gender term is used deliberately here.
41 https://dailyhistory.org/How_did_universities_develop.
42 https://dailyhistory.org/How_did_universities_develop.
43 Mark Cartwright, "Medieval Trades," in *World History Encyclopedia*. Last modified December 6, 2018. www.worldhistory.org/Medieval_Trades/.
44 www.britannica.com/topic/education/Lay-education-and-the-lower-schools.
45 www.britannica.com/topic/education/Europe-in-the-Middle-Ages.
46 www.britannica.com/topic/education/Nonscholastic-traditions.
47 www.britannica.com/topic/education/The-early-English-humanists.
48 Max Weber, *The Protestant Ethic and the Spirit of Capitalism* (London: Routledge, 1930), p. 38.
49 https://dailyhistory.org/How_did_universities_develop.
50 www.britannica.com/topic/education/The-new-scientism-and-rationalism.
51 Comenius is considered the "father of modern education." Paul Nash, *Models of Man: Explorations in the Western Educational Tradition* (Malabar, FL: Krieger Publishing Company, 1968/1983), pp. 203–232.
52 www.britannica.com/topic/education/The-new-scientism-and-rationalism.
53 www.britannica.com/topic/education/The-new-scientism-and-rationalism.
54 www.britannica.com/topic/education/The-Calvinist-Reformation#ref47555.
55 Barbara Thayer-Bacon, *Relational Ontologies* (New York: Peter Lang Publishing, Inc., 2017).
56 This paragraph is paraphrased from Thayer-Bacon, *Relational Ontologies*, p. 144.
57 Thayer-Bacon, *Relational Ontologies*, p. 136.
58 Adam Unwin & John Yandell, *Rethinking Education: Whose Knowledge is it Anyway?* (Ottawa: New Internationalist Publications and Between the Lines, 1998).
59 Nancy Kober and Diane Stark Rentner, *History and Evolution of Public Education in the US* (Washington DC: Center on Education Policy, George Washington University, 2020). https://files.eric.ed.gov/fulltext/ED606970.pdf.
60 Anthony Paré, "Foreword," in Unwin and Yandell, *Rethinking Education*, p. 7.
61 Ivan Illich, *Deschooling Society* (New York: Harper & Row, 1970), p. 17.
62 Illich, *Deschooling Society*.
63 Ivan Illich, "Needs," in Wolfgang Sachs, *The Development Dictionary: A Guide to Knowledge as Power* (London: Zed Books, 1992), p. 94.
64 Illich, "Needs," p. 94.
65 Through the enclosure movement, people no longer had public access to land, natural resources, or cultural resources (the commons), through which they could make a living.
66 Arvinsas Simon, "Ivan Illich on the Age of Disabling Professions," *Public Philosophy*, Blog, January 7, 2021. https://publicphilosophy.ca/2021/01/07/ivan-illich-on-the-age-of-disabling-professions/.
67 Ernesto Sirolli, *Ripples in the Zambezi* (Perth, Australia: Institute for Science & Technology Policy, Murdoch University, 1999).
68 Sirolli, *Ripples in the Zambezi*, p. 8.
69 Ingemar Fägerlind and Lawrence Saha, *Education and National Development: A Comparative Perspective* (Amsterdam: Elsevier, 1989), p. 97.
70 Elizabeth Kolbert, *The Sixth Extinction: An Unnatural History* (New York: Picador, 2014).
71 Martin Carnoy and Henry Levin, *The Limits of Educational Reform* (London: Longman Publishing, 1976), p. 7.
72 Robert McChesney and John Nicols, *People Get Ready: The Fight Against a Jobless Economy and a Citizenless Democracy* (New York: Nation Books, 2016), p. 37.

73 See Tiffany Prete and Elizabeth A. Lange, "Indigenous Voices and Decolonizing Lifelong Education," *International Journal of Lifelong Education*, Vol. 40, No. 4 (July/August 2021). https://doi.org/10.1080/02601370.2021.1968240.
74 Gustavo Esteva, "Development," in Wolfgang Sachs (Ed.), *The Development Dictionary* (London: Zed Books, 1992), pp. 6–25.
75 Gustavo Esteva and Madhu Suri Prakash, *Grassroots Postmodernism: Remaking the Soil of Cultures* (London: Zed Books, 1998), pp. 12–13.
76 Wolfgang Sachs, *Planet Dialectics: Explorations in Environment and Development* (London: Zed Books, 1999), p. 67.
77 Majid Rahnema, "Poverty," in Wolfgang Sachs (Ed.), *The Development Dictionary* (London: Zed Books, 1992), p. 168.
78 https://infed.org/mobi/ivan-illich-deschooling-conviviality-and-lifelong-learning/.
79 Trevor Harrison and Jerrold Kachur (Eds.), *Contested Classrooms: Education, Globalization, and Democracy in Alberta* (Edmonton: University of Alberta Press and Parkland Institute, 1999), p. xvii.
80 David Harvey, *A Brief History of Neoliberalism* (Oxford: Oxford University Press, 2005).
81 Donna Chovanec and Alexandra Benitez, "The Penguin Revolution in Chile: Exploring Intergenerational Learning in Social Movement," *Journal of Contemporary Issues in Education*, Vol. 3, No. 1 (2008), pp. 39–57.
82 Chovanec and Benitez, "The Penguin Revolution in Chile," p. 44.
83 Chovanec and Benitez, "The Penguin Revolution in Chile."
84 Stephen Ball, *Education plc: Understanding Private Sector Participation in Public Sector Education* (London: Routledge, 2007), p. 24.
85 Ball, *Education plc*, p. 13.
86 Ball, *Education plc*, p. 14.
87 Harrison and Kachur, *Contested Classrooms*, p. xvii.
88 Rosemary Deem, "'New Managerialism' and Higher Education: The Management of Performances and Cultures in Universities in the United Kingdom," *International Studies in Sociology of Education*, Vol. 8, No. 1 (1998), p. 47. https://doi.org/10.1080/0962021980020014.
89 Deem, "'New Managerialism' and Higher Education," p. 53.
90 Deem, "'New Managerialism' and Higher Education," p. 56.
91 Ball, *Education plc*, p. 40.
92 Ball, *Education plc*, p. 41.
93 The term 'New Managerialism' is generally used to refer to the adoption by public sector organizations of organizational forms, technologies, management practices, and values more commonly found in the private business sector. Deem, "'New Managerialism' and Higher Education," p. 47.
94 Harrison and Kachur, *Contested Classrooms*, p. xxi.
95 McChesney and Nicols, *People Get Ready*, p. 2.
96 James Galbraith cited in McChesney and Nicols, *People Get Ready*, p. 3.
97 Jeremy Rifkin, *The End of Work: The Decline of the Global Labor Force and the Dawn of the Post-Market Era* (New York: A Jeremy P. Tarcher/Putnam Book, 1995), p. xv.
98 Dana Stuchul, Madhu Suri Prakash, and Gustavo Esteva, "From Fear to Hope: Learning from BIPOC in Hard Times—Covid 19, Climate Collapse and Racial Violence," *International Journal of Lifelong Education*, Vol. 40, No. 4 (2021), p. 416.
99 Stuchul, Prakash, and Esteva, "From Fear to Hope," p. 416.
100 Karl Polanyi, *The Great Transformation: The Political and Economic Origins of our Time* (Boston: Beacon Press, 1944/2001). Scarcity and economic determinism are also described in Esteva, "Development," p. 19.
101 Joseph Stiglitz, "Foreword," in Polanyi, *The Great Transformation*, p. viii.
102 Jean Robert, "Production," in Wolfgang Sachs (Ed.), *The Development Dictionary* (London: Zed Books, 1992), p. 177.
103 Robert, "Production," p. 179.

104 Robert, "Production," p. 177.
105 Carolyn Merchant, *The Death of Nature* (New York: Harper & Row), pp. 171–172.
106 Robert, "Production', p. 181.
107 Peter Brown and Geoffrey Garver, *Right Relationship: Building a Whole Earth Economy* (San Francisco: Berrett-Koehler Publishers Inc., 2009).
108 Harrison and Kachur, *Contested Classrooms*, p. xix.
109 Vandana Shiva, *Oneness vs the 1%* (White River Junction, VM and London: Chelsea Green Publishing, 2020).
110 Rahnema, "Poverty," p. 167.
111 Rahnema, "Poverty," p. 168.
112 Just one example here is Bruce Pascoe, *Dark Emu: Aboriginal Australia and the Birth of Agriculture* (Melbourne: Scribe, 2018).
113 Greene, *Releasing the Imagination*, p. 43.
114 Randall Bass and J. W. Good, "*Educare* and *Educere*: Is a Balance Possible in the Educational System," *Educational Forum*, Vol. 68, No. 2 (2004), pp. 161–168. ERIC: EJ724880.
115 Michael Hardt and Antonio Negri, "Foreword," in William Fisher and Thomas Ponniah (Eds.), *Another World is Possible: Popular Alternatives to Globalization at the World Social Forum* (Halifax: Fernwood, 2003), p. xvii.
116 Fisher and Ponniah, *Another World Is Possible.*
117 Vandana Shiva in Ponniah and Fisher, "Introduction: The World Social Forum and the Reinvention of Democracy," in *Another World Is Possible*, p. 1.
118 Fisher and Ponniah, "Overview: The Affirmation of Civil Society and Public Space" in *Another World is Possible*, p. 193.
119 Francine Mestrum, "Reinventing the World Social Forum: How Powerful an Idea Can Be," *Open Democracy*/ISA RC-47: Open Movements, November 18, 2017. https://opendemocracy.net/francine-mestrum/reinventing-world-social-forum-how-powerful-idea-can-be.
120 Francine Mestrum, *Ten Years of World Social Forum*. www.cetri.be/Ten-years-of-World-Social-Forum?lang=fr.
121 Bernard Charlot and Paul Bélanger, "Education: Conference Synthesis," in William Fisher and Thomas Ponniah (Eds.), *Another World is Possible: Popular Alternatives to Globalization at the World Social Forum* (Halifax: Fernwood Publishing, 2003), p. 203.
122 Charlot and Bélanger, "Education," p. 204.
123 Charlot and Bélanger, "Education," p. 206.
124 Charlot and Bélanger, "Education," p. 206.
125 Washington, *Demystifying Sustainability*, p. 195.
126 Two are Charles Eisenstein, *The More Beautiful World Our Hearts Know is Possible* (Berkeley, CA: North Atlantic Books, 2013) and Christian de Quincey, *Radical Nature: The Soul of Matter* (Rochester, VM: Park Street Press, 2002/2010).
127 David Korten, *The Great Turning* (Bloomfield, CT: Kumarian Press and Berrett-Koehler, 2006).
128 Lester Brown, *The Great Transition* (New York; WW Norton and Company, 2015); Mauro Bonaiuti, *The Great Transition* (New York: Routledge, 2014). See also https://neweconomics.org/ and https://greattransition.org/about/what-is-the-great-transition.
129 Spretnak, *Relational Reality*, p. 6.
130 Jeremy Lent, *The Patterning Instinct* (Buffalo, NY: Prometheus Books, 2017).
131 Joanna Macy and Chris Johnstone, *Active Hope: How to Face the Mess We're in without Going Crazy* (Novato, CA: New World Library, 2012), p. 5.
132 Arturo Escobar, *Designs for the Pluriverse: Radical Interdependence, Autonomy, and the Making of Worlds* (Durham: Duke University Press, 2018), p. 142.
133 Jane Jacobs, *Dark Age Ahead* (Toronto: Random House Canada, 2004).
134 Jacobs, *Dark Age Ahead*, p. 17.
135 Michael Rostovtzeff, *Rome* (London: Oxford University Press, 1973), p. 318.

136 Rostovetzeff, *Rome*, pp. 320–321.
137 Jacobs, *Dark Age Ahead*, p. 21.
138 Ervin Laszlo, *The Whispering Pond: A Personal Guide to the Emerging Vision of Science* (Boston: Element, 1996). See diagram in Chapter 6.
139 Elizabeth A. Lange, "Transforming Transformative Education through Ontologies of Relationality," *Journal of Transformative Education*, Vol. 16, No. 4 (2018), pp. 280–301.
140 Jared Diamond, *Collapse: How Societies Choose to Fail or Succeed* (New York: Penguin Books, 2005), pp. 11–15.
141 Michael Hardt and Antonio Negri, *Empire* (Cambridge: Harvard University Press, 2000), p. xii.
142 Ronald Wright, *A Short History of Progress* (Toronto: House of Anansi Press, 2004), p. 3.
143 Wright, *A Short History of Progress*, p. 5.
144 Frederique Apffel-Marlin, *Subversive Spiritualities: How Rituals Enact the World* (Oxford: Oxford University Press, 2011), p. 43. Exclusive language in original.
145 Max Weber, *The Protestant Ethic and Spirit of Capitalism* (London: Routledge, 1930/1992).
146 Carolyn Merchant, *The Death of Nature: Women, Ecology, and the Scientific Revolution* (New York: Harper & Row, 1980/1983), p. xvi.
147 Wright, *A Short History of Progress*, p. 131.
148 Diamond, *Collapse* quoted by Frederique Apffel-Marlin, *Subversive Spiritualities*, p. 52.
149 Stuchul, Prakash, and Esteva, "From Fear to Hope," p. 416.
150 David Bollier and Silke Helfrich, *Free, Fair and Alive – The Insurgent Power of the Commons* (Gabriola Island, CA: New Society, 2019).
151 Carol Gould, *Marx's Social Ontology* (Cambridge: MIT Press, 1978), p. 7. Gould suggests that in abstract equality, individuals are aware of themselves as individuals and their rights "beings in and for themselves." In the move towards concrete equality, individuals maintain this awareness but then self-consciously reroot themselves in their social and natural relations, overcoming the alienated relations of capitalism and liberal democracy. Abstract, fragmented liberal notions create disunity, whereas concrete equality is a differentiated unity, a communal individuality.
152 Richard Tarnas, *The Passion of the Western Mind* (New York: Ballantine Books, 1991), p. xiii.
153 Tarnas, *The Passion of the Western Mind*, p. xiii.
154 Bollier and Helfrich, *Free, Fair and Alive*.

8

OUR GREAT WORK

Reimagining Education and Our Future

Civilization Transition

> The volume of education . . . continues to increase, yet so do pollution, exhaustion of resources, and the dangers of ecological catastrophe. If still more education is to save us, it would have to be education of a different kind: an education that takes us into the depth of things . . . Education cannot help us as long as it accords no place to metaphysics . . . More education can help us only if it produces more wisdom.[1]

DOI: 10.4324/9781003159643-8

We are at a historic moment when our choices impact the longevity of the human species. We are in a civilization transition, a transformation of not only,

> a given economic model (neoliberal capitalism), . . . a set of cultural traits inimical to life on the planet (say, rampant individualism and consumerism), high-level policy reform (e.g., more comprehensive climate change protocols), geopolitical power struggles for re- and de-Westernization, or the ever-growing military-industrial complex.[2]

Our very "civilizational model,"[3] that of a "modern colonial capitalist patriarchal[4] world system,"[5] is now in question. Naomi Klein asserts that "the full reality of this [climate] crisis will change everything."[6] It *needs* to change everything.

And so, it is. A global web of flourishing creativity is redesigning our way of perceiving, thinking, knowing, doing, and being. The multidimensional nodes of this web are in small, humble local places; yet, they are reweaving our "entire way of life and a whole style of worldmaking."[7] Those in these nodes have asked the "what if?" questions to reimagine ways of learning and educating as part of new ways of being. These are not only exciting times but also ones of trepidation and hard work, as we make choices regarding the future of human survival.

We can certainly go down the pathway of a technology-mediated world, where what is human and what is machine become less clear, and profound ethical issues come into sharper focus. Arturo Escobar explains,

> At the other end of the spectrum there lies the dream, held by the flashy techno-fathers of the moment, of a posthuman world wholly created by Man. This is the world, for instance, of synthetic biology, with its gene-centric view of life, of booming techno-alchemies for genetic enhancement and the prolongation of life; of robotics, cyborgian fantasies, space travel, nanotechnology, unlimited 3-D printing, and much more; of the bizarre geoengineering schemes concocted in corporate boardrooms as solutions to climate change; and of those advocating for the "Great Singularity," a technologically induced transformation "when humans transcend biology," in which life would finally be perfected, perhaps as in the world-without-mothers of artificial intelligence fictions . . . where women's ability to give life is finally completely usurped since wo/man is wholly created by man through the machine.[8]

Critical philosopher Herbert Marcuse[9] astutely observed that we no longer easily recognize the increasing totalitarian impulses hidden within scientific rationality and its technology progeny. The mechanisms of social control and "democratic unfreedom"[10] continue to expand. We are mystified because we do not peer behind the machine to those that wield this power, profit from it, and those who have paid with their lives and labour.[11] Continuing down this pathway will complete the deathmaking that is at the core of Western civilization . . . or, we can choose Life.

As statistician and economist Fritz Schumacher explains in the chapter opening, reweaving a way of being embedded in Life necessitates educators to get to the depth of things. This epochal transformation of our civilizational model involves resculpting the human for a new way of being human.[12] It is reinventing Western ways of living and knowing alongside restoring nonwestern ways of living and knowing to a respected place. This chapter examines education in relation to the social change process of alternative worldmaking.[13] From Chapter 7, it continues to address the crisis *of* education, repurposing education in relationship to society and the living world. It offers seeds of possibilities for reimagining education and learning, far beyond schooling. It offers seeds of a new imaginary for our future.

Our Great Work and Education

Educators have a purpose to fulfil in crafting life-giving forms of education that can rise to the civilization transition ahead of us. As demonstrated throughout this book, it means revisiting the long arc of history. It also means fostering active hope. As Rebecca Solnit explains, hope

> [I]s not the belief that everything was, is, or will be fine. The evidence is all around us of tremendous suffering and tremendous destruction. . . . [Hope is] the belief that what we do matters even though how and when it may matter, who and what it may impact, are not things we can know beforehand.[14]

Going to the depths of things means going into the very roots of Western thinking and being, the history of its development, and then unpacking our perceptual and material ways of being. This educational task will help enable the transition towards life-giving communities. It also means moving into relationing[15] ways of being that centre humility, wisdom, and relatedness once again. This is what I mean by Transformative Sustainability Education. This will require much unlearning and relearning.

Reweaving all the strands of ideas and values, language and concepts, cultural perceptions and imagery, as well as material practices enacted in all cultural, economic, and political structures is to undertake a "comprehensive transformation"[16] as Charlene Spretnak calls it. *We* are the transitional generation that stands in this liminal zone between two epochs. It is our Great Work as educators to assist in midwifing this transition in the most life-giving way possible.

We know that the Machine Story, particularly of the past 300+ years, is founded on a "worldview of four disconnects:"[17] estranging us from the natural world, erasing place-based traditions and spiritualities, detaching individuals from community and extended family, and uprooting Land-oriented ways of living. The Machine Story has been a process of alienation for humans—their bodies from their minds, their self from the social world, their being and knowing from the physical natural world and spiritual worlds, and their labour from their work and the products of their work.[18]

The New Science is catalysing rethinking in most disciplines. This Relationality view is a counterpoint to the foundations upon which our educational systems have been predicated as well as to the impact of the global proliferation of an industrial model of schooling. As Schumacher described previously, existing forms of education reinforce the very principles that are compromising life on Earth. Cultural historian and theologian Thomas Berry agrees, "Our educational institutions need to see their purpose not as training personnel for exploiting the Earth but as guiding students toward an intimate relationship with the Earth."[19]

Berry defines Great Work as that which connects our journey as a human species to the much longer journey of the Earth and the cosmos. Connecting individual lives to the destiny of the universe is "the Great Work of a people."[20] Frederique Apffel-Marglin describes this task before us in this beautiful passage inspired by the Indigenous cultures she has lived in.

> Before the triumph of Modernity . . . We were wealthy in our human and other-than-human communities. There was an abundance of beings to accompany us in our earthly journey. The multifarious beings of this world taught us to share the bounty of the world with them; they taught us the gestures of reciprocity; they taught us to fear greediness and accumulation. They taught us that the wealth of the plant beings, the tree beings, the water beings, the soil beings, the mineral beings, was not only ours, was not there for the sole purpose of satisfying our needs. They had their own reason for existing, their own requirements, and their own agency. We needed to ask permission, to share, to give back, and to give thanks. These very gestures made us aware that we were only one strand in an immense tapestry that wove the pattern of life on this earth.[21]

To engage the task of repurposing education, Berry suggests that we need a "vision of the future sufficiently entrancing that it will sustain us in the transformation of the human project that is now in process."[22] It is moving towards a consciousness that can break through the noise and distractions of the consumer society.[23] For Berry, this is a vision for an "Ecozoic Era,"[24] beyond the Anthropocene[25] into what others call an ecological civilization.[26]

How do we stretch beyond, living step by step into this new era? In review, Chapters 2 and 3 described the larger historical forces at work throughout Western civilization and the Modern Age, especially the Machine Story. These chapters also recounted the rise of social and environmental movements which contested different aspects of the Machine Story. Chapters 4 and 5 discussed the development of environmental education and then sustainability education that is nested within this larger history. Chapter 6 described the new emerging story from both quantum physics and the life sciences, comprising the Relationality revolution. Relationality is a new story and way of being that can address the crisis *in* education with its industrial structures. Chapter 6 introduced 15 principles of cosmo-onto-epi-axiological[27] stretching,[28] collaboratively developed with Katie E Ross, Joy Kcenia Polanco O'Neil, as a basis for reimagining education. See the original article for more description and exemplars, as well as my website for learning activities.[29]

Chapter 7 described the story of Modern Education, necessary to understand if we wish to reimagine education and engage an *educere*[30] form of learning. Every culture[31] relies on education to convey their social imaginary or Big Story of who they are and their way of life. Stretching beyond our current story *is* the journey of transformation, learning into sustainable or rather life-giving societies.

This chapter sketches new imaginaries related to the principles. These are my own interpretations of these principles, including possibilities for practice.[32] I go far afield to nonwestern sources, particularly Indigenous sources, to find inspiration for creating an educational container for seeds of possibilities. What do/did these principles look like in intact cultures? While few cultures have been untouched by the Modern way, perhaps we can see glimpses of potential while supporting the nonwestern reclamation of diverse Old Ways.[33] As *Tewa* Gregory Cajete suggests, there are "rudiments of eco-cosmologies of former times" still available.[34] These inspirational elements, "this new flowering is also a reflection of a new way of education for life's sake that must be nurtured once again."[35]

Decolonization, Dragonfly Seeing, and the Pluriverse

> The time will soon be here when my grandchild will long for the cry of a loon, the flash of a salmon, the whisper of spruce needles, or the screech of an eagle. But s/he will not make friends with any of these creatures and when the heart aches with longing s/he will curse me.[36]
>
> *Chief Dan George, Coast Salish*

Colonialism is ongoing, and so truth telling, reconciliation, restoration, indigenizing, and decolonizing are key elements of Transformative Sustainability Education. Indigenous leaders say that nonindigenous people cannot skip ahead to reconciliation without first hearing and recognizing the truth of deliberately attempted genocide, as well as the ongoing arrogance, ruthless extractivism, and duplicitous speech.[37] Truth telling comes first.

There have been multiple calls to decolonize education,[38] environmentalism,[39] sustainability,[40] systems thinking,[41] "development,"[42] regenerative agriculture,[43] epistemology,[44] research,[45] and transformation.[46] Decolonizing is the "dismantling of imperialism" by decentralizing power in ways that welcome non-White and non-Euro-American cultures, values, and knowledges.

One example, say Norwegian geographer Irmelin Gram-Hanssen and her colleagues, is climate change.

> Climate change has been conceptualized as a form and a product of colonization. In this perspective, it becomes important to base climate change adaptation and transformation efforts on decolonizing practices and imaginaries. A central aspect of decolonization is contained in the Indigenous conceptualization of relationality.[47]

By foregrounding the rights, traditions, and knowledges of Indigenous and other nonwestern peoples, working towards a *relationality* worldview, we can overcome colonizer imaginaries that keep alive the thought patterns that discount and denigrate and have created a culture driving climate change. It is recognizing Indigenous "Aboriginal Rights, Treaty Rights, and our Inherent Right to the land."[48]

Further, "Indigenous knowledge can reveal truths not visible with White, Eurocentric approaches"[49] within both environmental and sustainability education. *Potawatomi* Robin Wall Kimmerer says that Indigenous "unique cultural perspectives on environmental stewardship are greatly needed."[50] Gram-Hanssen and her colleagues utilize the concept of "right relation" to profile the "obligation to live up to the responsibilities involved when taking part in a relationship—be it to other humans, other species, the land, or the climate."[51] Reconciliation is walked when "both parties commit to truth and respect, which then grows into trust."[52] It is a commitment to recognition and respectful coexistence.

The work of Indigenous peoples has been and remains the work of cultural survival and reclamation of knowledge and cultural traditions while building bridges to Western knowledge. *Anishinaabe* Melanie Goodchild asks a key question: "How do you incorporate multiple ways of knowing, in a respectful way?"[53] Indigenous scholars call for "an approach that privileges and honours Indigenous intellectual traditions emanating from spiritual wisdom."[54] They call for an "equal but differentiated"[55] approach to knowing, knowledge, and scholarship. Kimmerer calls it "braiding" or "two ways of knowing." She explains,

> [B]oth Western science and Traditional ecological knowledge are methods of reading the land . . . in different ways. . . . From as far back as I can remember, I had this notion of plants as companions and teachers, neighbours and friends . . . As an aspiring botany major, I was pressured to adopt the scientific worldview . . . Later . . . I was invited to sit among Indigenous knowledge holders who understood plants as beings with their own songs and sensibilities . . . I am [now] able to stop speaking of plants as objects . . . Ideally the two ways of knowing can reinforce one another.[56]

Standing Rock Sioux Vine Deloria Jr also called "Indigenous knowledge the 'intellectual twin to science'" as does *Tewa* Gregory Cajete who sees these knowledges in a "circle of relationship."[57] A similar concept, "two-eyed seeing," is from *Mi'kmaq* Chief Charles Labrador. He says, "Go into a forest, you see the birch, maple, pine. Look underground and all those trees are holding hands. We, as people, must do the same." It is building on the strengths of both Western and Indigenous knowledges, not through merging them, but by "a weaving back and forth between knowledges in which each strand is necessary to the process."[58] *Cayuga* Chief Jacob Thomas says that each knowledge system is a tool to guide our cultural vessels in sharing the "everflowing River of Life."[59] Each way of knowing is held in a

> mutual understanding of peace, friendship, and respect . . . we are all traveling down the river of life together, but with each people in their own vessel with their own beliefs, languages, customs, and governments . . . We are to take care of this river as all of our survival depends on a healthy river.[60]

Each way of knowing and learning has their unique approach. The Western way is experiment, analysis, generalization, and critique through science and social science with "transmission" through education. Indigenous ways of knowing are spiritually derived and approached through contemplation, close observation, listening, and learning from the local natural world. As Kimmerer argues, both are theoretical, but they emerge from different assumptions, one from neutral objectivity by asking "How does it work?" and one from responsibility by asking "Who are you?"[61]

"The binocularity of this [two-eyed seeing] guiding principle means that by engaging the overlapping perspective of each 'eye,' a wider, deeper, and more generative 'field of view'" emerges.[62] While some of the more egregious values and practices in modern society require transformation, the Western way of knowing has strengths which can be balanced with many other ways of knowing and learning, not just Indigenous ways.

Like a dragonfly sees through a compound eye of thousands of facets, providing a spherical field of vision, a growing number of scholars recognize a *pluriverse* of alternatives.[63]

> [A] 'pluriverse': a world where many worlds fit, as the Zapatistas of Chiapas put it. All people's worlds should co-exist with dignity and peace without being subjected to diminishment, exploitation, and misery. A pluriversal world overcomes patriarchal attitudes, racism, casteism, and other forms of discrimination. Here, people re-learn what it means to be a humble part of 'nature,' leaving behind narrow anthropocentric notions of progress based on economic growth.[64]

It is the work of making room for a pluriverse of voices and sensing into a pathway for recreating new forms of educating, appropriate to each unique context.[65] Yet,

I readily acknowledge the difficult process of unlearning and relearning, struggling with my own colonized assumptions and embodied ways that most often grant me privilege, implanted by Western society and perpetuated through contemporary educational systems. My very patient and tenacious Indigenous Teachers,[66] my Earth Teachers, and my ceremonial experiences have planted seeds in me which are whispering, calling me to stretch into another way of being.

So, I am humbly sharing with you a small part of my journey to understanding, with the proviso that my understandings are partial and limited by my Western framework which I am working to decolonize, as well as a recognition of the depth, complexity, and diversity of Indigenous and Othered[67] knowledges. It has been a search for broad patterns across multiple knowledge systems to inform a Relationality framework that harmonizes with Life in all its forms, as the central pivot for our future. My goal is to honour Indigenous knowledge as being important to a future that is healthy, just, and wise and I endeavour to share this in the words of Indigenous people as much as possible, while acknowledging that they are the Teachers of this knowledge and pointing you towards them. My hope is that by holding up these written and spoken words, the spirit behind these words can resonate directly and work their way into collective understandings.

Theft of Knowledge, Theft of Biodiversity, and Right Relations

The esteemed Ugandan development education professor, Catherine Odora Hoppers, directly links theft of territory with theft of knowledge and biopiracy. Extractivism has pillaged material resources and people through slavery and then "systematically erased knowledge systems." So now, "bioprospecting still renders invisible the fact of prior use, prior knowledge, and prior rights associated with biodiversity."[68] Still today, "original economic and ecological systems disappear . . . whether for medical or agricultural use."[69] People are losing the materials that are their sources of income, food, and traditional medicines through expropriation and then patenting as proprietary knowledge, with "little or no compensation,"[70] a double loss. With these losses, she asserts that "From the perspective of biodiversity, it has become clear that a major threat to the sustainability of natural resources is the erosion of people's knowledge."[71] Indigenous peoples around the world have traditionally been protectors of land, water, and species, whose knowledge has developed from within their relations to these beings and elements.

Thus, Portuguese legal scholar Boaventura de Sousa Santos declares that it is time to end "cognitive empire,"[72] the epistemological face of colonization. "There is no social justice without cognitive justice."[73] Of importance to educators, Santos suggests one of the largest shifts is this shift past the stranglehold of Northern epistemology to make room for epistemologies of the South. It is ending *all* forms of piracy. Indian physicist and activist, Vandana Shiva, echoes that "what the industrial economy calls 'growth' is really a form of theft from nature and people,"[74] including intensive, industrial agriculture and criminalizing seed-saving and seed-sharing which undermine biological, cultural, and knowledge diversity.

Such an epistemological transformation makes many other transformations possible. It is making room and respecting the restoration of Other, particularly these indelible links between land, culture, language, and knowledge. Hoppers emphasizes that

> Indigenous knowledge and innovation systems must therefore be sustained through active support to the communities who are keepers of this knowledge, custodians of their ways of life, their languages, their social organization, and the environments in which they live.[75]

As educators, enacting cognitive justice is accepting and nurturing the legitimacy of nonWestern knowledge systems. It is accepting that all cultures contribute scientific knowledge of universal value.[76] Only "building bridges,"[77] says *Mi'kmaq* scholar Marie Battiste, between multiple living knowledges, cultures, and languages will provide a richness of alternatives for our future. Hoppers concludes, "The best possibility for IKS [Indigenous Knowledge Systems] wisdom and insights to be a shared resource is the creation of a better future for all humankind, stressing complementarity between knowledge systems."[78]

After hundreds of years of genocide and forced assimilation, Indigenous and Traditional peoples worldwide are reclaiming their ways of knowing, their Traditional storied knowledge, and ceremonial practices to inform their existence and relations once again, healing from generations of wounding. For *Mi'kmaq scholar* Marie Battiste,

> [D]ecolonizing is not a process generated only for Indigenous students [. . . but] forging a path of critique of current Eurocentric education and its practices while also researching and writing about a newer agenda of Indigenous science and humanities, both of which must be part of a global reclamation for education scholarship.[79]

Indigenizing education means to create space for Indigenous voices, knowledges, pedagogies, protocols, and research methodologies, as part of enacting cognitive justice. It means to respectfully centre, validate, honour, and protect Indigenous rights, worldviews, language, stories, and protocols.[80] It is learning to see and value multiplicities, difficult in a world that tends to monocultures. Goodchild talks about co-inquiry as sitting in the "sacred space of noninterference *in between epistemologies*,"[81] which is a healing space, a space of friendship.

In Canada, Indigenizing education means "every subject at every level is examined to consider how and to what extent current content and pedagogy reflect the presence of Indigenous/Aboriginal peoples and their valid contribution of

Indigenous knowledge."[82] Peter Cole, from *Douglas* or *Xa'xtsa* First Nation, brilliantly invokes this through his poetic writing of Indigenous becoming,

> [I]n the knots and weavings of storytelling indigenousness can become it by spinning weaving knitting knotting darning suturing and acknowledging the context of the practices and the sources of materials.[83]

Constructive transcultural dialogue is crucial. What is shared here is to prod imaginal knowing. It is an encouragement for people in all contexts and origins to dig deep into their own heritage(s) and their own indigenous roots, perhaps lost millennia ago. It is relearning relations with the Land where we currently find ourselves, while acknowledging our, perhaps multiple, Land origins. These are pathways towards life-giving societies, unique in each place. In terms of settlers who are guests on the land where they reside, *Haudenosaunee*, Diane *Kahontakwas* Longboat, explains,

> We are happy to share our sacred fire with you because at this fire is the essence of life, of who the Creator is. If you make your offerings, you make your prayers, have your fast, your vision quest, or whatever, we'll help you with that, but you've gotta do your work to find out who is the Creator and what does the Creator want you to do in your life, how do you activate that spiritual mandate that is in your life. We'll help you with it but in that journey of your healing, you need to go back to your homelands, walk in the places of your ancestors, and that will change you forever. Because that is where you belong and we are sharing this land with you, and we also have a duty to share with you how to respect and honour these homelands, and you need to live with those natural laws and those spiritual laws that govern Turtle Island [North America] . . . we're interested in if you will adhere to these natural laws and spiritual laws.[84]

Returning to the Big Questions

> The challenge to both men and women is to invent new myths. People are changed, not by intellectual convictions or ethical urgings, but by transformed imaginations.[85]

In an era of epochal shift, it is vital to remember the existential or big questions that every culture and/or society answers through their myths or Big Stories, disseminated through educational processes and structures. They are cosmological, ontological, axiological, and epistemological questions.

- How did the universe come about; what is our place in it as humans? What is the purpose of the non-human and more-than-human?[86]
- What is reality; what is real, what is illusion or abstraction?

- What is existence; what are life and death; is there a purpose to existence?
- What does it mean to be human as well as nonhuman; what is consciousness?
- What are our values; what does it mean to live a good life; how ought we live together; what is a good culture and/or society?
- What does it mean to know; how can we come to knowledge—even about these questions; what are legitimate and reliable sources of knowledge?[87]

Each epoch and culture have their own, often multiple, responses to these questions. It is the complex interplay between the questions, especially how they are framed and what is posed as legitimate responses, which concerns us as educators. These are the questions being reconsidered anew, with polyvalent responses, discussed in the next section, which educators and learners can consider together.

Seeds for Reimagining Education

The following themes are offered as seeds, little packets that hold the matrix of imaginal potential, from long and mixed ancestry. The seed is sleeping yet dreaming of what could be.[88] The sacred potential within is diverse, but the conditions in which it is seeded shape what emerges and develops. Seeds contain both death and life within them. To live, the seed itself must die and give way, enabling the sprout to root and develop.[89] So, something must die and let go, for the potential to live and become itself. The soil, of self and society, needs to be tilled, increasing the readiness, warmth, and fertility, so it is ready to welcome the seeds. The following themes or seeds are images of what may be possible, knowing that what emerges will be unexpected yet responsive to those who share a collective yearning for a life-giving way of being.

An Alive Cosmology

We need to return to "In the beginning . . ." as our way of being begins in cosmology. Big stories or core myths convey an understanding of the universe and reality. They are carried in smaller substories, reinforcing the deepest values, meaning frameworks, and aspirations of a culture.[90] Myth is understood as certain repeated patterns, part of a "Big Story" as described by Charles Eisenstein.[91] Regenerating a new cosmology, one that situates us in the matrix of a living cosmos, is vital, as landscape shapes mindscape and vice versa. *Haudenosaunee* Dan Longboat emphasizes that

> it's looking at the whole thing, looking at the whole system the way our ancestors did, the seen and the unseen, the past, present, and future, the spirit, the earth; all of a sudden, that's a whole system, that's what we need to bring back.[92]

Currently, our "inscape" shapes "landscape."[93] "The pathway of sensorial impression to material interference is strewn with an imagery that makes the inscape a template for the reshaping of the landscape."[94] Our imagery is a cosmos that is a dead, clockwork universe. Humans have been considered the only species in the universe with true intelligence (ability to reason), self-consciousness (awareness of self), and sentience (ability to feel and sense). Whatever was not the human mind, including the human body, was "mechanism."[95] Paradoxically, through mechanistic science which separated humans from the natural world, we are now led into a stunning view of the deep connectiveness and dynamism of the cosmos.

Mi'kmaq Murdena Marshall explains,

> In connectiveness, the three letters *-ive* introduce the idea of action, of tending toward a state, especially in a regular or lasting way . . . Thus the word connectiveness augments the state of being connected with the action of becoming connected, an important distinction for a verb-based language like Mi'kmaq.[96]

Thus, we exist within an active cosmos. It is a view of our "planet and its creatures [as] a single self-regulating system that is in fact a great living being."[97] Earth has a propensity towards Life. Life is naturally self-producing and self-renewing, a symphony of processes and beings who constantly regenerate the conditions for life. In fact, "the essence or process of the whole living being . . . [is] a self-creating autopoietic system."[98]

Deep History

Advocating a telling of Deep History, Thomas Berry says, "We need to think of the planet as a single, unique, articulated subject to be understood in a story *both* scientific and mythic."[99] He argues for a "cosmology of Gaia as well as a biology of Gaia,"[100] for Gaia is a "communion of Subjects not just a collection of objects."[101]

> We need a spirituality that emerges out of a reality deeper than ourselves, a spirituality that is as deep as the Earth process itself, a spirituality that is born out of the solar system and even out of the heavens beyond the solar system. For it is in the stars that the primordial elements take shape in both their physical and psychic aspects. Out of these elements the solar system and Earth took shape, and out of Earth, ourselves.[102]

The science story now tells of the birth of our solar system four and a half billion years ago, as in Figure 8.1. This story, beautifully written by Thomas Berry with mathematician and evolutionary cosmologist Brian Swimme, is called *The Universe Story*.[103] Since the birth of the cosmos, they say, Earth has been in process—a dance of solids, liquids, gases, gravitation, and electromagnetism. From the simplest of cells has come the sophistication, complexity, and artistry of our abundant planet and all its intricate and connected life forms.

FIGURE 8.1 Pinwheel of Star Birth

Credit: NASA. Reprint permitted.

Working from this story, professor of Philosophy, Cosmology and Consciousness, Sean Kelly, identifies eight thresholds or transformative moments in the evolution of Earth, which are, the:

1. great flaring forth as the origin of the universe,
2. formation of stars and galaxies,
3. formation of heavier chemical elements,
4. formation of Earth and the solar system,
5. emergence of life,
6. birth of *Homo sapiens*,
7. Agricultural Revolution, and the
8. Modern revolution.[104]

Kelly suggests that we are on the cusp of another threshold, a ninth threshold. While civilizational collapse is one possibility, he suggests that we may be entering a Second Axial Age. Coined by philosopher Karl Jaspers, the "First Axial Age" referred to a pivotal historical time between 800 and 300 BCE, when all the main religions and philosophies emerged. Each of these main religions and philosophies, however, considered themselves to be the one source of "the only truth or the highest truth," a form of universalist thinking that was rooted in specific ethno-linguistic cultures,[105] despite their spread. The divisions and contestations between these religions, cultures, and their cosmologies have led to thousands of years of war, death, and erasure as well as the need for the modern secular state.[106]

Now, however, we see that Gaia, from the soaring grandeur of mountain spires to the most delicate intricacy of mosses,[107] is "the ground of what we all share in common."[108] Earth is the "concrete universality"[109] that we all share and that binds us as human and nonhuman family. Humans (*Homo sapiens*) have only been a part of this 4.5-billion-year Earth journey for 300,000 years. It was only 11,000 years ago when early agriculture began. It was only 300 years ago that industrialization began, 70 years since the beginning of space travel, and 50 years since computers became ubiquitous.[110] After all this time, we have come full circle to realize, once again, that the "story of Gaia is sacred, because it tells of our common origin and will include our shared destiny."[111] This destiny is not just of an individual ethno-cultural group of People, but of the whole human family. We are here together.

Earth-Honouring Spiritualities

Honouring our cosmos by learning this new Gaia cosmology and our human presence in it should be a core part of reimagining education. It is the cosmic dimension of learning "all our relations."[112]

Pedagogical inspiration is available for telling the *Universe Story*, which can be beautifully shaped into a meditative walk.[113] Such a storytelling of Deep History blends awe and reverence with the science of the universe. It is a reflection on the magnificence of how our living, breathing planet came to be—a song of praise to its complexity and mystery.

The story of the cosmos is a foundational part of earth-honouring spiritualities. Such spirituality, as a direct experience of the numinous or sacred, is generally not contrary to religious traditions, each with their unique paths to the sacred. The *Faith for Earth* initiative of the UN Environmental Programme (UNEP),[114] together with the World Council of Religious Leaders[115] and Parliament of World's Religions,[116] state that our "flourishing global interfaith movement . . . is increasingly bringing people together to protect and sustain life on Earth."[117] That said, this ought not eclipse the many origin stories of specific Peoples. These are the mythical stories they were given in their local places since time immemorial, their unique consciousness in the universe. As Joseph Campbell suggests, these stories overlap, with shared archetypes, remarkable and diverse, part of the human story, and another form of differentiated unity.

Yet, ecofeminist theologian Heather Eaton expresses caution.

> The cosmos cannot be seen as a backdrop to the human drama, or as a context, an unfolding, a progression, or a potential. It is more a becoming; not linear and determined, but creative and dynamic, yet, seemingly, with an orientation. As the universe develops, it becomes more: more complex, interactive, entwined, vibrant, and intense. . . . the expansion of human consciousness into the cosmos is also the universe and Earth becoming conscious in humanity.[118]

We have a human role in this living, dynamic universe, as one node of self-consciousness that is the universe reflecting on itself.[119] Our expanding consciousness is part of the expanding universe. Quiet, mystical moments in the natural world, walking the Deep History of the cosmos and the Earth, through the "Universe Story" pondering with awe all that surrounds us as a matrix of life, feeds the spirits of learners, which I have witnessed repeatedly. These slow moments of wonder need to be cultivated by educators once again. Such moments anticipate a respect way of being.

Cosmic Principles: Balancing the Feminine and Masculine

An Earth-based cosmology is also present in ancient myths, such as those about the Great Goddess or Mother Goddess, revered throughout the Palaeolithic, Neolithic, and Early Bronze ages. Through new archaeological interpretations of these eras,[120] the earliest humans honoured the regeneration principle that gave them ongoing life. This was symbolized in the Great Goddess or Venus Figure, as seen in Figure 8.2 who was "considered the source of all life."[121]

Many agree that the earliest organizations of humans were gynecocratic, where the principle of "mother-right" was recognized as being primary.[122] This is seen in the beautiful statuary of women councils, sometimes with 13 wise women and their attendants as seen in Figure 8.3.[123] These images were honouring the principle of life-givingness that flows through all human and other earthly forms.

This has echoed down the ages to the International Council of the Thirteen Indigenous Grandmothers. These women Elders were brought together for the first time in 2004.[124] Their Council has become a new global "alliance of prayer, education, and healing for our Mother Earth—for all Her inhabitants, for all the children, and for the next seven generations."[125] According to *Laguna Pueblo* literary scholar, Paula Gunn Allen,

> Surely this is the time referred to in ancient prophecy: the time when . . . the Grandmother goddesses return and Native Americans lead the world into a new age of peace, balance, harmony, and respect for all that is.[126]

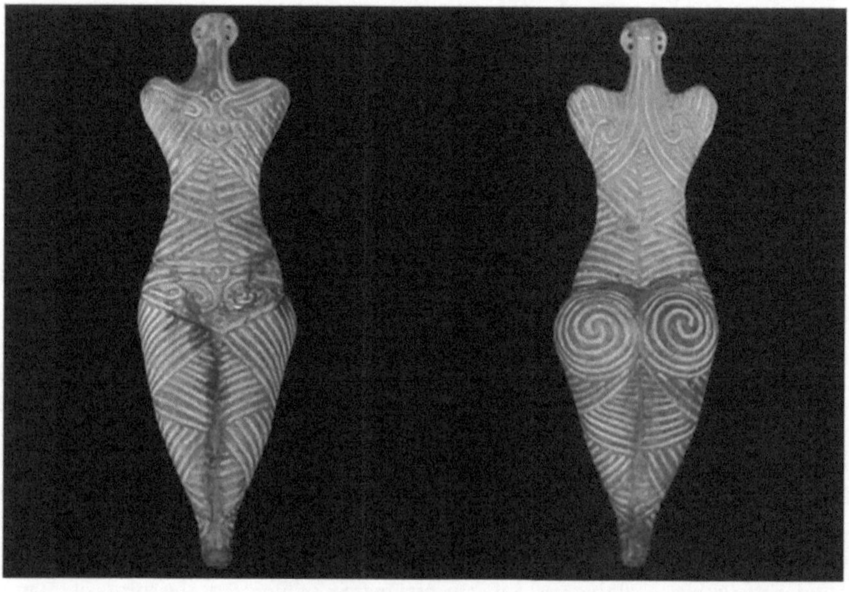

FIGURE 8.2 Venus Figurine, Front and Back, Cucuteni, Drăguşeni, 4050–3900 BCE

Credit: Reproduced with permission of the Botoşani County Museum, Romania.

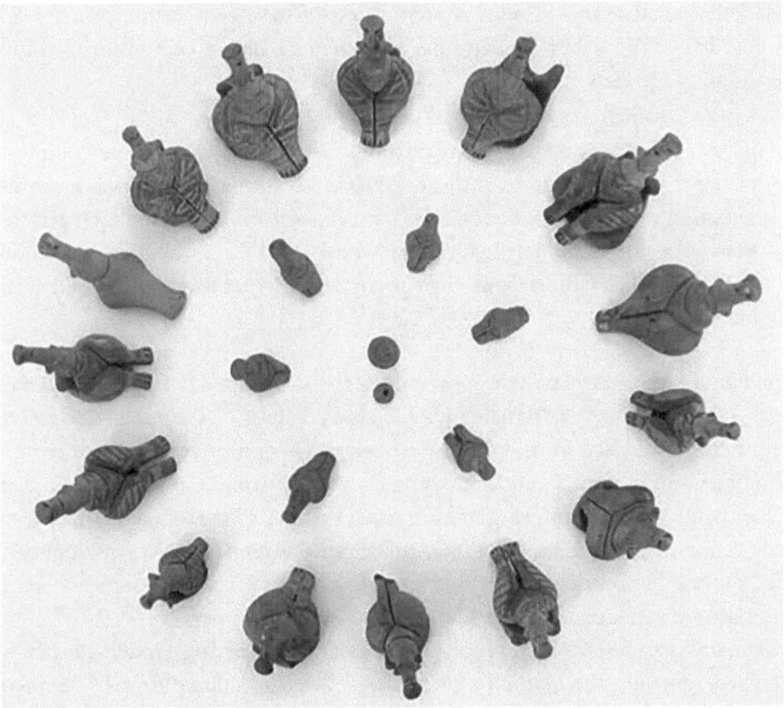

FIGURE 8.3 "The Council of the Goddess," Set of Twenty-One Figurines and Thirteen Chairs, Poduri-Dealul Ghindaru, Pre-Cucuteni II, 4900–4750 BCE

Credit: Used with permission of Princeton University Press, from *The Lost World of Old Europe: The Danube Valley, 5000-3500 BC,* edited by David W. Anthony with Jennifer Y. Chi, 2010, pp. 114–115; permission conveyed through Copyright Clearance Center, Inc.

French historian of pre-Christian civilizations, Jean Markale, has found evidence of what he calls "the vast mother" on many continents. Riane Eisler describes how this was lost through a radical reversal in which the worship of the Great Earth Goddess in agricultural Old Europe (chalice cultures) shifted to the Father Sky God of the early Indo-Europeans who were pastoral (sword cultures).[127] Markale contends that since this time, there has been an ongoing contestation between the Feminine and Masculine principles in many cultures. This contestation has been very violent at times, such as the witch burnings of the Inquisition or the erasure of a Feminine principle in some religions.

Markale traces the many historical faces of the Goddess in the Near East, Middle East, and Far East as well as throughout Christian Europe. Despite many attempts to demonize and eliminate the Goddess and suppress all her complexity, metamorphoses, and symbolism, she persists. As the primal matrix of regeneration, the Feminine cosmic principle of both femininity and maternity, as well as the Masculine principle, are needed. Although, he advises, the Divine Feminine will always be culturally shaped, named, and heard.[128]

Markale says that this ancient consciousness of balanced cosmic principles has been lost in the West but is resurging—a remembering of our historical human beginnings.[129] He elaborates:

> [D]ating back to the beginning of time, [there exists] a feminine component of the divine, sometimes anthropomorphized into an individual, sometimes remaining a pure abstraction. This existence of a great universal feminine Principle, virgin and fertile, original Womb of all things, will seem absolutely logical to those familiar with the laws of analogy that form the basis of eternal harmony.[130]

Joseph Campbell agrees on the importance of the Feminine Divine, particularly for women who are bereft of "female mythological models."[131] Campbell declares that this resuscitation is vital for men too. We need the Feminine principle for women's mythological quest for self-understanding as well as for men, as it is part of themselves and aids "patient fostering of each other's growth."[132] He maintains that "we are the 'ancestors' of an age to come, the unwitting generators of its supporting myths, the mythic models that will inspire its lives. In a very real sense, therefore, this is a moment of creation."[133]

For many decades, women scholars have been resuscitating Goddess myths and archetypes, revealing the Feminine principle in diverse multiple forms.[134] Ecofeminist theologian Rosemary Radford Ruether traces the emergence and evolution of new movements and cultural symbolism related to the Divine Feminine.[135] She concludes that perhaps instead of personifications of divine energy, "the divine is seen as the matrix of life-giving energy that is in, through, and under all things, sustaining and renewing life."[136] In this way we can hold, in common, a vision of a "life-giving community from many strands of tradition, culture, and history."[137] Berry agrees, "The sacred community must now be considered the integral community of the entire universe and, more immediately, the integral community of the planet Earth."[138]

Thus, the Feminine principle is being languaged once again. Ancient myths featuring her many faces have been rewritten for current audiences or remembered back from disuse. Charlene Spretnak says this enables "engagement with the Goddess in symbol, myth, and ritual as participatory fields of relation [which] encourage the expression of one's unique gifts while evoking a sense of one's larger self, the fullness of our being."[139]

The recovery of myths and archetypes for those across the full spectrum of gender and sexuality is also needed, in respect and full acknowledgement of the diversity of humanity. Again, many Indigenous cultures recognize two-spirited people and other gender identifications and sexual lifeways within their own traditions, social roles, and moral voices.[140] Gunn Allen explains that "recent scholarly work reveals the universal or nearly universal presence of homosexuality and lesbianism among tribal peoples, the special respect and honour often accorded gay men and women."[141] She explains that the white Christian colonizers reshaped "the ritual status of all women and those males [and females] labelled 'deviant,'"[142] which connects misogyny with homophobia. Traditionally, there has been a more open sexuality and a positive acceptance of a range of gender identity in Indigenous societies, considered a spiritual gift to the society, lived out through spiritual guidance.

Cultural historian Riane Eisler calls these gender reconciliations the Partnership Principle, where the Masculine and Feminine acknowledge and value each other, as vital elements present in every human.[143] Similarly, writer Frank Waters, with *Cheyenne* ancestry, describes the cosmic dualities that are commonly present across the symbolism and dances of the *Navajo* and *Pueblo*, the ancient mystery plays of the Greeks and Egyptians, and Buddhist cosmology.[144] A higher consciousness, sought out through mythopoetic rituals, can express and move beyond the life-giving tensions between the dualities of sun/moon, day/night, male/female, and winter/summer. In other words, these principles ought to have meaning in their relation with the other, not in overcoming the other, like a unique fertilized egg resulting from female and male contributions. Many traditional cosmologies consider there to be a Oneness at the heart of all dualities, only accessible through spiritual discipline. Relearning the rhythms, powers, and balancing of dualities, residing within the Oneness, is part of a newly emerging cosmo-ontology.

And so, in these reclaimed myths, we come face to face with the great cosmic principles and our place within this vastness. Ritual tellings of such ancient myths and the restoration of still-existent cultural myths, particularly relating to the scientific and mythic stories of Gaia, can help learners deepen into an understanding of a balanced cosmos, including the balancing of various knowledges. All people need to see themselves in the Sacred and the Sacred in them.

Sacred Geography, Sacred Architecture

The certainties of the modern age no longer cohere. Rather than perceiving the universe as separate objects moving in mechanical patterns, quantum science now conveys the unbroken wholeness underlying the universe as a complex network of energy fields. In this webbed understanding of the universe, there is a plasticity of time, space, and matter.[145] The universe is one of impermanence, interchangeability, unpredictability, connected by a substrate of continual flowing energy.

> Several [scientists] thought again about a few equations that had always been subtracted out in quantum physics. These equations stood for the Zero Point Field—an ocean of microscopic vibrations in the space between things. If the Zero Point Field were included in our conception of the most fundamental nature of matter, they realized, the very underpinning of our universe was a heaving sea of energy—one vast quantum field. If this were true, everything would be connected to everything else like some invisible web.[146]

Quasi-instant communication among the parts through this energy-sea, called nonlocality or nonlocal coherence, illustrates the relationality and energy at the heart of the universe. All beings and forces, cosmos and Earth are entangled.

> On our most fundamental level, living beings, including human beings, were packets of quantum energy constantly exchanging information with this inexhaustible energy sea. . . . At our most elemental, we are not a chemical reaction, but an energetic charge. Human beings and all living things are a coalescence of energy in a field of energy connected to every other thing in the world. This pulsating energy field is the central engine of our being and our consciousness.[147]

This matrix of the cosmos as life-giving energy flowing through all things and in all places is prominent in most Indigenous cosmologies. *Blackfoot* Leroy Little Bear states that "The Native American paradigm is comprised of and includes ideas of constant motion and flux, existence consisting of energy waves, interrelationships, all things being animate, space/place, renewal, and all things being imbued with spirit."[148] This cosmic energy flows through every individual as well. Little Bear says that "What Native Americans refer to as 'spirit' and energy waves are the same thing."[149] Meditating on this matrix puts humans, their spirit, and psyche into harmony with cosmic energies.[150]

Standing Rock Sioux (*Lakota*) theologian and historian, Vine Deloria Jr, explains that the North American Indigenous "sweat lodge is built in the shape of a sphere and reproduces, as closely as possible, the cosmos as a whole."[151] In it, through ceremony,

> we learn that sacred places represent the power [of the cosmos] by showing us that we can become a part of a preexisting set of relationships. Each tribe has its own center and its own boundaries marked out by reference to existing landscape features.[152]

Over history, many Traditional Peoples built structures—in caves, homes, or sacred sites—reproducing and honouring the directions of the cosmos. Acknowledging the cardinal directions and cosmic principles in their living and ceremonial structures provides both an orientation within the sensory world as well as pays "respect to the powers of the universe."[153]

As *Piikani* Elders (of the *Blackfoot Confederacy*) Dr. Reg Crowshoe and Geoff Crow Eagle explain, the Blackfoot Tipi Circle model informs all their teaching, governance processes, community life, and spiritual elements and rituals, as in Figure 8.4. The tipi was part of the ongoing respect conversation with the living world as well as the spirit world.

> Many Elders believe that there is an oral circle, or system, that exists across all the First Nations of North America, and that we are all part of. Our Piikani

FIGURE 8.4 Tipi Doors Face East

Credit: Paul Goble, *Tipi: Home of the Nomadic Buffalo Hunters* (Bloomington: World Wisdom, Inc., 2007), p. 10. Reproduced with permission from World Wisdom, Inc.

> Blackfoot systems are only a part of that greater circle. We are honoured to share with you our Piikani perspective, as part of that great circle that unites us all.[154]

In the very formation of their living or ceremonial structures, the cosmic matrix of relations and energy is shaping the inhabitants. The circle as cosmological geography recognizes a parallel reality between the seen and unseen worlds, nesting realities.

> [The circle] has no beginning and no end; it is a form that always comes back around to its beginning, integrating its past, constantly enlarging its circumference and deepening its dimensions through its connection with the unconscious and the sacred.[155]

Indigenous peoples build their structures as sacred architecture in which every element connects to their living cosmology, past, and future.[156] Tipi encampments, common to most of the migratory tribes of the North American Great Plains, were also in a circular geometry, with the horizon embraced by the sky as the third layer of circle. Jungian analyst Joan Golden-Alexis explains,

> The circle, however, embraces the wholeness central to the Plains Indian culture; it depicts the unity of the universe—the unity between history and destiny, the unity between present and memory, the unity between above and below, between the conscious and unconscious, and the mundane and sacred.[157]

Tipi structures act as spiritual containers filled with symbols, so that occupants always recognize they are part of this spacious sacred geography. "The floor of the tipi is the earth, the walls [are] the sky, and the poles are the trails that lead to the spirit world."[158] Golden-Alexis then describes the central point in such sacred architecture.

> These zones of power are connected through a vertical axis termed the axis mundi, the cosmic axis or world tree. The Native American sees this axis as an intimate part of [human] nature and a path ultimately available . . . to connect with this cosmic energy.[159]

In addition to a circle cosmology, "being in balance" is a core cosmological idea, shared by many other Peoples as well. Buryat-Mongolian shaman Sarangerel says their practices aim at

> keeping negative and positive actions in balance. While one has the intention of leading a good and benevolent life, one can accept one's own flaws as well, because balance, rather than absolute goodness or perfection, is the ideal.[160]

The cosmic principles are visualized in the geometric shape of the circle, "as everything has a circular motion"[161] from the sun, moon, other planets, and the seasons. Each direction is generally associated with an element—such as fire, earth, water, and air—and a colour or sometimes plants, animals, and minerals. The four cardinal directions of east, south, west, and north (or six, with the inclusion of up/spirit world and down/Earth world) are fundamental in many Indigenous worldviews. The doors of the tipi were located in the East, to be greeted by the Sun's first rays. The tipis were oriented to the heavens in other ways—to stars or constellations, spirit dreams, and, as illustrated in Figure 8.5, dreams of where to find the buffalo.

FIGURE 8.5 Inside the Blackfoot Black Buffalo Lodge

Credit: Paul Goble, *Tipi: Home of the Nomadic Buffalo Hunters* (Bloomington: World Wisdom, Inc., 2007), p. 38. Reproduced with permission from World Wisdom, Inc.

This is similar to African *Bantu* philosophy, which is "based upon the belief that life is a vital unity and that the human being is only a point on the cosmic circle of life."[162]

In describing another of the oldest religious and cultural traditions, Sarangerel says that they too consider the cosmos and world to be vibrantly alive, with a "spirit" or living energy flowing through all beings and things. For nomadic pastoralist Mongolians, since 3000 BCE, their traditional homes or yurts are called a *ger*.[163] The *ger* is constructed in a sacred circle with a latticework of logs and poles traditionally covered in skins or felt made from sheep's wool and cotton, connected by cowhide straps as well as horsehair and camel hair. Geography professor Batchuluun Yembuu explains that it no longer contains iron and thus is considered the most ecological housing, given biodegradability, heat exchange qualities, and ability to stay cool in hot summers and warm in cold winters.[164]

Sarangerel describes that, generally, the centre smoke hole denotes the four directions related to colours, sky spirits, and elements. The *ger* is structured as multiple circles, a correlation to their perceptions of the universe, as in Figures 8.6 to 8.10.

> The ger is not only the center of the universe but also a microcosm within it. In fact, it is a map of the universe at large, and the vault of the heavens is reflected in the arched shape of the interior of the ger roof.[165]

Inside, each part of the structure has a related "spatial zone" for genders, tools, and activities, which coheres as a map of the universe, acknowledging the sacred principles within the whole. Traditionally, movement in the *ger* would generally be sunwise in relation to the movements of the heavens.

Yembuu explains that the door is south-facing so that the *ger* is always in relation to the movement of the sun. The *toono* or opening in the roof acts as a sundial, orienting the inhabitants by sun and shadow. Daily work tasks would depend on the changing

FIGURE 8.6 Inside a New *Ger*

Credit: Dr Batchuluun Yembuu, Mongolian National University of Education. Photo reproduced with permission.

FIGURE 8.7 Mongolian *Ger* Frame

Credit: Dr Batchuluun Yembuu, Mongolian National University of Education. Photo reproduced with permission.

sunlight, from guiding the grazing of cattle, watering or settling of the cattle, milking, and salting. The *toono* is divided into 12 parts which correlate to the 12 zodiac signs: Rat, Ox, Tiger, Rabbit, Dragon, Snake, Horse, Sheep, Monkey, Rooster, Dog, and Pig. For instance when the sunlight reflects off the top of the *ger's* roof, between 4 and 6 am, it is the hour of tigers and rabbits. When the sun rises, between 6 am and 10 am, the morning sun rays descend into the *ger* illuminating the wall, and it is the hour of dragons and serpents. The hour of horse and sheep is from 10 am to noon when the sunlight penetrates onto the floor of the *ger*, in the north.

FIGURE 8.8 A New *Ger* for Married Couple

Credit: Dr Batchuluun Yembuu, Mongolian National University of Education. Photo reproduced with permission.

Similarly, this cosmology is reflected in the *hogans*[166] and sand paintings of the *Navajo*, as well as in the tipis, medicine wheels, and sacred hoops of the *Plains Cree*. Golden-Alexis explains,

> [T]he American Indian resonates with this inaudible cosmic pulse, responsive to both time and place . . . life is not a shallow horizontal stream flowing out of an impatiently forgotten past through the present into a future one is hurrying to reach. Rather, it has a vertical dimension that cups past and

FIGURE 8.9 Mongolian *Gers*

Credit: Dr Batchuluun Yembuu, Mongolian National University of Education. Photo reproduced with permission.

> future in a timeless present. This allows entrance into something ancient and primordial—a fluid space, which brings the American Indian's sacred symbols into consciousness. These symbols then take on a defining reality, penetrating [one's] character, providing . . . an essential dignity.[167]

Like the traditional Mongolians, the pole is understood to touch the "Pole Star . . . the sky nail that holds up the heavens."[168] In these ways, Indigenous people constantly moved in a sacred universe, whether inside or outside their habitations. The circular space is fluid, energy ever flowing within and without, circles within circles. One is constantly connected to, and participating in, the visible and invisible world.

For different purposes or by different peoples, square or rectangular habitations are also built, still in relation to cosmology. For the *Navajo*, square buildings acknowledge the four corners of their world, which are the four mountains sacred to them. For the Maya, their world is "four-pointed, four-sided, four-bordered"[169] as they understand their Land having four corners. As education scholar Cora Weber-Pillwax explains, the Woodland Cree and other Northern bush peoples embody a networked "trails model of consciousness and identity formation," with meaningful places bestowed with markers, as markers of events or personal and group development embedded in geographic markers.[170] So, cosmologies evoke geographical symbolism and geography evokes cosmology expressed in physical space, including architecture.

Sacred Time

Vine Deloria Jr and Paula Gunn Allen both explain how North American Indigenous cosmology connects to quantum physics. While the "magic" of shamans was renown, many stories which Deloria recounts are explainable through the new quantum understandings of time, space, consciousness, and matter. Physicist Fred Alan Wolf explains the connection in one of Vine Deloria Jr.'s books,

> Today our [scientific] position is close to the one discovered by basic tribal peoples. The concept of universal energy in our language might be called "universal quantum wave function" or "matter wave" or "probability wave of quantum physics." This "wave" pervades everything, and like the universal energy, it resists objective discovery. It appears as a guiding influence in all that we observe.[171]

Some have claimed that Indigenous peoples have few concepts related to the future, but this is a misunderstanding. Rather, the future correlates with the past and present. As Westerners are fixated on the "mysteries of matter,"[172] they have maintained a unilineal concept of time with a discrete past, present, and future. Yet, John Mbiti explains that in African cosmology, macro-time "embraces all the past events that link the beginning of things to the present unfoldment of events in the universe . . . [including] the period of myth."[173] In this way, the past, present, and potential of the future, including the ancestors, the mythological world, and community perpetuation, co-exist.

This is also present in the Australian Indigenous concept of "Dreaming." Australian Indigenous people exist within a matrix of songlines—the most basic understanding is that these songs tell the mythological stories of the Land. The stories are sung while walking, addressing the mythological events that occurred, creating landmarks, or leaving footprints of the Ancestors that became spirit-dwelling places. The rhythms of the songs are living narratives corresponding to the movement of the walker and the contour of the Land. The songs guide the people from one place to another, connecting them to the Dreamtime, the spiritual plane of existence. Thus, the musical phrase is a map reference, and the music is a memory bank for finding one's way around. It also constitutes one's identity by linking past, present, and future within the moment. *Apalech* Tyson Yunkaporta better explains that this is a "supra-rational interdimensional ontology endogenous to custodial ritual complexes."[174] So, time crosses dimensions, evoked in ritual and other levels of consciousness, linking all within the present moment.

Such perspectives on time have given rise to all ways of keeping time, from cosmological calendars to simple observances related to night/day and the seasons. Wheels of the Year are derived from astronomical movements, which generally govern the time and places of festivals and ceremonies as well as the seasonal activities of agricultural, pastoral, hunting, and gathering people. These wheels followed the solar movements, lunar movements, and/or other planetary movements. Such wheels have been common among many, including the Celtic, Germanic, Slavic, Hellenic, Roman, Mayan as well as Chinese Peoples, which are another way of placing humans within the workings of the universe.

In an ancient example, the Babylonians, Greeks, and then Romans and pre-Christian Germanic peoples named the days of the week by the names of the planets or mythological figures: Sunday for the Sun, Monday for the Moon, Tuesday for Mars, Wednesday for Mercury, Thursday for Thor, Friday for Frigg as wife of Odin, and Saturday for Saturn.[175] In this ancient Mediterranean cosmology, the names of the months are associated with gods as well as with specific powers of planets. For instance, January was named by the Ancient Romans for the god Janus, a two-faced god which symbolizes retrospection as well as anticipation, a time of transition. The Latin word *jānua* means doorway or entrance, evoking the ceremonial gateways into and out of Rome as new beginnings. The planet Saturn is the dominant force, a slowly revolving planet which generates energy for a path of rules, responsibility, and discipline.

For many Indigenous peoples, the "Medicine Wheel"[176] or Circle of Life is a physical expression of their ontology or way of being. It is typically divided by a cross in the middle, creating four equal quadrants, often further divided. There are multiple meanings for the quadrants, some protected and private. Overall, the spatial and temporal geography were replicated in living space as well as replicated in the body and in the community. For instance the four directions resonate with the four stages of life, four aspects of a human (mind, body, spirit, and emotions), and four races of humans, just to name a few.[177]

In another example, Claudine Louis explains the *Omisimaw* Leadership Model,[178] which reclaims knowledge passed down generations of Plains Cree or *Nêhiyawak* women. Using the Medicine Wheel, there are four stages in a woman's life and four parts to the self. Briefly, the Girl-Child sits in the east full of life and play; the Sacred Woman sits in the south holding responsibility for the continuance of the community in multiple ways; the Warrior Woman sits in the west as the visionary leader, and the Wise Woman sits in the North as a source of great wisdom.[179] In this way, the cosmic circle and Earthly realities are reflected in one's own body and psyche. Part of one's growth, says Louis, is learning from the Medicine Wheel, through ceremony, visions, dreams or mentoring by Elders, so that the meanings become personal to one's life journey.

In conclusion, to lessen the grandeur of the cosmos that we live in is to experience a loss of soul. Berry emphasizes that,

> This is especially a loss for children, for it is from the stars, the planets, and the moon in the heavens as well as from the flowers, birds, forests, and woodland creatures of Earth that some of their most profound inner experiences originate.
> Berry, *The Sacred Universe*, p. 133.

I take adult learners out into the natural world in most of my teaching, watching the night skies or the sun's travel across the sky, perhaps leisurely exploring by a creek or seashore, meandering in deep untouched snow, or laying in deep rustling grasses watching clouds. Often participants return to this sense of grandeur that they experienced as children, but which children now experience less and less. These are just some of the ways we can experience being fundamentally nested within these magnificent realities and sense of natural time, rather than isolated monads who are the random outcome of cosmic events. It begins with the development of a new cosmo-ontology that encapsulates the scientific origin story and the mythic stories.

Implications for Reimagining Education

As *Tewa* Gregory Cajete says, "Cosmology is the contextual foundation for . . . a grand guiding story . . . [giving] rise to philosophy, values, and action, which in turn form the foundation of a society's guiding institutions."[180] In this diverse world, many cosmologies already co-exist. Yet, can an Earth-based cosmology complement as well as bind us into one Earth community with respect for diversity? The cosmological possibilities lead us to some questions for reimagining education.

BOX 8.1 COSMOLOGY—IMPLICATIONS FOR REIMAGINING EDUCATION

Imagine...

What if we had living and learning places and/or spaces that rooted us, educators and learners from age 1 to 100, in the mythic Deep History of the cosmos as well as the spatial cosmology of the planets and forces of Gaia?

What is communicated through sacred geography, architecture, and timekeeping before anything is ever said? How might a circle cosmology be manifested?

What if we participated in an Earth-based spirituality that saw these cosmic workings as sacred and revered, orienting our physical movements and psychic orientation?

Out of a sacred Deep Time cosmology, what kind of human civilization might unfurl as a result?

A Relational, Participatory Universe

Evolution biologist Elisabet Sahtouris and chemist James Lovelock say that such a grand cosmic story requires humans to "recognize their lack of maturity in believing we can manage the planet" rather than follow its lead.[181] Drawing deeply from her Greek heritage, Sahtouris sees the Gaian sensibility as coming full circle back to Greece, from where Western civilization began and the myth of Gaia originated.

The human response to the cosmos, she says, should be EarthDancing, dancing as a basic form of human expression. For her, EarthDancing should be the nature of human participation in cosmic processes—adaptative, innovative, lightfooted, and rhythmic, continually finding new steps and configurations that work harmoniously. Such participation as a dance partner in the cosmos is appropriate, she says, as participation is one of the central organizing principles of the universe,[182] contrary to passive observation, intrusive intervention, or the violent reaping undertaken by Western civilization. Learning the language and patterns of Gaia is part of EarthDancing or, rather, human responsiveness as a new ontology. Following Earth's creativity, we too can engage in the creativity principle, redesigning human habitation and ways of being that follow the dancing patterns of life.

For Sahtouris, we belong in this cosmos. She explains that this means holding two perceptions simultaneously: that we are part of a "single being," the cosmos, as well as made of self-bounded cells, a single organism.[183] David Bollier and Silke Helfrich, who promote the power of the *commons*, use the term "nested-I"[184] to mean overcoming the isolated individual identity and seeing our entanglement in the whole. We are differentiated in unity, meaning we are distinctive while simultaneously profoundly related, what Carol Gould calls "communal individuality."[185] Various contemplative pedagogies are available to assist in a perceptual experience of the nested-I.[186]

Reorganizing education in a way that follows the patterns of life and creates conditions for participatory "intra-acting"[187] in the flow of life is part of a relational ontology. Educational philosopher Barbara Thayer-Bacon talks about the weft and warp of ontology and epistemology, especially in education. "We cannot divorce ourselves from epistemological and ontological questions for they form the very weaving of the nets we use to catch our everyday concerns and give them meaning."[188] She explains that woven together, ontology and epistemology overcome dualist thinking. Together, they comprise our net of understanding. They are taught from the moment babies are born.

> Before they take their first step or say their first word, they have been bathed in the sounds, smells, tastes, and touches of their family and community. Already we have been naming their experiences and giving meaning to them. And they influence us as well.[189]

Thayer-Bacon turns to several nonwestern sources of nondualist thought as well to see how this weaving of ways of being and ways of knowing might give life rather than continue to destroy life.

She acknowledges the underlying Ocean of energy, that we emerge from, float on, and sink back into at the end our lives, undergirding all reality. She describes that we throw our nets overboard, and it is our concepts that bring meanings up to us, letting other meanings slip through. Each of these onto-epistemological nets is different, culturally shaped, part of the radical pluralism of meaning systems across the world. Each is partial and limited. Each sees only what their conceptual and perceptual frameworks capture in their net. In addition, there is much that cannot be articulated or fully captured, only experienced as implicit knowing,[190] perhaps enacted through the arts.

In trying to move from an ontology "based on Science, Judeo-Christian, and Reason,"[191] Thayer-Bacon, like many, is looking outside this self-referencing system. Two of Thayer-Bacon's sources for relational ontologies are Indigenous ontologies as well as Buddhist ontology. As she explains, ontologies have generally been conveyed through spiritualities or religions, providing an explanation of reality. In secular Western civilization, we largely depend on both ideologies and scientific theories, sometimes alongside religion, to explain and give meaning to reality, often holding these in tension. Thayer-Bacon suggests Indigenous peoples, who have had "longtime-developed relational nets,"[192] can be instructive and assist in becoming more aware of one's own net.[193]

Cherokee Marilou Awiakta emphasizes that cultural stories, which comprise these nets of meaning making, are highly contextual, evoking a particular culture on a particular land base in communication with the spirit world in that place.[194] Nevertheless, it is important to explore the nature of these mythologic stories, while honouring them as the compasses that Land and Spirit gave to particular people. The way forward is *not* appropriating the myths of nonwestern and Traditional cultures, but to let them resonate and lead us into resuscitating or recreating myths deep within our own ancestries, appropriate to the land where we find ourselves as well as the lands we came from. This is the work of both colonizer descendants and settler descendants.

Stories and the Mythologic Mind

In her study of comparative religion, Karen Armstrong asserts that "Human beings have always been mythmakers"[195] because we are "meaning-seeking creatures."[196] Now, however, "mythical thinking has fallen into disrepute."[197] As the Modern mind is so literal, myths have been dismissed as tall tales, fairy tales, fanciful fables, or imaginary worlds, considered irrational and unreal. It is a hard task to revivify the mythological mind.

For the most part, Armstrong says, traditional myths are "an art form"[198] that speak to the numerous planes of existence, profile timeless human fears and longings as well as convey universal experiences such as suffering, death, pleasure, and bliss, all part of the human condition.

Myths are not history or factual. They were never meant to be universal, given their contextuality. They do, however, pierce to the core of reality, offering "timeless truths"[199] that reach into both the soul and spirit. They share universal truths using local or ethnic symbolism that speak to a People.[200] Myths can have a transformative impact, hence their importance to ontology, shifting ways of seeing reality. There is never just one version of the same myth, as it continuously changes to speak to new conditions[201] and new generations. If it fails to do this, then the myth dies, and the shell that is left has little contextual framework from which to breathe it back into life.

Redeveloping a spherical, mythologic mind out of a flattened, linear mindscape is a key aspect of transforming and reorienting our onto-epistemology towards Life. Drawing from her Hungarian and Latina heritage as a *cantadora* or "keeper of the old stories," Clarissa Pinkola Estés suggests that "the old knowing is long overdue."[202] She calls this a process of "psychic-archeological digs,"[203] in reinterpreting myths and facilitating a visceral "knowing of the soul," part of healing from the Machine mind.[204]

Myth offers a spiritual container for the deepening understanding of the human experience. A mythological mind is in touch with the instinctive psyche. A mythological mind enacts deep meanings through daily activities, nourishes the development of wisdom over a lifetime, and participates ritually in the sacred, living universe.

Importantly for Westerners, it enlarges the sense of self beyond base self-interests and the demands of the ego, towards a storied life with dignity and larger purpose.

A storyteller from the Appalachians, Marilou Awiakta, tells some of the *Cherokee* creation story of *Selu*, the Corn-Mother. These are the stories she often heard as a child, when Big Stories are seeded in young minds. The delight of, and learning process in, stories, she says, is that they create images for the eye, sounds for the ear, and rhythms for the heart.[205] As listeners sit with the story over time, it works as a compass, ready when its wisdom is called for. In Indigenous communities, some creation stories take over a week for a full ritual telling[206] and years to evolve deeper levels of understanding.[207]

Indigenous Origin Stories, like many religious texts, convey the origins, development, and history of a People within the aliveness of their Earth place, within the cosmos. It conveys their way of apprehending reality. These Big Stories are spiced with human, nonhuman, and otherworldly characters with myriad interactions. They are not only teaching ethics or ways of relationing, but also they set up a vibrating energy field that teaches at deep, embodied levels. For this reason, Origin Stories have traditionally been oral. They are considered alive, carriers of power, responsive to each moment when and where they are told. *Sto:lo* Jo-ann Archibald says that stories "take on their own life" and "become the teacher" as they awaken powerful and creative healing energy.[208]

Thus, traditionally, one facet of education has been through continual oral history and storytelling, so all members of a community are constantly in a learning relation. Education has traditionally been village based, carried out by adult mentors and Elders[209] who are specific knowledge holders. Just like seed, the story is planted by Elders in the minds of young children to later bear fruit and mature wisdom. These stories hold central teachings, including the idea that the natural world holds the Original Instructions for each being and the purposes of, and shape of, good relationing between the human and natural worlds.

The stories encapsulate instructions, yet go far beyond simple lessons or spiritual imperatives to guide living. Myths were the original transformative process, where the myth transports one out of the here and now "toward some ineffable experience."[210] These stories have untold depths, which cannot be understood through the surface mind or with just one hearing.[211] This is different from understanding stories literally or metaphorically—stories as "moral lessons." *Mayan* shaman Martín Prechtel says writing stories down petrifies them—they need to remain oral and respond to the people, locale, and moment.[212] He continues, offering just a glimpse into this way of thinking:

> We owe our lives to the story it tells because it tells us into life. Then when we die, we're still in the story and die for a good reason, and the story doesn't die for our having kept it alive by the way we live.[213]

In this way, Big Stories contain "wisdom of the deep mysteries of life."[214] Often, these Big Stories and associated art forms embody "a whole philosophy of life."[215]

The symbolism, metaphors, relations, plot lines, and energy speak to certain people in a specific material circumstance and historical era. Generally speaking, says Joseph Campbell, "deities are personifications of the energies of nature."[216] Stories from different places may contain the same gods/powers, called by multiple names. Or there may be different gods/powers all acknowledged in one locale. In all cases, myth connects the spiritual, human, and natural worlds, recognizing that the power and energy that course through all three have the same source, reverberating through each other.

Thus, stories have been vital teaching forms. On the surface, they convey types of knowledge, a moral order to live by, a remembered communal past, and an identity for a sense of belonging as well as teaching at deep levels. There are many kinds of stories: stories of origin; stories about the nature of reality; stories that include quests, initiations, maturation, and death/rebirth processes; stories about morality, conduct, and character; stories of balance and harmonizing of body, spirit, and mind; and stories of learning to "walk in a good way," being in good relations, just to name a few. Many of these stories have been lifted out of their natural context, desiccating their meaning. Yet, says Prechtel,

> A real mythologic story is never about something, it's a schematic for how to do something. The story is woven into your work, your landscape, your being. Maybe you don't remember it, but its telling remembers you back to life.[217]

Thus, stories are immortalized in song, art, dance, and poetry—from cave paintings and pictographs to distinctive music, dance, ceremonial regalia, regional dress, and daily tools. The sacred as found in a place is woven through these storied forms, "membering" humans into their place. You live the mythology in all that you do. In part, this explains the tragedy of the theft of Indigenous land and belongings and the urgent need for the repatriation of land and ceremonial and other stolen items for all Indigenous peoples. This is one concrete manifestation of what is meant by "decolonization."[218]

So, myths are so much more than "lessons." They are a form of mythologic thinking that speaks to dimensions beyond the material. In fact, Campbell suggests that myth is a secret opening to the energies of the cosmos. "This is an important point about symbols: they do not refer to historical events; they refer *through* historical events to spiritual or psychological principles and power that are of yesterday, today, and tomorrow, and that are everywhere."[219]

Ritual and Reenchantment

Family therapist and ceremonialist Linda Neale suggests that it is time to resuscitate our ceremonial roots. Imagine, she says, "a society in which the sacred is not separate from everyday life."[220] Sociologist Max Weber coined the term disenchantment in the Modern age to mean the progressive loss of cosmic meaning to a rationalized form of living. In Traditional societies, myth together with ritual and ceremony have been tightly woven pedagogically, enchanting life and community daily with deeper meaning, learning, and higher purpose. *Chippewa* Sun Bear explains,

> When humans participate in ceremony, they enter sacred space. Everything outside of that space shrivels in importance. Time takes on a different dimension. Emotions flow more freely. The bodies of participants become filled with the energy of life, and this energy reaches out and blesses the creation around them.[221]

The primary purpose of myth, ritual, and ceremony is contact with the Sacred, pondering life meaning, seeking guides for moral action and/or healing, and remembering individual and community identity. As Cajete suggests,

> Native cosmologies . . . and relationships . . . must be remembered and re-remembered through art, ceremony, story, ways of community and personal experience. We human beings are forgetful creatures, and we need cosmology, philosophy, and lived experience at a personal and communal level to remember our life-sustaining relationships.[222]

Ritual consists of repeated symbolic actions with spiritual meanings.[223] Rituals can include postures, movements, bodily motions, sounds, and multiple levels of consciousness. Space and time are transformed into a "locus of power"[224] as symbols are activated and a spiritual experience facilitated. Ritual can be individual or collective. An example of an individual ritual is offering prayers of gratitude to the rising sun and the whole matrix of life each morning. When collective, myths speak through ritualized tellings and doings, such as honouring ancestors. Myth may also explain the rituals that are carried out in a ceremony, or the ceremony may enact a myth through multiple rituals.[225]

Of the African *Dagara* tribe, Malidoma Patrice Somé asserted that "where ritual is absent, the young ones are restless or violent, there are no real elders, and the grownups are bewildered. The future is dim."[226] For him, ritual is "anti-machine" in that it is "not compatible with the rapid rhythm" of industrialism, which in part is why societies "touched by progress" abandon rituals.[227] It is also why so many are spiritually searching and why other kinds of rituals prevail, such as political rituals or sports rituals, in place of spiritual ritual that reenchants.

Ceremony is typically a set of participatory actions that are witnessed by the cosmos. Ceremony has a performative and participatory aspect, as a re-enactment of a cosmological story, a conducting of important rites of passage, or facilitating a direct experience of the Sacred. As Neale describes, ceremony is most often for "keeping the world going"[228] in celebrating and participating in the spiritual, celestial, and Earth movements. The human purpose, then, is participation in this great rhythmic symphony of life. Theologian Matthew Fox exclaims that "The great mysteries, the great moments of our lives as individuals and as community members, have always been marked by ceremony."[229] Thus, ceremony has always been collective among humans and the cosmos.[230]

A ceremony is a full bodied, intellectual, emotional, psychological, and spiritual experience, penetrating levels of consciousness to the deep unconscious where new meanings are tapped and released. Ceremony is the original ritualized transformative process. Ceremony typically features death/rebirth, quest/return, or creation/dissolution motifs[231] as pathways of transformation. Ceremonies are structured and collectively engaged, so that life passages or healing are explicitly acknowledged. One type of ceremony that only has faint echoes within the educational system is coming of age or initiation ceremonies.

Rites of Passage: Becoming—Over the Life Walk

Philosopher Charles Taylor suggests that to find some idea of 'the good', we must understand our lives in narrative form as a quest.[232] Some myths are specifically reserved for meaning quests and important life passages, but often speak across multiple stages. Ceremonies can translate a stage of growth into a collective experience.

The first stage of adulthood has often been ritualized as a coming of age.[233] Even within adulthood, whether moving into young adulthood, elderhood, or into spiritual, governance, or social leadership positions, the transition to a new status or role is often embedded in ceremony. For instance, mentors might be offered to the newly married, the newly with children, those becoming part of a council of leaders, or those reaching elder status, offering teachings to assist the transition. These are some of the oldest forms of adult and lifelong education. Education for children would lay the preparation for these moments.

In initiation ceremonies, participants pass from one "universe" into another, as a profound changing of one's life.[234] Initiations are meant to be difficult. They can be times of testing before sacred knowledge can be passed on and a new stage of life embraced. As Paula Gunn Allen says, "Spiritual discipline is the hallmark of any ritual path."[235] Each step of the initiation process is an opportunity for a transformative experience. Thus, transformation on a lifelong spiritual path is understood as "perpetual self-renewal," always in process.[236] There are four general steps in an initiation: a crisis or dramatic loss; entering liminal space, despair, darkness or lostness out of which develops knowledge and wisdom; a moment of epiphany; then a slow restoration to a (new) sense of self and into community.[237]

Finding Face, Footing, Heart, and Kin

In many cultures, upon birth, each child is welcomed by the community as an important Being, a gift to the community. Each child is understood as containing all the knowledge that they require for their Life Walk.[238] The purpose of education, an *educere* approach, is assisting young people in "remembering" themselves, with the assistance of one-on-one guides or mentors. The mentor "sees" the mentee—a form of deep recognition—and works to awaken their dormant knowledge.

Each child is understood as bringing something of value, a gift, that must be coaxed out into the open, for the benefit of the community.[239]

Cajete says that the *Tewa* understanding of education is " 'breathing in life' or to 'be with life.' "[240]

> The understanding and perspectives that native people have about the life process of learning are coded within their stories, their designs and traditions of art and dance, within their languages, communities, and expressions of science and technology.[241]

Cajete explains that one central Mexican *Nahutl* poem lays out the tasks of a "true and rightful education:" first to find one's face or identity; second to find one's heart which is the divine spirit as well as source of desire or passion that motivates; third is to find one's foundation or footing which is their work or vocation; and fourth is to find one's kin relations—to self, family, clan, tribe, place, Land, natural world, and cosmos.[242] Then you become a whole person, with harmony between inner and outer realities. This ought to be the task of all education and educators in some way. Malidoma Somé emphasizes this sense of service to the larger community,

> For when we see with spiritual eyes, we remain in service to nature; we see nature as the originator of our tools, and we know that our tools or our technology must be used in harmony with nature's design and purposes, which are to maintain and serve the individual and community.[243]

In these ways, education is not an acquisition for the purpose of individual development, with calculative personalities demanding the delivery of goods (marks and know how) for monetary exchange or as "my right." Neither is education the task of fitting humans as tools into industrial cogs. Rather, knowledge and knowing is "becoming" into one's responsibility to the larger community and cosmos, a giving rather than a taking. It is an evocation of what lies inside, passing through challenging thresholds experienced throughout one's life, to benefit the human and the greater-than-human communities.

The 4Rs: Respect, Reciprocity, Restraint, and Response-abilities

Within such a cosmo-ontology, one basic task of education is to teach the 4Rs of respect, reciprocity, restraint, and response-abilities. Learning is not only studying pre-existing "texts" of meaning and "disciplines" with bodies of knowledge. It is also unfolding one's own meaning in dialogue with the natural world and all other beings.

The foundation of learning ought to be one of respect. The etymology of the word "respect"[244] is from the Old French and Latin *respectus*, meaning to regard or observe, as the act of looking back often. When we hold respect for a being or force, we esteem them and are entranced. We want to know them better, remaining in their company. We offer "especial attention" or doting interest. In other words, we understand ourselves as being in relation to this being or force, learning from them. Thus, we are continually drawn towards them. It is a constant returning to deep beholding, watching with great admiration.

Unfortunately, in the Modern world, what tends to garner our respect are money and status and those with other kinds of power in social hierarchies. This informs most of our current storytelling, including entertainment and news. People who are outside of these privileges need to "earn" our respect. In other words, we do not begin from the position of respect, unless we have something to gain. This is part of the calculative personality.

Similarly, we approach learning with a grasping mind demanding that we know, so it becomes part of our consumptive mindset. Erich Fromm explains that the grasping mode is part of an acquisitive society, the having mode, which is never satiated.[245] In this view, teachers need to make the knowing easy for consumption as their part in the "contractual" obligation with the learner. Individual power comes from being someone who knows something, a knower, a collector of knowings, with knowledge in the bank which can be traded for something else. Brazilian Paulo Freire called this "banking education"[246] where the knowledge is deposited from the teacher to the student who as "objects" of education passively regurgitate the knowledge on an exam in exchange for the receipt of marks. Knowledge in this way is a "thing," a commodity, not living, and not creative. This is exactly what we mean by a "knowledge society," knowledge that can be bought and sold in the global market. In this way, Freire says, we are creating "beings for another"[247] in a prescriptive cookie cutter process which is dehumanizing, rather than for the unique purpose they are to fulfil within their communities or, what he calls, "the vocation of becoming more fully human."[248]

Starting with an attitude of respect begins by understanding ourselves in relation to all living beings and forces. In this way, we are part of all we study, and they are a part of us. We approach the learning delicately, as we wish to be well in relation. We want to get closer, to be let into the mystery like a lover or a moth circling a flame. Parker Palmer calls these dynamics "subject-centered" education where a question, observation, being, or force (not an object to be disassembled or a discipline of pre-established knowledge to be consumed) is placed in the middle of a learning circle and approached carefully, with great regard.

> If we want a community of truth in the classroom, a community that can keep us honest, we must put a third thing, a great thing, at the center of the pedagogical circle. When student and teacher are the only active agents, community easily slips into narcissism, where either the teacher reigns supreme or students can do no wrong. A learning community that embodies both rigor and involvement will elude us until we establish a plumb line that measures teacher and students alike—as great things can do. . . . The subject-centered classroom is characterized by the fact that the third thing has a presence so real, so vivid, so vocal, that it can hold teacher and students alike accountable for what they say and do. . . . the power of a subject that transcends our self-absorption and refuses to be reduced to our claims about it.[249]

While I would replace the word "thing" with being, force, or query, Palmer is able to illustrate that there is a reciprocity relation among learners, educators, and the subject of focus, where each is not discrete. We are not studying a dead world, but one that is alive and active. In this way, we show restraint, not harming while seeking to understand, utilizing multiple modalities. Similar to Nobel Prize winner Barbara McClintock and her study of corn,[250] Kimmerer says that in studying plants, she approaches them as companions and teachers. This helps shape our responsibility—both our "ability to respond" appropriately within the relation and the nature of our response in ways that are restrained and nonviolating.

David Bohm offers another pedagogical pathway. He developed what has become known as Bohmian dialogue in the latter years of his life. It is a process whereby a group begins to explore a topic, their initial positions, and their contrasting values and then naturally evolves to explore deeper personal assumptions and reactiveness. What appears into view are the deeper structures of consciousness and societal programming, often held collectively.[251]

He felt that fostering a "nonjudgmental curiosity" about one's thinking is to "see things as freshly and clearly as possible."[252] In the process of dialogue, it becomes clear that reactivity and defensiveness undermine relations, particularly collaboration and friendship. Relations are central to dialogue, contrary to the debate of ideas, which often violates relation. Bohm himself describes what emerges naturally through this process:

> A new kind of mind thus begins to come into being which is based on the development of a common meaning that is constantly transforming in the process of the dialogue. People are no longer primarily in opposition, nor can they be said to be interacting, rather they are participating in this pool of common meaning which is capable of constant development and change . . . Going further along these lines would open up the possibility of transforming not only the relationship between people, but even more, the very nature of consciousness in which these relationships arise.[253]

Bohmian dialogue centres on the understanding of consciousness as collective thought and collective communication. Bohmian dialogue is constituted by:

- listening in a different way,
- suspending assumptions momentarily,
- identifying representations that we take as reality rather than as re-presentations,
- proprioception or becoming aware of thought as well as of one's body as sources of knowledge, and
- learning to follow the flow of meaning in both.

Proprioception can lead us to reflective discernment, which is of vital importance when dealing with the paradoxes that currently confound us. These are just two brief examples of bringing the 4Rs into a reimagined education.

On Being Orphans and Becoming Re-memberers

For Westerners who have been orphaned from their ancestry(ies) and thus a sense of deep belonging, education can take the form of retuning one's sensibilities to older stories and older traditions. It is respecting the stories and traditions of others, while reclaiming *your own* ancestral traditions, whatever may remain or however they have been mingled. It is regrowing traditions in beautiful, multidimensional ways in relation to the land, forces, and beings where you find yourself.

"The People,"[254] as most Indigenous people understand themselves, are remembered as being placed upon their land by the Spirit world. At some point, all of us were indigenous to a particular place. However, many of these histories have been lost through multiple Empires and waves of colonization. My own Indigenous roots were lost during the Roman Empire when the Romans successively conquered Europe, and then St Boniface Christianized the eastern Germanic tribes, calling us pagans. The Ancient Germanic culture, with its rich stories, mythologies, poems, runes, songs, and histories, was collected, and then the first Holy Roman Emperor Charlemagne ordered them to be burned at his death in 814 CE. Even in the 20th century, I grew up with warnings to avoid anything with pagan roots, divorcing me from my deepest cultural and spiritual roots. There are only a few remnants left of this ontology, found in Scandinavian texts or in our veins as blood memory.[255]

We can become Re-memberers, remembering our own indigenous roots from our ancestral home(s). We may find that we have blood or body memory[256] or can tap other extrarational sources of knowing. Those who still reside in their ancestral lands, perhaps despite waves of colonization, can relearn their history and re-place themselves within their ancestral home by re-membering their ancestral relations, further enlivening these relations. We do this alongside Indigenous people working hard to prevent loss of their stories, songs, rituals, and traditions.

Implications for Reimagining Education

This ontological discussion leads to additional imaginal questions in relation to education.

BOX 8.2 ONTOLOGY—IMPLICATIONS FOR REIMAGINING EDUCATION

Imagine...
What if we saw ourselves as participating in the creativity and rhythm of the cosmic story—as EarthDancing? How might we view ourselves differently?

If we saw ourselves as "nested-I's" or as communal individuals, emerging from the underlying sea of energy and returning to it, how would our daily behaviours change?

What kinds of meanings do our "nets" currently bring up for us (from the science, Judeo-Christian, and reason net)? What kinds of meanings could our "nets" bring up if we were open to Relationality?

How might we redevelop a mythologic mind? Where might we find artful Big Myths authentic to ourselves—the kinds of mythologic stories that teach us cosmic principles, wisdom, and deep belonging? Where might we find these storytellers of Old? How might we be the Re-memberers who access this needed knowledge through multiple ways of knowing?

How might we engage learners of any age in meaningful rituals that begin to foster a sense of contact with the cosmos, the Sacred, and their purpose within the larger community?

What kind of transformative ceremonies might be created that engage adults, from adolescence to elderhood, in the important transitions and passages throughout adult life?

How might we revalue the birth of each child and assist them in their Life Walk? How might we help them discern their face, footing, heart, and kin matrix? What kind of teaching and learning—in the spirit of *educere*—would this evoke?[257]

How might our learning manifest the 4Rs of respect, reciprocity, restraint, and response-abilities and thereby reembody our relations?

Kinship Ethics: A New Moral Order

> We are as much alive
> as we keep the earth alive.[258]
>
> Chief Dan George, Coast Salish

Indigenous Origin Stories often include Original Instructions which are instructions for relationing well in a connective human and natural context. Gregory Cajete says that the *Lakota* phrase *mitakuye oyasin* means "we are all related."[259] This phrase and its meaning are shared broadly among Turtle Island Indigenous peoples, spoken in an array of languages. Rather than individual rights, this understanding enables kinship with all other beings, with a sense of responsibility toward the "entire world around them."[260]

Original Instructions, First Teachings, Covenants

Joseph Campbell suggests that myths are ritual roadmaps, where an individual or group is "performing an act . . . in accord with the order of the universe."[261]

Whether hunting, planting, herding, trucking, serving food, or doing office work, there can be ritual actions related to Original Instructions, First Teachings, or Covenants. These have been understood as being received from the spirit world for living in harmony with the natural, human, and spirit worlds. Rituals provide humanity "a way" for enacting respect, honour and "giving back" to that which gives life to humans.

Anishinaabe/Cree/Métis/Norwegian professor, Melissa Nelson, defines Original Instructions as

> the literal and metaphorical instructions, passed on orally from generation to generation, for how to be a good human being living in reciprocal relation with all of our seen and unseen relatives. They are natural laws, that, when ignored, have natural consequences.[262]

In other words, there is a basic pact between people, place, and the spirit world to enact principles that ensure the continued regeneration of life. The warning about consequences speaks about our current reality of climate change and the damages of extractivism, all brought on by activities motivated by human arrogance, profit, and wealth seeking.

For Indigenous peoples, *Seneca* Chief Oren Lyons (of the *Haudenosaunee* Confederacy) says

> Our instructions, and I'm talking about for all human beings, our instructions are to get along. Understand what these laws are. Get along with laws and support them and work with them. We were told a long time ago that if you do that, life is endless. It just continues on and on in great cycles of regeneration, great powerful cycles of life regenerating and regenerating and regenerating. If you want to tinker with that regeneration, if you want to interrupt it, that's your choice, but the results that come back can be very severe because again, the laws are absolute.[263]

This is a beautiful definition of deep sustainability.

Pacts or covenants existing in other religious systems include the Ten Commandments, with similarities across the Torah, Bible, and Quran, and other guides such as the moral precepts of Buddhism. However, these systems do not tend to consider other species as equal partners in the web of life. Indigenous cultures have understood "in profound and elegant ways that 'we are all related.'"[264] It is time to expand our ethics or axiological scope, including within religious systems, which many theologians and religious leaders are now addressing.[265]

Further, ethical systems have common attributes for human behaviour across Indigenous cultures.[266] Cajete suggests that these ethics relate to the quality of person, their depth of being, and personal ideals. They also include communal ideals and spiritual ideals.[267] One of the primary concerns was one of "communal harmony in rituals, deeds, and thoughts."[268] One rich example is the *Ojibway* Seven Teachings, which are:

- Love (a gift from Eagle),
- Respect (a gift from Buffalo),
- Courage (a gift from Bear),
- Honesty (a gift from Sasquatch),[269]
- Wisdom (a gift from Beaver),
- Humility (a gift from Wolf), and
- Truth (a gift from Turtle).[270]

These gifts, understood as manifested in the spirit of various animals or forces, are connoted by the capitalization of the name. By referring to these animals as Spirit Animals, constantly reminds one of their kinship and therefore their responsibilities to the human and the more-than-human. Cajete suggests that forgetting these relations and principles "has led to the ecological crisis we see today."[271]

From his Christian background, Thomas Berry concurs that "This discontinuity between the human and the nonhuman breaks the covenant of the universe, the covenant whereby every being exists and has . . . value in relation to the great universe community."[272] Columbian anthropologist Arturo Escobar agrees and suggests that what is needed once again is an acknowledgement of the

> relational matrix [. . . which is] a politics for an other civilization that respects, and builds on, the interconnectedness of all life, based on a spirituality of the Earth, and that nourishes community because it acknowledges that love and emotion are important elements of knowledge and of all of life.[273]

For Escobar, it is our "act of emotioning" that is the biological basis for existence, given that "our desires . . . determine the kinds of worlds we create."[274] This is a call to give up our need to control and amass, which informs our desires for wealth, status, and power. He says that it is recognizing that "a civilization based on the love of life is a far better option . . . than the project of death."[275] For him also, this revives a matristic culture, respecting and honouring women, their leadership, and the Feminine cosmic principle. It revives human dignity, as the quality of relationing which overcomes human ego and the illusion of separateness. This is hard personal work needing deep attentiveness and reflexivity, as Western impulses operate in the opposite direction.

Escobar calls for a "transition imagination" to enable us to move towards this kind of flourishing, which he calls *Buen Vivir* or well-being, unique in each place.[276] This includes moving past instrumentalized forms of education where people are trained to surrender to "an altogether powerful economic, political, and technocratic elite,"[277] towards vernacular educations unique in each place, and towards developing a web of learning commons, both discussed in subsequent sections.

Land as Teacher

> The land, language, and stories are not lost. The people don't lose them. They're there. It's the people who get lost.[278]

The word "Indigenous" in Latin means "born of the land" or "springs from the land," which is the essence of Indigenous identity,[279] as Land is the home that holds a community. Land also holds the bones of ancestors over millennia. Many say that Land also holds wisdom, and if we enter into respectful relations, it can teach us. So, knowledge and knowing occur within specific eco-spiritual relations, orienting Indigenous people in space-time.[280] Indigenous education engages an explicit cosmology conveying a very specific and profound connection between a land, a people, and the spirit world, unique to that place.[281] *Mohawk* Sandra Styres explains that to Indigenous people, the Land is "living and emergent."[282] Land, Country, or Territory is capitalized, as

> all my Elders say that it is Land that holds all knowledge of life and death and is a constant teacher . . . the Land constantly speaks. It is constantly communicating. We survived and thrived by listening to its teachings—to its language—and then inventing human words to retell its stories to our succeeding generation.[283]

Land is understood as a living being where one is a self-in-relationship with Land, with attendant responsibilities. The goal is to stay in a direct dialogue with the natural world, as Land and all entities are living.

Language derives from Land. In fact, Pretchel says that "intact natural people" live *for* their Land, instead of *off* it or *on* it.[284] The Maya can never begin speaking without first "talking into a mental field of vision" that emanates from the storied Land.[285] He says that each Land has distinct sounds, and so humans have distinct sounds in their language, a tonal ecology that emanates from their *natureculture* matrix. The Traditional Maya keep their world alive through the "beauty of the motion of their speech," feeding the spirits of the land.[286]

The Land not only constitutes human identity but is also an important referent, as Indigenous people and many "Oldtime" people observe the patterns, cycles, and events—from cosmic movements to animal migrations, cycles of plant life, and seasons. Regularities observed over a long period of time are not "laws" but patterns and rhythms, as the only "constant is change."[287] They watch for the nuances and subtleties as you would in a human loved one. Land is not wilderness needing taming. Land in a well state is "full of life energy," says Cajete.[288] The "real test of living" is establishing harmonious relations.[289] He says that "[t]hrough the seeking, making, sharing, and celebrating of these natural relationships, [Indigenous people] came to perceive themselves as living in a sea of relationships."[290]

Thus, a matrix of relations emerges from within an ecosystem, people, language, historical events, and cultural customs.[291] All life forms are part of the spiritual family, where humans perceive themselves as living within a big Life process. The physical features are part of "its mythic and spiritual meaning,"[292] and this shapes the character of the people and their way of being. In fact, to know you, Indigenous people ask what Land you come from, meaning what is your matrix of relations?

Your relations include ancestral relations. One *Plains Cree* Elder explains: "We don't speak in lines. We speak in circular motion thinking, . . . our sound system is a neural pathway to our memory [. . . that] we inherited throughout cellular memory. Indigenous people learn through listening."[293] In *Sand Talk*, Yunkaporta describes this as following the songlines or "maps of story carrying knowledge along the lines of energy . . . webbed throughout the traditional lands of the First Peoples."[294] Thus, the Land carries the vital life forces of the ecological matrix begetting a language which carries this force, emphasizing the importance of language protection.

Words carry power,[295] and so syllables carry certain vibrations that echo the land and waters. Martín Pretchel explains the importance of language is "speaking deliciously" to the Land, as part of ritual honouring, maintaining a "courtesy relationship."[296] Cajete agrees that "Native cultures talk, pray, and chant the landscape into their being. This is the animating power of language inherent in the spoken word that connects the breath of each person to the breath of the world."[297]

Further, Land is an extension of one's body. Each child given to a tribe is considered to embody a divine spirit and the spirit of Earth, the joining of a spirit and soul.[298] Over time, a group of people become unique as they are "formed to some extent by the distinct climate, soil, geography, and living things of a place" and the spirit energy that runs through it.[299] Land is thus a map of one's soul, so that the natural world participates in the "ensoulment" of humans.[300] Just as humans witness the many faces and temperaments of the Land, so these are parts of themselves. The emergent archetypes are personified in stories and characters. In this way, their relationship with land is a "co-creative relationship."[301] This reverence for Land "allowed for survival,"[302] which is something we need to consider at this pivotal moment, also called "literacies of the Land."[303]

Land education is different from place-based education. Sandra Styres differentiates among space, place, and Land/Territory/Country. Space is a "continuous area or expanse that is free, available or unoccupied,"[304] more an abstraction. Place is a gathering of "experiences and histories, even languages and thoughts . . . and the trajectories of inanimate things,"[305] a concrete entity. People are "emplaced" by their interactions through the depth of place names, stories, songs, and teaching that develop from "deeply intimate and spiritual expressions of our connections to Land."[306] As Edward Casey suggests, "we are not only *in* places, but *of* them."[307] Thus, place still refers to a physical geography as well as to all the beings and ecologies therein.

Land is all this, but, in addition, is also "the underlying . . . Indigenous philosophical construct."[308] In this way, Land not only embodies both space and place but also all the "principles, philosophies and ontologies and making of place."[309] For this reason, Land is "a sentient and conscious being."[310]

> For those who want to live in deeply sacred and intimate relationship to Land must understand that it first and foremost requires a respectful and consistent acknowledgement of whose traditional lands we are on, a commitment to journeying—a seeking out and coming to an understanding of the stories and knowledges embedded in those lands, a conscious choosing to live in intimate, sacred, and storied relationships with those lands and not the least of which is an acknowledgement of the ways one is implicated in the networks and relations of power that comprise the tangled colonial history of the lands one is upon.[311]

Yunkaporta warns that, particularly in nonindigenous settings, there "is always a polite acknowledgement of connection to the land rather than true engagement. It is always about the *what*, and never about the *how*. "[312] He says that sinking into "Indigenous pattern-thinking, connective thinking" requires a different way of being in space, place, and Land. This is often not well expressed in the English language, except through stories, imagery and "yarning."[313] If we take on this deep challenge to go beyond politeness, as *Mohawk* Taiaiake Alfred says, colonials "will be unable to function as colonials"[314] moving from denial, disrespect, or romanticization to more appropriate and respectful ways of being in relation.

Languaging Relationality and Animacy

> Remove just and it from your vocabulary—they are intended to trivialize things and make them less alive and smaller than you. Instead learn to use language that makes the world bigger and more alive![315]

Moving towards life-giving societies requires an honest examination of our present language, concepts, cultural perceptions, and imagery in everyday language as well as in science and social theory.[316] Indigenous scholars and teachers consistently describe the difficulty of understanding their relational cosmology and ontology through the English language. *Potawatomi* Robin Wall Kimmerer[317] is most eloquent on the limitation of language by identifying how the English language structures thinking in terms of linearity and object-drivenness.

Kimmerer described her frustration as she was learning her native language in all its complexity. This gave way, however, with her discovery that Indigenous languages are verb-based, not noun-based.[318] While the English language calls all beings and things which are not human an "it," meaning an object, Indigenous language uses grammar that conveys kinship (subject–subject relations) and animacy (living entities as constantly in motion). The only thing that is called an "it" in Indigenous languages is a human-made "thing." Living beings and forces cannot be "things." Kimmerer says, as all beings and elements are alive, it is disrespectful to call them "it" just as you would not call any member of your family an "it." Thus, in some cultures, the Sun may be referred to as Father or Brother, the Earth as Mother, the Moon as Sister or Grandmother, and the Four Winds or Thunderers as Grandfathers, as this form of address establishes a kinship relation. The word is capitalized as it is expressive of both a cosmic principle and the quality of aliveness of that entity. Such language structure allows for kinship thinking.

Furthermore, transformed grammar can express animacy, as "a world of being, full of unseen energies that animate everything."[319] It acknowledges the multitudes of nonhuman languages in which humans are embedded, from the language of the wind to leaves to birds. From the language of a lake or a stone, instead of seeing an object, the language structure recognizes subjects speaking with each other. In sum, it acknowledges the animacy of the world. In Potawatomi language, 70% of words are verbs. In this way, a fox is "streaking redness," and an ocean bay is "being a bay held by shores." The word combinations convey being-in-action. In English, only 30% of the language are verbs, and they grant agency only to humans.

Indigenous languages often do not have gendered distinctions, hard to describe in English. Kimmerer summarizes the impact, especially on children.

> When we tell them that a tree is not a who, but an it, we make that maple an object; we put a barrier between us, absolving ourselves of moral responsibility and opening the door to exploitation. Saying it makes a living land into "natural resources." If a maple is an it, we can take up the chain saw. If a maple is a her, we think twice.[320]

I remember well when my 4-year-old daughter could not understand how a horse or the air could be "owned." Her instincts were such that property relations, which create "it" relations, made no sense. So, Kimmerer says that a part of our work, natural to children, is learning to speak the language of kinship and animacy. This can be done by learning an Indigenous language or other old languages that keep these understandings alive. It can also be done by shifting the English language away from "it" language towards "verbing." The terms emotioning or relationing are using such process-based languaging.

In sum, Gregory Cajete describes that "Native languages invest their homeland with their presence through the active verb-based process of 'talking the land.' That is, naming its places, singing its virtues, and telling its stories."[321] As he suggests, we can begin to "pay special attention to the way the event or place they are describing is in a perpetual state of motion."[322] Animacy and verb languaging convey Life.

Intelligibility of All Life

> Using he or she in place of it isn't to humanize the animals, but animalize the humans. This isn't anthropocentrically referring to the world as human, but reidentifying humans as part of a living world.[323]

Consciousness exists at the most primary level of existence, says philosopher Christian de Quincey.[324] Thus, all living systems and entities experience and have an intelligibility. Participating and relating within this greater consciousness are the participatory aspects of the cosmos, becoming a quantum self within the coherent, conscious whole. De Quincey explains,

> [W]e will need to cultivate an alternative epistemology beyond the faculties of rational analysis and conceptual understanding . . . developed in many non-Western traditions, such as Taoism, Buddhism, Hinduism, and Shamanism. The spiritual or mystical practices of such traditions employ nonrational participatory epistemologies that enable them to make contact with "alternative realities" . . . and to develop worldviews based on ontologies radically different from the Modern Western paradigm of materialism and mechanism.[325]

He asserts that these models can be sources of inspiration. Thus, healing the *nature-culture* split and *bodymind* schism requires different forms of knowing and relating.

Some traditions, as de Quincey says, identify body centres of knowing, not just reason (head), but also emotion (heart), intuition (gut), and will (solar plexus).[326] In fact, according to biophysicists and neuroscientists, we have three brain centres, if you think in terms of neural cells.[327] The heart has approximately 40,000 neural cells, similar in construction to those in the brain, but these neural cells make up 60–65% of the cells in the heart. It is 100,000 times stronger electrically than the brain, acting like the sun around which our body revolves.[328] So it is a powerful centre for receiving and processing information, as the heart sends more messages to the brain than vice versa. In this way, the heart generates the most powerful electromagnetic field in your body, a heart quantum field. Thus, compassion is the most fundamental practice towards all living beings.[329]

De Quincey goes on to say that process philosophy assumes "that the fundamental ontology is not substance but events with *experience*."[330] Thus, "the ultimate entities are essentially *feelings*,"[331] or felt experience. Consciousness, as a quality of being, "would be an emergent property of experience."[332]

> The body literally incorporates the world . . . experientially, through acts of physical prehension—acts that take into the moment of experience the completed forms of prior moments. . . . This process view of the mind, creatively integrating input from the body and brain through a hierarchy of levels, provides a solution to the long-standing problem in philosophy of mind.[333]

In other words, we are embodied consciousness. "Nature, the cosmos—*matter itself*—is inherently and thoroughly meaningful, purposeful, and valuable in and for itself. Nature, we must see, is sacred."[334]

This leads to a new paradigm:

> [a] sense of the cosmos where all bodies, including human bodies, resonate and participate in the web of cosmic matter, matter that is intrinsically sentient and capable of nonlocal communication of feelings, where one body's being literally incorporates the being of other bodies both local and nonlocal.[335]

Traditional Indigenous perceptions of animals and plant life are similar, where "all matter" has the capacity for feeling and experiencing. There is a "way of knowing the presence of consciousness in other beings."[336]

As Cajete suggests, "our human intelligence [is] a microcosm of the intelligence of the Earth and the universe—with the heart and mind."[337] As the world is constantly in flux, Native science does not attempt to firmly categorize, but rather understand the nature of each thing.[338] As de Quincey suggests, the heart, instincts, intuition, and rational mind each are sources of knowing. *Cree* Patricia Steinhauer-Hill agrees that intelligences of the body, mind, heart, and spirit are all vital, requiring balance, particularly the balancing of heart and mind.[339]

Each animal, bird, insect as well as every type of plant has its own "spirit" or living energy. The goal is to learn the internal logic of that life form, as part of being in a relationship that co-creates. Other beings and entities can be observed and studied—the sounds, motions, habits, special qualities and powers, and growth patterns—a form of mentoring in how to work and communicate with the life form and their way of being in that place.[340] Individuals can develop special relations with some species, even hearing them "speak." In this view, the natural world is as much a teacher of new ways of knowing and being as other humans. We come to see our bodily boundaries as perforated, pulsing energy and open to communication within the web of life.[341]

Learning, then, is coming to know the intelligibility of our world. It is presencing—bringing one's attention into the present. It is "intra-acting with the intelligibility of the world, which goes beyond the contents of our minds toward felt participation and presencing[342] in a constantly creative universe."[343]

As described in Chapter 6, de Quincey says we can do this through the four gifts of knowing: the philosopher's gift of rationality, the scientist's gift of the senses, the shaman's gift of participatory engagement, and the mystic's gift of communion. Integrating these ways of knowing and levels of consciousness brings much fuller, richer forms of learning.

This understanding also nuances ideas of "rewilding as environmental education."[344] The world in its naturalness should not be thought of as being "wild" as defined from within the colonial, dualist view. This view engages in dualist differentiating of what is wild or tame/domesticated, uncivilized/savage or civilized, and urban/suburban/superurban or rural and wilderness. Rewilding echoes the original conservation movement that emerged from and for colonialism in protecting spaces from its own extractivist dynamics, rather than from people indigenous to the area who may have resided there sustainably for millennia.

The "wild" is better understood as the matrix of each being or entity going about their natural business, wherever they are found. It is ensuring that the natural world can go about its business in areas untouched by Western humans or touched respectfully. Such areas need significant expansion until the Western world has learned restraint. What also makes the "wild" is the energy or living spirit that flows through all entities, the spark of life animating them. In response, Marc Bekoff calls for "rewilding our hearts," which is rebuilding the compassion required for coexistence with our animal and plant kin and for creating conditions for ecological repair.[345]

Rather than thinking of nature as co-teacher, it might be useful to think of the natural world as co-learner, where there is reciprocal agency. *We* need rewilding so that our spirits and ways of being return to a state of vitality that has not been groomed, anesthetized, and comfortized through a consumer society. As the Land contributes to our ensoulment, so the state of the Land reflects the state of our souls.

If we pull on one strand within a matrix to study, we cannot forget all the relations comprising the natural world we are engaging. Mapping these relations and looking for patterns are ways of knowing which do not let us forget the relational matrix. Learners gain an understanding of "the set of essential relationships and principles [their web of life] which parallel those of transformational and holistic education."[346] Tapping other levels of consciousness and ways of knowing in learner explorations as well as leaving room for nonlinearity, emergence, and spontaneity are important in rewilding pedagogies. Each modality of exploration unveils another facet of the whole.

Becoming Makers

> In a corrupt society you learn how to survive by taking.
> In a good society you learn how to make beauty by hand and
> farm food in beautiful clothing. In the best of societies
> you learn to bestow the wonderfulness you never received
> on a time of hope beyond your own.[347]

Learning occurs from within activities, not from an objective distance as a spectator, as one participates in the natural world as a porous being. Learning that involves traditional handwork, particularly in an intergenerational context, can help to resuscitate a variety of traditions and tools appropriate to contexts. Not only do students of all ages learn basic skills long forgotten within various cultures, also they understand how cultural traditions and ways of living and working emerge from communication and reverence within a natural context. It probes the

4Ds of dress, dance, diet, and dialect to resonate much deeper—understanding the kinship and animacy from which they emerged.

As Tyson Yunkaporta describes, Aboriginal knowledge in what is now Australia "endures because everybody carries a part of it, no matter how fragmentary."[348] He describes hand gestures, making, and oral culture exchanges through yarning as "a traditional form we have always used to create and transmit knowledge."[349] As a way to express and enact knowledge for his book, he carved bark shields and used sand talk or images drawn on the ground, "all part of a big story or meta-story that connects and extends all over Australia through massive songlines in the earth and sky."[350] Aboriginal knowledge uses pattern-thinking as a "framework for the understandings needed in the co-creation of sustainable systems."[351]

Similarly, Batchuluun Yembuu describes that the Traditional Ecological Knowledge (TEK) of the Mongolian pastoral nomads is quickly disappearing despite being "one of the last surviving nomadic civilizations . . . and one of the most sustainable ways of life in the world."[352] The globalization of education in the 1990s intensified the adoption of a Western model of schooling and English language training for the employment of Mongolians in the global economy. Traditional ways are now "museumized," increasingly shared only at a university dedicated to culture and the arts. Even more destructive, they are commercialized for sale. In this way, the intergenerational knowledge and language which were passed systematically from one generation to the other, in families and communities, are being lifted out of their natural and spiritual matrix of Life. Students are increasingly alienated from their traditions, considering these traditions "backward." Many modernized Mongolians are now leaving Mongolia altogether to pursue employment elsewhere. This "learning to leave" as the outcome of schooling has been noted in other places too, such as rural Canada.[353]

The task, Yembuu says, is reknitting this rich context back into place with the Land, while honouring Western and Indigenous epistemologies equally. She sees some reclamation occurring in schools, but it is particularly strong in community-based education programmes. While Elders do not pass their knowledge to their own descendants, they have an opportunity to pass it onto the interested younger generations. This is vital as the "forgotten culture and knowledge of nomads cannot ever be restored."[354]

As the Canadian anthropologist Wade Davis has been arguing, the ethnosphere and biosphere are integrally linked.[355] Only when these ways are once again revered as being rich examples of life-giving societies—societies that privilege harmony as a core principle—and when honest thought is given to bridging these epistemological systems, will we move towards a new era of truly sustainable societies.

Implications for Reimagining Education

This axiological discussion leads us to some imaginal questions in relation to reimagining education.

BOX 8.3 AXIOLOGY—IMPLICATIONS FOR REIMAGINING EDUCATION

Imagine...
As part of a transition imagination, what might a life-oriented ethical system—a covenant or Great Instructions for living on Earth—look like?

What should comprise basic Relationality ethics across all humans and nonhuman life? How can these be contextually grown? How can we help evoke humans of greater depth through educational practices?

How might we feed the spirit of Land, our sea of relations? How might we learn to feel the pulse of the Land as part of our own bodies, as basic felt experience?

What if we used all our centres of knowing, experiencing the consciousness of all other beings and sensing their flowing spirit or life energy?

What if we removed the word "it" from all our languaging about the living world? What if we removed the word "thing" as much as possible from our languaging regarding living entities and forces? What if we devised new languaging to illustrate beings-in-motion?

What if we became makers once again, learning original human skills, such as weaving, ropemaking, felting, knitting, beadmaking, tracking, wild harvesting, gardening, pottery, firemaking, and carving, just to name a few? How might this re-establish our relations with each other and with the living world? How might this meet the many needs of children and adults for active learning? How might this reinstitute the original multitasking, listening, or relationing while engaged in making?[356]

Epistemology as Process, as Wisdom-Seeking

Process Learning and Teaching

Rivers and water give us a sense of flowing—a process way of learning and knowing, which flows with the animacy and relationality of the natural world, a flowing dance of life. As Capra and Luisi assert, "mind and consciousness are not things but processes."[357] "Cognition is a process of knowing,"[358] which occurs in most living entities. "Primary consciousness" emerges from within "basic perceptual, sensory, and emotional experience."[359] In other words, basic sensing and feeling are part of the endless processing or phenomenal consciousness of any living entity.[360] Some of this knowing "nonconsciously maintains the state of the body"[361] for survival. In this way, "cognition involves the entire process of life."[362]

For us, this form of cognition is not located in the brain but exists throughout our human body. Not only basic body processes but also emotions play a critical role as "complex patterns of chemical and neural responses that have specific regulatory functions," providing a range of "survival-oriented behaviours."[363] So a part of process learning is learning from this range of sensory apparatus, valuing the role of bodily sensing and the connections between consciousness, emotions, and the physical body.

Auroville in India[364] is an international community whose schools provide a learning environment where learners can

> become conscious of their own perceptions and abilities so that they may become self-aware, self-directed individuals. . . . The programme works by first bringing the individual into a state of receptivity in which s/he can better "listen" to the many and varied inputs s/he is continuously receiving.[365]

Thus, the process of being alive *is* learning, adapting, and responding to one's environment. So, "to live is to know."[366]

As process-relational philosopher Robert Mesle elaborates, knowing is also worldmaking—participating and evoking, not acting as an objective, passive observer who then represents phenomena. "Cognition, then, is not a representation of an independently existing world but rather a continual bringing forth of a world through the process of living."[367]

The broadly assumed Judeo-Christian perspective in the EuroAmerican worldview has been that the future exists and that we are living into this ongoing story. While "the future may be unknown to us . . . it exists—is already fully actual—for God."[368] In this Western view, "God" is the "ultimate unchanging reality."[369] If one could rise above to God's perspective, an Archimedean point, then one could see the story ahead. This is a "static view of time" where the ultimate reality or ending of the story is already known from a transcendent viewpoint.

In a process view, "the future does not exist." Nothing is preordained or known but waiting to happen. "The entire universe [is] bursting into existence in each moment."[370] Therefore, the continual flow of our experiencing and sensing is creative in each moment. In this way, instead of concrete entities, "the world is composed of events and processes."[371] As Bohm explains, "knowledge is an integral part of the total flux of process" so that one is "becoming of knowledge" or continuously dipping into the everchanging, everflowing stream of knowledge.[372]

Further, relatedness "goes all the way down to the roots of reality."[373] Because we are processes not things and, as processes, integrally linked to all other processes, we are constantly "Becoming." This view is similar to many Eastern spiritualities, where "time and change are illusions."[374] Process theologian David Ray Griffen suggests that the Divine "is best understood in terms of relatedness and process rather than as an unchanging, static Being unaffected by the world."[375] Mesle argues that this integration of metaphysics, religion, science, and ancient worldviews has the potential to generate a shared vision that we can live into.

In terms of education, Swedish educational researcher Simon Ceder suggests that this "relationality . . . is the point of departure"[376] for educators. "Relations are the very condition for education"[377] as they are constantly making us and constituting us. In this view, learning is inherent in living. Thus, learning emerges from within events, happenings, and doings as we relate to other people, the natural world, physical objects, and texts.[378] Education is the process of learning within the intelligibility of all that is around us in which we are embedded.[379] Therefore, the learning place is the whole cosmic order and Earth community.[380] From within these relations, we can see the multitude of ways we are attracted through our relations.[381] This relationing, or attraction, is the starting point for learning.[382]

Our relationing begins with our educational intentions, intention being ideas with purpose. This establishes the way in which we engage and how our attention is focussed. Our intentions set up a force field that can bring us into creative, organizing energy, as it courses through the focus of our attention. We are drawn into a mutuality where learners as subjects and the subject of our attention are responding to each other, unveiling facets and unfolding meanings. This is the matrix of creative possibilities from which learning emerges.

Physicist Karen Barad[383] says that subjects do not precede their interaction but emerge through their intra-acting, as various possibilities become materialized. As we intra-act, we can study agency as manifested in performances, enactments, or doings-in-action.[384] We do not engage in reductionism but in exploring multiple facets of becomingness. What we learn is just one small part of the Big Story that is constantly unfolding and re-enfolding.

As energy flow is the substrate of the universe and substance of life, following energy is part of learning. Educators have rarely addressed the role of energy in the learning process—how it is generated, how it impacts learning, and how it is an enabler of new practices. The very intentions we take into the learning relation, which have their own energy, shift and shape the interchange. Yet, the intention is also to allow things to take their own course rather than bend consciousness. This can allow for emergence or spontaneity of learning, leading to new levels of understanding.

Scientists now see rhythms of wavelike movements in the cosmos, bringing them to reconsider the Renaissance notion of "the harmonies of nature" or "the music of the spheres." In this way, it is following the "vibration resonating through the instrumentation of created forms"[385] as a process of knowing. For instance while controversial, the

> superstring theory . . . postulates that the fundamental energy that enlivens everything in the universe may be compared to the vibrating energy that occurs when we move the bow over a musical string, the music being the "voiced" language of the silent energy.[386]

Looking for movement, rhythms, and intra-actional patterns will likely lead to learning much differently from reducing matter to successively smaller bits and observing "static matter."

"Morality is tied up with breadth of vision"[387] in that we now witness intra-connectedness, seeing that our actions affect every other element and entity, even in our learning process. We slow down our learning, so we become thoughtfully, respectfully part of this flowing dance of life.

Beyond primary or core consciousness, humans have reflective consciousness. If experience and consciousness go all the way down, then

> elementary drops of feeling are organized into successively more complex forms, like molecules and cells and animal bodies, central nervous systems and brains, the complexity of those feelings will increase until it crosses a crucial threshold into conscious self-awareness such as you are having right now.[388]

With core consciousness, our feelings are primary. With the reflective consciousness, we narrate our experience, out of which emerges a more stable concept of self. We form an idea about ourselves over time, creating an autobiographical self.[389] From this form of consciousness, emerges "mental images . . . values, beliefs, goals and strategies"[390] which make us human and out of which human community springs. Using metaphors, imagery, and other conceptual structures, we language our experience. Through language, behaviour, and social relationing, we bring forth a world with others, one that is constantly moving.

Religious studies professor Christopher Bache advocates working with fields of consciousness in educational settings. Our minds are consciousness fields which are open and porous systems.

> While not losing their individual integrity, human minds "flow" and "merge" with other minds to form larger wholes—group fields. Though generated by individual minds, these fields have a life of their own that persists even after the minds that contributed to them have moved on to other activities.[391]

If there is an underlying collective consciousness, our individual minds rest within it. Learning can be accelerated for individuals by activating the learning that resides within the collectivity. While our "minds are distinct, they are not private."[392] In this way, educators can work with these group fields: "preparing the field, nourishing the field, using visualization exercises, closing the field . . . [as well as] connecting the fields."[393]

Bache calls the intentions he sets for teaching, the "seed catalyst" of the group field, much like how I have structured this book. I have written it with my full attentive consciousness and with intention that the seeds travel and resonate in your consciousness. In setting intentions and planting seeds, the educator then gets out of the way of the energy field, so that spontaneous happenings can occur within the learning community.

In nourishing the group field, the educator's "presence enters the room"/space,[394] meeting learners, trying to deeply understand their underlying themes, relying on images, and sense impressions. Educators are constantly developing or finding activities and ideas that can elevate the learning among them. The shifts in energy and learner feedback can be assessed, while staying aligned with the energy field.

Like Bache, I use learner visualization exercises to activate latent potential in the group field which settles bodies, opens learning pathways, guides attention, and synergizes divergence that may have been conflictual in the past. Using other levels of consciousness in this way can contribute to conditions for transformative processes. Various visual and other art forms are used for this purpose, too, inviting learners into being "makers." Using such activities, they remain open to the unfoldings of the implicate order,[395] sometimes experiencing quite remarkable insights and learning. Closing the field includes various rituals such as review games, inspirational material, offering thanks to each other, gifting, or closing reflective circles. As Bache describes, the "circle of learning between student and teacher is a flow,"[396] coming back to where one began.

Coming together in a learning place activates meaningfulness from deep within an "intelligent universe,"[397] if openness exists. Learners may continue some of the practices utilized with the group, as an energetic bridge to other parts of their lives. Or they may continue sharing learnings which keep them connected at the energetic level. Often, the full manifestation of such learning happens long after the educational gathering ends, vibrating and gestating. Bache concludes, "When the spark of connection jumps in my classroom, it feels as if a living consciousness that I am part of uses my knowledge to accomplish a purpose that lies beyond my personal horizon."[398] I too have felt this—an intelligence beyond my own, working the connectiveness for the greater good.

Wisdom-Seeking

As mentioned previously, Gregory Cajete talks about the process of education as " 'breathing in life' or to 'be with life.' "[399] In the *Tewa* way, "the very nature of any true educational process is developmental and transformational."[400] In other words, as Freire says, it is the process of "becoming human"[401] or, rather, moving into deeper and deeper levels of becoming human. Echoing this, Capra and Luisi suggest that in ancient languages, soul or spirit was considered "breath." For them "describing cognition as the breath of life seems to be the perfect metaphor."[402]

Spirit, as in the spirit that moves in all things or life energy, is considered the "ultimate moving force and source of all life."[403] In good education, learners are moved by their life energy onto a path that seeks for meanings behind meanings, deepening the understanding of life. As Cajete indicates, we successively learn who our relations are, how to get along with each other, how to engage the natural world respectfully, identify the skills gifted to us, and what we are called to contribute during our existence. Humans are always called towards knowledge but, even more deeply, towards wisdom as the harvest of a life well lived. For some ancient cultures, wisdom was "the principal thing."[404]

Wisdom, explains neuropsychiatrist Dilip Jeste and science writer Scott LaFee, has generally been considered a religious and philosophical concern.[405] However, through new findings about cognition, brain function, consciousness, and life, there are now new scientific bases for this discussion. Of course, smartness or high levels of intelligence do not necessarily relate to wisdom. Neither does age necessarily relate to wisdom, as wisdom can manifest across various ages, or not at all.

The term *homo sapiens* is Latin for "wise man [/human]"[406] which Jeste and Lafee suggest needs to be foregrounded as part of our needed evolution. Through comparative research across religions and cultures, they found that the concept of wisdom has remained stable across millennia. There are also significant similarities across cultures and religions.

As discussed in Chapter 6, using neuroscience research, they discovered some cognitive and biological roots of wisdom, particularly areas of the brain responsible for various attitudes and capacities, from emotional regulation to empathy. They suggest that emotion and cognition, feeling and thinking, must be *balanced* for wisdom to emerge.[407] Further, they say "[r]eflection—the fixing of thoughts on a particular topic with careful consideration—is foundational to wisdom."[408] Moreover, "becoming wiser is a process."[409] Not only does it emerge over life experience as a practical result of navigating challenges and moral conundrums, but biologically based wisdom can also be increased through the plasticity of the brain. Finally, wisdom is not only individual, but societal wisdom can be fostered as well. These, then, are aspects that educators can foster in learning contexts. As described in Chapter 6, the six primary attributes of wisdom they identified are: prosocial attitudes and behaviours, emotional stability with happiness, balancing decisiveness with acceptance of uncertainty, self-reflection and understanding, social decision-making with pragmatic knowledge for living, and spirituality.

FIGURE 8.10 Agia Sophia Church/Temple, Kardamyli, Greece

Credit: Permission to reprint from 123RF, Standard Licence.

Harkening back to the Ancient Greeks, they discuss *sophia* as theoretical wisdom and *phronesis* as practical wisdom. Based on their research, they offer a variety of insights and practices in the quest for these forms of wisdom. In particular, they discuss the relation between individual and societal wisdom. They make the case that global society is both less and more wise today, based on happiness measures and human development indexes. Typically, they observe that if a society makes wise choices and is protective of their members, the need for individual wisdom is less. The corollary is that with less societal wisdom, the more the need for individual wisdom. Given the declining level, quality, and breadth of education in many Western nations, Jeste and LaFee believe that there is less opportunity for the development of either individual or societal wisdom.

For them, education and wisdom are directly correlated, not as part of educational outcomes and achievement, but as one context where the capacities for wisdom can be developed.

> We must find ways to infuse the education of our young with lessons that promote the elements of wisdom: compassion and other prosocial behaviours, self-reflection, emotional regulation, openness to divergent perspectives, and ability to make good decisions.[410]

Education is about the journey to full personhood, becoming people of dignity who live lives of respectfulness and who make, as well as who can counsel, wise choices, as we move into an increasingly uncertain future. After posing some implications, let us turn to the last part of this discussion, education in relation to social change or, rather, processes for worldmaking.

Implications for Reimagining Education

This axiological discussion leads us to some imaginal questions in relation to reimagining education.

BOX 8.4 EPISTEMOLOGY—IMPLICATIONS FOR REIMAGINING EDUCATION

Imagine…

What if we engaged education as a process way of learning and cognition as breathing in life, as important as breathing in oxygen? What if learning entailed a slowing down and receptivity to the intelligence of all life around us?

If to live is to know, and our relations are the very condition of education, how might we perceive and focus on constant learning and relationing rather than acknowledging learning as only that which is prescribed and accumulated under the auspices of an educational institution? How might we break open these boundaried conceptions? What needs to be unlearned and relearned?

What if a learning space was perceived as an energetic force field? What if a learning space was perceived as a field of consciousness, not confined to individual minds? How might we approach the learning process differently?

What if we understood learning not only as developmental but also transformative, towards challenges that evoke full personhood?

What if the principal thing was evoking wisdom? What if educators more explicitly engaged in wise mentoring?

How might we transform educating and learning towards cosmic principles such as the multiple faces of balance, adhering to natural laws, and respect? What if we balanced *phronesis* and *sophia* every day in learning?

How might educators embrace their Great Work of connecting individual lives to the universe, thereby contributing to the transformation of human habitation on Earth?

Pluriverse Worldmaking: The Alchemy of Change

The Right Kind of Love

The right kind of love,
the silent, deep, and lasting one, pleases the Creator.
Sometimes there may be darkness
in our hearts that make us dread the future.
Young people should never overlook that this love
is working silently in a thousand ways.
Because of it we can have confidence
in the years that lie ahead.
Where this love lives
beauty grows.[411]

Chief Dan George, Coast Salish

From One-Dimensionality to Pluriversality

Over time, the Western civilizational model has flattened perceptual abilities, created linear minds, and stultified the energy of life. As Herbert Marcuse[412] intuited, we have become one-dimensional beings where we no longer see the irrationality and reprehensible nature of our current societal organization. While promising freedom from want for all, yet only delivering comfort to some and extravagant wealth to very few, human freedom and dimensionality have been desiccating. For Marcuse, part of one-dimensionality is the loss of discontent and opposition which is needed for fuelling a dialectical process of social change. So, one pathway of social change is dissent and resistance to dominant societal forms. It is reflective in the many new global social movements since the turn of the millennium, from the anti-globalization movement to the #MeToo, Black Lives Matter, Arab Spring, Occupy, and Idle No More/Land Back/Defend Life movements. As Naomi Klein claims, these movements are completing the social movement work of the 19th and 20th centuries.

Marcuse, writing during the existential nuclear moment in the 1960s but which is applicable in this existential environmental moment, foresaw two tendencies: one pathway "explodes the society,"[413] and the other contains the seeds of qualitative, transformative change. He argued for the Great Refusal, a "'NO' to repression and domination" in all its forms and a YES to "life instincts over aggressive instincts.[414] Similar to Paulo Freire, he says it is vital to both denounce the existing system as well as announce the new one,[415] keeping announcing and denouncing in balance. Many now believe that creating new social forms, amid the dying forms, is where creative energy should be concentrated, as part of this transitional time. Yet, as Antonio Gramsci warned, "the crisis consists precisely in the fact that the old is dying and the new cannot be born; in this interregnum a great variety of morbid symptoms appear."[416]

Both resistance and visioning are occurring around the world in thousands of communities that are marginalized from advanced industrial society or have become the disposable ones for whom there is no place. These are the places where immense creativity is emerging. Columbian development theorist Arturo Escobar turned away from the "development enterprise," citing that it has been "literally scrapping the vernacular design and endogenous practices that for centuries had nourished, for better or worse, the lives of millions throughout the centuries." He calls "Western forms" the deliberate design of "elimination."[417] For him, what is needed is a reverse elimination process, eliminating the "structures of unsustainability" while investing energy in the "creation of new, nonexploitative forms of life."[418]

Originating from Mexico and India, respectively, Gustavo Esteva and Madhu Suri Prakash assert that Modernity is "being transformed by the non-modern majorities into opportunities for regenerating their own traditions, their cultures, their unique indigenous and other non-modern arts of living and dying."[419] Escobar, Esteva, and Prakash together call this pathway "vernacular design"[420] as opposed to global thinking monocultures, or what Escobar calls the "One-World World."[421] Vernacular here means "native, indigenous, not of foreign origin or of learned formation,"[422] where living and learning are indistinguishable in ways that "widen the competencies for living."[423]

Rather than submitting to the flattening processes of extractive capitalism and neoliberal globalization, Escobar, like many, advocates an ontological turn towards Relationality that includes "communal forms of autonomy."[424] This transition process is about walking away from Western forms and engaging "plural ways of making the world." Zapatistas see it as plural ways of thinking and plural ways of being which create a "pluriverse," where the many are welcomed.

Escobar suggests that social change needs to go "deeper than capitalism" and needs to go broader than structural change. It involves the entire constellation of ideas, materials, forms, structures, and systems,[425] as manifested in cosmology, ontology, axiology, and epistemology. It is process-based with an emergent quality that cannot be predicted. Embracing such imaginal and creative dimensions is what generates this energetic pluriverse of alternatives. These alternatives begin from very different premises, and they begin in the small, local, and sometimes forgotten places.[426]

Emergence is "interaction among living agents [which] unexpectedly produces novel and more complex organization at larger scales."[427] This is more an alchemical process of change, of perturbations that create increasing amounts of instability until there is a slow dissolution of forms and accumulating losses. This shakes loose the old frameworks, enabling transformative energy to penetrate. A cleansing of illusions and an unlearning of habits occur on a social scale. Then, at the tipping point of maximum instability, with no energy or resources left to pursue business as usual, a spaciousness emerges, and a moment of bifurcation is reached. A digestion process breaks down ideas, habits, and social forms, reworking them into new forms. In this moment of liminality, intuitive leaps and even civilizational leaps[428] can occur, emerging as a reconstituted form with a new level of self-organization, a new dynamic balance that can generate life-giving cultures or, rather, cultures of perpetual self-renewal.

Revitalizing the Commons

In studying the hidden and forgotten places where new social forms are emerging, Escobar, Esteva, and Prakash as well as Bollier and Helfrich all see a revitalization of the commons. This has many forms, from degrowth, postdevelopment, relocalization, commoning, conviviality, transition communities, vernacular communities, and *Buen Vivir*[429] (good living or collective well-being initially from Andean communities). In commoning, "[e]very community practices the design of itself," as a form of autonomy that is life-enhancing for their region.[430]

Many Latin American examples were stimulated by the 1994 Chiapas Revolution in Mexico, begun in opposition to the rise of neoliberalism and the North American Free Trade Agreement (NAFTA). As a largely Indigenous *Mayan* region, NAFTA spelled the end to agrarian reform and democratic reform for them, alongside an intensification of extractivism. Rather than submit, they rebelled and called for their independence. Alternating between armed and diplomatic approaches, they declared their freedom of governance, establishing 32 autonomous municipalities. In 2005, they released their Sixth Declaration[431] which was their manifesto of their vision for Mexico and all dispossessed everywhere. As Esteva and Prakash ask, "What can others learn from a provincial movement of desperately impoverished and oppressed Third World peasants struggling for their cultures, shamed and silenced for five centuries?"[432] It turns out—quite a lot.

The Zapatistas solidified the extensive international support they received and began to re-establish "self-governance following their own indigenous traditions."[433] They were deliberately escaping One-World thinking which is part of the "Global Project." They established radical democracy through very localized, participatory processes. They declared economic autonomy resisting the "continual interference by market or state,"[434] devising their own economic processes. They proclaimed cultural self-determination and established decentralized mutual aid in many social spheres, from medical care to education. They began unlearning colonialism and relearning the ways of knowing, processes, and skills they had lost over the colonial years.

They have resisted becoming an "enlightened vanguard" or a "unique ideology." Rather, they stand as an example of a People taking back their living functions and processes while "evolving their own cultural notions of 'a good life.'"[435] In Summer 2021, they[436] sailed to Spain as a symbolic "reverse invasion," seen in Figure 8.11. They renamed the European continent SLUMIL K'AJXEMK'OP or "Rebel Land, Land that does not yield."[437]

Instead of carrying a flag and gun, however, they were carrying their *Declaration for Life*,[438] encouraging other marginalized groups who wished to reclaim their traditions and defend their Land (see Figure 8.12). In this symbolic way, they are turning the Doctrine of Discovery and the claim of *terra nullius* (empty land) on its head.[439] In their words:

> Down there, in the cities and the haciendas, we did not exist. Our lives had less value than their machines or animals. We were like stones, like weeds

FIGURE 8.11 Zapatista Travel for Life

Credit: Photo reprinted with the permission of Enlace Zapatista and Progressive International.

FIGURE 8.12 Delegation of the Zapatista Invasion

Credit: Photo reprinted with the permission of Enlace Zapatista and the Progressive International.

> growing by the side of the road. We had no word. We had no face. We had no name. We had no tomorrow. We did not exist.
>
> Today, thousands of struggles of the five continents struggle themselves here, in the mountains of the Mexican southeast, for life and against death . . . to announce a tomorrow of inclusion and tolerance . . . The beginning of the construction of a new and good world, that is, a world where all worlds fit [*quepan*].[440]

Many yearn for such meaningful community, given that the loss of community bonds was one of the early erasures of the Modern capitalist age. As Jane Jacobs says, we have lost even the memory of this loss, as we do not even know what we don't know. The tools, processes, and capacities need to be reinvented, inspired by those reclaiming their traditions from shreds of memory.

At a personal level, this is a learning of conviviality, learning to rely on others and be reliable, instead of relying on money, called "money-lite commoning."[441] This will require living relationally with a focus on social and natural wealth but material "enoughness." This will require living consciously in terms of slowing down and being mindful of one's rhythms and relations. This will require living simply with both a sense of inner and outer fulfilment that goes beyond material things, busyness, overcommitment, and debt towards finding abundance in beauty, elegant living, simple pleasures, being a maker and grower, and giving back. This will require living spiritually by shrinking one's ego needs and cravings and developing ritual practices of awe, wonder, and thankfulness. This requires developing a thick skin to detractors. This means challenging existing notions and language around beauty, success, independence, freedom, identity, work, money, wealth, ownership, and security, mostly based on the framework of individualism. This means countering language and tendencies towards hierarchy, competition, and all forms of aggression and violence, mostly based on predatory competitive capitalism. This means questioning concepts and language around strength, weakness, power, anger, sex, intimacy, consent, control, safety, shame, and fear, mostly based on patriarchy.

Collectively, it means the "recrafting of communal forms of knowing, being and doing,"[442] which lay the groundwork for communality. Bollier and Helfrich identify many global places where "commoning" is occurring. They have created a framework for commoning[443] on the basis of a triad of new practices in social life, peer governance, and provisioning (see Figure 8.13). The central premise is "primarily about creating and maintaining relationship."[444]

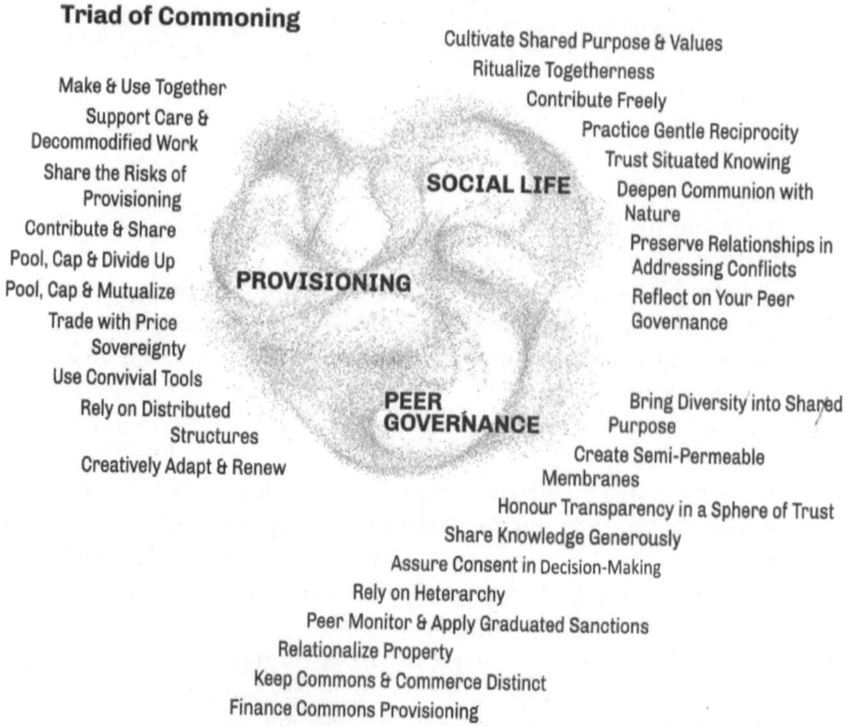

FIGURE 8.13 Triad of Commoning

Credit: Artist: Merce M. Tarres. Authors: Jack Bollier and Silke Helfrich, *Free, Fair and Alive: The Insurgent Power of the Commons.* Gabriola Island, BC: New Society Publishers. Reprinted with permission of the artist and under a Peer Production Licence.

It requires relearning 27 dynamics and practices of collective governance, working together, and relocalizing the meeting of community needs, which are needed to work through the sticky parts of commoning. Just as the Zapatistas assert, commoning is context-specific. While it is normal in many Indigenous cultures, in Western industrialized settings, it requires conscious and mindful self-organization.

Matristic Cultures and the Partnership Ethic

Both Marcuse and Escobar,[445] alongside feminist scholars, suggest one key pathway of change: that of addressing the 5,000 to 7,000-year-old system of patriarchy. The term patriarchy refers to social systems organized around the idea of father-rule, in which wealth, power, and privilege in family and society accrue to males over females. This usually includes patrilineal descent as the mechanism whereby name, wealth, property, and status are passed to sons in relation to the father's bloodline. Androcentrism operates to give males and their actions primary attention and importance.

Claudia von Werlhof,[446] among many, many ecofeminist scholars,[447] asserts that the present ontological shift includes a turn away from "patriarchal alchemy"[448] or the mode of knowing that is a "war against life" manifested in a conquering mindset; violence by hand or machine; and exploitation of women, children, and Earth—our Mother.[449] German developmental psychologist and educational researcher Gerda Verden-Zöller and Chilean biologist Humberto Maturana suggest that the Western civilizational culture arose "in the midst of a patriarchal culture centered on the emotioning of appropriation, control, mistrust, and arrogance . . . sexual and racial discrimination, and war.[450] They expand at length:

> Through the continuous expansion of patriarchality, religious doctrines and political philosophies replaced spiritual experiences as the basis for seeing and understanding human existence, and the natural world became more and more an alien realm that had to be dominated and submitted to linear causal control. As a consequence, daily life progressively lost its spiritual presence as the awareness of its cosmic connectedness was lost, desacralized by philosophical, religious, and scientific theories that placed a transcendental object or essence (such as God, truth, reality, matter or energy) at the center of all reflections about humanness, rather than the awareness of the systemic coherences of life.[451]
>
> Nevertheless, the very same expansion of the reflective thinking that has opened our desires for power and for domination over everything, and has alienated us from ourselves, opens for us the possibility for the conservation of loving humanness, if we indeed want it. We do not have to do everything that is possible for us to do, we can choose. We do not have to live all the realities that we may create, they are not equally desirable if one has self-respect and social consciousness.[452]

Related to Riane Eisler's work, Verden-Zöller and Maturana illustrate how the pastoral patriarchal cultures, who became the Indo-Europeans, invaded the matristic cultures of Old Europe which had a different biological emotioning pattern, that of trust, love, participation, gratitude, respect, and acceptance. They suggest that the matristic culture was not fully eclipsed, however, but preserved domestically between women and their children. They call for a (re)turn to "matristic cultures" where this configuration of emotioning, living, and relating exists once again and gives children primary importance.[453]

Thomas Berry's work converges with this, in that he considers the Earth as "the maternal principle out of which we are born and from which we derive all that we are and all that we have."[454] He says that we have lived a very long motherless period, partly due to the failure of Christianity to "recognize the Earth's spiritual qualities."[455] For him, the rising "importance of the feminine or maternal principle is . . . the emergence of a new age of human culture [which] will necessarily be expressed through the symbol of woman [. . . and] this mother-child relationship."[456]

Laguna Pueblo professor Paula Gunn Allen affirms that "traditional tribal lifestyles were more often gyneocratic than not" by which she means "spirit-centered, women-focused worldviews."[457] These systems are characterized by

> nurturing, pacifist, and [nonviolent] males, and self-defining, assertive, decisive women. . . . Tribal gynocracies prominently feature even distribution of goods among all members of the society on the grounds that First Mother enjoined cooperation and sharing on all her children.[458]

As educators, exploring and attuning to the Feminine and Masculine as cosmic principles can assist in profoundly shifting human relationing.

Reimagining Our Future, Reimagining Learning

A Web of Vernacular Learning Commons

> For years I labored with the idea of reforming the existing institutions of society, a little change here, a little change there. Now I feel quite differently. I think you've got to have a reconstruction of the entire society, a revolution of values.[459]
>
> Martin Luther King, Jr

What do pluriversality, commoning, and matristic cultures mean for learning? They mean, as educators, we are being given permission to think beyond the box of schooling. Learning has had a multitude of forms over the existence of humanity. While schooling is important for the existing system, it does not have to be the only way into the future. Further, it leaves out so many, while simultaneously crushing other forms of learning and ways of being. It will likely only ever "permit" small spaces for introducing a new cosmology, ontology, axiology, and epistemology. Nevertheless, teachers and educators can begin experimenting and find ways of engaging this new view of reality, this "OntoShift".[460] Many footholds have been provided throughout this book for your creativity.

If our task is to protect life, enhance life, and generate life-giving societies, we need to find imaginal spaces for worldmaking outside of schooling, finding pathways that include community-organized education and convivial forms of learning. This requires building up a parallel web of "learning commons" to the institutions of schooling and higher education. These learning commons will be "vernacular" to their places but connected to each other by intention and design, as a living web. Just as much of sustainability education has been most fruitful in nonformal and informal sites, such as afterschool programmes, NGOs, civil society organizations, and continuing education,[461] additional sites can be created for engaging communities and learners of all ages.

David Selby calls vernacular learning "place-based learning that is rooted in intimacy and a porous exchange with our natural surroundings."[462] It is "locally grounded learning" that connects us to the "spirit of place."[463] "A deep sense of place aris[es] out of a focus on the small, the particular and the specific [which] carries within itself the potential to link the learner to the universals in human experience."[464] The task will be unique to each place and based on the amount of penetration by Western social forms as well as the memory available for reclaiming and restoring alternative cultural forms. The Venezuelan environmental justice movement calls this awakening the "sleepy knowledge" of understandings and lived diversities outside of Western forms.[465]

Convivial forms of education refer to accessing knowledge and design ideas for creating tools and structures that "strengthen creativity and self determination."[466] It is where there is free sharing of learning according to local requirements, a way of learning that pre-existed schooling. Continued schooling is needed but should be completed by lifelong learning that is available in the commons outside of property relations, for designing a respect-based way of being and life-giving way of doing in a particular locale. It provides for "enlivenment" so that a sense of aliveness and vitality rather than passiveness and resignation is fostered. It is learning needed for commoning and regeneration rather than consumption, the basis of sustainable societies.

Finally, it requires a hospitality to diversity. As religion scholar Martin Marty explains, when faiths and other belief systems collide, what we must risk is hospitality to the Other.[467] This is not changing who we are but being open to dialogue to "act and speak with integrity and to regard others as doing so."[468] It is engaging the other in open-ended ways beyond doctrines and creeds. Parker Palmer encourages,

> Inviting the stranger into our private space, whether that be the space of our own home or the space of our personal awareness and concern. And when we do so, some important transformations occur. Our private space is suddenly enlarged; no longer tight and cramped and restricted, but open

> and expansive and free. And our space may also be illumined . . . Hospitality to the stranger gives us a chance to see our own lives afresh, through different eyes.[469]

While acknowledging that there will competing ways of life with contradictory beliefs and values, we can avoid culture and race wars by practicing hospitality in the most "public zones of life."[470] Dana Stuchul, Madhu Suri Prakash, and Gustavo Esteva agree that "the practice of hospitality, then, becomes the seedbed of hope" as fears can surrender to hope.[471] They draw from Ivan Illich who says that "I do think that if I had to choose one word to which hope can be tied, it is hospitality."[472] Hospitality comes from friendship, where one recovers "threshold, table, patience, listening, and from there generating seedbeds for virtue and friendship on the one hand—on the other radiating out for . . . (the) rebirth of community."[473] Stuchul, Prakash, and Esteva conclude that "The lifelong learning we propose is simple, yet profoundly sophisticated—to learn again to live together, to recover and to extend traditions of hospitality. Such learning cannot but occur over our lifetimes, cannot but be worthy of our living."[474]

Sprouting Seeds of Hope

Gregory Cajete concludes that "Choosing a life of seeking life is the central message of Indigenous cosmologies, and also reflects the deepest aspects of human nature."[475] We need to be "recreate[d] as 'people of hope.'"[476] Vaclav Havel, too, speaks hope into what is being birthed:

> Something is being born . . . [W]e are in a phase when one age is succeeding another, when everything is possible. . . . The only real hope of people today is probably a renewal of our certainty that we are rooted to the earth and at the same time to the cosmos . . . [I]t logically follows that in today's multicultural world, the truly reliable path to coexistence, to peaceful coexistence and creative cooperation, must start from what is at the root of all cultures and what lies infinitely deeper in human hearts and minds than political opinion, convictions, antipathies, or sympathies: It must be rooted in self-transcendence. Transcendence is a hand reached out to those close to us, to foreigners, to the human community, to all living creatures, to nature, to the universe; transcendence as a deeply and joyously experienced need to be in harmony, even with what we ourselves are not, what we do not understand, what seems distant from us in time and space, but with which we are nevertheless mysteriously linked because, together with us, all this constitutes a single [reality]. Transcendence is the only real alternative to extinction. Something is being born.[477]

It is a hard time to be alive. Some days, the grief and tangible losses threaten to engulf one, from climate change to biodiversity loss to human desperation. But the thread of hope unfolded through such imaginal possibilities provides images of living with the depth of which we are capable, where the living world is continuously regenerating strength to speak and be heard, where human communities are revitalizing mythic consciousness of origins and cosmic principles, and where our societies are becoming cultures of life-giving ways, stretching towards harmony within the full matrix of cosmic and Earthly life.

Hope resides in the (re)weaving of loose threads back into the web of life. Educators and learners filled with such longings will know that something *is* being born and something is "be-coming" from the seeds woven into the knots throughout the web of Life. In many hearts and places, the seeds are finding nourishing soil, their protective shells perishing to give way to tender new growth. Keep them protected, keep them nourished. One day, we may find we have lived into the prophecies of a new day.

Notes and References

Chapter line drawing by Canadian artist Erin Marie Lange Christensen. www.erinchristensen.com. Reproduced with permission.

1 E. F. Fritz Schumacher (1997), *'This I Believe' and Other Essays* (Dartington: Green Books, 1974). Fritz Schumacher, *Small is Beautiful: A Study of Economics as if People Mattered* (London: Penguin Group; ABACUS Edition by Sphere Books, 1973), pp. 76, 66.
2 Arturo Escobar, *Designs for the Pluriverse: Radical Interdependence, Autonomy, and the Making of Worlds* (Durham: Duke University Press, 2018), p. ix.
3 Escobar, *Designs for the Pluriverse*, p. ix.
4 Patriarchy here means a society governed by father-rule, in which wealth, power, and privilege in family and society accrue to males over females.
5 Ashish Kothari, Ariel Salleh, Arturo Escobar, Federico Demaria, and Albertao Acosta, *Pluriverse: A Post-Development Dictionary* (New Delhi: Tulika Books and Authorsupfront Publishing Services, 2019), p. xxiii.
6 Naomi Klein, *This Changes Everything* (London: Allen Lane, 2014), p. 4.
7 Escobar, *Designs for the Pluriverse*, p. x.
8 Escobar, *Designs for the Pluriverse*, p. 17.
9 Herbert Marcuse, *One-Dimensional Man* (Boston: Beacon Press, 1964), p. 1.
10 Marcuse, *One-Dimensional Man*, p. 1.
11 Marcuse, *One-Dimensional Man*, p. 189.
12 The phrase "reinventing the human' was used by Thomas Berry, *The Great Work: Our Way into the Future* (New York: Bell Tower, 1999), p. 159 cited by Escobar in *Designs for the Pluriverse*, pp. 139, 143.
13 Worldmaking here refers to how "language is a means for co-creating the world" as vocabularies shape perceptual realities. See David Bollier and Silke Helfrich, *Free, Fair and Alive: The Insurgent Power of the Commons* (Gabriola Island: New Society Press, 2019).
14 Rebecca Solnit, *Hope in the Dark: Untold Histories, Wild Possibilities* (Chicago: Haymarket Books, 2016).
15 I thank Katie Ross as a model for transitioning the English language from a noun language to a verb language, as a process approach. See Katie E. Ross, *Transforming the Ways We Create Change: Experiencing and Cultivating Transformative Sustainability Learning* (Sydney: University of Technology Sydney, 2021). https://opus.lib.uts.edu.au/handle/10453/149105. I give thanks to Joy Kcenia Polanco O'Neil for her writing on "becoming" as a process approach.
16 Charlene Spretnak, *Relational Reality: New Discoveries of Interrelatedness That Are Transforming the Modern World* (Topsham, ME: Green Horizon Books, 2011).
17 Spretnak, *Relational Reality*, p. 16.
18 Based on Spretnak, *Relational Reality*, p. 15 and also Bertell Ollman, *Alienation: Marx's Conception of Man in Capitalist Society* (Cambridge: University Press, 1971).
19 Thomas Berry, *The Great Work: Our Way into the Future* (New York: Bell Tower, 1999), p. x.

20 Berry, *The Great Work*, p. 1.
21 Frederique Apffel-Marlin, *Subversive Spiritualities: How Rituals Enact the World* (Oxford: Oxford University Press, 2011), pp. 3–4.
22 Berry, *The Great Work*.
23 Edmund O'Sullivan, *Transformative Learning: Educational Vision for the 21st Century* (Toronto: University of Toronto Press and London: Zed Books, 1999).
24 Thomas Berry describes the Ecozoic era as the time when humans are "present to the Earth in a mutually enhancing manner," *The Great Work*, p. 55.
25 The Anthropocene is understood as the era in which human actions are the key force impacting planetary well-being. Carl Folke, "Respecting Planetary Boundaries and Reconnecting to the Biosphere," in World Watch Institute (Ed.), *Is Sustainability Possible?* (Washington, DC: Island Press, 2013).
26 The term ecological civilization has been used by Herman Greene, *The Promise of Ecological Civilization* (Anoka, MN: Process Century Press, 2015) as cited in Escobar, *Designs for the Pluriverse*, p. 144.
27 The concept of cosmo-onto-axi-epistemological transformation refers to the changes needed in cosmology or the understanding of the universe, ontology or the understanding of reality and being, axiology or the moral and ethical system, and epistemology or the understanding of knowledge and knowing. See Katie E. Ross, *Transforming the Ways We Create Change: Experiencing and Cultivating Transformative Sustainability Learning* (Sydney: University of Technology Sydney, 2021 https://opus.lib.uts.edu.au/handle/10453/149105.
28 I am indebted to Dr Katie Ross for this concept of stretching, as applied to these spheres of inquiry. See Katie E. Ross, *Transforming the Ways We Create Change: Experiencing and Cultivating Transformative Sustainability Learning* (Sydney: University of Technology Sydney, 2021) https://opus.lib.uts.edu.au/handle/10453/149105. She borrowed the concept from Matthew Harmin, M. J. Barrett, and Carolyn Hoessler, "Stretching the Boundaries of Transformative Sustainability Learning: On the Importance of Decolonizing Ways of Knowing and Relations with the More-than-human," *Environmental Education Research*, Vol. 23, No. 10 (2016), pp. 1489–1500. https://doi.org/10.1080/13504622.2016.1263279.
29 Original source: Elizabeth A. Lange, Joy Kcenia Polanco O'Neil, and Katie E. Ross, "Educating during the Great Transformation: Relationality and Transformative Sustainability Education," *Studies in Adult Education and Learning*, Vo. 27, No. 1 (2021), pp. 23–46. https://doi.org/10.4312/as/9692.
30 The Latin concept *educere* means to lead out or evoke the wisdom held within by the learner through questioning, thinking, and creating. From Randall Bass and J. W. Good, "*Educare* and *Educere*: Is a Balance Possible in the Educational System," *Educational Forum*, Vol. 68, No. 2 (2004), pp. 161–168. ERIC: EJ724880.
31 Culture refers to "The ways of thinking, the ways of acting, and the material objects that together form a people's way of life." This includes material and nonmaterial culture. From John Macionis and Linda Gerber, *Sociology*, Eighth Canadian Edition (Toronto: Pearson Education, Inc., 2012), p. 58.
32 See the works of Joy Kcenia Polanco O'Neil (www.researchgate.net/profile/Joy-Polanco-Oneil) and Katie E Ross (www.researchgate.net/profile/Katie-Ross-19#) for their own interpretations and explorations of these principles as well as for their unique research directions.
33 The term Old Ways refers to the ways of living of First Peoples, wherever found. As an example, see Elizabeth Marshall Thomas, *The Old Ways: A Story of the First People* (New York: Picador, 2006).
34 Cajete, *Native Science*, p. 267.
35 Cajete, *Native Science*, p. 263.
36 Chief Dan George (and Helmut Hirnschall), *My Spirit Soars* (Surrey, BC: Hancock House Publishers, 1982/1989), p. 9. Inclusive language added.
37 Lisa Korteweg and Connie Russell, "Editorial: Decolonizing + Indigenizing = Moving Environmental Education towards Reconciliation," *Canadian Journal of Environmental Education*, Vol. 17 (2012), pp. 5–14.

38 Marie Battiste, *Decolonizing Education: Nourishing the Learning Spirit* (Saskatoon: Purich Publishing Limited, 2013).
39 Jazmin Murphy, "Decolonizing Environmentalism," *Yes! Magazine*, September 15, 2020. www.yesmagazine.org/environment/2020/09/15/conservation-decolonize-environmentalism.
40 Tim Frandy, "Indigenizing Sustainabilities, Sustaining Indigeneities: Decolonization, Sustainability, and Education," *Journal of Sustainability Education*, June 17, 2018. www.susted.com/wordpress/content/indigenizing-sustainabilities-sustaining-indigeneities-decolonization-sustainability-and-education_2018_06/.
See also the whole decolonization issue and Noel Gough, "Decolonising Sustainability: Subverting and Appropriating Mythologies of Social Change," *South African Journal of Environmental Education*, Vol. 18 (1998), pp. 3–13.
41 Melanie Goodchild, "Relational Systems Thinking: That's How Change Is Going to Come, from Our Earth Mother," *Journal of Awareness-Based Systems Change*, Vol. 1, No. 1 (2021), p. 79.
42 Kothari et al., *Pluriverse*, p. xvii.
43 Arty Mangan and A-dae Romero-Briones, *Decolonizing Regenerative Agriculture: An Indigenous Perspective*, Bioneers. https://bioneers.org/decolonizing-regenerative-agriculture-indigenous-perspective/.
44 Boaventura de Sousa Santos, *The End of the Cognitive Empire: The Coming of Age of Epistemologies of the South* (Durham: Duke University Press, 2018).
45 Linda Tuhiwai Smith, *Decolonizing Methodologies: Research and Indigenous Peoples* (London: Zed Books, 1999/2006.
46 Irmelin Gram-Hanssen, Nicole Schafenacker, and Julia Bentz, "Decolonizing Transformations through 'Right Relations'," *Sustainability Science*, Vol. 17 (2022), pp. 673–685. https://link.springer.com/article/10.1007/s11625-021-00960-9.
47 Gram-Hanssen et al., "Decolonizing Transformations through 'Right Relations'," p. 673.
48 Jocelyn Cheechoo, Ooshkahneekwayweuk, "Living What I Love Most," in Leanne Simpson (Ed.), *Lighting the Eighth Fire: The Liberation, Resurgence, and Protection of Indigenous Nations* (Winnipeg: Arp Books, 2008), p. 146.
49 Murphy, "Decolonizing Environmentalism," p. 6.
50 Leath Tonino, "Two Ways of Knowing: Robin Wall Kimmerer on Scientific and Native American Views of the Natural World," *The Sun Magazine*, April 2016, p. 2. www.thesunmagazine.org.issues/484/two-ways-of-knowing.
51 Gram-Hanssen et al., "Decolonizing Transformations through 'Right Relations'," p. 673.
52 Goodchild, "Relational Systems Thinking," p. 83.
53 Goodchild, "Relational Systems Thinking," p. 79.
54 Goodchild, "Relational Systems Thinking," p. 94. See also Fritjof Capra and Pier Luigi Luisi, *The Systems View of Life: A Unifying Vision* (Cambridge: Cambridge University Press, 2014), p. 70.
55 Goodchild, "Relational Systems Thinking," p. 81.
56 Kimmerer in Tonino, "Two Ways of Knowing."
57 Goodchild, "Relational Systems Thinking," p. 81.
58 Marilyn Iwama, Murdena Marshall, Albert Marshall, and Cheryl Bartlett, "Two-Eyed Seeing and the Language of Healing in Community-Based Research," *Canadian Journal of Native Education*, Vol. 32, No. 2 (2009); ProQuest p. 3.
59 Goodchild, "Relational Systems Thinking," p. 82.
60 Goodchild, "Relational Systems Thinking," p. 82.
61 Kimmerer in Tonino, "Two Ways of Knowing," p. 4.
62 Iwama et al., "Two-Eyed Seeing and the Language of Healing in Community-Based Research," pp. 4–5.
63 The term pluriversity here relates to the term pluriverse, from the Zapatistas, meaning where one world among many exists. See Escobar, *Designs for the Pluriverse*.
64 Kothari et al., *Pluriverse*, p. xxviii.
65 Kothari et al., *Pluriverse*, p.

66 Any errors in understanding and explanation are my sole responsibility. The word Teachers, as in the word Elders, is capitalized when it is a formalized relation in which one is recognized and respected for their knowledge and responsibilities as part of this role.
67 "Othering" refers to the process whereby an individual or groups of people attribute negative characteristics to other individuals or groups of people that set them apart as representing that which is opposite to them. Poul Rohleder, "Othering," in Thomas Teo (Ed.), *Encyclopedia of Critical Psychology* (New York: Springer, 2014). https://doi.org/10.1007/978-1-4614-5583-7_414.
68 Catherine Odora Hoppers, "Research on Indigenous Knowledge Systems: The Search for Cognitive Justice," *International Journal of Lifelong Education*, Vol. 40, No. 4 (2021), p. 312.
69 Hoppers, "Research on Indigenous Knowledge Systems," p. 312.
70 Hoppers, "Research on Indigenous Knowledge Systems," p. 312.
71 Hoppers, "Research on Indigenous Knowledge Systems," p. 315.
72 Boaventura de Sousa Santos, *The End of the Cognitive Empire: The Coming of Age of Epistemologies of the South* (Durham: Duke University Press, 2018).
73 Santos, *The End of the Cognitive Empire*, p. 6.
74 Vandana Shiva, *Stolen Harvest: The Hijacking of the Global Food Supply* (Cambridge: South End Press, 2000), pp. 1–2.
75 Hoppers, "Research on Indigenous Knowledge Systems."
76 Cajete, *Native Science*.
77 Marie Battiste, "Introduction: Unfolding the Lessons of Colonization," in Marie Battiste (Ed.), *Reclaiming Indigenous Voice and Vision* (Vancouver: UBC Press, 2000), p. xvii.
78 Hoppers, "Research on Indigenous Knowledge Systems," p. 324.
79 Marie Battiste, *Decolonizing Education: Nourishing the Learning Spirit* (Saskatoon: Purich Publishing, 2013), p. 14.
80 Korteweg and Russell, "Editorial," p. 7.
81 Goodchild, "Relational Systems Thinking," p. 99.
82 Marlene Brant Castellano, *Indigenizing Education* (EdCan Network, June 10, 2014), p. 1. www.edcan.ca/articles/indigenizing-education/#:~:text=The%20change%20I%20would%20propose,valid%20contribution%20of%20Indigenous%20knowledge.
83 Peter Cole, "Coyote and Raven Talk about Indigenizing Environmental Education: Or Reconfiguring the Shenangigans of Otis O'Dewey Esquire," *Canadian Journal of Environmental Education*, Vol. 17 (2012), pp. 15–29.
84 Diane *Kahontakwas* Longboat in Goodchild, "Relational Systems Thinking," p. 91.
85 Madonna Kolblenschlag as cited in Tom Hampson and Loretta Whalen, *Tales of the Heart: Affective Approaches to Global Education* (New York: Friendship Press, 1991), p. 2.
86 Frederique Apffel-Marglin suggests that the nonhumans are other living beings and material forces and that the other-than-human world are the spirits and otherworldly elements who "continue to be loved or feared or both, continue to accompany human beings in all their activities." Apffel-Marlin, *Subversive Spiritualities*, p. 6.
87 Elizabeth Ann Lange, *Living Transformation: From Midlife Crisis to Restoring Ethical Space* (Edmonton: University of Alberta, 2001), UMI Dissertation Services from ProQuest Dissertations Publishing, 2001, NQ68960.
88 I wish to thank Nova Poirier for her teaching on seeds.
89 For this teaching on seeds, I thank Martín Prechtel. For more, see *The Unlikely Peace at Cuchumaquic: The Parallel Lives of People as Plants: Keeping the Seeds Alive* (Berkeley: North Atlantic Books, 2012).
90 Wright, *A Short History of Progress*, paraphrased from p. 4.
91 The use of Big Story not only refers to the usage of Charles Eisenstein, but also invokes Indigenous philosophies. See *The More Beautiful World our Hearts Know is Possible* (Berkeley: North Atlantic Books, 2013).
92 Dan Longboat in Goodchild, "Relational Systems Thinking," p. 90.

93 Pierre Dansereau, *Inscape and Landscape*, Massey Lectures (Toronto: Canadian Broadcasting Corporation, 1973).
94 Dansereau, *Inscape and Landscape*, p. iii.
95 Thomas Berry, *The Sacred Universe: Earth, Spirituality, and Religion in the Twenty-First Century* (New York: Columbia University Press, 2009), p. 131.
96 Iwama et al., "Two-Eyed Seeing and the Language of Healing in Community-Based Research," p. 5.
97 Elisabet Sahtouris, *EarthDance: Living Systems in Evolution* (San Jose: iUniversity Press, 2000), p. xvii.
98 Sahtouris, *EarthDance,* pp. 71, 72.
99 Berry, *The Sacred Universe*, p. 112 (my emphasis).
100 Berry, *The Sacred Universe*, p. 103.
101 Thomas Berry, *Evening Thoughts: Reflecting on Earth as Sacred Community* (San Francisco: Sierra Club, 2006), p. 149.
102 Berry, *The Sacred Universe*, p. 74.
103 Brian Swimme and Thomas Berry, *The Universe Story: From the Primordial Flaring Forth to the Ecozoic Era, A Celebration of the Unfolding of the Cosmos* (New York: HarperSanFrancisco, 1992). For young people, see Brian Swimme, *The Universe Is a Green Dragon: A Cosmic Creation Story* (Santa Fe, NM: Bear & Company Publishing, 1984).
104 Sean Kelly, "Gaia and the Second Axial Age," in Sam Mickey, Mary Evelyn Tucker, and John Grim (Eds.), *Living Earth Community: Multiple Ways of Being and Knowing* (Cambridge: Open Book Publishers, 2020). https://doi.org/10.11647/OBP.0186.
105 Robert Bellah as cited by Kelly, "Gaia and the Second Axial Age," p. 208.
106 See Ronald Johnston, "Chapter 7: Religion and Politics" for a description of relationships between religions and states in *Religion in Society: A Sociology of Religion*, Eighth Edition (Upper Saddle River, NJ: Pearson Prentice Hill, 2007), pp. 145–176.
107 Robin Wall Kimmerer, *Gathering Moss: A Natural and Cultural History of Mosses* (Corvallis: Oregon State University Press, 2003).
108 Kelly, "Gaia and the Second Axial Age," p. 209.
109 Kelly, "Gaia and the Second Axial Age," p. 209.
110 I have created a version of this story for educators, called *A Cosmic Story*, which is available on my website at www.elizabethlange.ca/.
111 Kelly, "Gaia and the Second Axial Age," p. 212.
112 I was taught that the Cree word for "all our relations" is *Wahkohtowin* by my Teacher, Claudine Louis. For more, see Claudine C. Louis, "Indigenous Ways of Knowing: Returning to the Women Fire *Kiweytotahk Iskwew Iskotayow*," in Susan Brigham, Robert McGray, and Kaela Jubas (Eds.), *Adult Education and Lifelong Learning in Canada: Advancing a Critical Legacy* (Toronto: Thompson Educational Publishing Inc, 2021), pp. 4–14.
113 See www.elizabethlange.ca/teaching resources.
114 https://fore.yale.edu/project/UNEP-Faith-for-Earth-Initiative.
115 www.millenniumpeacesummit.org/.
116 https://parliamentofreligions.org/.
117 https://parliamentofreligions.org/faith-earth-call-action.
118 Heather Eaton, "The Human Quest to Live in a Cosmos," in Sam Mickey, Mary Evelyn Tucker, and John Grim (Eds.), *Living Earth Community: Multiple Ways of Being and Knowing* (Cambridge: Open Book Publishers, 2020), p. 219. https://doi.org/10.11647/OBP.0186.
119 Eaton, *The Human Quest to Live in a Cosmos*, paraphrased from p. 220.
120 See Marija Gimbutas, *The Living Goddesses* (Berkeley: University of California Press, 1999); Merlin Stone, *When God was a Woman* (Orlando: A Harvest Book, Harcourt, Inc., 1976); Riane Eisler, *The Chalice and the Blade: Our History, Our Future* (New York: HarperSanFrancisco, 1987); Marija Gimbutas, *The Goddesses and Gods of Old Europe* (Berkeley: University of California Press, 1974/2007).

121 Jean Markale, *The Great Goddess: Reverence of the Divine Feminine from the Paleolithic to the Present* (Rochester: Inner Traditions International), p. 10.

122 Campbell, *Goddesses*, p. 266. Generally, mother-right refers to societies where the Feminine cosmic principle is primary. The spectrum includes phenomena such as female leadership or matriarchy, to female or matrilineal descent, to notions of the sacred as maternal.

See also Paula Gunn Allen, *The Sacred Hoop: Recovering the Feminine in American Indian Traditions* (Boston: Beacon Press, 1986/1992).

123 David Anthony (Ed.), *The Lost World of Old Europe: The Danube Valley, 5000–3500 BC* (New York: The Institute for the Study of the Ancient World at New York University and Princeton University Press, 2010), pp. 114–115.

124 The 13 Indigenous Grandmothers are: Agnes Baker Pilgrim (Elected Chairperson) *Takelma*, Confederated Tribes of Siletz, Grants Pass, Oregon, USA; Aama Bombo, *Tamang*, Nepal; Margaret Behan, *Arapaho/Cheyenne*, Montana, USA; Rita Pitka Blumenstein, *Yup'ik*, Alaska, USA; Julieta Casimiro, *Mazatec*, Huautla de Jimenez, Mexico; Maria Alice Campos Freire, Amazonian Rainforest, Brazil; Flordemayo, *Mayan*, Highlands of Central America/New Mexico; Tsering Dolma Gyaltong, Tibetan; Clara Shinobu Iura, Amazonian Rainforest, Brazil; Beatrice Long Visitor Holy Dance, *Oglala Lakota*, Black Hills, South Dakota, USA; Rita Long Visitor Holy Dance, *Oglala Lakota*, Black Hills, South Dakota, USA; Mona Polacca, *Hopi/Havasupai/Tewa*, Arizona, USA; and Bernadette Rebienot, *Omyene*, Gabon, Africa.

125 Carol Schaefer, *Grandmothers Counsel the World: Women Elders Offer Their Vision for our Planet* (Boston: Trumpeter, 2006). See also *Grandmothers Wisdom: Reverence for All Creation* (New York: MA Creative, 2019). The Grandmothers Wisdom Project is at www.macreativeinc.com/ and www.grandmotherscouncil.org/.

126 Allen, *The Sacred Hoop*, p. x.

127 Riane Eisler, *The Chalice and the Blade: Our History, Our Future* (New York: HarperSanFrancisco, 1987).

128 Markale, *The Great Goddess*, p. 3.

129 Markale, *The Great Goddess*, p. 42.

130 Markale, *The Great Goddess*, pp. 30–31. The last sentence he quoted is from Grillot de Givry.

131 Joseph Campbell, *Goddesses: Mysteries of the Feminine Divine* (Novato, CA: New World Library, 2013), p. xiii.

132 Campbell, *Goddesses*, p. xiv.

133 Campbell, *Goddesses*, p. xiv.

134 Two such examples are Jean Shinoda Bolen, *Goddesses in Everywoman: A New Psychology of Women* (New York: Harper Colophon Books, 1984) or Carol Christ, *A Serpentine Path: Mysteries of the Goddess* (Cleveland: The FAR Press, 2016).

135 Rosemary Radford Ruether, *Goddesses and the Divine Feminine* (Berkeley: University of California Press, 2005).

136 Ruether, *Goddesses and the Divine Feminine*, p. 308.

137 Ruether, *Goddesses and the Divine Feminine*, p. 308.

138 Berry, *The Sacred Universe*, p. 133.

139 One example of Charlene Spretnak, *Lost Goddesses of Early Greece: A Collection of Pre-Hellenic Myths* (Boston: Beacon Press, 1978/1992), p. xii.

140 For more, see Gilbert Herdt, "Coming of Age and Coming Out Ceremonies across Cultures," in Michael Kimmel, Amy Aronson, and Amy Kaler (Eds.), *The Gendered Society Reader*, Canadian Edition (Don Mills, ON: Oxford University Press, 2008), pp. 45–60.

141 Gunn Allen, *The Sacred Hoop,* p. 198.

142 Gunn Allen, *The Sacred Hoop,* pp. 196–200.

143 Riane Eisler, *Tomorrow's Children: A Blueprint for Partnership Education for the 21st Century* (Boulder, CO: Westview Press, 2000).

144 Frank Waters, *Masked Gods: Navaho and Pueblo Ceremonialism* (Chicago: Sage Books and The Swallow Press, 1950).
145 Allen, *The Sacred Hoop*.
146 Lynne McTaggart, *The Field: The Question for the Secret Force of the Universe* (New York: Harper, 2008), p. xxvii.
147 McTaggart, *The Field*, pp. xxvii, xxiv.
148 Leroy Little Bear, "Foreword," in Gregory Cajete (Ed.), *Native Science: Natural Laws of Interdependence* (Santa Fe: Clear Light Publishing, 2000), p. x.
149 Leroy Little Bear, "Foreword," p. x.
150 Campbell, *Goddesses*, p. 14.
151 Vine Deloria Jr., *The World We Used to Live In: Remembering the Powers of the Medicine Man* (Golden, CO: Fulcrum Publishing, 2006), p. 202.
152 Deloria Jr., *The World We Used to Live In*, p. 202.
153 Deloria Jr., *The World We Used to Live In*, p. 202.
154 Elders Dr. Reg Crowshoe and Georr Crow Eagle, *Piikani Blackfoot Teaching*, n.d. www.fourdirectionsteachings.com/transcripts/blackfoot.html.
155 Joan Golden-Alexis, *The Native American Sense of the Sacred: The Tipi as a Collective Transformational Vessel*. ARAS Connections, No. 1 (2015), p. 4. Part of the quote is a reference to Frank Waters, *Book of the Hopi* (New York: Penguin Group, 1977), p. 8.
156 Golden-Alexis, *The Native American Sense of the Sacred*, p. 4.
157 Golden-Alexis, *The Native American Sense of the Sacred*, p. 8.
158 Golden-Alexis, *The Native American Sense of the Sacred*, pp. 23–24.
159 Golden-Alexis, *The Native American Sense of the Sacred*, p. 9.
160 Sarangerel, *Riding Windhorses: A Journey into the Heart of Mongolian Shamanism* (Rochester, VT: Destiny Books, 2000), p. 3.
161 Sarangerel, *Riding Windhorses*, p. 8.
162 Sulayman Nyang, "Reflections on Traditional African Cosmology," *New Directions*, Vol. 8, No. 1 (1980), pp. 28–32.
163 Dr Batchuluun Yembuu, personal correspondence, June 21, 2022.
164 Dr Batchuluun Yembuu, personal correspondence, June 21, 2022.
165 Sarangerel, *Riding Windhorses*, pp. 10–11.
166 A *hogan* is a traditional *Navajo* dwelling made of logs and packed mud, generally in a circular shape.
167 Golden-Alexis, *The Native American Sense of the Sacred*. Inclusive language added.
168 Sarangerel, *Riding Windhorses*, p. 14.
169 Waters, *Masked Gods*, p. 170.
170 Cora Weber-Pillwax, *Identity Formation and Consciousness with Reference to Northern Alberta Cree and Metis Indigenous Peoples* (Edmonton: University of Alberta, 2003), p. 174.
171 Vine Deloria Jr., *The World We Used to Live In*, p. 195.
172 John Mbiti, *Concepts of God in Africa* (New York: Praeger, 1970), cited in Nyang, "Reflections on Traditional African Cosmology," p. 32.
173 Mbiti In Nyang, "Reflections on Traditional African Cosmology," p. 31.
174 Tyson Yunkaporta, *Sand Talk: How Indigenous Thinking Can Save the World* (Melbourne: Text Publishing, 2019), p. 22.
175 Berry, *The Sacred Universe*, pp. 142–143.
176 The "Medicine Wheel" was named such by European settlers; the original names for this structure are sacred and protected. Jamie Oxendine, *Medicine Wheel: Comparison in Life*. www.powwows.com.
177 Oxendine, *Medicine Wheel*, p. 3.
178 Louis, "Indigenous Ways of Knowing," pp. 4–14.
179 Louis, "Indigenous Ways of Knowing," pp. 9–10.
180 Cajete, *Native Science*, p. 58.
181 James Lovelock, "Foreword," in Sahtouris, *EarthDance*, p. xiii.

182 Lyall Watson, *Gifts of Unknown Things: A True Story of Nature, Healing, and Initiation from Indonesia's Dancing Island* (Rochester, VT: Destiny Press, 1991) cited in Sahtouris, *EarthDance*, p. iii.
183 Sahtouris, *EarthDance*, p. 81.
184 Bollier and Helfrich, *Free, Fair and Alive*, p. 44.
185 Carol Gould, *Marx's Social Ontology* (Cambridge: MIT Press, 1978).
186 Jin Lin, Tom Culham, and Sachi Edwards, *Contemplative Pedagogies for Transformative Teaching, Learning and Being* (Charlotte: Information Age Publishing, Inc., 2019). Another source is from Aloka Marti and Joan Sala, *Awareness through the Body: A Way to Enhance Concentration, Relaxation and Self-knowledge in Children* (Auroville: Sri Aurobindo International, 2018).
187 The term intra-acting, from Karen Barad, is distinguished from interaction, to mean that it is not humans who hold agency always with the potential to execute it, but rather that reality emerges from a dynamism of forces. The implication is that agency flows through the world "through which part of the world makes itself differentially intelligible to another part of the world." Karen Barad, *Meeting the Universe Halfway: Quantum Physics and the Entanglement of Matter and Meaning* (Durham: Duke University Press, 2007), p. 140. See Joy Kcenia Polanco O'Neil's work in terms of application within the field of education.
188 Barbara Thayer-Bacon, *Relational Ontologies* (New York: Peter Lang, 2017).
189 Thayer-Bacon, *Relational Ontologies*, p. 138.
190 For more on implicit knowing, see David Bohm, *Unfolding Meaning* (London: Routledge, 1985) or *Wholeness and the Implicate Order* (London: Routledge, 1980) and Alexis Shotwell, *Knowing Otherwise: Race, Gender, and Implicit Understanding* (University Park, PN: The Pennsylvania State University Press, 2011).
191 Thayer-Bacon, *Relational Ontologies*, pp. 33–34.
192 Thayer-Bacon, *Relational Ontologies*, p. 34.
193 Thayer-Bacon, *Relational Ontologies*, p. 34.
194 Marilou Awiakta, *Selu: Seeking the Corn-Mother's Wisdom* (Golden, CO: Fulcrum Publishing, 1993).
195 Karen Armstrong, *A Short History of Myth* (Toronto: Vintage Canada, 2006), p. 1.
196 Armstrong, *A Short History of Myth*, p. 2.
197 Armstrong, *A Short History of Myth*, p. 2.
198 Armstrong, *A Short History of Myth*, p. 7.
199 Armstrong, *A Short History of Myth*, p. 11.
200 Campbell, *Primitive Mythology: The Masks of God* (New York: Arkana, 1959/1991), p. 461.
201 Armstrong, *A Short History of Myth*, p. 11.
202 Clarissa Pinkola Estes, *Women Who Run with the Wolves: Myths and Stories of the Wild Woman Archetype* (New York: Ballantine Books, 1992/1995), p. 2.
203 Estes, *Women Who Run with the Wolves*, p. 1.
204 Estes, *Women Who Run with the Wolves*, p. 7.
205 Awiakta, *Selu*, p. 9.
206 Dr. Claudine Louis, personal communication.
207 One example of layered origin stories is Martin Prechtel's *The Disobedience of the Daughter of the Sun: A Mayan Tale of Ecstasy, Time, and Finding One's True Form* (Berkeley: North Atlantic Books).
208 Jo-Ann Archibald, *Indigenous Storywork: Educating the Heart, Mind, Body, and Spirit* (Vancouver: UBC Press, 2008), p. ix.
209 When the word Elders is capitalized, it refers to a formalized relation in which one is recognized and respected as a knowledge carrier with communal responsibilities as part of their Elder role. See Cajete, *Native Science*, p. 71.
210 Campbell, *Primitive Mythology*, p. 462.
211 Awiakta, *Selu*, p. 16.

212 Marín Prechtel, *Workshop Four* (Ojo Caliente, New Mexico, November, 2018).
213 Martín Prechtel, *Rescuing the Light* (Berkeley: North Atlantic Books, 2021).
214 Joseph Campbell, *Goddesses* (Novato, CA: New World Library, 2013), p. 15.
215 Gregory Cajete, "A Pueblo Story for Transformation," in Edmund O'Sullivan and Marilyn Taylor (Eds.), *Learning toward an Ecological Consciousness: Selected Transformative Practices* (New York: Palgrave Macmillan, 2004), p. 106.
216 Campbell, *Goddesses*, p. 13.
217 Prechtel, *Rescuing the Light*, p. 63.
218 Eve Tuck and Wayne Yang, "Decolonization Is Not a Metaphor," *Decolonization: Indigeneity, Education & Society*, Vol. 1, No. 1 (2012).
219 Joseph Campbell, *Goddesses*, p. 53. Emphasis added.
220 Linda Neale, *The Power of Ceremony: Restoring the Sacred in Our Selves, Our Families, Our Communities* (Portland: Eagle Spirit Press, 2011), p. 12.
221 Sun Bear in Neale, *The Power of Ceremony*, p. 11.
222 Cajete, *Native Science*, p. 82.
223 Meredith McGuire, *Religion: The Social Context* (Belmont, CA: Wadsworth), p. 17.
224 McGuire, *Religion*, p. 17.
225 While I am drawing from Neale and using McGuire's sociological definitions here, *Dagara* Malidoma Patrice Somé from Africa understands ritual and ceremony in the reverse, where ritual (individual, family, and community) includes symbol and ceremony. See his book *Ritual: Power, Healing, and Community* (New York: Penguin Compass, 1993).
226 Somé, *Ritual*, p. 12.
227 Somé, *Ritual*, pp. 19–20.
228 Neale, *The Power of Ceremony*, p. 23
229 Matthew Fox, "Foreword: A Return to Ceremony, a Return to Healing," in Neale, *The Power of Ceremony*, p. xi.
230 See proviso in endnote 237.
231 Mircea Eliade, *Rites and Symbols of Initiation: The Mysteries of Birth and Rebirth* (Thompson, CN: Spring Publications, 2017).
232 Charles Taylor cited in Maxine Greene, *Releasing the Imagination: Essays on Education, the Arts, and Social Change* (San Francisco: Jossey Bass, 1995), p. 75.
233 One example is in Malidoma Somé, *The Healing Wisdom of Africa: Finding Life Purpose through Nature, Ritual, and Community* (New York: Jeremy P. Tarcher/Putnam, 1998).
234 Neale, *The Power of Ceremony*.
235 Paula Gunn Allen, *Grandmothers of the Light* (Boston: Beacon Press, 1991).
236 Will McWhinney and Laura Markos, "Transformative education: Across the Threshold," *Journal of Transformative Education*, Vol. 1, No. 1 (2003), pp. 16–37.
237 McWhinney and Markos, "Transformative Education," p. 24.
238 Frank Waters, *Masked Gods: Navaho and Pueblo Ceremonialism* (Chicago: The Swallow Press Inc., 1950), p. 9.
239 Somé, *The Healing Wisdom of Africa*, pp. 102–103.
240 Cajete, "A Pueblo Story for Transformation," p. 103.
241 Cajete, "A Pueblo Story for Transformation," p. 105.
242 Cajete, "A Pueblo Story for Transformation," paraphrased from p. 110.
243 Malidoma Somé, *The Healing Wisdom of Africa: Finding Life Purpose through Nature, Ritual, and Community* (New York: Jeremy P. Tarcher/Putnam, 1998), p. 65.
244 www.etymonline.com/word/respect.
245 Erich Fromm, *To Have or to Be?* (New York: Continuum, 1999).
246 Paulo Freire, *Pedagogy of the Oppressed* (New York: Seabury Press, 1970), pp. 59–60.
247 Freire, *Pedagogy of the Oppressed*, p. 34.
248 Freire, *Pedagogy of the Oppressed*, p. 28.
249 Parker Palmer, *The Courage to Teach: Exploring the Inner Landscape of a Teacher's Life* (San Francisco: Jossey-Bass Publishers, 1998), pp. 116–117.

250 Evelyn Fox Keller, *A Feeling for the Organism: The Life and Work of Barbara McClintock* (New York: Henry and Holt Company, 1983).
251 Lee Nichol, "Foreword," in David Bohm (Ed.), *On Dialogue* (London: Routledge, 1996), p. ix.
252 Nichol, "Foreword," p. ix.
253 David Bohm, *Unfolding Meaning* (London: Routledge, 1987), p. 175.
254 Most Indigenous people refer to themselves as "The People" in their own language.
255 Cora Weber-Pillwax, "It's in the Blood: Theory and Praxis of Lifelong Indigenous Education," *International Journal of Lifelong Education*, Vol. 40, No. 4 (2021), pp. 395–414. https://doi.org/10.1080/02601370.2021.1966528.
256 Weber-Pillwax, "It's in the Blood," pp. 395–414.
257 I will be developing pedagogical materials to assist in these learning explorations. Please check back regularly on my website: www.elizabethlange.ca/.
258 George, *My Spirit Soars*, p. 52.
259 Cajete, *Native Science*, p. 178.
260 Cajete, "Look to the Mountain," p. 11.
261 Campbell, *Goddesses*, p. 5.
262 Melissa Nelson, *Original Instructions: Indigenous Teachings for a Sustainable Future* (Rochester: Bear & Company, 2008), p. 3.
263 Chief Oren Lyon, "Listening to Natural Law," in Nelson, *Original Instructions*, p. 24.
264 Cajete, *Native Science*, p. 178.
265 See, for instance, the World Council of Religious Leaders. www.millenniumpeacesummit.org/ and Parliament of World's Religions. https://parliamentofreligions.org/.
266 Gregory Cajete, "'Look to the Mountain:' Reflections on Indigenous Ecology," in Gregory Cajete (Ed.), *A People's Ecology: Explorations in Sustainable Living* (Santa Fe, NM: Clear Light Publishers, 1999), p. 5.
267 Cajete, *Native Science*, p. 260.
268 Nyang, "Reflections on Traditional African Cosmology," p. 32.
269 A sasquatch or Big Foot is an elusive giant, hairy primate that is reported to live in the Pacific Northwest as well as in other remote areas, including down the spine of the Rocky Mountains. The name sasquatch comes from *Halkomelem* (the *Salishan* language of southwestern British Columbia), written as *sésqəc* (www.merriam-webster.com/dictionary/Sasquatch). *See John Zada, In the Valleys of the Noble Beyond: In Search of the Sasquatch (Vancouver: Greystone Books, 2019)* for more descriptions and understandings of sasquatch from the Pacific Northwest of North America.
270 "The Seven Teachings." www.southernnetwork.org/site/seven-teachings. Accessed April 27, 2021
271 Cajete, "Look to the Mountain," p. 12.
272 Berry, *The Sacred Universe*, p. 138.
273 Escobar, *Designs for the Pluriverse*, p. 12.
274 Escobar, *Designs for the Pluriverse*, p. 13.
275 Escobar, *Designs for the Pluriverse*, p. 14.
276 Escobar, *Designs for the Pluriverse*, p. 4.
277 Escobar, *Designs for the Pluriverse*, p. 10.
278 Martín Prechtel, *Rescuing the Light: Quotes from the Oral Teachings of Martín Prechtel* (Berkeley: North Atlantic Books, 2021), p. 36.
279 Cajete, "Look to the Mountain," p. 4.
280 Cajete, "Look to the Mountain," p. 7.
281 Sandra Styres, "Literacies of Land: Decolonizing Narratives, Storying and Literature," in Linda Tuhiwai Smith, Evelyn Tuck, and K. Wayne Yang (Eds.), *Indigenous and Decolonizing Studies in Education: Mapping the Long View* (New York: Routledge, 2019), pp. 24–37.
282 Styres, "Literacies of Land," pp. 24–25.
283 Jeanette Armstrong, "Land speaking" in Simon Ortiz (Ed.), *Speaking for the Generations: Native Writers on Writing* (Tuscon: University of Arizona Press, 1998), pp. 175–194 cited in Styres, "Literacies of Land," pp. 25–26.

284 I cite this from a previous article, Elizabeth A. Lange, "Riverspeaking: The Spiraling of Transformative and Restorative Learning toward Kinship Ethics," in Peter Blaze Corcoran, Joseph Weakland, and Arjen Wals (Eds.), *Envisioning Futures for Environmental and Sustainability Education* (Wageningen: Wageningen Academic Publishers, 2017), p. 35.
285 Martin Prechtel, *The Unlikely Peace at Chuchumaquic* (Berkley: North Atlantic Books, 2012), p. 73. Originally quoted in Elizabeth A. Lange, "Riverspeaking toward Kinship Ethics," p. 35.
286 Martin Prechtel, *The Unlikely Peace at Chuchumaquic* (Berkley: North Atlantic Books, 2012), p. 72. Originally quoted in Elizabeth A. Lange, "Riverspeaking toward Kinship Ethics," p. 34.
287 Leroy Little Bear, "Foreword," p. xii.
288 Cajete, *Native Science*, p. 180.
289 Cajete, *Native Science*, pp. 180, 179.
290 Cajete, *Native Science*, p. 178.
291 Marie Battiste, *Indigenous Knowledge and Pedagogy in First Nations Education* (Ottawa: Indian and Northern Affairs Canada, 2002).
292 Cajete, *Native Science*, p. 183.
293 Elder Kinuwa, 1998, in Patricia Steinhauer-Hill, *Kihkipiw: A Cree Way* (Edmonton: University of Alberta, 2008), p. 19.
294 Yunkaporta, *Sand Talk*, p. 2.
295 Nyang, "Reflections on Traditional African Cosmology," p. 30.
296 Martin Prechtel. Workshop Series. (Ojo Caliente, New Mexico, 2017–2020).
297 Cajete, *Native Science*, p. 184.
298 This is a generality, as each tribal peoples have their own ontology of spirit and soul, with much variation and complexity.
299 Cajete, *Native Science*, p. 187.
300 Cajete, *Native Science*, p. 186.
301 Cajete, *Native Science*, p. 187.
302 Sarangerel, *Riding Windhorses*, p. 22.
303 Styres, "Literacies of the Land," p. 29.
304 Styres, "Literacies of Land," p. 26.
305 Edward Casey, "How to Get from Space to Place in a Fairly Short Stretch of time: Phenomological Prolegomena," in Steven Feld and Keith Basso (Eds.), *Senses of Place* (Santa Fe: School of American Research Press, 1996), pp. 13–32 cited in Styres, *Literacies of Land*, p. 26.
306 Styres, "Literacies of Land," p. 27.
307 Edward Casey cited in Styres, "Literacies of Land," p. 27.
308 Styres, "Literacies of Land," p. 27.
309 Styres, "Literacies of Land," p. 27.
310 Styres, "Literacies of Land," p. 27.
311 Styres, "Literacies of Land," p. 27.
312 Yunkaporta, *Sand Talk*, p. 19.
313 Yunkaporta, *Sand Talk*, p. 21, 19. Yarning and yarning circles are a dialogue form where knowledge and stories are shared (passed on) in an informal, meandering way, built from honest, respectful relations. This traditional form of interweaving knowledge has been used for centuries in Aboriginal and Torres Strait Islander cultures (Australian continent).
314 Taiaiake Alfred, "Restitution is the Real Pathway to Justice for Indigenous Peoples," in *Responsibility and Renewal: Canada's Truth and Reconciliation Journey* (Ottawa: Aboriginal Healing Foundation, 2009), pp. 179–187 cited in Styres, *Literacies of Land*, p. 30.
315 Prechtel, *Rescuing the Light*, p. 43.
316 Esteva, *Designs for the Pluriverse*.
317 See her lovely book, *Braiding Sweetgrass*, 2013.
318 Robin Wall Kimmerer, "Learning the Grammar of Animacy," in *Braiding Sweetgrass* (Minneapolis, MN: Milkweed Editions, 2013), pp. 48–59.

319 Kimmerer, *Braiding Sweetgrass*, p. 49.
320 Kimmerer, *Braiding Sweetgrass*, p. 57. Emphasis added for clarity.
321 Cajete, *Native Science*, p. 184.
322 Cajete, *Native Science*, p. 184.
323 Prechtel, *Rescuing the Light*, p. 40.
324 Christian de Quincey, *Radical Nature: The Soul of Matter* (Rochester: Park Street Press, 2002/2010).
325 De Quincey, *Radical Nature*, p. 187.
326 De Quincey, *Radical Nature*, p. 187.
327 Peter Levine, *In an Unspoken Voice: How the Body Releases Trauma and Restores Goodness* (Berkeley: North Atlantic Books, 2010).
328 Information and related practices found in Andrea Cohen, *Practicing the Art of Compassionate Listening* (The Compassionate Listening Project, 2017), pp. 34–35. Her sources include Daniel Siegel, Antonio Damasio, Eugene Gendlin, and Christian Huygens.
329 For important compassion practices, see *The Compassionate Listening Project* or teachers such as Thich Nhat Hanh and Pema Chödrön.
330 De Quincey, *Radical Nature*, p. 189.
331 De Quincey, *Radical Nature*, p. 189.
332 De Quincey, *Radical Nature*, pp. 195–196.
333 De Quincey, *Radical Nature*, p. 238.
334 De Quincey, *Radical Nature*, p. 260.
335 De Quincey, *Radical Nature*, p. 261.
336 De Quincey, *Radical Nature*, p. 261.
337 Cajete, *Native Science*, p. 71.
338 Cajete, *Native Science*, p. 72.
339 Patricia Steinhauer-Hill, *Kihkipiw: A Cree way* [Unpublished manuscript] (Edmonton: University of Alberta, 2008), p. 43.
340 Cajete, *Native Science*, pp. 150–151.
341 Cajete, "Look to the Mountain," p. 10.
342 For presencing practices, see Olen Gunnlaugson, *Dynamic Presencing: A Transformative Journey into Presencing Mastery, Leadership and Flow* (Vancouver: Trifoss Business Press, 2020). This work is partly built upon Theory U by Otto Scharmer. See Otto Scharmer, Addressing the Blind Spot of our Time: An Executive Summary of Theory U: Leading from the Future as it Emerges. https://solonline.org/wp-content/uploads/2016/12/Exec_Summary_Sept19-Theory-U-leading-from-the-emerging-future-copy.pdf.
343 Lange, O'Neil, and Ross, "Educating During the Great Transformation," pp. 23–46.
344 The Crex Crex Collective. "Six Touchstones for Wild Pedagogies in Practice". Finalized chapter found in Bob Jickling, Sean Blenkinsop, Nora Timmerman, and Michael Sitka Sage, *Wild Pedagogies: Touchstones for Re-Negotiating Education and the Environment in the Anthropocene* (London: Palgrave Macmillan, 2018). See also https://wildpedagogies.com/Publications.
345 Marc Bekoff, *Rewilding our Hearts: Building Pathways of Compassion and Coexistence* (Novato: New World Library, 2014).
346 Gregory Cajete, "A Pueblo Story for Transformation," in Edmund O'Sullivan and Marilyn Taylor (Eds.), *Learning toward an Ecological Consciousness: Selected Transformative Practices* (New York: Palgrave Macmillan, 2004), pp. 103–113.
See also Anne Phillips, *Holistic Education: Learning from Schumacher College* (Dartington, Devon: Green Books, 2008).
347 Prechtel, *Rescuing the Light*, p. 43.
348 Yunkaporta, *Sand Talk*, p. 14.
349 Yunkaporta, *Sand Talk*, pp. 16–17.
350 Yunkaporta, *Sand Talk*, pp. 17–18.
351 Yunkaporta, *Sand Talk*, p. 18.

352 Batchuluun Yembuu, "Intergenerational Learning of Traditional Knowledge through Informal Education: The Mongolian Context," *International Journal of Lifelong Education*, Vol. 40, No. 4 (2021), pp. 339–358. https://doi.org.10.1080/02601370.2021.1967488.
353 See, for instance, Michael Corbett, *Learning to Leave: The Irony of Schooling in a Coastal Community* (Halifax: Fernwood Publishing, 2007).
354 Yembuu, "Intergenerational Learning of Traditional Knowledge through Informal Education," p. 355.
355 Wade Davis, *The Wayfinders: Why Ancient Wisdom Matters in the Modern World* (Toronto: Anansi Press, 2009).
356 I wish to respectfully thank Martín Pretchel for his workshops exemplifying this kind of learning.
357 Capra and Luisi, *The Systems View of Life*, p. 252.
358 Capra and Luisi, *The Systems View of Life*, p. 252.
359 Capra and Luisi, *The Systems View of Life*, p. 260.
360 Michel Bitbol in Capra and Luisi, *The Systems View of Life*, p. 266.
361 Capra and Luisi, *The Systems View of Life*, p. 269.
362 Capra and Luisi, *The Systems View of Life*, p. 254.
363 Capra and Luisi, *The Systems View of Life*, p. 269.
364 I thank Dr Katie Ross for introducing me to Auroville. See: Bem Le Hunte, Katie Ross, Suryamayi Clarence-Smith, and Aditi Rosegger, "Lessons from Utopia: Reflections on Peak Transformative Experiences in a University Studio in Auroville, India," in Aliki Nicolaides et al. (Eds.), *The Palgrave Handbook of Learning for Transformation* (Cham: Springer Nature Switzerland, 2022), pp. 863–881. https://doi.org/10.1007/978-3-030-84694-7.
365 Aloka Marti and Joan Sala, *Awareness through the Body* (Auroville: Sri Aurobindo International Institute of Education Research, 2019), p. 7. Inclusive language added.
366 Citing Humberto Maturana and Francisco Varela in Capra and Luisi, *The Systems View of Life*, p. 256.
367 Capra and Luisi, *The Systems View of Life*, p. 256.
368 Robert Mesle, *Process-Relational Philosophy: An Introduction to Alfred North Whitehead* (West Conshohocken, PA: Templeton Press, 2008), p. 5.
369 Mesle, *Process-Relational Philosophy*, p. 5.
370 Mesle, *Process-Relational Philosophy*, p. 7.
371 Mesle, *Process-Relational Philosophy*, p. 8.
372 Bohm, *Wholeness and the Implicate Order*, p. 63.
373 Mesle, *Process-Relational Philosophy*, p. 8.
374 Mesle, *Process-Relational Philosophy*, p. 6.
375 Theologist David Ray Griffen cited in Mesle, *Process-Relational Philosophy*, p. 6.
376 Simon Ceder, *Cutting through Water: Towards a Posthuman Theory of Educational Relationality* (Lund University, 2015), p. 20. www.academia.edu/23696265/Cutting_Through_Water_Towards_a_Posthuman_Theory_of_Educational_Relationality_Cutting_Through_Water.
377 Simon Ceder, *Cutting through Water*, p. 27 cited in Elizabeth A. Lange, "*Living* Transformation: The Alchemy of Change in an Epochal Shift," in Aliki Nicolaides, Petra Buergelt, Saskia Eschenbacher, Marguerite Welch, Yabome Gilpin-Jackson, and Mitsunori Misawa (Eds.), *The Palgrave Handbook on Learning for Transformation* (Palgrave Macmillan, 2022), p. 736.
378 Lange, "*Living* Transformation," pp. 733–749.
379 Lange, "*Living* Transformation," p. 737.
380 Ceder, *Cutting through Water*, p. 13.
381 Lange, "*Living* Transformation," p. 737.
382 See also Simon Ceder, *Towards a Posthuman Theory of Educational Relationality* (London: Routledge, 2019).
383 I thank Joy Kcenia Polanco O'Neil for introducing me to Karen Barad's work.
384 Barad, *Meeting the Universe Halfway*.

385 Diarmuid O'Murchu, *Quantum Theology: Spiritual Implications of the New Physics* (New York: The Crossroad Publishing Company, 2004), p. 54.
386 O'Murchu, *Quantum Theology*, p. 55.
387 Mesle, *Process-Relational Philosophy*, p. 18.
388 Mesle, *Process-Relational Philosophy*, p. 37.
389 Capra and Luisi, *A Systems View of Life*, p. 261.
390 Capra and Luisi, *A Systems View of Life*, p. 260.
391 Christopher Bache, *The Living Classroom: Teaching and Collective Consciousness* (New York: SUNY Press, 2008), p. 98.
392 Bache, *The Living Classroom*, p. 154.
393 Bache, *The Living Classroom*, p. 99.
394 Bache, *The Living Classroom*, p. 148.
395 David Bohm, *Wholeness and the Implicate Order* (London: Routledge, 1980).
396 Bache, *The Living Classroom*, p. 145.
397 For more on clusters of vibrations as part of an intelligence cosmos, see Ervin Laszlo, *The Intelligence of the Cosmos: Why are We Here—New Answers from the Frontiers of Science* (Rochester, VT: Inner Traditions, 2017).
398 Bache, *The Living Classroom*, p. 152.
399 Cajete, "A Pueblo Story for Transformation," p. 103.
400 Cajete, "A Pueblo Story for Transformation," p. 103.
401 Cajete, "A Pueblo Story for Transformation," p. 104.
402 Capra and Luisi, *The Systems View of Life*, p. 256.
403 Capra and Luisi, *The Systems View of Life*, p. 256.
404 Dilip Jeste with Scott LaFee, *Wiser: The Scientific Roots of Wisdom, Compassion, and What Makes Us Good* (Boulder, CO: Sounds True, 2020), p. 4.
405 Jeste with LaFee, *Wiser*.
406 Jeste with LaFee, *Wiser*, p. xv.
407 Jeste with LaFee, *Wiser*, p. 23. Emphasis is mine.
408 Jeste with LaFee, *Wiser*, p. xv.
409 Jeste with LaFee, *Wiser*, p. xv.
410 Jeste with LaFee, *Wiser*, p. 279.
411 George, *My Spirit Soars*, p. 86.
412 Marcuse, *One-Dimensional Man*.
413 Marcuse, *One-Dimensional Man*, p. xlvii.
414 Arnold Farr, "Herbert Marcuse," in Edward N. Zalta (Ed.), *The Stanford Encyclopedia of Philosophy* (Summer 2021 Edition). https://plato.stanford.edu/archives/sum2021/entries/marcuse/.
415 Paulo Freire, "Education, Liberation, and the Church," *Religious Education*, Vol. 79, No. 4 (1984), pp. 524–548.
416 Antonio Gramsci cited in Kothari et al., "Introduction: Finding Pluriversal Paths," p. xxiv.
417 Escobar, *Designs for the Pluriverse*, pp. 6–7.
418 Escobar, *Designs for the Pluriverse*, p. 7.
419 Gustavo Esteva and Madhu Suri Prakash, *Grassroots Post-Modernism: Remaking the Soil of Cultures* (London: Zed Books, 1998).
420 Escobar, *Designs for the Pluriverse*. Also Esteva and Prakash, *Grassroots Post-Modernism*.
421 Escobar, *Designs for the Pluriverse*, p. 66.
422 Gustavo Esteva, Dana Stuchul, and Madhu Suri Prakash, "From a Pedagogy for Liberation to Liberation from Pedagogy," in Chet Bowers and Frederique Apffel-Marlin (Eds.), *Rethinking Freire: Globalization and the Environmental Crisis* (Mahwah, NJ: Lawrence Erlbaum Associates, Publishers, 2005), p. 25.
423 Esteva et al., "From a Pedagogy for Liberation to Liberation from Pedagogy," p. 28.
424 Escobar, *Designs for the Pluriverse*, p. x.
425 Escobar, *Designs for the Pluriverse*, p. xvii.

426 Escobar, *Designs for the Pluriverse*, p. 7.
427 David Bollier and Silke Helfrich, *Free, Fair and Alive: The Insurgent Power of the Commons* (Gabriola Island, BC: New Society Press, 2019), p. 78.
428 Klein, *This Changes Everything*.
429 Bollier and Helfrich, *Free, Fair and Alive*, p. 148.
430 Escobar, *Designs for the Pluriverse*, p. 5.
431 https://schoolsforchiapas.org/library/sixth-declaration-lacandona-jungle/.
432 Esteva and Prakash, *Grassroots Post-Modernism*, p. 6.
433 Esteva and Prakash, *Grassroots Post-Modernism*, p. 35.
434 Esteva and Prakash, *Grassroots Post-Modernism*, p. 37.
435 Esteva and Prakash, *Grassroots Post-Modernism*, p. 46.
436 The delegation was: Lupita, 19, Ximena, 25, Carolina, 26, Yuli, 37, Marijose, 39, Darío, 47, and Bernal, 57.
437 Jerome Baschet, *Progressive International*, June 3, 2021. https://progressive.international/wire/2021-06-30-the-zapatista-invasion-has-begun/en.
438 https://enlacezapatista.ezln.org.mx/2021/01/01/part-one-a-declaration-for-life/. See footnote 436 for the names of the delegation.
439 Eva Mackey, *Unsettled Expectations: Uncertainty, Land and Settler Decolonization* (Halifax: Fernwood Publishing, 2016).
440 Speech by Mayor Ana Maria for the Intercontinental Encounter for Humanity and Against Neoliberalism, July 27, 1996, cited in Escobar and Prakash, *Grassroots Post-Modernism*, pp. 173–174.
441 Bollier and Helfrich, *Free, Fair and Alive*, p. 83.
442 Escobar, *Designs for the Pluriverse*, p. 16.
443 Bollier and Helfrich, *Free, Fair and Alive*, p. 98.
444 Bollier and Helfrich, *Free, Fair and Alive*, p. 101.
445 Richard Tarnas, *The Passion of the Western Mind: Understanding the Ideas That Have Shaped our World View* (New York: Ballantine Books, 1991).
446 Claudia von Werlhof, *The Failure of Modern Civilization and the Struggle for a "Deep Alternative": On "Critical Theory of Patriarchy as a New Paradigm"* (Bern: Peter Lang, 2011).
447 Riane Eisler, *The Chalice and the Blade: Our History, Our Future* (New York: HarperSanFrancisco, 1987); Rosemary Radford Ruether, *Gaia & God: An Ecofeminist Theology of Earth Healing* (New York: HarperSanFrancisco, 1992); Greta Gaard, *Ecofeminism: Women, Animals, Nature* (Philadelphia: Temple University Press, 1993); Thomas Berry, *The Sacred Universe: Earth, Spirituality and Religion in the Twenty-First Century* (New York: Columbia University Press, 2009); Carolyn Merchant, *The Death of Nature: Women, Ecology and the Scientific Revolution* (San Francisco: Harper & Row, 1980), to name a few.
448 Cited in Escobar, *Designs for the Pluriverse*, p. 10.
449 Paraphrased from Escobar, *Designs for the Pluriverse*, pp. 12–13, who is citing von Werlhof.
450 Humberto Maturana and Gerda Verden-Zöller, *The Origin of Humanness in the Biology of Love* (Thorverton, UK: Imprint Academic, 2012), pp. 87, 209.
451 Maturana and Verden-Zöller, *The Origin of Humanness in the Biology of Love*, pp. 209–210.
452 Maturana and Verden-Zöller, *The Origin of Humanness in the Biology of Love*, pp. 210–211.
453 Humberto Maturana and Gerda Verden-Zöller, *The Origin of Humaness in the Biology of Love*.
454 Berry, *The Sacred Universe*, p. 69.
455 Berry, *The Sacred Universe*, p. 69.
456 Berry, *The Sacred Universe*, pp. 76–77.
457 Gunn Allen, *The Sacred Hoop*, p. 2.
458 Gunn Allen, *The Sacred Hoop*, p. 3.
459 Martin Luther King Jr in Gary May, "A Revolution of Values: Martin Luther King Jr and the Poor People's Campaign," January 18, 2015. https://billmoyers.com/2015/01/18/revolution-values/.

460 Bollier and Helfrich, *Free, Fair and Alive*, p. 29.
461 UNESCO, *Review of Contexts and Structures for Education for Sustainable Development* (Paris: Division for the Coordination of United of United Nations Priorities for Education, 2009), cited in Elizabeth A. Lange, "Transformative Sustainability Education: From Sustainababble to a Civilization Leap," in Marcela Milana et al. (Eds.), *The Palgrave International Handbook on Adult and Lifelong Education and Learning* (London: Palgrave Macmillan, 2018), pp. 397–421.
462 David Selby, "Education for Sustainable Development, Nature and Vernacular Learning," *C•E•P•S Journal*, Vol. 7, No. 1 (2017), pp. 9–27.
463 Selby, "Education for Sustainable Development, Nature and Vernacular Learning," p. 16.
464 Selby, "Education for Sustainable Development, Nature and Vernacular Learning," p. 19.
465 Budd Hall, "A River of Life: Learning and Environmental Social Movements," *Interface: A Journal for and about Social Movement*, Vol. 1, No. 1 (2009), pp. 46–78.
466 Bollier and Helfrich, *Free, Fair and Alive*, p. 77.
467 Martin Marty, *When Faiths Collide* (Malden, MA: Blackwell Publishing, 2005).
468 Marty, *When Faiths Collide*, p. 145.
469 Parker Palmer, *The Company of Strangers: Christians and the Renewal of America's Public Life* (New York: Crossroad, 1981), p. 69 cited in Marty, *When Faiths Collide*, p. 138.
470 Marty, *When Faiths Collide*, p. 148.
471 Dana Stuchul, Madhu Suri Prakash and Gustavo Esteva, "From Fear to Hope: Learning from BIPOC in Hard Times—Covid 19, Climate Collapse & Racial Violence," *International Journal of Lifelong Education*, Vol. 40, No. 4 (2021), pp. 415–426. https://doi.org/10.1080/02601370.2021.1936234.
472 Ivan Illich cited in Stuchul, Prakash and Esteva, "From Fear to Hope," p. 425.
473 Ivan Illich cited in Stuchul, Prakash and Esteva, "From Fear to Hope."
474 Stuchul, Prakash, and Esteva, "From Fear to Hope," p. 425.
475 Cajete, *Native Science*, p. 268.
476 Cajete, *Native Science*, p. 268.
477 Vaclav Havel, "Something is Being Born." Remarks presented by Vaclav Havel on July 4, 1994, at Independence Hall, on the occasion of his receiving the Philadelphia Liberty Medal. *New Age Journal* (September–October 1994), pp. 31–45 cited in Cajete, *Native Science*, pp. 264–265.

INDEX

Note: Page numbers in *italics* indicate a figure and page numbers in **bold** indicate a table on the corresponding page.